This book generalizes the classical theory complex unit circle or on the real line to o _____ ___ functions whose poles are among a prescribed set of complex numbers.

The first part treats the case where these poles are all outside the unit disk or in the lower half plane. Classical topics such as recurrence relations, numerical quadrature, interpolation properties, Favard theorems, convergence, asymptotics, and moment problems are generalized and treated in detail. The same topics are discussed for the different situation where the poles are located on the unit circle or on the extended real line. In the last chapter, several applications are mentioned including linear prediction, Pisarenko modeling, lossless inverse scattering, and network synthesis.

This theory has many applications in both theoretical real and complex analysis, approximation theory, numerical analysis, system theory, and in electrical engineering.

Adhemar Bultheel is a professor in the Computer Science Department of Katholieke Universiteit Leuven. In addition to coauthoring several books, he teaches introductory courses in analysis and numerical analysis for engineering students and an advanced course in signal processing for computer science and mathematics.

Pablo González-Vera is a professor in the Faculty of Mathematics at La Laguna University, Canary Islands. He teaches numerical analysis in mathematics, introductory courses in calculus for engineering, as well as advanced courses in numerical integration for physics and mathematics.

Erik Hendriksen is currently a researcher with the Department of Mathematics at the University of Amsterdam. He teaches introductory courses in analysis and linear algebra for students in mathematics and physics and advanced courses in functional analysis.

Olav Njåstad is a professor in the Department of Mathematical Sciences of the Norwegian University of Science and Technology. He is currently teaching introductory and advanced courses in analysis.

CAMBRIDGE MONOGRAPHS ON APPLIED AND COMPUTATIONAL MATHEMATICS

Series Editors
P. G. CIARLET, A. ISERLES, R. V. KOHN, M. H. WRIGHT

5   Orthogonal Rational Functions

The *Cambridge Monographs on Applied and Computational Mathematics* reflects the crucial role of mathematical and computational techniques in contemporary science. The series publishes expositions on all aspects of applicable and numerical mathematics, with an emphasis on new developments in this fast-moving area of research.

State-of-the-art methods and algorithms as well as modern mathematical descriptions of physical and mechanical ideas are presented in a manner suited to graduate research students and professionals alike. Sound pedagogical presentation is a prerequisite. It is intended that books in the series will serve to inform a new generation of researchers.

Also in this series:

A Practical Guide to Pseudospectral Methods, *Bengt Fornberg*

Dynamical Systems and Numerical Analysis, *A. M. Stuart and A. R. Humphries*

Level Set Methods, *J. A. Sethian*

The Numerical Solution of Integral Equations of the Second Kind, *Kendall E. Atkinson*

Orthogonal Rational Functions

ADHEMAR BULTHEEL   PABLO GONZÁLEZ-VERA
ERIK HENDRIKSEN   OLAV NJÅSTAD

CAMBRIDGE
UNIVERSITY PRESS

CAMBRIDGE UNIVERSITY PRESS
Cambridge, New York, Melbourne, Madrid, Cape Town, Singapore, São Paulo, Delhi

Cambridge University Press
The Edinburgh Building, Cambridge CB2 8RU, UK

Published in the United States of America by Cambridge University Press, New York

www.cambridge.org
Information on this title: www.cambridge.org/9780521115919

First published 1999
This digitally printed version 2009

A catalogue record for this publication is available from the British Library

Library of Congress Cataloguing in Publication data

Orthogonal rational functions / Adhemar Bultheel ... [et al.].
   p.   cm. – (Cambridge monographs on applied and computational
mathematics; 4)
Includes bibliographical references.
ISBN 0-521-65006-2 (hb)
1. Functions, Orthogonal.   2. Functions of complex variables.
   I. Bultheel, Adhemar.   II. Series.
QA404.5.075   1999
515'.55 – dc21
                                                        98-11646
                                                        CIP

ISBN 978-0-521-65006-9 hardback
ISBN 978-0-521-11591-9 paperback

Contents

vii

List of symbols

$\mathbb{C}$ the complex plane $\mathbb{C} = \{z = \operatorname{Re} z + \mathbf{i} \operatorname{Im} z\}$

$\overline{\mathbb{C}}$ the Riemann sphere $\overline{\mathbb{C}} = \mathbb{C} \cup \{\infty\}$

$\mathbb{R}, \overline{\mathbb{R}}$ the real line $\mathbb{R} = \{z \in \mathbb{C} : \operatorname{Im} z = 0\}$,

$\mathbb{Z}$ the integers

$\mathbb{U}, \mathbb{L}, \mathbb{H}$ $\mathbb{U} = \{z \in \mathbb{C} : \operatorname{Im} z > 0\}$, $\mathbb{L} = \{z \in \mathbb{C} :$

$\operatorname{Im} z < 0\}$, $2\mathbb{H} = \{z \in \mathbb{C} : \operatorname{Re} z > 0\}$

$\mathbb{D}, \mathbb{T}, \mathbb{E}$ $\mathbb{D} = \{z \in \mathbb{C} : |z| < 1\}$, $\mathbb{T} = \{z \in \mathbb{C} : |z| = 1\}$,

$\mathbb{E} = \{z \in \mathbb{C} : |z| > 1\}$

$\mathbb{O}, \partial\mathbb{O}, \mathbb{O}^e$ $\mathbb{O}$ region in $\mathbb{C}$: $\mathbb{D}$ or $\mathbb{U}$

$\partial\mathbb{O}$ boundary of $\mathbb{O}$: $\mathbb{T}$ or $\mathbb{R}$

$\mathring{\mu}$ normalized measure on $\overline{\partial\mathbb{O}}$:

$d\lambda, d\mathring{\lambda}$ normalized Lebesgue measures:

$\langle f, g \rangle_{\mathring{\mu}}$ inner products:

$\hat{z}$ reflection in the boundary:
$\hat{z} = 1/\overline{z}$ for $\mathbb{D}$, $\hat{z} = \overline{z}$ for $\mathbb{U}$ 20

$f_*(z)$ substar conjugate: $f_*(z) = \overline{f(\hat{z})}$ 20

$\alpha_k, k \geq 0$ basic interpolation points:
$\alpha_k \in \partial\mathbb{O}$ for boundary case 257
$\alpha_k \in \mathbb{O}$ otherwise. 43

$A_n, \hat{A}_n$ $A_n = \{\alpha_1, \ldots, \alpha_n\}$, $\hat{A}_n = \{\hat{\alpha}_1, \ldots, \hat{\alpha}_n\}$ 44

$A, \hat{A}$ $A = \{\alpha_1, \alpha_2, \ldots\}$, $\hat{A} = \{\hat{\alpha}_1, \hat{\alpha}_2, \ldots\}$ 44

A_n^w, A_n^0 $A_n^w = \{w, \alpha_1, \ldots, \alpha_n\}$ 133
$A_n^0 = \{\alpha_0, \alpha_1, \ldots, \alpha_n\}$ 44

$\mathbb{O}_0, \mathbb{O}_0^e$ $\mathbb{O}_0 = \mathbb{O} \setminus A_n$, $\mathbb{O}_0^e = \mathbb{O}^e \setminus A_{n*}$ 241

α_0 special point: $\alpha_0 = 0$ for $\mathbb{D}$, $\alpha_0 = \mathbf{i}$ for $\mathbb{U}$........19
in the boundary case: $\alpha_0 = -1$ for $\mathbb{T}$, $\alpha_0 = \infty$
for $\mathbb{R}$ 257

$\varpi_w(z), \varpi_i(z)$ $\varpi_w(z) = 1 - \overline{w}z$ for $\mathbb{D}$, $\varpi_w(z) = z - \overline{w}$ for $\mathbb{U}$,
$\varpi_i = \varpi_{\alpha_i}$ 19

$\pi_n(z)$ $\pi_n(z) = \prod_{k=1}^{n} \varpi_k(z)$ 44

$D(t, z), E(t, z)$ $D(t, z)$ Riesz–Herglotz–Nevanlinna kernel:
$D(t, z) = (t + z)/(t - z)$ for $\mathbb{D}$....................27
$D(t, z) = -\mathbf{i}(1 + tz)/(t - z)$ for $\mathbb{U}$...............27
$E(t, z) = 1 + D(t, z)$ 83

$\mathcal{C}, \mathcal{B}$ $\mathcal{C}$ positive real functions:
$\mathcal{C} = \{f \in H(\mathbb{O}) : f(\mathbb{O}) \subset \mathbb{H}\}$....................23
$\mathcal{B}$ bounded analytic functions:
$\mathcal{B} = \{f \in H(\mathbb{O}) : f(\mathbb{O}) \subset \mathbb{D}\}$ 23

$\mathcal{A}$ $\mathcal{A} = \{[\Delta_1 \ \Delta_2] : \Delta_1, \Delta_2 \in H(\mathbb{O}),$
$\Delta_2(z) \neq 0, z \in \mathbb{O}, \Delta_1/\Delta_2 \in \mathcal{B}\}$ 141

$\Omega_\mu(z) \in \mathcal{C}$ Riesz–Herglotz–Nevanlinna transform:
$\Omega_\mu(z) = \mathbf{i}c + \int D(t, z) \, d\mu(t), c \in \mathbb{R}$ 27

$C(t, z)$ Cauchy kernel:
$C(t, z) = [\varpi_0(\alpha_0)\varpi_{z*}(t)]^{-1}$ for $\mathbb{O}$,
$C(t, z) = t/(t - z)$ for $\mathbb{D}$,
$C(t, z) = 1/[2\mathbf{i}(t - z)]$ for $\mathbb{U}$ 23

$P(t, z)$ Poisson kernel:
$P(t, z) = [\varpi_z(z)/\varpi_0(\alpha_0)]/[\varpi_z(t)\varpi_{z*}(t)]$ for $\mathbb{O}$....27
$P(t, z) = (1 - |z|^2)/|t - z|^2$ for $\mathbb{D}$ if $t \in \mathbb{T}$,........27
$P(t, z) = \mathrm{Im}\, z/|t - z|^2$ for $\mathbb{U}$ if $t \in \mathbb{R}$ 27

M_w Möbius transform:
$M_w(z) = (z - w)/(1 - \overline{w}z)$ 25

Introduction

This monograph forms an introduction to the theory of orthogonal rational functions. The simplest way to see what we mean by orthogonal rational functions is to consider them as generalizations of orthogonal polynomials.

There is not much confusion about the meaning of an orthogonal polynomial sequence. One says that $\{\phi_n\}_{n=0}^{\infty}$ is an orthogonal polynomial sequence if ϕ_n is a polynomial of degree n and it is orthogonal to all polynomials of lower degree. Thus given some finite positive measure μ (with possibly complex support), one considers the Hilbert space $L_2(\mu)$ of square integrable functions that contains the polynomial subspaces $\mathcal{P}_n$, $n = 0, 1, \ldots$. Then $\{\phi_n\}_{n=0}^{\infty}$ is an orthogonal polynomial sequence if $\phi_n \in \mathcal{P}_n \setminus \mathcal{P}_{n-1}$ and $\phi_n \perp \mathcal{P}_{n-1}$. In particular, when the support of the measure is (part of) the real line or of the complex unit circle, one gets the most widely studied cases of such general orthogonal polynomials. Such orthogonal polynomials appear of course in many different aspects of theoretical analysis and applications. The topics that are central in our generalization to rational functions are moment problems, quadrature formulas, and classical problems of complex approximation in the complex plane.

Polynomials can be seen as rational functions whose poles are all fixed at infinity. For the orthogonal rational functions, we shall fix a sequence of poles $\{\gamma_k\}_{k=1}^{\infty}$, which, in principle, can be taken anywhere in the extended complex plane. Some of these γ_k can be repeated, possibly an infinite number of times, or they could be infinite. However, the sequence is fixed once and for all and the order in which the γ_k occur (possible repetitions included) is also given. This will then define the n-dimensional spaces of rational functions $\mathcal{L}_n$ that consist of all the rational functions of degree n whose poles are among $\gamma_1, \ldots, \gamma_n$ (including possible repetitions). We then consider $\{\phi_n\}_{n=0}^{\infty}$ to be a sequence of orthogonal rational functions if $\phi_n \in \mathcal{L}_n \setminus \mathcal{L}_{n-1}$ and $\phi_n \perp \mathcal{L}_{n-1}$.

1

There are two possible generalizations, depending on whether one generalizes the polynomials orthogonal on the real line or the polynomials orthogonal on the unit circle. The difference lies in the location of the finite poles that are introduced in the rational case. In the case of the circle, the pole at infinity is outside the closed unit disk. There, it is the most natural choice to introduce finite poles that are all outside the closed unit disk. This guarantees that the rational functions are analytic at least inside the unit disk, which allows us to transfer many properties from the polynomial to the rational case. Moreover, if the poles are not on the circle, then we avoid difficulties that could arise from singularities of the integrand in the support of the measure.

If the support of the measure is, however, contained in the real line, then the pole at infinity may be in the (closure of) the support of the measure. The most natural generalization is here to choose finite poles that are on the real line itself, that is, possibly in the support of the measure for which orthogonality is considered.

Of course one can by a Cayley transform map the unit circle to the (extended) real line and the open unit disk to the upper half plane. Since this transform maps rational functions to rational functions, it makes sense to consider the analog of the orthogonal rational functions on the unit circle with poles outside the closed disk, which are the orthogonal rational functions orthogonal on the real line with poles in the lower half plane. Conversely, one can consider the orthogonal rational functions with poles on the unit circle and that are orthogonal with respect to a measure supported on the unit circle as the analog of orthogonal rational functions on the real line with poles on the real line.

The cases of the real line and the unit circle, which are linked by such a Cayley transform, are essentially the same and can be easily treated in parallel, which we shall do in this monograph. The distinction between the case where the poles are outside or inside the support of the measure is, however, substantial. We have chosen to give a detailed and extensive treatment in several chapters of the case where the poles are outside the support. The case where the poles are in the support (which we call the boundary case) is treated more compactly in a separate chapter.

This brief sketch should have made clear in what sense these orthogonal rational functions generalize orthogonal polynomials. Now, what are the results of the polynomial case that have been generalized to the rational case? As we suggested above, we do not go into the details of all kinds of special orthogonal polynomials by imposing a specific measure or weight function. We do keep generality by considering arbitrary measures, but we restrict ourselves to measures supported on the real line or the unit circle. In that sense we are not as general as the "general orthogonal polynomials" in the book of Stahl and Totik [193].

Orthogonal polynomials have now been studied so intensely that many different and many detailed results are available. It would be impossible to give in one volume the generalizations of all these to the rational case. We have opted for an introduction to the topic and we give only generalizations of classical interpolation problems of Schur and Carathéodory type, of quadrature formulas, and of moment problems. There is a certain logic in this because interpolation problems are intimately related to quadrature formulas and these quadrature formulas are an essential tool for solving the moment problems.

These connections were made clear and were used explicitly in the book by Akhiezer [2], which treats "the classical moment problem." To some extent we have followed a similar path for the rational case.

First, we derive a recurrence relation for the orthogonal rational functions. In our setup, this is mainly based on a Christoffel–Darboux type relation. In the boundary case, this recurrence generalizes the three-term recurrence relation of orthogonal polynomials; in the case where the poles are outside the support of the measure, this is a generalization of the Szegő recurrence relation.

To describe all the solutions of the recurrence relation, a second, independent solution is considered, which is given by the sequence of associated functions of the second kind.

These functions of the second kind appear as numerators and the orthogonal rational functions as denominators in the approximants of a continued fraction that is associated with the recurrence relation. The continued fraction converges to the Riesz–Herglotz–Nevanlinna transform of the measure and the approximants interpolate this function in Hermite sense. This is the interpolation problem that we alluded to. It is directly related to the algorithm of Nevanlinna–Pick, which is a (rational) multipoint generalization of the Schur algorithm that relates to the polynomial case.

A combination of the orthogonal rational functions and the associated functions of the second kind give another solution of the recurrence relation called the quasi-orthogonal or para-orthogonal functions in the boundary or nonboundary case respectively. It can be arranged that these functions have simple zeros that are on the real line or on the unit circle. These zeros are used as the nodes of quadrature formulas. In the nonboundary case, such n-point quadrature formulas are optimal in the sense that corresponding weights can be chosen in such a way that the quadrature formulas have the largest possible domain of validity. For the boundary case, these quadrature formulas are "nearly optimal" in general. Their domain of validity has a dimension one less than the optimal one. However, when the zeros of the orthogonal rational functions themselves happen to be a good choice, then the quadrature formula is really optimal. In the polynomial case, this corresponds to Gaussian quadrature formulas on the real line or Szegő quadrature formulas for the circle.

These quadrature formulas are an essential tool in the construction of a solution of the moment problems. These moment problems are rational generalizations of the polynomial case that correspond to the Hamburger moment problem in the case of the real line and the trigonometric moment problem in the case of the circle.

Two other aspects are important or are at least closely connected to the solution of these moment problems. First, there is the well-known fact that, as n goes to ∞, the polynomial spaces $\mathcal{P}_n$ become dense in the Hardy spaces H_p. A similar result will only hold for the spaces $\mathcal{L}_n$ under certain conditions for the poles. Second, there is the general question of asymptotics for the orthogonal rational functions, for the interpolants, for the quadrature formulas, etc., when n tends to infinity. Such results were extensively studied in the polynomial case. We shall devote a large chapter to their generalizations.

After this general introduction, let us have a look at the roots of this theory, at the applications in which it was used, and let us have a closer look at the technical difficulties that arise by lifting the polynomial to the rational case. Since the central theme up to Chapter 10 is the generalization of results related to Szegő polynomials, orthogonal on the unit circle, let us take these as a starting point.

The particularly rich and fascinating theory of polynomials orthogonal on the unit circle needs no advertising. These polynomials are named after Szegő since his pioneering work on them. His book on orthogonal polynomials [196] was first published in 1939, but the ideas were already published in several papers in the 1920s. The Szegő polynomials were studied by several authors. For example, they play an important role in books by Geronimus [94], Freud [87], Grenander and Szegő [102], and several more recent books on orthogonal polynomials.

It is also in Szegő's book that the notion of a reproducing kernel is clearly introduced. Later on these became a studied object of their own. The book by Meschkowski [148] is a classic. In our exposition, reproducing kernels take a rather important place and the Christoffel–Darboux summation formula, which expresses the nth reproducing kernel in terms of the nth or $(n+1)$st orthogonal polynomials (in our case rational functions), is used again and again in many places throughout our monograph.

Szegő's interest in polynomials orthogonal on the unit circle was inspired by the investigation of the eigenvalue distribution of Toeplitz forms, an even older subject related to coefficient problems as initiated by Carathéodory [48, 49] and Carathéodory and Fejér [50] and further discussed by F. Riesz [184, 185],

Gronwall [103], Schur [189, 190, 191], Hamel [107], and many others. This problem is closely related to the trigonometric moment problem. Indeed, the matrices of the Toeplitz forms are Gram matrices because for the inner product

$$\langle f, g \rangle = \int f(t)\overline{g(t)}\, d\mu(t),$$

where μ is a measure on the unit circle; we have $\langle z^k, z^l \rangle = \int t^{k-l} d\mu(t) = \mu_{l-k}$, where $\mu_k = \int t^{-k} d\mu(t)$ are the moments of the measure μ. Thus if the measure μ is positive, then the Gram matrix should be positive definite, and because it is Hermitian, this means that its eigenvalues should be positive. The converse is also true: Given a Hermitian Toeplitz matrix, then there will be a positive measure for which the entries of this matrix are the moments if the matrix is positive definite. Another way of putting this is to say that the function $\Omega_\mu(z) = \sum_{k=0}^{\infty} \mu_k z^k$ is a Carathéodory function. This means that it is analytic in the unit disk and it has a positive real part there. Since this is an infinite-dimensional problem, it can not be checked by a finite number of computations. A computational procedure consists basically in approximating Ω_μ by some rational Ω_n that fits the first $n+1$ Fourier coefficients of Ω_μ and letting n range over the natural numbers $n = 0, 1, 2, \ldots$. These Ω_n turn out to be related to ψ_n / ϕ_n, where ϕ_n is the nth Szegő polynomial and ψ_n is the associated polynomial of the second kind. Both of these are solutions of the recurrence relation for the orthogonal polynomials and they appear as successive approximants in continued fractions. Thus checking the positivity of the infinite Toeplitz matrix comes down to checking the positivity of all its leading principal submatrices, or, equivalently, checking whether Ω_μ is a Carathéodory function reduces to checking that Ω_n are Carathéodory functions for all $n = 0, 1, \ldots$. The Carathéodory coefficient problem is in a sense an inverse of this: Given the coefficients $\mu_0, \mu_1, \ldots, \mu_n$, can these be extended to a sequence such that $\sum_{k=0}^{\infty} \mu_k z^k$ is a Carathéodory function, or, equivalently, such that the corresponding Toeplitz matrix with entries μ_{i-j}, $(\mu_{-k} = \overline{\mu_k})$ is positive definite, or, equivalently, such that there is a positive measure μ on the unit circle for which these μ_k are the moments $\mu_k = \int t^{-k} d\mu(t)$, $n = 0, 1, 2, \ldots$.

Schur solved an equivalent problem [189, 190, 191]. By mapping the right half plane to the unit disk, functions with a positive real part are mapped to functions bounded by 1 or Schur functions. So the problem is reduced to checking whether a given function is a Schur function. This can be done recursively and the papers by s–like algorithm to actually check if the given coefficients (moments) correspond to a bounded analytic function. The algorithm produces some coefficients (Schur coefficients) that turned out later to be exactly the complex conjugates of the coefficients that appeared in the recurrence relation for the orthogonal polynomials as formulated by Szegő.

It was Pick who first considered an interpolation problem as a generalization of the coefficient problems of Carathéodory [176, 177, 178]. Nevanlinna was not aware of Pick's work when he developed the same theory in a long memoir in 1919 [153]. See also his later work [154, 155, 156]. Nevanlinna also gave an algorithm that directly generalized the algorithm given by Schur.

Since then, these problems and a myriad of generalizations played an important role in several books, such as in Akhiezer [2], Kreĭn and Nudel'man [131], and Walsh [200], and more recently in Donoghue [71], Garnett [92], Rosenblum and Rovnyak [186], Ball, Gohberg, and Rodman [22], Bakonyi and Constantinescu [19], etc.

Some of the more recent interest in this subject was stimulated by the work of Adamyan, Arov, and Kreĭn [4, 5] and most of all by their fundamental papers [6, 7]. We should also mention Sarason's paper [188], which had great influence on some developments made in later publications. These results relate the theory to operator theoretic methods for Hankel and Toeplitz operators.

Besides this, there is also a long history where the same theory is approached from several application fields. Grenander and Szegő themselves discussed the application in the theory of probability and statistics [102]. But one finds also the applications in the prediction theory of stationary stochastic processes in work by Kolmogorov [129] and Wiener [201]. Some benchmark papers on this topic are collected in [125]. The book by Wiener contained a reprint from Levinson's celebrated paper [134], which is in fact a reformulation of the Szegő recursions. Other engineering applications include network theory (see, e.g., Belevitch [23] and Youla and Saito [204]), spectral estimation (see Papoulis [174] for an excellent survey), maximum entropy analysis as formulated by Burg [47] (see the survey paper [133]), transmission lines and scattering theory as studied by Arov*, Redheffer [183], and Dewilde and Dym [60, 62], digital filtering (see the survey of Kailath [124]), and speech processing (see [144] or the tutorial paper by Makhoul [143]).

It is from these engineering applications that methods for inverting and factorizing Toeplitz or related matrices also emerged (see [190]), and people are now even using these ideas for designing systolic arrays for the solution of a number of linear algebra problems [24]. The linear algebra literature in this connection has a complete history of its own, which we shall not mention here. Most of it was devoted to Toeplitz and Hankel matrices or related matrices that appear in connection with a theory of Schur–Szegő. However, only recently have

---

* D. Z. Arov. Darlington's method in the study of dissipative systems. *Dokl. Akad. Nauk SSSR,* **201** (1971), 559–562. (In Russian). English translation: *Soviet Physics-Doklady* **16** (1972), 954–956.

people started looking at matrices that are related to interpolation problems.

We could go on like this and probably never be complete in summing up all the application fields and this is without ever touching all the related generalizations of this theory that were obtained recently or the analog theory that has been developed for the complex half plane instead of the unit circle or the continuous analog of Wiener–Hopf factorization. We just stop here by referring to a survey paper on the applications of Nevanlinna–Pick theory by Delsarte, Genin, and Kamp [57].

In all these papers on theory and applications, the approach of the Nevanlinna–Pick theory from the point of view of the orthogonal functions has not been fully put forward. In this monograph, we try to give an approach to the theory that is an immediate generalization of the theory of Szegő for orthogonal polynomials. This theory is related to the interpolation theory of Pick and Nevanlinna like the Szegő theory was related to the Schur and Carathéodory–Fejér coefficient problems.

By a Cayley transform, the complex unit circle is mapped to the extended real line and its interior to the upper half plane. Thus there is a natural analog of this theory for the real line. There are, however, technical differences that make the transformation not always trivial. Therefore we shall treat both cases in parallel.

A central role in the theory is played by moment problems. We give some details because it illustrates very well the difference between the case of the circle and the case of the line. For the unit circle, we have the classical trigonometric moment problem. In that case one defines the moments μ_k for $k \in \mathbb{Z}$ as

$$\mu_k = \int t^{-k} d\mu(t), \quad k \in \mathbb{Z}, \ t = e^{i\theta}.$$

By giving $\mu_{-k}, k = 0, 1, 2, \ldots$, one has defined a linear functional on the set of polynomials and the problem is to find the measure μ on the unit circle that represents this functional. Note that when μ is a real positive measure, then $\mu_k = \bar{\mu}_{-k}$, so if the Toeplitz moment matrix $G = [\mu_{k-l}]$ is Hermitian and positive definite, it is completely characterized by half of these moments, namely by the moments $\mu_k, k = 0, 1, \ldots$ with a nonnegative index. Thus the previous moment problem where only the moments $\mu_{-k}, k = 0, 1, \ldots$, or equivalently only the moments μ_k for $k = 0, 1, 2, \ldots$, are defined, is equivalent to the moment problem where all the $\mu_k, k \in \mathbb{Z}$ are prescribed. The working instrument for solving this trigonometric moment problem is the set of orthogonal Szegő polynomials obtained by orthogonalizing the basis $\{1, t, t^2, \ldots\}$. The orthogonality is with respect to the inner product given by $\langle P, Q \rangle = M\{P(z)\overline{Q(1/\bar{z})}\}$, where the linear functional M is defined by $M\{t^k\} = \mu_{-k}, k = 0, 1, 2, \ldots$. Note that the inner product is well defined for arbitrary polynomials P and Q because the argument $P(z)\overline{Q(1/\bar{z})}$ is a Laurent polynomial and, as we remarked a moment ago, if the linear functional is defined on the set of polynomials, then it is automatically defined on the set of Laurent polynomials.

For the rational generalization of this problem we consider a sequence of points $\alpha_1, \alpha_2, \ldots$ all inside the unit disk and define the moments

$$\mu_{kl} = \int \frac{d\mu(t)}{\omega_k(t)\pi_l^*(t)}, \quad k, l = 0, 1, \ldots,$$

where for $k \geq 1$

$$\omega_k(z) = z^{-k} \prod_{i=1}^{k} (1 - \overline{\alpha}_i z), \quad \pi_l^*(z) = \prod_{i=1}^{l} (z - \alpha_i)$$

while $\omega_0 = \pi_0^* = 1$. The rational generalization of the trigonometric moment problem is to define a linear functional on the space $\mathcal{L}_\infty = \text{span}\{\omega_0^{-1}, \omega_1^{-1}, \omega_2^{-1}, \ldots\}$ by prescribing the moments $\mu_{k0}, k = 0, 1, 2, \ldots$. Note that if all $\alpha_i = 0$, then

$$\mu_{k0} = \int t^k d\mu(t) = \mu_k, \quad t = e^{i\theta}.$$

Thus, in that case the moment problem reduces to the classical trigonometric moment problem. Moreover, if $G = [\mu_{kl}]$ is Hermitian positive definite, then it is obvious that

$$\overline{\mu}_{k0} = \int \frac{d\mu(t)}{\overline{\omega_k(t)}} = \int \frac{d\mu(t)}{\pi_k^*(t)} = \mu_{0k}, \quad k \geq 0.$$

Thus, by partial fraction decomposition, it is seen that the knowledge of μ_{k0}, $k = 0, 1, 2, \ldots$ also gives the moments $\mu_{kl}, k, l = 0, 1, \ldots$. Again, it is sufficient to prescribe only the moments $\mu_{k0}, k = 0, 1, \ldots$. To solve this rational generalization, the role of the orthogonal polynomials in the classical trigonometric moment problem is taken over by orthogonal functions obtained by orthogonalization of the basis $\{1, \omega_1^{-1}, \omega_2^{-1}, \ldots\}$. These are the orthogonal rational functions that are studied in this monograph. This also explains why it is called a rational generalization. Note that at any point in the discussion we can replace all the interpolation points α_k by zero and recover at any moment the corresponding result of the polynomial case. In this respect it is a natural generalization of the Szegő theory.

For the unit circle, both in the polynomial and the rational case, we are in a convenient situation where it is sufficient to have a functional defined on $\mathcal{L}_\infty = \text{span}\{\omega_0^{-1}, \omega_1^{-1}, \omega_2^{-1}, \ldots\}$, and we immediately have defined the functional on a larger set, which allows us to define orthogonality. Indeed, in the rational case we should be able to evaluate integrals of the form $\int f(t)\overline{g(t)}\,d\mu(t), t = e^{i\theta}$ for $f, g \in \mathcal{L}_\infty$. This requires the existence of moments μ_{kl}, which is guaranteed because the μ_{k0} also define the μ_{kl}.

Let us now consider the Hamburger moment problem. Here we define the moments

$$\mu_k = \int t^k d\mu(t), \quad t \in \mathbb{R}, \ k \in \mathbb{Z}.$$

Again, prescribing the μ_k for $k = 0, 1, 2, \ldots$ defines a linear functional on the set of polynomials and the problem is to find a positive measure μ on the real line such that it matches these prescribed moments for $k = 0, 1, 2, \ldots$. The instruments in this case are again the orthogonal polynomials obtained by orthogonalization of the basis $\{1, t, t^2, \ldots\}$. However, the inner product now generates a moment matrix $G = [\mu_{k+l}]$, which is a Hankel matrix. It is no longer true that the μ_k for $k = 0, 1, \ldots$ also give the values of the moments $\mu_{-k}, k = 1, 2, \ldots$. The problems where the moments μ_k are prescribed for $k = 0, 1, 2, \ldots$ (the moment problem for the set of polynomials) and where all the moments $\mu_k, k \in \mathbb{Z}$ are given (the moment problem in the set of Laurent polynomials) are not equivalent anymore. The first problem is called the (classical) Hamburger moment problem, whereas the second is the strong Hamburger moment problem. To solve the strong Hamburger moment problem, one has to deal with orthogonal Laurent polynomials instead of the usual polynomials [119, 51]. We may note that in the classical Hamburger moment problem, there is no problem when talking about orthogonality because the inner product is defined as $\langle f, g \rangle = M\{f(x)g(x)\}$ with the linear functional defined by $M\{x^k\} = \mu_k, k = 0, 1, 2, \ldots$, and since the product of two polynomials is again a polynomial, it is sufficient to define the functional on the set of polynomials. A similar observation concerning the Laurent polynomials holds for the strong Hamburger moment problem.

The rational generalization of the Hamburger moment problem is to consider the moments

$$\mu_{k0} = \int \frac{d\mu(t)}{\omega_k(t)}, \quad k = 0, 1, \ldots,$$

where

$$\omega_k(z) = z^{-k} \prod_{i=1}^{k} (1 - z/\alpha_i), \quad k \geq 1, \ \omega_0 = 1,$$

and the α_i are points on the (extended) real line. Note that when all $\alpha_i = \infty$, this reduces to the classical Hamburger moment problem. The orthogonal polynomials are replaced by the orthogonal rational functions obtained by orthogonalization of $\{1, \omega_1^{-1}, \omega_2^{-1}, \ldots\}$. There is a considerable complication that did not occur in any of the previous situations. Indeed, the product of two rational functions from $\mathcal{L}_\infty = \text{span}\{1, \omega_1^{-1}, \omega_2^{-1}, \ldots\}$ is not in $\mathcal{L}_\infty$ anymore (unless there is a special choice of the α_i). Thus, to be able to consider orthogonality here, we should also have a functional defined on $\mathcal{L}_\infty \cdot \mathcal{L}_\infty$ and this requires moments that are in general not obtainable from the μ_{k0} alone. Therefore, we need moments of the form

$$\mu_{kl} = \int \frac{d\mu(t)}{\omega_k(t)\omega_l(t)}, \quad k, l = 0, 1, 2, \ldots.$$

Note that when there is only a finite number of different α_i that are cyclically repeated (the so-called cyclic case) then it does hold that the product of rational functions in $\mathcal{L}_\infty$ is again in $\mathcal{L}_\infty$ and the previously sketched difficulties do not occur.

For the strong Hamburger moment problem, the rational equivalent is to prescribe the moments

$$\mu_{kl} = \int \frac{d\mu(t)}{\omega_k(t)\pi_l(t)}, \quad k, l = 0, 1, \ldots,$$

with ω_k as before and

$$\pi_l(z) = \prod_{i=1}^{l} (z - 1/\alpha_i), \quad l \geq 1, \ \pi_0 = 1.$$

This is possible, if we are willing to introduce a more complicated notation to formulate this as a problem where the special case of the strong Hamburger moment problem appears by choosing alternatively $\alpha_{2k} = 0$ and $\alpha_{2k+1} = \infty$. The problem of orthogonality for the functions in these spaces is even more difficult since this requires even more complicated moments to be defined. In this monograph, we shall not consider this rational generalization of the strong Hamburger moment problem.

Even for the classical Hamburger moment problem, the rational generalization gives yet other complications with regard to the point at ∞ that do not occur in the polynomial case. Indeed, if μ is a solution of a polynomial moment problem, then, since all polynomials, except the constants, tend to ∞ at ∞, there can not be a mass at ∞ if the moments are finite. However, in the rational case, all the rational functions can have finite values at ∞ and hence a mass at infinity is perfectly possible for a solution of the moment problem. Therefore, the integrals should be taken over the extended real line $\overline{\mathbb{R}} = \mathbb{R} \cup \{\infty\}$, rather than over $\mathbb{R}$.

Note also that the situation of the real line and the complex unit circle are considerably different in the respect that the natural generalization of the polynomial case requires the points α_i to be inside the disk (to be able to recover the polynomial case by setting them all equal to 0) whereas for the real line case, these points are all on the extended real line (to be able to set them all equal to ∞).

Also, integrals of the form $\int f(t) \, d\mu(t)$, where f has poles at $\alpha_1, \alpha_2, \ldots$, are definitely more complicated when the α_i are in the support of the measure than when they are not.

Rational generalizations of the Hamburger moment problem (called extended moment problems or multipoint moment problems) were considered in Refs. [157], [158], [160], [164], and [165] where only a finite number of different α_i is used. See also Refs. [137] and [138]. Orthogonal rational functions occur first in the work of M. M. Djrbashian [66, 65, 67, 68, 69, 70] (see also Ref. [150]) and in Ref. [30].

This brief discussion of the moment problem shows that there is a distinction to be made between the case of the real line and the case of the unit circle. In fact when the Cayley transform is put into play, there are four cases to be considered:

1. supp (μ) is in $\mathbb{T}$, all $\alpha_i \in \mathbb{D}$.
2. supp (μ) is in $\mathbb{T}$, all $\alpha_i \in \mathbb{T}$.
3. supp (μ) is in $\mathbb{R}$, all $\alpha_i \in \mathbb{U}$.
4. supp (μ) is in $\mathbb{R}$, all $\alpha_i \in \mathbb{R}$.

Cases 1 and 3 are related by a Cayley transform and so are 2 and 4. Chapters 2–10 will be devoted to cases 1 and 3. Chapter 11 discusses similar results for cases 2 and 4 to which we shall refer as the boundary case. It is perfectly possible to combine the two cases, as for example in Ref. [62], where the α_i are allowed to be inside the closed unit circle, thus combining cases 1 and 2.

Let us conclude this introduction with an outline of the text. Since the previous more technical introduction introduced some terminology, we can give it in some detail.

In Chapter 1, we start with preliminaries from complex analysis and some properties of reproducing kernels. It also includes some generalities on positive real functions, that is, analytic functions in the unit disk with positive real part, also known as Carathéodory functions since they appeared in the Carathéodory–Fejér problem. We give also the relation with the analytic functions bounded by 1. The latter are also known as Schur functions because it was Schur who used these functions in his algorithm to solve the Carathéodory–Fejér problem. A last important tool in our analysis are the J-contractive matrices studied by Potapov. They are introduced in Section 1.5.

It is then time to be more specific and we introduce the fundamental spaces of rational functions that will generalize the spaces of polynomials as they appear in the Szegő theory. They are defined and discussed in Chapter 2 in some detail.

Rather than starting with the recurrence for the orthogonal functions themselves, it will turn out to be easier to start with the reproducing kernels, which we do in Chapter 3. This chapter also contains the important Christoffel–Darboux relations.

It will become clear in Chapter 4, when we give the recurrence for the orthogonal functions, that, compared with the reproducing kernels, they are somewhat less simple to handle, since the recurrence can not be easily described in terms of a J-unitary recursion. It is however possible to get some recurrence that generalizes the Szegő relations, and one can also define functions of the second kind (Section 4.2). As in Szegő's theory, these functions of the second kind appear as another independent solution of the recurrence for the orthogonal functions, exactly like the Szegő polynomials of the second kind. We also mention in this chapter how the coupled Szegő type recursions can be transformed into three-term recurrence relations, which are associated with continued fractions.

The relation to quadrature on the unit circle is given in Chapter 5. It gives quadrature formulas that, like the Gaussian quadrature on the real line, have a maximal domain of validity. However, the zeros of the orthogonal functions on the circle are all inside the unit disk and they are thus not well suited to be used as abscissas. One can construct para-orthogonal functions that do have unimodular zeros and are missing the usual orthogonality conditions just by not being orthogonal to the constants. This deficient orthogonality causes the dimension of the domain of validity for the quadrature formulas to be one lower than the dimension in the classical case of Gaussian formulas. Also, the domain of validity is not a polynomial space but a space of rational functions. Of course, similar results hold for the half plane and they are formulated in parallel.

In Chapter 6, we relate several results to rational interpolation problems in the class of Carathéodory and Schur functions. We also give the algorithm of Nevanlinna–Pick, which is related to the recurrence for the reproducing kernels, and also a similar algorithm related to the recurrence of the orthogonal functions.

Some density problems of the functions in $\mathcal{L}_\infty$, that is, the completeness problem of the rational basis functions of $\mathcal{L}_\infty$, in the spaces L_p and H_p or in

$L_2(\mu)$ and $H_2(\mu)$ are discussed in Chapter 7. This forms an introduction to the chapter on convergence but is also used in the Favard theorem formulated in the next chapter.

Favard theorems for the spaces of rational functions are possible, exactly as in the polynomial case. In the polynomial case, these theorems state that if one has polynomials satisfying a three-term recurrence relation, then they are orthogonal with respect to some measure. The same holds for the orthogonal rational functions where we may also replace the three-term recurrence by a recurrence as discussed in Chapters 3 and 4. We give constructive proofs for such theorems. Also, the kernels satisfy a recurrence relation and one may be interested in a Favard type theorem for the kernels, that is, if some kernel functions satisfy a typical recurrence relation, will they be reproducing kernels for nested spaces with respect to some inner product? Even in the polynomial case this problem has not been solved. We do not succeed in finding an easy characterization of the type of recurrence that will guarantee this Favard type result, but we give some indications of what can be obtained.

Chapter 9 starts by giving a generalization of the Szegő problem, which can be solved by a limiting approximation process in our rational function spaces. This is done in Section 9.1. We could formulate it as finding the projection of z^{-1} onto the space $H_2(\mu)$, that is, the space of polynomials closed in the $L_2(\mu)$ metric. A crucial fact will then be to find out when the space $H_2(\mu)$ is not only spanned by the polynomials, which is the original Szegő approach, but also when it is spanned by the rational functions under certain conditions. Some further convergence results are discussed in Section 9.2. We give asymptotics for ϕ_n^* in Section 9.3. We have to require that the points α_i stay away from the boundary. However, such results can be obtained under weaker conditions for which we introduce in Section 9.5 the orthogonal polynomials with respect to a varying measure as studied by López [140], Stahl and Totik [193], and others.

The convergence results are given in Section 9.6 and subsequent sections. There are strong and weak convergence results in norm, locally uniform convergence, and results about ratio and root asymptotics. They are obtained under various conditions that hold for the measure [Szegő ($\int \log \mu' > -\infty$) or Erdős–Turán ($\mu' > 0$ a.e.)] and under several conditions on the point set $\alpha_1, \alpha_2, \ldots$: It can be assumed to be compactly included in the disk or the half plane, or its Blaschke product can be assumed to diverge or a Carleman type of condition can be assumed.

In Chapter 10 the moment problem is discussed, in particular, the classical theory of nested disks is generalized.

In Chapter 11, we treat the boundary situation where the α_i are on the boundary, that is, on the unit circle or on the extended real line. The complication there is that the α_i can be in the support of the measure. At a somewhat quicker pace, the orthogonal rational functions are introduced and their recurrence relations, Christoffel–Darboux type relations, quadrature formulas, the moment problem, interpolation properties, and convergence results are discussed.

Finally, in Chapter 12, we give some of the direct applications of the theory that was developed in the previous chapters.

1

Preliminaries

In this chapter we shall collect the necessary preliminaries from complex analysis that we shall use frequently. Most of these results are classical and we shall give them mostly without proof.

We start with some elements from Hardy functions in the disk and the half plane in Section 1.1.

The important classes of analytic functions in the unit disk and half plane and with positive real part are called positive real for short and are often named after Carathéodory. By a Cayley transform, they can be mapped onto the class of analytic functions of the disk or half plane, bounded by one. This is the so-called Schur class. These classes are briefly discussed in Section 1.2.

Inner–outer factorizations and spectral factors are discussed in Section 1.3.

The reproducing kernels are, since the work of Szegő, intimately related to the theory of orthogonal polynomials and they will be even more important for the case of orthogonal rational functions. Some of their elementary properties will be recalled in Section 1.4.

The 2×2 J-unitary and J-contractive matrix functions with entries in the Nevanlinna class will be important when we develop the recurrence relations for the kernels and the orthogonal rational functions. Some of their properties are introduced in Section 1.5.

1.1. Hardy classes

We shall be concerned with complex function theory on the unit circle and the upper half plane. The complex number field is denoted by $\mathbb{C}$. We use the following notation for the unit circle, the open unit disk, and the complement

15

of the closed unit disk:

$$\mathbb{T} = \{z : |z| = 1\}, \qquad \mathbb{D} = \{z : |z| < 1\}, \qquad \mathbb{E} = \{z : |z| > 1\}.$$

The upper bar denotes complex conjugation when appropriate or closures when it concerns sets (e.g., $\overline{\mathbb{D}} = \mathbb{D} \cup \mathbb{T}$ is the closed unit disk and $\overline{\mathbb{C}} = \mathbb{C} \cup \{\infty\}$ is the Riemann sphere). The real axis is denoted as $\mathbb{R}$. Real and imaginary parts of a complex number z are indicated by $\operatorname{Re} z$ and $\operatorname{Im} z$ respectively: $z = \operatorname{Re} z + \mathbf{i} \operatorname{Im} z$, $\mathbf{i}$ is reserved for the unit on the imaginary axis, and the open right half plane is denoted as

$$\mathbb{H} = \{z : \operatorname{Re} z > 0\}.$$

We also need the upper and lower half plane, which we denote as

$$\mathbb{U} = \{z \in \mathbb{C} : \operatorname{Im} z > 0\} \quad \text{and} \quad \mathbb{L} = \{z \in \mathbb{C} : \operatorname{Im} z < 0\}.$$

Furthermore, we denote the real line as $\mathbb{R} = (-\infty, \infty)$, its closure as $\overline{\mathbb{R}} = \mathbb{R} \cup \{\infty\}$, and $\overline{\mathbb{U}} = \mathbb{U} \cup \mathbb{R}$ and $\mathbb{C} = \mathbb{U} \cup \mathbb{L} \cup \mathbb{R}$.

We shall give a uniform treatment for the disk case ($\mathbb{D}$) or the half plane case ($\mathbb{U}$) as much as possible. Therefore we shall use the notation

$$\mathbb{O} = \mathbb{D} \text{ or } \mathbb{U}, \qquad \partial\mathbb{O} = \mathbb{T} \text{ or } \mathbb{R}, \qquad \mathbb{O}^e = \mathbb{E} \text{ or } \mathbb{L}.$$

We first mention that the disk $\mathbb{D}$ and the half plane $\mathbb{U}$ can be conformally mapped onto each other by the Cayley transform τ. We set

$$z = \tau(w) = \mathbf{i}\frac{1-w}{1+w}; \qquad w = \tau^{-1}(z) = \frac{\mathbf{i}-z}{\mathbf{i}+z}, \quad w \in \mathbb{D}, \ z \in \mathbb{U}.$$

The mapping τ is a one-to-one map of the unit disk $\mathbb{D}$ onto the upper half plane $\mathbb{U}$, -1 onto ∞, and $\mathbb{T} \setminus \{-1\}$ onto $\mathbb{R} = (-\infty, \infty)$. Thus the boundary $\mathbb{T}$ is mapped onto the *extended* real line $\overline{\mathbb{R}}$. Thus we note the slight discrepancy in notation for our two cases: $\partial\mathbb{O} = \overline{\partial\mathbb{O}} = \mathbb{T}$ for the disk, but for the half plane $\partial\mathbb{O} = \mathbb{R}$ (not including ∞) whereas $\overline{\partial\mathbb{O}} = \overline{\mathbb{R}}$.

By $\mathcal{P}_n$ we mean the set of polynomials of degree at most n. The set of complex functions holomorphic on X are denoted by $H(X)$.

Let μ be a positive measure on $\mathbb{T}$, whose support is an infinite set. Sometimes it is characterized by a distribution function $\psi(t) = \int_0^t d\mu$, which has an infinite number of points of increase. If $t = e^{\mathbf{i}\alpha} \in \mathbb{T}$ is a point of discontinuity of the distribution function, then $\mu(\{t\})$ is the *concentrated mass* or *point mass* at t. A similar thing can be said about a measure supported on the real line or on any measure space with (positive) measure μ.

The metric spaces $L_p(X, \mu)$, $0 < p \le \infty$ are well known [187, Chap. 3]. In our case $X = \overline{\partial\mathbb{O}}$. The normalized Lebesgue measure for $\mathbb{T}$ is denoted by $\lambda : d\lambda(\theta) = (2\pi)^{-1}d\theta$. For the real line $\mathbb{R}$, this becomes $d\lambda(x) = \pi^{-1}dx$. If $\mu = \lambda$, we just write $L_p(\overline{\partial\mathbb{O}})$ instead of $L_p(\overline{\partial\mathbb{O}}, \lambda)$. Also, at other occasions, we shall drop μ from the notation when $\mu = \lambda$. When we want to stress the difference between the cases $\partial\mathbb{O} = \mathbb{T}$ and $\partial\mathbb{O} = \mathbb{R}$, we write this explicitly (e.g., $L_p(\mathbb{T})$ or $L_p(\mathbb{R})$), but when something is true for both, regardless of the fact $\partial\mathbb{O} = \mathbb{T}$ or $\mathbb{R}$, we also drop $\partial\mathbb{O}$ from the notation. For example, it is well known that $L_p(\mu)$ are complete metric spaces for $1 \le p \le \infty$ and that $L_2(\mu)$ is a Hilbert space.

The inner product in $L_2(\mu)$ is denoted by

$$\langle f, g \rangle_\mu = \int_{t \in \mathbb{T}} f(t)\overline{g(t)}\, d\mu(t) \quad \text{or} \quad \langle f, g \rangle_\mu = \int_{x \in \mathbb{R}} f(x)\overline{g(x)}\, d\mu(x).$$

Depending on the case, our integrals will be over $\mathbb{T}$ or $\mathbb{R}$ in some form or another and we shall not mention this explicitly. So, with an obvious abuse of notation, we shall take the freedom to write the previous integrals as

$$\int f\bar{g}\, d\mu = \int f(t)\overline{g(t)}\, d\mu(t).$$

We shall need many results from H_p theory. These Hardy classes H_p are the next thing we introduce. We consider the disk first. Different equivalent definitions of the Hardy classes for the disk can be given. We use the following: $H_p(\mathbb{D})$ is the set of functions $f \in H(\mathbb{D})$ for which the subharmonic functions $|f|^p$ have an harmonic majorant in $\mathbb{D}$ [92, p. 51; 76, Chap. 10]. This means that

$$\sup_{r<1} \int |f(rt)|^p\, d\lambda(t) = \|f\|_p^p < \infty$$

whereas $H_\infty(\mathbb{D})$ is just the set of bounded analytic functions in $\mathbb{D}$. This definition of H_p classes is conformally invariant. Conformally invariant means that if we replace in that definition $\mathbb{D}$ by $\mathbb{U}$, we get some corresponding H_p classes for the upper half plane. We shall assume that $H_p(\mathbb{U})$ is defined as such. Thus $H_p(\mathbb{U})$ is the set of analytic functions in the upper half plane such that for $0 < p < \infty$

$$\sup_{y>0} \int_{-\infty}^{\infty} |f(x + \mathbf{i}y)|^p\, dx = \|f\|_p^p < \infty$$

whereas $f \in H(\mathbb{U})$ is in $H_\infty(\mathbb{U})$ if and only if

$$\sup_{z \in \mathbb{U}} |f(z)| = \|f\|_\infty < \infty.$$

More generally, these definitions could be used to define H_p classes in any plane domain or Riemann sphere, but we shall not need this. The term conformally invariant is, however, misleading, since one might think that whenever $f \in H_p(\mathbb{D})$ and τ is the Cayley transform as given above, then $f \circ \tau$ is in $H_p(\mathbb{U})$. Unfortunately, this is only true for $p = \infty$, but not for a general $p < \infty$. The classes thus obtained turn out to be too big as one can simply check. For example, $H_1(\mathbb{D})$ contains the constant functions, while obviously $H_1(\mathbb{U})$ does not. To get most of the H_p properties we need, the more restrictive definition we gave above, in terms of subharmonic functions having an harmonic majorant, should be adopted for $H_p(\mathbb{U})$, $p < \infty$.

The following alternative definitions are equivalent with the previous ones. They are classical and can be found, for example, in [92, p. 51], [130, p. 159], and [76, p. 197]. They do use the Cayley transform and the $H_p(\mathbb{D})$ spaces but with a twist:

$$H_\infty(\mathbb{U}) = \{f : f \circ \tau \in H_\infty(\mathbb{D})\},$$
$$H_p(\mathbb{U}) = \{(z+i)^{-2/p} f(z) : f \circ \tau \in H_p(\mathbb{D})\}, \quad 0 < p < \infty.$$

The presence of the extra factor $(z+i)^{2/p}$ in the definition of $H_p(\mathbb{U})$ can be explained as follows. Let us temporarily use an extra argument for a measure to indicate where it is supported. For example, $\lambda(\theta, P) = (2\pi)^{-1}\theta$ is the normalized Lebesgue measure for the interval $P = (-\pi, \pi]$. We have identified this previously with the measure $\lambda(t, \mathbb{T})$ on the unit circle where $t = e^{i\theta}$ by setting $\lambda(e^{i\theta}, \mathbb{T}) = \lambda(\theta, P)$. The normalized Lebesgue measure on the line $\mathbb{R}$ is indicated as $\lambda(x, \mathbb{R}) = \pi^{-1}x$. Similar notational conventions are used for other measures. Later on we shall drop the extra indication, and it should be clear from the context which particular measure is meant. Transforming the Lebesgue measure $\lambda(\theta, P) = \lambda(t, \mathbb{T})$ from $(-\pi, \pi)$ or $\mathbb{T}$ to $\mathbb{R}$ using the relation

$$t = e^{i\theta} = \frac{i-x}{i+x}, \quad x \in \mathbb{R}, \ t \in \mathbb{T}, \ \theta \in (-\pi, \pi), \tag{1.1}$$

results in [130, p. 143]

$$d\lambda(\theta, P) = d\lambda(t, \mathbb{T}) = \frac{d\theta}{2\pi} = \frac{dx}{\pi(1+x^2)} = \frac{d\lambda(x, \mathbb{R})}{1+x^2}.$$

Thus the conformal map of the Lebesgue measure on $\mathbb{T}$ corresponds to the measure

$$d\mathring{\lambda}(x, \mathbb{R}) = \frac{d\lambda(x, \mathbb{R})}{1+x^2}$$

on $\mathbb{R}$. Note that $1 + x^2 = |x + \mathbf{i}|^2$ if $x \in \mathbb{R}$, which explains the extra factor in the definition of $H_p(\mathbb{U})$. We shall use the circle to symbolize the unit disk. For the unit disk, the objects with a circle are the same as those without a circle, but for the half plane, the objects with a circle will refer to objects that are obtained by conformally transforming the corresponding objects for the disk. So we shall also use the notation $\mathring{H}_p(\mathbb{U})$ to mean $\mathring{H}_p(\mathbb{U}) = \{f = g \circ \tau : g \in H_p(\mathbb{D})\} = (z + \mathbf{i})^{2/p} H_p(\mathbb{U})$, or more explicitly

$$f \in \mathring{H}_p(\mathbb{U}) \Leftrightarrow (z + \mathbf{i})^{-2/p} f(z) \in H_p(\mathbb{U}) \Leftrightarrow f \circ \tau \in H_p(\mathbb{D}).$$

This will suffice thus far for the definition of the Hardy classes.

Later on, the following expressions will play an important role. We define

$$\varpi_w(z) = 1 - \overline{w}z \text{ for } \mathbb{D} \quad \text{and} \quad \varpi_w(z) = z - \overline{w} \text{ for } \mathbb{U}. \tag{1.2}$$

Furthermore, for a fixed sequence of points α_i, we shall use the abbreviation $\varpi_i(z) = \varpi_{\alpha_i}(z)$. The point α_0 is special. In the first part (Chapters 2–10), it is always defined as

$$\alpha_0 = 0 \text{ for } \mathbb{D} \quad \text{and} \quad \alpha_0 = \mathbf{i} \text{ for } \mathbb{U}.$$

Hence we have

$$\varpi_0(z) = 1 - \overline{\alpha}_0 z = 1 \text{ for } \mathbb{D} \quad \text{and} \quad \varpi_0(z) = z - \overline{\alpha}_0 = z + \mathbf{i} \text{ for } \mathbb{U}.$$

This notation will be essential in the rest of the text. Notice that for the disk

$$\varpi_z(z) = \varpi_z(z)/\varpi_0(\alpha_0) = \varpi_0(\alpha_0)\overline{\varpi_z(z)} = 1 - |z|^2,$$

whereas for $\mathbb{U}$

$$\varpi_z(z) = 2\mathbf{i} \operatorname{Im} z, \quad \varpi_z(z)/\varpi_0(\alpha_0) = \operatorname{Im} z, \quad \varpi_0(\alpha_0)\overline{\varpi_z(z)} = 4 \operatorname{Im} z.$$

Hence we can characterize the sets $\mathbb{O}$, $\partial\mathbb{O}$, and $\mathbb{O}^e$ as follows:

$$\partial\mathbb{O} = \{z \in \mathbb{C} : \varpi_z(z) = 0\},$$

$$\mathbb{O} = \{z \in \mathbb{C} : \varpi_z(z)/\varpi_0(\alpha_0) > 0\} = \{z \in \mathbb{C} : \varpi_0(\alpha_0)\overline{\varpi_z(z)} > 0\}, \tag{1.3}$$

$$\mathbb{O}^e = \{z \in \mathbb{C} : \varpi_z(z)/\varpi_0(\alpha_0) < 0\} = \{z \in \mathbb{C} : \varpi_0(\alpha_0)\overline{\varpi_z(z)} < 0\}.$$

With this notation, we can define $d\mathring{\lambda}$ by

$$d\mathring{\lambda}(t) = d\mathring{\lambda}(t, \partial\mathbb{O}) = \frac{d\lambda(t, \partial\mathbb{O})}{|\varpi_0(t)|^2},$$

which puts $d\overset{\circ}{\lambda} = d\lambda$ for $\partial\mathbb{O} = \mathbb{T}$ and gives the previous definition with $1 + x^2$ in the denominator for $\partial\mathbb{O} = \mathbb{R}$. Similarly, we shall use $\overset{\circ}{H}_p = \overset{\circ}{H}_p(\mathbb{O})$ to mean $H_p(\mathbb{D})$ for $\mathbb{O} = \mathbb{D}$ and $\overset{\circ}{H}_p(\mathbb{U})$ when $\mathbb{O} = \mathbb{U}$.

We note that $H_p(\mathbb{O}) \subseteq \overset{\circ}{H}_p(\mathbb{O})$ with equality for $\mathbb{O} = \mathbb{D}$. It is well known that $H_p(\mathbb{O})$ is a Banach space for $1 \le p \le \infty$.

The Nevanlinna class $N(\mathbb{O})$ is the class of functions f for which the sub-harmonic function $\log^+ |f(z)| = \max(\log |f(z)|, 0)$ has a harmonic majorant. This class $N(\mathbb{O})$ contains all spaces $H_p(\mathbb{O})$ for $0 < p \le \infty$. It can be characterized as the class of functions that are the ratio of bounded analytic functions:

$$f \in N \Leftrightarrow f = g/h; \quad g, h \in H_\infty, \quad h \text{ has no zeros in } \mathbb{O}.$$

This characterization comprises the contents of a theorem by F. and R. Nevanlinna [76, p. 16]. It is known that each function $f \in N$ has a nontangential limit to the boundary $\partial\mathbb{O}$ a.e. and $\log |f| \in L_1(\partial\mathbb{O}, \overset{\circ}{\lambda})$, unless $f \equiv 0$ [76, p. 17], [186, p. 85]. Moreover, for $p < q$ we have the inclusions $H_q \subset H_p \subset N$ and $H_p \subset \overset{\circ}{H}_p \subset N$.

The operation of taking the complex conjugate of a function defined on the boundary $\partial\mathbb{O}$ is extended to the whole Riemann sphere $\overline{\mathbb{C}}$ by the involution operation

$$f_*(z) = \overline{f(\hat{z})}, \quad \hat{z} = 1/\bar{z} \text{ for } \mathbb{T};$$
$$f_*(z) = \overline{f(\hat{z})}, \quad \hat{z} = \bar{z} \text{ for } \mathbb{R}.$$

Note that for $z \in \partial\mathbb{O}$, $f_*(z)$ is just $\overline{f(z)}$.

The Hardy and Nevanlinna classes of analytic functions in $\mathbb{O}^e$ are indicated by a prime, for example,

$$H'_p = \{f : f_* \in H_p\} \quad \text{and} \quad N' = \{f : f_* \in N\}.$$

The transformation between the Lebesgue measure for the circle and the line, which we gave above, can be generalized for other positive measures. Let $\mu(\theta, P)$ be a measure on $P = (-\pi, \pi]$. We identify this with a measure $\mu(t, \mathbb{T})$ on $\mathbb{T} = \{e^{i\theta} : -\pi < \theta \le \pi\}$. With the above correspondence (1.1) among θ, t, and x, we introduce $\overset{\circ}{\mu}(x, \mathbb{R}) = \mu(\tau^{-1}(x), \mathbb{T})$, which has to be understood in the following sense. Suppose $E \subset \mathbb{T}$ is a $\mu(\cdot, \mathbb{T})$-measurable subset of the unit circle; then the Cayley transform τ will map this onto $E' = \tau(E) \subset \mathbb{R}$. We define a measure $\overset{\circ}{\mu}(\cdot, \mathbb{R})$ on $\mathbb{R}$ by $\overset{\circ}{\mu}(E', \mathbb{R}) = \mu(E, \mathbb{T})$. Moreover, it will be convenient to also define the measure $\mu(E', \mathbb{R}) = \int_{E'}(1 + x^2)\, d\overset{\circ}{\mu}(x, \mathbb{R})$. Note that a possible point mass at $t = -1$, or equivalently at $\theta = \pi$, will result in a point mass at infinity for the transformed measure $\overset{\circ}{\mu}(x, \mathbb{R})$. We also remark that a finite positive measure $\mu(t, \mathbb{T})$ on $\mathbb{T}$ corresponds to a finite positive measure $\mathring{\mu}(x, \overline{\mathbb{R}})$ on $\overline{\mathbb{R}}$. However, if

$$\int d\mathring{\mu}(x, \overline{\mathbb{R}}) = \int \frac{d\mu(x, \overline{\mathbb{R}})}{1 + x^2} < \infty$$

then $\int d\mu(x, \overline{\mathbb{R}})$ need not be finite! If one wants μ to be finite on $\overline{\mathbb{R}}$, then one should impose a stronger condition on the original measures on $\mathbb{T}$. Only for some (not all) finite measures on $\mathbb{T}$ do we find that the corresponding μ is finite on $\overline{\mathbb{R}}$.

In the sequel, measures $\mathring{\mu}$ and μ on $\overline{\partial \mathbb{O}}$ will be related by

$$d\mu(t) = |\varpi_0(t)|^2 \, d\mathring{\mu}(t).$$

As before, $\mathring{\mu} = \mu$ on $\mathbb{T}$, whereas $d\mathring{\mu}(x) = (1 + x^2)^{-1} d\mu(x)$ on $\overline{\mathbb{R}}$.

Let the Lebesgue decomposition of μ be $\mu = \mu_a + \mu_s$ with μ_a satisfying

$$d\mu_a = \omega \, d\lambda, \tag{1.4}$$

the *absolutely continuous part*: $\mu_a \ll \lambda$. The function $\mu' = d\mu_a/d\lambda = \omega \in L_1$ is a weight function. The remaining μ_s is the *singular part* w.r.t. λ : $\mu_s \perp \lambda$. Obviously $\mu' = d\mu/d\lambda = d\mathring{\mu}/d\mathring{\lambda}$.

Define the *moments* of a positive measure $\mathring{\mu}$ on $[-\pi, \pi]$ as the Fourier–Stieltjes coefficients

$$c_k = \int e^{-ik\theta} \, d\mathring{\mu}(\theta), \quad k \in \mathbb{Z}. \tag{1.5}$$

Clearly $c_{-k} = \overline{c}_k$ and $|c_k| \leq c_0$ for a positive measure μ. The computations will simplify substantially if we suppose the measure to be normalized. This means that we divide out c_0, which is always possible since it is not zero, and we shall thus set $c_0 = 1$ from now on, which is no restriction of the generality. In other words, we work with the normalized measure that satisfies $\int d\mathring{\mu} = 1$.

For a function belonging to $H_1(\mathbb{D})$, the Fourier coefficients are defined as

$$\hat{f}_k = \int t^{-k} f(t) \, d\lambda(t).$$

It is known [76, p. 38] that $H_p(\mathbb{D})$, $1 \leq p \leq \infty$ is precisely the class of functions in $H(\mathbb{D})$ whose boundary function is in $L_p(\mathbb{D})$ and whose Fourier coefficients vanish for $k < 0$. One has for $f \in H_p(\mathbb{D})$

$$f(z) = \sum_{k=0}^{\infty} \hat{f}_k z^k, \quad z \in \mathbb{D}.$$

For the real axis, the Fourier transform of $f \in H_1(\mathbb{R})$ is

$$\hat{f}(x) = \int e^{-\mathrm{i}xt} f(t) \, d\lambda(t)$$

(to be evaluated as a Cauchy principal value integral), and the class $H_p(\mathbb{U})$ is precisely the class of functions in $H(\mathbb{U})$ whose boundary value belongs to $L_p(\mathbb{R})$ and whose Fourier transform vanishes for $x < 0$ a.e. One has

$$f(z) = 2 \int_0^\infty e^{\mathrm{i}zt} \hat{f}(t) \, dt, \quad z \in \mathbb{U}.$$

The latter is known as the Paley–Wiener theorem in the case $p = 2$.

Concerning Cauchy integrals we recall the following facts. Consider the case of the unit circle first [76, p. 39]. Let μ be a complex measure of bounded variation. Then a Cauchy–Stieltjes integral is of the form

$$F(z) = \int \frac{t \, d\mu(t)}{t - z}.$$

We get a Cauchy integral if $d\mu$ is replaced by $\psi \, d\lambda$, $\psi \in L_1(\mathbb{T})$. This represents two analytic functions: one in $\mathbb{D}$ and one in $\mathbb{E}$. The function $F(z)$, $z \in \mathbb{D}$ is in $H_p(\mathbb{D})$ for all $p < 1$. If $F \equiv 0$ in $\mathbb{E}$, then $F \in H_1(\mathbb{D})$. If $F \in H_p(\mathbb{D})$ for some p, $1 \leq p < \infty$, then ψ is the boundary function for F.

There are some consequences: When

$$D(t, z) = \frac{t + z}{t - z} \quad \text{and} \quad f \in L_1(\mu)$$

then

$$\int D(t, z) f(t) \, d\mu(t)$$

is an analytic function in $\mathbb{D}$. Also, for any function f analytic in $\mathbb{D}$ and integrable on $\mathbb{T}$,

$$\int f(t) \, d\lambda(t) = f(0). \tag{1.6}$$

For the real line, we have [76, p. 195] $f \in H_p(\mathbb{U})$, $1 \leq p < \infty$ iff (if and only if)

$$f(z) = \frac{1}{2\mathrm{i}} \int \frac{\psi(t)}{t - z} \, d\lambda(t), \quad \psi \in L_p(\mathbb{R}), \ z \in \mathbb{U},$$

and the integral is zero for $z \in \mathbb{L}$. The function ψ is the boundary function of f.

For example, if $f \in \overset{\circ}{H}_2(\mathbb{U})$, then it follows that $\overset{\circ}{f}(z) = f(z)/(z + \mathbf{i}) \in H_2(\mathbb{U})$, so that

$$\int f(t)\, d\overset{\circ}{\lambda}(t) = \int \frac{\overset{\circ}{f}(t)}{t - \mathbf{i}}\, d\lambda(t) = f(\mathbf{i}). \tag{1.7}$$

Hence this is zero iff f vanishes in $\mathbf{i}$.

Putting (1.6) and (1.7) together, we get: If $f \in \overset{\circ}{H}_2(\mathbb{O})$ and $f(\alpha_0) = 0$, then

$$\int f(t)\, d\overset{\circ}{\lambda}(t) = 0. \tag{1.8}$$

In order to have a uniform notation for the disk and the half plane, we define the Cauchy kernel as

$$C(t, z) = \frac{1}{\varpi_0(\alpha_0)\varpi_{z*}(t)} = \begin{cases} \dfrac{t}{t - z} & \text{for } \mathbb{D}, \\[2mm] \dfrac{1}{2\mathbf{i}(t - z)} & \text{for } \mathbb{U}. \end{cases} \tag{1.9}$$

Hence we have for $f \in H_1(\mathbb{O})$

$$\int C(t, z) f(t)\, d\lambda(t) = \begin{cases} f(z) & \text{for } z \in \mathbb{O}, \\ 0 & \text{for } z \in \mathbb{O}^e. \end{cases}$$

1.2. The classes $\mathcal{C}$ and $\mathcal{B}$

The class $\overline{\mathcal{C}}$ of *positive real* functions, also known as the class of *Carathéodory functions*, will be introduced now, as well as the closed unit ball $\overline{\mathcal{B}}$ in H_∞, which corresponds to the class of *Schur functions*.

We shall use the notation $\mathbb{H} = \{z \in \mathbb{C} : \operatorname{Re} z > 0\}$ for the (open) right half plane.

The class $\overline{\mathcal{C}}$ of Carathéodory functions is defined as follows:

$$\overline{\mathcal{C}} = \overline{\mathcal{C}}(\mathbb{O}) = \{f \in H(\mathbb{O}) : f(\mathbb{O}) \subset \overline{\mathbb{H}} = \mathbb{H} \cup \mathbf{i}\mathbb{R}\}. \tag{1.10}$$

The class $\overline{\mathcal{B}}$ of *bounded analytic* functions or *Schur* functions is defined as

$$\overline{\mathcal{B}} = \overline{\mathcal{B}}(\mathbb{O}) = \{f \in H(\mathbb{O}) : f(\mathbb{O}) \subset \overline{\mathbb{D}} = \mathbb{D} \cup \mathbb{T}\}. \tag{1.11}$$

Since $\overline{\mathcal{B}}$ can be regarded as the closed unit ball in $H(\mathbb{O})$, we have chosen the notation $\overline{\mathcal{B}}$ for it.

However, in the sequel, we shall work most of the time with slightly smaller classes:

$$\mathcal{C}(\mathbb{O}) = \{f \in H(\mathbb{O}) : f(\mathbb{O}) \subset \mathbb{H}\} \tag{1.12}$$

and

$$\mathcal{B}(\mathbb{O}) = \{f \in H(\mathbb{O}) : f(\mathbb{O}) \subset \mathbb{D}\}. \tag{1.13}$$

Note that because $\overline{\mathcal{B}}(\mathbb{O})$ consists of analytic functions, it follows by the maximum modulus principle that a function $f \in \overline{\mathcal{B}}(\mathbb{O})$ can only take a value that is 1 in modulus when it is evaluated on the boundary $\partial\mathbb{O}$, unless it is a constant function of modulus 1. Thus $\mathcal{B}(\mathbb{O})$ merely excludes the unimodular constant functions from $\overline{\mathcal{B}}(\mathbb{O})$. Similarly, $\mathcal{C}(\mathbb{O})$ merely excludes the constant functions with values on the imaginary axis from $\overline{\mathcal{C}}(\mathbb{O})$. The classical *Schwarz* lemma for functions in $\overline{\mathcal{B}}(\mathbb{D})$ reads

Lemma 1.2.1 (Schwarz's lemma). Suppose $f \in \overline{\mathcal{B}}(\mathbb{D})$ and $f(0) = 0$. Then

$$|f'(0)| \le 1 \quad and \quad |f(z)| \le |z|, \quad z \in \mathbb{D}. \tag{1.14}$$

Equality holds if and only if $f(z) = c\,z$ with $|c| = 1$.

Proof. This is a classical result and we are not going to prove it here. See, for example, Ref. [46, p. 191] or [92, p. 1]. □

Note that for $f \in \overline{\mathcal{B}}(\mathbb{D}) \setminus \mathcal{B}(\mathbb{D})$, it can never be true that $f(0) = 0$; thus the lemma actually gives a statement about functions in $\mathcal{B}(\mathbb{D})$.

A *Möbius transform* is a linear fractional transform that conformally maps the unit circle/disk onto itself. It has the general form

$$M_{a,b} : z \mapsto \frac{az + b}{\overline{b}z + \overline{a}}, \quad |b| < |a|, \tag{1.15}$$

or, equivalently,

$$M_\alpha : z \mapsto \eta \frac{z - \alpha}{1 - \overline{\alpha}z}, \quad \alpha \in \mathbb{D}, \ |\eta| = 1. \tag{1.16}$$

Note that M_α is the most general conformal map of this type that transforms α into the origin. The unit circle $\mathbb{T}$ is transformed into itself. The inverse transformation is given by

$$M_\alpha^{-1} : w \mapsto \frac{w/\eta + \alpha}{1 + \overline{\alpha}w/\eta}. \tag{1.17}$$

Clearly M_α itself is a function from the class $\mathcal{B}(\mathbb{D})$. Usually, since η is not relevant, it is put equal to 1.

To give an invariant form of the Schwarz lemma, we recall our notation (1.2) and we set

$$\zeta_w(z) = \frac{\varpi_w^*(z)}{\varpi_w(z)} = \begin{cases} (z-w)/(1-\overline{w}z) & \text{for } \mathbb{D}, \\ (z-w)/(z-\overline{w}) & \text{for } \mathbb{U}. \end{cases} \quad (1.18)$$

A form of the Schwarz lemma that is invariant can now be formulated:

Theorem 1.2.2. *Let* $f \in \overline{\mathcal{B}}$ *and* $z, w \in \mathbb{O}$. *Then*

$$|M_{f(w)}(f(z))| \le |\zeta_w(z)|, \quad z \ne w \quad (1.19)$$

and in case $z = w$, *we get*

$$\frac{|f'(z)|}{1 - |f(z)|^2} \le \frac{1}{\varpi_z(z)}, \quad z \in \mathbb{O}. \quad (1.20)$$

Equality holds if and only if f *is a Möbius transformation.*

Proof. This result also is classical for $\mathbb{D}$. See, for example, Ref. [46, p. 192] or [92, p. 2]. For $\mathbb{U}$, the result is easily obtained by a Cayley transform. □

We note:

1. The expression

$$\rho(z, w) = |M_w(z)| = \left| \frac{z-w}{1-\overline{w}z} \right|, \quad z, w \in \mathbb{D}$$

is invariant under Möbius transformations. It is called the *pseudo-hyperbolic distance* between z and w and it forms a metric in $\mathbb{D}$. In the case of the disk, ζ_w coincides with M_w, and thus inequality (1.19) can be written as $\rho(f(z), f(w)) \le \rho(z, w)$, and it implies that $f \in \mathcal{B}(\mathbb{D})$ is Lipschitz continuous w.r.t. the pseudo-hyperbolic distance.
2. The second form (1.20) is the limiting case of the first one (1.19) for $z \to w$. The following property forms the basis of the Nevanlinna–Pick algorithm.

Theorem 1.2.3. *Let* M_α *be a Möbius transform as defined in (1.16) and* ζ_w *be as defined in (1.18).* $\mathcal{B} = \mathcal{B}(\mathbb{O})$.

1. *Let* $f \in \mathcal{B}$ *and* $\alpha \in \mathbb{D}$. *Then* $M_\alpha(f) \in \mathcal{B}$. *More precisely:*

$$M_\alpha(\mathcal{B}) = \mathcal{B}. \quad (1.21)$$

2. *If $f \in \mathcal{B}$ and $w \in \mathbb{O}$, then*

$$M_{f(w)}(f)/\zeta_w \in \mathcal{B}. \qquad (1.22)$$

3. *If $f \in \mathcal{B}$ and $f(w) = 0$ for some $w \in \mathbb{O}$, then $f/\zeta_w \in \mathcal{B}$.*

Proof.

1. Since $M_\alpha \in \mathcal{B}$ and the composition of functions in $\mathcal{B}$ is also in $\mathcal{B}$, we find that $M_\alpha(\mathcal{B}) \subset \mathcal{B}$. Hence $\mathcal{B} \subset M_\alpha^{-1}(\mathcal{B})$. But since $M_\alpha^{-1} = M_{-\alpha}$ (take $\eta = 1$ in (1.16), without loss of generality) we also have $M_{-\alpha}(\mathcal{B}) \subset \mathcal{B}$. Thus

$$\mathcal{B} \subset M_\alpha^{-1}(\mathcal{B}) = M_{-\alpha}(\mathcal{B}) \subset \mathcal{B},$$

   so that equality holds.
2. This is a rewriting of the invariant form of Schwarz's lemma.
3. This is a special case of 2 because $f(w) = 0$.   □

The link with class $\mathcal{C}$ functions can be made as follows. The *Cayley transform*

$$c : z \mapsto \frac{1 - z}{1 + z} \qquad (1.23)$$

is a one-to-one map of $\mathbb{D}$ onto $\mathbb{H}$ and of $\mathbb{T}$ onto $i\overline{\mathbb{R}}$ (-1 is mapped onto ∞). The following result is now simple to see.

Theorem 1.2.4. *The following relations between class $\mathcal{C}$ and class $\mathcal{B}$ exist.*

1. *The Cayley transform c of (1.23) is a one-to-one map of $\mathcal{C}$ onto $\mathcal{B}$. That is,*

$$c(\mathcal{B}) = \mathcal{C} \quad and \quad c(\mathcal{C}) = \mathcal{B}. \qquad (1.24)$$

2. *For the extended classes, define $\overline{\mathcal{B}}' = \overline{\mathcal{B}} \setminus \{-1\}$, that is, we exclude the constant function $f \equiv -1$. Then*

$$c(\overline{\mathcal{B}}') = \overline{\mathcal{C}} \quad and \quad c(\overline{\mathcal{C}}) = \overline{\mathcal{B}}'. \qquad (1.25)$$

3. *More generally, let $\gamma \in \mathbb{H}$, $\eta \in \mathbb{T}$, and $f, g \in H(\mathbb{O})$ be related by*

$$\eta \frac{f - \gamma}{f + \overline{\gamma}} = g \Leftrightarrow \frac{g\overline{\gamma} + \eta\gamma}{\eta - g} = f.$$

Then $f \in \mathcal{C}$ iff $g \in \mathcal{B}$.

Proof.

1. If $f \in B$, then $|f| < 1$ in $\mathbb{O}$ so that $|1 + f| > 0$ in $\mathbb{O}$. Hence $c(f) \in H(\mathbb{O})$ and conversely, if $f \in C$, then $1 + f$ has strictly positive real part in $\mathbb{O}$ and therefore $1 + f$ does not vanish. Thus again, $f \in H(\mathbb{O})$. The rest follows from the one-to-one map given by the Cayley transform.
2. Here we have to exclude $f(z) \equiv -1$ because then the transform would fail.
3. This is proved along the same lines as 1.

This concludes our proof. $\qquad\qquad\qquad\qquad\qquad\qquad\qquad\qquad\qquad\qquad\qquad\qquad$ □

We give now integral expressions for functions in C, the well-known Riesz–Herglotz–Nevanlinna representation of class C functions.

We start with the case of the disk. Therefore we introduce the *Riesz–Herglotz kernel*

$$D(t, z) = \frac{t + z}{t - z}. \tag{1.26}$$

To a positive measure on $[0, 2\pi]$ (i.e., on $\mathbb{T}$), we associate the C function $\Omega_\mu(z)$ in $\mathbb{D}$:

$$\Omega_\mu(z) = \mathrm{i}c + \int \frac{e^{\mathrm{i}\theta} + z}{e^{\mathrm{i}\theta} - z} \, d\mu(\theta), \quad c \in \mathbb{R}, \; z \in \mathbb{D}. \tag{1.27}$$

This function is analytic in $\mathbb{D}$ and belongs to H_p for all $p < 1$ [76, p. 34] and hence it has a nontangential limit a.e. The constant c is the imaginary part of $\Omega_\mu(0) = 1 + \mathrm{i}c$. This integral representation of C functions is called the *Riesz–Herglotz* representation for functions of class C. Conversely, every C function can be represented in this form. The relation between μ and Ω_μ is one to one except for the constant c. Since μ is uniquely defined by Ω_μ, we shall refer to it as the *Riesz–Herglotz measure* for Ω_μ.

Note that the real part of the kernel $D(t, z)$, $t \in \mathbb{T}$, $z \in \mathbb{D}$ is given by

$$P(t, z) = \mathrm{Re}\, D(t, z) = \frac{1 - |z|^2}{|t - z|^2}, \quad t \in \mathbb{T}. \tag{1.28}$$

It is the *Poisson kernel* for $\mathbb{D}$. It features in the *Poisson–Stieltjes integral*, which represents the (positive) real part of Ω_μ:

$$\mathrm{Re}\, \Omega_\mu(z) = \int \mathrm{Re}\, D(t, z) \, d\mu(t) = \int P(t, z) \, d\mu(t)$$

$$= \int \mathrm{Re}\, \frac{t + z}{t - z} \, d\mu(t) = \int \frac{1 - |z|^2}{|t - z|^2} \, d\mu(t), \quad z \in \mathbb{D}.$$

This is obviously positive since the integrand on the right is positive. By Fatou's theorem [116, p. 34], this also has a radial limit given by

$$\lim_{r\uparrow 1} \operatorname{Re} \Omega_\mu(re^{i\theta}) = \mu'(\theta) \quad \text{a.e.} \tag{1.29}$$

Here μ' is the density of the absolutely continuous part of μ in its Lebesgue decomposition, and at the discontinuous points, it can be replaced by the symmetric derivative, that is,

$$\mu'(\theta) = \lim_{h\to 0} \frac{\mu((\theta - h, \theta + h))}{2h}.$$

See Ref. [76, p. 4].

The relation $\operatorname{Re} D(t, z) = P(t, z)$ for $t \in \mathbb{T}$ can be generalized to a relation for $t \in \mathbb{C}$ by defining

$$P(t, z) = \frac{1}{2}[D(t, z) + D(t, z)_*] = \frac{t(1 - |z|^2)}{(t - z)(1 - \bar{z}t)}, \tag{1.30}$$

where the substar is w.r.t. the variable t.

Using a conformal mapping from $\mathbb{D}$ to $\mathbb{U}$, we see that the Riesz–Herglotz kernel for $\mathbb{D}$ transforms into the *Nevanlinna kernel* for $\mathbb{U}$:

$$D(t, z) = \frac{1}{i}\frac{1 + tz}{t - z}. \tag{1.31}$$

The Poisson kernel for $\mathbb{D}$ transforms into

$$P(t, z)(1 + t^2),$$

where now $P(t, z)$ is the Poisson kernel for $\mathbb{U}$, which is defined as

$$P(t, z) = \frac{\operatorname{Im} z}{|t - z|^2}, \quad t \in \mathbb{R}. \tag{1.32}$$

Now we do not have $P(t, z) = \operatorname{Re} D(t, z)$, but instead

$$\frac{1}{2}[D(t, z) + D(t, z)_*] = P(t, z)(1 + t^2) = \frac{1}{2i}\frac{z - \bar{z}}{(t - z)(t - \bar{z})}(1 + t^2).$$

Note that with our previous definitions, we can give an invariant expression for the Poisson kernel that catches both the disk and the half plane case:

$$P(t, z) = \frac{\varpi_z(z)}{\varpi_0(\alpha_0)}\frac{1}{\varpi_z(t)\varpi_{z*}(t)}, \tag{1.33}$$

whereas

$$\frac{1}{2}[D(t,z) + D(t,z)_*] = P(t,z)\varpi_0(t)\varpi_{0*}(t)$$

is also invariant.

For functions in $\mathcal{C}(\mathbb{U})$, we have the following *Nevanlinna representation*:

$$\Omega_\mu(z) = \mathbf{i}c + \int D(t,z)\,d\mathring{\mu}(t), \quad c \in \mathbb{R}, \ z \in \mathbb{U}, \qquad (1.34)$$

where $\mathring{\mu}$ is a finite positive measure for $\overline{\mathbb{R}}$. Recall that our integral is over the extended real line $\overline{\mathbb{R}}$. If there is a point mass $b \geq 0$ at infinity, we could split it off explicitly and write

$$\Omega_\mu(z) = \mathbf{i}c - \mathbf{i}bz + \int_{\mathbb{R}} D(t,z)\,d\mathring{\mu}(t), \quad b \geq 0$$

[115, p. 588; 130, p. 144]. This is called its Nevanlinna representation.

If $\sup |y\Omega(\mathbf{i}y)|$ is bounded for $y > 1$, then μ will indeed be a finite measure and there will be no mass at infinity and the representation can be simplified even more to a Hamburger representation:

$$\Omega_\mu(z) = \frac{2}{\mathbf{i}} \int \frac{d\mu(t)}{t - z}.$$

See Refs. [115, p. 590] and [2, p. 92].

The correspondence between Ω_μ and μ is one to one except for the constant term. The measure $\mathring{\mu}$ will be called the *Nevanlinna measure* of Ω_μ.

The real part of Ω_μ is

$$\mathrm{Re}\,\Omega_\mu(z) = by + \int \mathrm{Re}\,D(t,z)\,d\mathring{\mu}(t), \quad z = x + \mathbf{i}y \in \mathbb{U}$$

$$= by + \int P(t,z)\,d\mu(t) > 0.$$

Consequently, Ω_μ is analytic in $\mathbb{U}$ with values in $\mathbb{H}$, which confirms that it is a positive real function.

Again by Fatou's theorem for the half plane [116, p. 123], we know that the nontangential limit of $\mathrm{Re}\,\Omega_\mu$ on the boundary converges a.e. to the density of the absolutely continuous part of the measure μ:

$$\lim_{y \downarrow 0} \mathrm{Re}\,\Omega_\mu(x + \mathbf{i}y) = \mu'(x)$$

[92, p. 29], [130, p. 146].

We shall in the rest of the text use Riesz–Herglotz representation, Riesz–Herglotz measure, etc. in the case of the unit circle $\mathbb{T}$, and Nevanlinna representation, Nevanlinna measure, etc. for the case of the real line $\mathbb{R}$. For the general case $\partial \mathbb{O}$, we use the adjective Riesz–Herglotz–Nevanlinna instead.

As it has been said before, in the sequel we assume that the measure is normalized by $\int d\mathring{\mu} = 1$ and we shall normalize the $\mathcal{C}$ function Ω_μ by $\Omega_\mu(\alpha_0) = 1$. This avoids the extra constant c and we get a strict one-to-one relation for $z \in \mathbb{O}$ between the positive measure $\mathring{\mu}$ and the $\mathcal{C}$ function Ω_μ:

$$\Omega_\mu(z) = \int D(t, z) \, d\mathring{\mu}(t),$$

$$\mathrm{Re}\,\Omega_\mu(z) = \int \mathrm{Re}\, D(t, z) \, d\mathring{\mu}(t) = \int P(t, z) \, d\mu(t).$$

When $\Omega_\mu \in H_1(\mathbb{D})$, the analysis simplifies considerably, because then μ is absolutely continuous so that the Fourier coefficients of μ are equal to the Taylor coefficients of Ω_μ, since indeed writing

$$D(t, z) = \frac{t + z}{t - z} = 1 + 2 \sum_{k=1}^{\infty} z^k t^{-k}, \quad z \in \mathbb{D}$$

gives

$$\Omega_\mu(z) = c_0 + 2 \sum_{k=1}^{\infty} c_k z^k, \quad c_0 = 1, \tag{1.35}$$

which converges locally uniformly in $\mathbb{D}$. Any positive real function Ω of $H_1(\mathbb{D})$ with $\Omega(\alpha_0) > 0$ can be characterized by

$$\Omega(z) = \int D(t, z) \mathrm{Re}\, \Omega(t) \, d\mathring{\lambda}(t). \tag{1.36}$$

Note that the converse is not true: the measure μ can be absolutely continuous without Ω_μ being an H_1 function.

The relation (1.36) holds for $\mathbb{D}$ and $\mathbb{U}$ simultaneously since we wrote $d\mathring{\lambda}$ instead of $d\lambda$. However, the relation (note λ and not $\mathring{\lambda}$)

$$\mathrm{Re}\,\Omega(z) = \int P(t, z) \mathrm{Re}\, \Omega(t) \, d\lambda(t)$$

holds for both $\mathbb{D}$ and $\mathbb{U}$.

This is a special case of the more general theorem [92, p. 61] that says that any $f \in H_1(\mathbb{O})$ can be recovered from its boundary function by a Poisson integral

$$f(z) = \int P(t, z) f(t) \, d\lambda(t), \quad z \in \mathbb{O}.$$

This formula also holds when f is replaced by its real part, Re f. Conversely, if μ is a finite complex measure such that the Poisson–Stieltjes integral

$$f(z) = \int P(t, z) \, d\mu(t)$$

is analytic in $\mathbb{O}$, then μ is absolutely continuous: $d\mu = f(t) \, d\lambda(t)$ with $f(t)$ the boundary function of $f(z)$.

1.3. Factorizations

It is a classical result [76, pp. 24,193] that every $f \in \mathring{H}_p(\mathbb{O}), 0 < p < \infty$ has a canonical inner–outer factorization. This means that there exists an essentially unique factorization

$$f \in \mathring{H}_p(\mathbb{O}) \Leftrightarrow f(z) = U(z) F(z) \tag{1.37}$$

with U an inner function and F an outer function in $\mathring{H}_p(\mathbb{O})$.

An *inner* function U is a function $U \in \mathcal{B}(\mathbb{O})$ with

$$|U(t)| = 1 \quad \text{a.e. on } \partial\mathbb{O}.$$

An outer function $F \in \mathring{H}_p(\mathbb{O})$ has the form

$$F(z) = e^{i\gamma} \exp \left\{ \int D(t, z) \log \psi(t) \, d\mathring{\lambda}(t) \right\}, \quad \gamma \in \mathbb{R}, \tag{1.38}$$

with

$$\log \psi \in L_1(\mathring{\lambda}) \quad \text{and} \quad \psi \in L_p(\mathring{\lambda}).$$

In (1.37), F is of the form (1.38) with $\psi = |f|$.

The inner–outer factorization holds also for a function $f \in H_p(\mathbb{O})$ when in the definition of the outer function, the condition $(|f| =) \psi \in L_p(\mathring{\lambda})$ is replaced by $\psi \in L_p(\lambda)$ [76, p. 194].

Since an outer function has no zeros in $\mathbb{O}$, its inverse will be in $H(\mathbb{O})$.

An example of an inner function is a Blaschke product. It is defined as

$$B(z) = \prod_n \zeta_n(z) \tag{1.39}$$

with

$$\zeta_n(z) = z_n \frac{\varpi_n^*(z)}{\varpi_n(z)} = \begin{cases} z_n \dfrac{z - \alpha_n}{1 - \overline{\alpha}_n z} & \alpha_n \in \mathbb{D} \text{ for } \mathbb{D}, \\[3mm] z_n \dfrac{z - \alpha_n}{z - \overline{\alpha}_n} & \alpha_n \in \mathbb{U} \text{ for } \mathbb{U}. \end{cases} \tag{1.40}$$

The convergence factors $z_n \in \mathbb{T}$ are defined as

$$z_n = \begin{cases} -\dfrac{|\alpha_n|}{\alpha_n} & \text{for } \mathbb{D}, \\[3mm] \dfrac{|1 + \alpha_n^2|}{1 + \alpha_n^2} & \text{for } \mathbb{U}. \end{cases} \tag{1.41}$$

For $\alpha_n = \alpha_0$, we set $z_n = 1$ by convention.

A Blaschke product converges iff

$$\sum (1 - |\alpha_n|) < \infty \quad \text{for } \mathbb{D},$$

$$\sum \frac{\operatorname{Im} \alpha_n}{1 + |\alpha_n|^2} < \infty \quad \text{for } \mathbb{U}.$$

In the case of the half plane, we may replace the convergence condition by $\sum \operatorname{Im} \alpha_n < \infty$ if we know that the moduli $|\alpha_n|$ are bounded [92, p. 56].

Any inner function has the form

$$U(z) = e^{i\gamma} B(z) S(z), \quad \gamma \in \mathbb{R},$$

where B is a Blaschke product and S is a singular inner function, which is of the form

$$S(z) = \exp\left\{ -\int D(t, z) \, d\nu(t) \right\},$$

where ν is a bounded, positive, singular ($\nu' = 0$ a.e.) measure.

In (1.37), the Blaschke factor of the inner factor U catches all the zeros α_n of f.

Inner functions in $\mathbb{O}$ have a *pseudo-meromorphic extension* across the boundary $\partial \mathbb{O}$ [73]. This means the following: Because $U \in H_p$, it is an analytic function in $\mathbb{O}$ and therefore $U_* \in H_p'$, and thus it is analytic in $\mathbb{O}^e$. Moreover, on the boundary $\partial \mathbb{O}$ since we have almost everywhere for any inner function $|U|^2 = 1$ or $U U_* = 1$, we can write $U = 1/U_*$ on $\partial \mathbb{O}$. In this way, U has an analytic extension to the whole complex Riemann sphere, where we have to exclude the poles $\hat{\alpha}_j$, $j = 1, 2, \ldots$ of course as well as the points of $\partial \mathbb{O}$ that are in the support of the singular measure ν generating the singular part of U. The nontangential limits from outside or inside $\mathbb{O}$ to the boundary $\partial\mathbb{O}$ coincide wherever they exist. See also Ref. [92, p.75 ff]. Douglas, Shapiro, and Shields [73] showed that a general function $f \in H_2$ has a pseudo-meromorphic extension across $\partial\mathbb{O}$ if there exists an inner function $U \in H_2$ such that on $\partial\mathbb{O}$ we have $\overline{U}f \in H_2'$ or, equivalently, if f can be factored as $f = h_*/U_*$ on $\partial\mathbb{O}$ with $h \in H_2$ and U inner in H_2. Again, the left-hand side has an extension to $\mathbb{O}$ and the right-hand side to $\mathbb{O}^e$, which defines f in the sphere $\overline{\mathbb{C}}$.

Let μ' be the density of the absolutely continuous part of the positive measure μ. Suppose the *Szegő condition* $\log \mu' \in L_1(d\overset{\circ}{\lambda})$, that is,

$$\int \log \mu'(t)\, d\overset{\circ}{\lambda}(t) > -\infty \qquad (1.42)$$

is satisfied, then we can define a *spectral factor* of μ' as

$$\sigma(z) = c \exp\left\{ \frac{1}{2} \int D(t, z) \log \mu'(t)\, d\overset{\circ}{\lambda}(t) \right\}, \quad z \in \mathbb{O},\ c \in \mathbb{T}. \qquad (1.43)$$

It is defined up to an arbitrary unimodular constant factor c. We shall refer to *the* spectral factor when we set $c = 1$. Note that then $\sigma(\alpha_0) > 0$. The spectral factor is an *outer* function in H_2. Outer implies that σ as well as $1/\sigma$ are both in H_2. See, for example, Ref. [187]. Since σ is in H_2, it has a nontangential limit that satisfies

$$\mu'(t) = |\sigma(t)|^2 \quad \text{a.e. } t \in \partial\mathbb{O}. \qquad (1.44)$$

Note also that we have

$$|\sigma(z)|^2 = \exp\left\{ \int P(t, z) \log \mu'(t)\, d\lambda(t) \right\}, \quad z \in \mathbb{O}.$$

As one can see from its definition, the spectral factor σ does not depend on the singular part of the measure. It is completely defined in terms of the absolutely continuous part. Recall that $d\mu_s = d\mu - \mu' d\lambda = d\mu - d\mu_a$. From the Szegő theory of orthogonal polynomials, we know that in the circle case $1/\sigma$ vanishes $d\mu_s$ a.e. if $\log \mu' \in L_1$ [87, p. 202]. The same is true for the real line: $1/\sigma = 0\, d\mu_s$ a.e. on $\partial\mathbb{O}$.

The condition $\log \mu' \in L_1$ is fundamental in the theory of Szegő for orthogonal polynomials on the unit circle. Szegő's theory has been extended beyond this condition if $\mu' > 0$ a.e. on $\mathbb{T}$ [152].

Suppose that the spectral factor σ has a pseudo-meromorphic extension across $\partial\mathbb{O}$. Then the relations

$$\mu'(z) = \sigma(z)\sigma_*(z) = \frac{1}{2}[\Omega_\mu(z) + \Omega_{\mu*}(z)], \qquad (1.45)$$

valid on $\partial\mathbb{O}$, can be extended to $\overline{\mathbb{C}}$.

Inner and outer functions are also related to (*shift*) *invariant subspaces* of H_2 [110]. One says that a subspace $M \subset H_2$ is S-invariant if $f \in M \Rightarrow Sf \in M$, where S is a *shift* operator. A shift operator is a partial isometry. For example, in $H_2(\mathbb{D})$, the multiplication with z is a canonical shift operation. For $H_2(\mathbb{U})$, usually $e^{i\gamma z}$ with $\gamma > 0$ is the shift operator. However, it is shown in Ref. [116, p. 107] that a subspace of $H_2(\mathbb{U})$ is invariant under multiplication with $e^{i\gamma z}$ iff it is invariant under multiplication with the *canonical shift* $(z - i)/(z + i)$ [186, p. 93]. Thus $\zeta_0(z)$ is the canonical shift operator for $\mathbb{O}$.

The classical theorem of Beurling–Lax [116, Chap. 7] says that M is a shift invariant subspace of $H_2 = H_2(\mathbb{O})$ iff there exists an essentially (up to a constant factor) unique inner function U of H_2 such that $M = U H_2$.

An outer function $F \in H_2$ can also be characterized by the fact that the set $\{S^k F\}_{k \geq 0}$ is dense in H_2.

1.4. Reproducing kernel spaces

In this section we recall some definitions and properties of reproducing kernel spaces. The necessary background can be found in Ref. [148].

Definition 1.4.1 (Reproducing kernel). *Let H be a Hilbert space of functions defined on X with inner product $\langle \cdot, \cdot \rangle$. Then we call $k_w(z) = k(z, w)$ a reproducing kernel if*

1. $k_w(z) \in H$ for all $w \in X$,
2. $\langle f, k_w \rangle = f(w)$ for all $w \in X$ and $f \in H$.

For example, $(1 - \overline{w}z)^{-1}$ is a reproducing kernel for $H_2(\mathbb{D})$ since we have [186, p. 15]

$$\langle f(t), 1/(1 - \overline{w}t) \rangle = f(w), \quad f \in H_2(\mathbb{D}), \quad w \in \mathbb{D},$$

and $(z - \overline{w})^{-1}$ is a reproducing kernel for $H_2(\mathbb{U})$ because [186, p. 92]

$$\langle f(t), 1/(t - \overline{w}) \rangle = f(w), \quad f \in H_2(\mathbb{U}), \quad w \in \mathbb{U}.$$

In both cases the inner product represents the Cauchy integral for the appropriate space $H_2(\mathbb{O})$.

It is a well known property [148] that if the Hilbert space is separable and $\{\phi_k\}_{k \in \Gamma}$ is an orthonormal basis, then the unique reproducing kernel is given by

$$k(z, w) = \sum_{k \in \Gamma} \phi_k(z) \overline{\phi_k(w)}.$$

These reproducing kernels can also be used to find best approximants in subspaces as the following property shows.

Theorem 1.4.1. *Let H be a separable Hilbert space and K a closed subspace with reproducing kernel $k_w(z) = k(z, w)$. Then the best approximant (w.r.t. the norm $\|\cdot\| = \langle \cdot, \cdot \rangle^{1/2}$) of $f \in H$ from K is given by*

$$h(w) = \langle f, k_w \rangle.$$

This h is the orthogonal projection of f onto K.

Proof. Suppose $\{\phi_k : k \in \Gamma'\}$ is an orthonormal basis for K. Extend this with $\{\phi_k : k \in \Gamma''\}$ such that $\{\phi_k : k \in \Gamma = \Gamma' \cup \Gamma''\}$ is an orthonormal basis for H. Then the kernel of K is given by $k_w = \sum_{k\in\Gamma'} \phi_k \overline{\phi_k(w)}$. Any element $f \in H$ can be expanded as

$$f = \sum_{k\in\Gamma} a_k \phi_k \quad \text{with } a_k = \langle f, \phi_k \rangle.$$

The best approximant from K is given by

$$h = \sum_{k\in\Gamma'} a_k \phi_k,$$

whereas

$$\langle f, k_w \rangle = \sum_{k\in\Gamma'} \langle f, \phi_k \rangle \, \phi_k(w) = \sum_{k\in\Gamma'} a_k \phi_k(w) = h(w).$$

This proves the theorem. □

With these kernels, it is also possible to solve a number of classical extremal problems in Hilbert spaces. We find in Ref. [148, p. 44] the following theorem.

Theorem 1.4.2. *Let H be a Hilbert space with reproducing kernel $k(z, w)$. Then all the solutions of the following problem:*

$$P^1(a, w): \quad \sup\{|f(w)|^2 : \|f\| = a, \ w \in X\}$$

are parametrized by

$$f = \eta \, a \, k(z, w)[k(w, w)]^{-1/2}, \quad |\eta| = 1.$$

The supremum is

$$|a|^2 k(w, w).$$

The problem

$$P^2(a, w): \quad \inf\{\|f\|^2 : f(w) = a, \; w \in X\}$$

reaches an infimum for

$$f = a\, k(z, w)[k(w, w)]^{-1}$$

and this solution is unique. The minimum reached is

$$|a|^2[k(w, w)]^{-1}.$$

Proof. This theorem was given in Ref. [148] for $a = 1$, but the introduction of a is trivial.   □

The problems $P^1(a, w)$ and $P^2(a, w)$ are related as dual extremal problems as can be found in Ref. [72, p. 133] in a much more general context of Banach spaces.

Problem $P^2(a, w)$ can be understood as the problem of finding the orthogonal projection in H of 0 onto the space $V = \{f \in H : f(w) = a\}$.

1.5. J-unitary and J-contractive matrices

We shall consider 2×2 matrices θ whose entries are functions in the Nevanlinna class N: $\theta = [\theta_{ij}] \in N^{2\times2}$. We consider such matrices that are unitary with respect to the indefinite metric

$$J = \begin{bmatrix} 1 & 0 \\ 0 & -1 \end{bmatrix} = 1 \oplus -1.$$

We mean that they satisfy

$$\theta^H J \theta = J \quad \text{on } \partial\mathbb{O}, \tag{1.46}$$

where the superscript ^{H} denotes complex conjugate transpose. If we define the substar conjugate for matrices as the elementwise substar conjugate of the transposed matrix:

$$\begin{bmatrix} \theta_{11} & \theta_{12} \\ \theta_{21} & \theta_{22} \end{bmatrix}_* = \begin{bmatrix} \theta_{11*} & \theta_{21*} \\ \theta_{12*} & \theta_{22*} \end{bmatrix},$$

then we can write (1.46) as

$$\theta_* J \theta = J \quad \text{on } \partial\mathbb{O}. \tag{1.47}$$

As we did for inner functions in $\mathcal{B}$, we can define a pseudo-meromorphic extension for such a θ-matrix. Indeed, it follows from (1.47) that $|\det \theta| = 1$ a.e. on $\partial \mathbb{O}$. Hence, θ is invertible on $\partial \mathbb{O}$ a.e. and therefore also in $\mathbb{O}$ a.e. From the relation

$$\theta_* = J\theta^{-1}J \quad \text{a.e. on } \partial \mathbb{O},$$

we can extend the right-hand side to $\mathbb{O}$ and hence we define also $\theta_*(z) = [\theta(\hat{z})]^H$ for $z \in \mathbb{O}$, which is equivalent with defining $\theta(y) = [\theta_*(\hat{y})]^H$ for $y \in \mathbb{O}^e$. Thus θ is defined on the sphere $\overline{\mathbb{C}}$.

We shall call the matrix functions satisfying

$$\theta_* J\theta = J \quad \text{a.e. in } \overline{\mathbb{C}}$$

J-unitary matrices and denote the set of these matrices as

$$\mathbb{T}_J = \{\theta \in N^{2\times 2} : \theta_* J\theta = J \text{ a.e.}\}.$$

We have the following properties for J-unitary matrices:

Theorem 1.5.1. *For elements of* $\mathbb{T}_J$ *the following relations hold:*

1. $\theta_1, \theta_2 \in \mathbb{T}_J \Rightarrow \theta_1\theta_2 \in \mathbb{T}_J$.
2. $\theta \in \mathbb{T}_J \Rightarrow |\det \theta| = 1$ *on* $\partial \mathbb{O}$.
3. $\theta \in \mathbb{T}_J \Rightarrow \theta^{-1} = J\theta_* J$.
4. $\theta \in \mathbb{T}_J \Rightarrow \theta J\theta_* = J$.
5. *If* $\theta = [\theta_{ij}] \in \mathbb{T}_J$, *then*
   (a) $\theta_{11*}\theta_{11} - \theta_{21*}\theta_{21} = \theta_{22*}\theta_{22} - \theta_{12*}\theta_{12} = 1$,
   (b) $\theta_{11*}\theta_{12} - \theta_{21*}\theta_{22} = \theta_{11*}\theta_{21} - \theta_{12*}\theta_{22} = 0$,
   (c) $\theta_{12*}\theta_{12} - \theta_{21*}\theta_{21} = \theta_{11*}\theta_{11} - \theta_{22*}\theta_{22} = 0$,
   (d) $(\theta_{11} + \theta_{12})_*^{-1}(\theta_{11} - \theta_{12})_* = (\theta_{22} + \theta_{21})^{-1}(\theta_{22} - \theta_{21})$.
6. *Let* $\theta = [\theta_{ij}] \in \mathbb{T}_J$ *and set* $a = \theta_{11} - \theta_{12}$; $b = \theta_{11} + \theta_{12}$; $c = \theta_{22} - \theta_{21}$; *and* $d = \theta_{22} + \theta_{21}$. *Then the following holds true*

$$\frac{1}{2}\left[\frac{a}{b} + \frac{a_*}{b_*}\right] = \frac{1}{bb_*} = \frac{1}{2}\left[\frac{c}{d} + \frac{c_*}{d_*}\right] = \frac{1}{dd_*}$$

$$ab_* + a_*b = cd_* + c_*d = 2.$$

Proof. Parts 1–3 are trivial to check. Part 4 follows from $\theta_* J\theta = J$ so that $J\theta_* = \theta^{-1}J$ and by multiplying with θ, we get $\theta J\theta_* = J$. Part 5 is just an explicitation of $\theta_* J\theta = J = \theta J\theta_*$. The equalities of part 5 then give an easy proof for part 6. $\quad\square$

An important example of a constant J-unitary matrix is

$$U_\rho = (1 - |\rho|^2)^{-1/2} \begin{bmatrix} 1 & -\rho \\ -\overline{\rho} & 1 \end{bmatrix}, \quad |\rho| \neq 1. \tag{1.48}$$

In fact, this example turns out to be almost the most general constant J-unitary matrix.

Theorem 1.5.2. *The most general constant* $\theta \in \mathbb{T}_J$ *is given by*

$$\begin{bmatrix} \eta_1 & 0 \\ 0 & \eta_2 \end{bmatrix} U_\rho, \tag{1.49}$$

with $|\eta_i| = 1$, $i = 1, 2$, *and* U_ρ, $|\rho| \neq 1$ *as given in (1.48).*

Proof. This is a matter of simple algebra. One can make use of the properties given in Theorem 1.5.1. □

A simple nonconstant matrix from the class $\mathbb{T}_J$ is given by the *Blaschke–Potapov factor* with a zero in $\alpha \in \mathbb{O}$:

$$B_\alpha = \begin{bmatrix} \zeta_\alpha & 0 \\ 0 & 1 \end{bmatrix} = \zeta_\alpha \oplus 1; \quad \zeta_\alpha(z) = \frac{z - \alpha}{\varpi_\alpha(z)}, \quad \alpha \in \mathbb{O}. \tag{1.50}$$

The matrices in $N^{2\times 2}$ that are also *J-contractive* in $\mathbb{O}$ form an important class we shall often need in this paper. J-contractive in $\mathbb{O}$ means

$$\theta^H J \theta \leq J \quad \text{a.e. in } \mathbb{O}.$$

By the inequality, we mean that $J - \theta^H J \theta \geq 0$, that is, this is positive semidefinite. The class of strictly J-contractive matrices is denoted by

$$\mathbb{D}_J = \{\theta \in \mathbb{T}_J; \theta^H J \theta < J \text{ a.e. in } \mathbb{O}\} \tag{1.51}$$

and its closure by

$$\overline{\mathbb{D}}_J = \{\theta \in \mathbb{T}_J; \theta^H J \theta \leq J \text{ a.e. in } \mathbb{O}\}. \tag{1.52}$$

Note that $\mathbb{D}_J$ and $\overline{\mathbb{D}}_J$ are closed under multiplication since for $x \in \mathbb{C}^2$, and $\theta_1, \theta_2 \in \overline{\mathbb{D}}_J$ it holds that

$$x^H J x - x^H \theta_2^H \theta_1^H J \theta_1 \theta_2 x \geq x^H \theta_2^H J \theta_2 x - x^H \theta_2^H \theta_1^H J \theta_1 \theta_2 x$$
$$\geq x^H \theta_2^H (J - \theta_1^H J \theta_1) \theta_2 x \geq 0.$$

For these matrices a number of additional properties can be proved. The following theorem is due to Dewilde and Dym [60, p. 448].

Theorem 1.5.3. *For* $\theta = [\theta_{ij}] \in \overline{\mathbb{D}}_J$ *the following hold:*

1. $\theta^H \in \overline{\mathbb{D}}_J$.
2. $\theta^H J \theta \geq J$ *a.e. in* $\mathbb{O}^e$.
3. $(\theta_{11} + \theta_{12})_*^{-1} \in \mathring{H}_2$.
4. $(\theta_{11} + \theta_{12})_*^{-1}(\theta_{11} - \theta_{12})_* \in C$.
5. $(\theta_{22} + \theta_{21})^{-1} \in \mathring{H}_2$.
6. $(\theta_{22} + \theta_{21})^{-1}(\theta_{22} - \theta_{21}) \in C$.
7. $(\theta_{11} + \theta_{12})_*^{-1}(\theta_{21} - \theta_{22})_*$ *is inner.*

Proof. Part 1 was shown in Potapov [180, p. 171]. We shall, however, give an explicit proof using an idea of Dym [77, p. 14].

Because $J - \theta^H J \theta \geq 0$ a.e. in $\mathbb{O}$, we find for the $(2, 2)$ element

$$-1 - [\bar{\theta}_{12} \quad \bar{\theta}_{22}]J[\theta_{12} \quad \theta_{22}]^T \geq 0.$$

Hence

$$|\theta_{22}|^2 \geq |\theta_{12}|^2 + 1 > 0.$$

Thus θ_{22}^{-1} is analytic in $\mathbb{O}$ and we can define

$$\Sigma_{11} = \theta_{11} - \theta_{12}\theta_{22}^{-1}\theta_{21},$$
$$\Sigma_{12} = \theta_{12}\theta_{22}^{-1},$$
$$\Sigma_{21} = -\theta_{21}\theta_{22}^{-1},$$
$$\Sigma_{22} = \theta_{22}^{-1},$$

which are all analytic in $\mathbb{O}$. Set $\Sigma = [\Sigma_{ij}]$ and define $P = (I + J)/2$ and $Q = (I - J)/2$. Then we can check easily that

(a) $P \pm Q\Sigma$ and $P \pm \Sigma Q$ are invertible in $\mathbb{O}$.
(b) $\theta = (P\Sigma + Q)(Q\Sigma + P)^{-1} = (P - \Sigma Q)^{-1}(\Sigma P - Q)$ in $\mathbb{O}$.
(c) $J - \theta(z_1)J\theta(z_2)^H = (P - \Sigma(z_1)Q)^{-1}(I - \Sigma(z_1)\Sigma(z_2)^H)(P - \Sigma(z_2)Q)^{-H}$
   for $z_1, z_2 \in \mathbb{O}$.
(d) $J - \theta(z_2)^H J\theta(z_1) = (P + Q\Sigma(z_2))^{-H}(I - \Sigma(z_2)^H\Sigma(z_1))(P + Q\Sigma(z_1))^{-1}$
   for $z_1, z_2 \in \mathbb{O}$.

From (d), we find that in $\mathbb{O}$, Σ is contractive iff θ is J-contractive, while we see similarly from (c) that Σ^H is contractive iff θ^H is J-contractive. Hence, to show that $\theta \in \overline{\mathbb{D}}_J \Leftrightarrow \theta^H \in \overline{\mathbb{D}}_J$, we only have to show that Σ is contractive iff Σ^H is contractive. This is a classical result. See, for example, Ref. [77, Lemma 0.1]. In fact $\Sigma\Sigma^H \leq I$ as well as $\Sigma^H\Sigma \leq I$ are equivalent with the singular values of Σ being bounded by 1.

(Note: The matrix Σ is called a scattering matrix and θ is called a chain scattering matrix. They describe the same scattering phenomenon by a rearrangement of the inputs and outputs. See Ref. [62] and Section 12.3.)

Part 2 follows from part 1 and the definition of $\mathbb{D}_J$, which imply that for $z \in \mathbb{O}$ it holds that a.e. $\theta(z)^{-1} J \theta(z)^{-H} \geq J$. Now make use of $\theta(z)^{-1} = J\theta(\hat{z})^H J$ to get

$$\theta(w)^H J \theta(w) \geq J \quad \text{with } w = \hat{z} \in \mathbb{O}^e. \tag{1.53}$$

Using part 1, we also have

$$\theta(\hat{z}) J \theta(\hat{z})^H \geq J, \quad z \in \mathbb{O}.$$

Then it follows from writing out the (1,1) element that

$$|\theta_{11*}|^2 - |\theta_{12*}|^2 \geq 1 \quad \text{a.e. in } \mathbb{O}. \tag{1.54}$$

Therefore, $(\theta_{11*} + \theta_{12*})$ is not zero and its inverse is analytic in $\mathbb{O}$.

Computing the real part of the expression of part 4, we get, using (1.54),

$$\text{Re}\left(\frac{\theta_{11*} - \theta_{12*}}{\theta_{11*} + \theta_{12*}}\right) = \frac{|\theta_{11*}|^2 - |\theta_{12*}|^2}{|\theta_{11*} + \theta_{12*}|^2} \geq \frac{1}{|\theta_{11*} + \theta_{12*}|^2} \geq 0.$$

From this, part 4 follows. Since the left-hand side in the previous expression is a harmonic majorant in $\mathbb{O}$ for the analytic function $|\theta_{11*} + \theta_{12*}|^{-2}$, it follows from Ref. [76, Theorem 2.12, p. 28] that part 3 is true.

Part 5 and 6 follow from the (2,2) element in much the same way as 3 and 4 followed from the (1,1) element in (1.53).

To prove the last part, note that we had for $z \in \mathbb{O}$ that $\theta_* J \theta_*^H \geq J$. So we get in $\mathbb{O}$:

$$[1 \quad 1]\theta_* J \theta_*^H [1 \quad 1]^T \geq [1 \quad 1]J[1 \quad 1]^T = 0$$

with equality on $\partial\mathbb{O}$. Working this out gives

$$1 - a\bar{a} \geq 0,$$

where $a = (\theta_{11*} + \theta_{12*})^{-1}(\theta_{21*} + \theta_{22*})$ and with equality on $\partial\mathbb{O}$. This identifies a as an inner function. $\square$

The following theorem describes a simple matrix from the class $\mathbb{D}_J$.

Theorem 1.5.4. *The most general first degree matrix in $\mathbb{D}_J$ with a zero at $z = \alpha \in \mathbb{O}$ is given by*

$$\begin{bmatrix} \eta_1 & 0 \\ 0 & \eta_2 \end{bmatrix} U_\rho B_\alpha U_\gamma,$$

with $\eta_1, \eta_2 \in \mathbb{T}$, U_ρ and U_γ constant J-unitary matrices as defined in (1.48) for ρ and $\gamma \notin \mathbb{T}$, and B_α a Blaschke–Potapov factor as in (1.50).

Proof. This is a classical result that can be found, for example, in Potapov [180, pp. 187–188]. □

A matrix in $N^{2 \times 2}$ that is J-contractive in $\mathbb{O}$ and J-unitary will be called J-inner, a terminology used by Dym [77]. The set of J-inner matrix functions is denoted as

$$\mathcal{B}_J = \mathbb{D}_J \cap \mathbb{T}_J$$

while we set

$$\overline{\mathcal{B}}_J = \overline{\mathbb{D}}_J \cap \mathbb{T}_J.$$

The matrix of Theorem 1.5.4 is a J-inner matrix.

Much more on J-unitary matrix functions and the work of Potapov can be found in the V. P. Potapov memorial volume [100].

2

The fundamental spaces

This chapter serves to introduce the spaces of rational functions with fixed poles in $\mathbb{O}^e$. In the case of the disk, these spaces generalize the spaces of polynomials. The latter are a special case if all poles are at infinity. For the half plane, the situation is similar. Also here, the polynomial case is recovered by letting the poles tend to infinity, although the fact that now $\infty \in \overline{\mathbb{R}}$, with $\mathbb{R} = \partial\mathbb{O}$, makes the situation less trivial. For $\partial\mathbb{O} = \mathbb{R}$, the polynomials appear in a much more natural way when the interpolation points are located on the boundary. This will be discussed later in Chapter 11.

We shall first discuss several equivalent characterizations of the spaces in Section 2.1, and in Section 2.2 we give several rules for doing calculus in these spaces. The results of the latter section are frequently used in the rest of the text. It requires some skill to perform the computations fluently and the reader is warmly recommended to have a close look at this section because these results are used in practically every subsequent section of this book. The last section of this chapter reconsiders, for the spaces of rational functions, several extremal problems that are related to the extremal problem that was given in general reproducing kernel spaces in Section 1.4. Their solutions can be described in terms of kernels or orthogonal functions.

2.1. The spaces $\mathcal{L}_n$

In this section we shall introduce the spaces $\mathcal{L}_n$, which are the fundamental spaces dealt with in Chapters 2–10.

We already defined the *Blaschke factors* $\zeta_n(z)$ in Section 1.3 (1.40–1.41). We recall the definitions

$$\zeta_i(z) = z_i \frac{\varpi_i^*(z)}{\varpi_i(z)}, \tag{2.1}$$

where depending on $\mathbb{O} = \mathbb{D}$ or $\mathbb{O} = \mathbb{U}$, the definitions of the factors are

$$\varpi_i(z) = 1 - \overline{\alpha}_i z, \qquad \varpi_i^*(z) = z - \alpha_i \quad \text{for } \mathbb{D},$$
$$\varpi_i(z) = z - \overline{\alpha}_i, \qquad \varpi_i^*(z) = z - \alpha_i \quad \text{for } \mathbb{U}.$$

There is a special point $\alpha_0 = 0$ for $\mathbb{D}$ and $\alpha_0 = \mathbf{i}$ for $\mathbb{U}$. For $\alpha_i = \alpha_0$, we always set $z_i = 1$ and for $\alpha_i \neq \alpha_0$, the normalizing factors are

$$z_i = -\overline{\alpha}_i/|\alpha_i| \qquad \text{for } \mathbb{D},$$
$$z_i = \left|\alpha_i^2 + 1\right|/\left(\alpha_i^2 + 1\right) \quad \text{for } \mathbb{U}.$$

Thus

$$\zeta_i(z) = -\frac{\overline{\alpha}_i}{|\alpha_i|} \frac{z - \alpha_i}{1 - \overline{\alpha}_i z} \qquad \text{for } \mathbb{D},$$

$$\zeta_i(z) = \frac{\left|\alpha_i^2 + 1\right|}{\alpha_i^2 + 1} \frac{z - \alpha_i}{z - \overline{\alpha}_i} \qquad \text{for } \mathbb{U}.$$

In what follows we shall always assume that an expression such as z_i, even if it does not appear in a Blaschke factor, will be 1 when $\alpha_i = \alpha_0$. The factor $\zeta_0(z) = z$ for $\mathbb{D}$ and $\zeta_0(z) = (z - \mathbf{i})/(z + \mathbf{i})$ for $\mathbb{U}$ is important since

$$1 - |\zeta_0(z)|^2 = \begin{cases} 1 - |z|^2 & \text{for } \mathbb{O}, \\ 4 \operatorname{Im} z/|z + \mathbf{i}|^2 & \text{for } \mathbb{U}, \end{cases}$$

and thus it is fairly easy to characterize the sets $\mathbb{O}$, $\partial\mathbb{O}$, and $\mathbb{O}^e$ as

$$\mathbb{O} = \{z \in \mathbb{C} : |\zeta_0(z)| < 1\},$$
$$\partial\mathbb{O} = \{z \in \mathbb{C} : |\zeta_0(z)| = 1\},$$
$$\mathbb{O}^e = \{z \in \mathbb{C} : |\zeta_0(z)| > 1\}.$$

Recall also (1.3), which gave an equivalent but more complicated characterization of these three parts of $\mathbb{C}$.

Next we define finite Blaschke products as

$$B_0 = 1 \quad \text{and} \quad B_n = \zeta_1 \cdots \zeta_n = B_{n-1}\zeta_n \quad \text{for } n \geq 1. \tag{2.2}$$

We then consider the spaces

$$\mathcal{L}_n = \operatorname{span}\{B_k : k = 0, 1, \ldots, n\}. \tag{2.3}$$

They will often be considered as subspaces of $L_2(\mu)$ but from time to time we shall also consider them as subspaces of $L_2(\mathring{\lambda})$ or some other space.

There are of course many equivalent ways to describe the spaces $\mathcal{L}_n$. One of them is to say that $\mathcal{L}_n$ is a space of rational functions whose poles are all in the prescribed set $\{\hat{\alpha}_i : i = 1, \ldots, n\} \subset \mathbb{O}^e$. Recall that $\hat{\alpha} = 1/\overline{\alpha}$ for $\mathbb{D}$ whereas $\hat{\alpha} = \overline{\alpha}$ for $\mathbb{U}$. Thus we may write

$$\mathcal{L}_n = \left\{ f = \frac{p(z)}{\pi_n(z)}, \pi_n(z) = \prod_{i=1}^{n} \varpi_i(z); \, p \in \mathcal{P}_n \right\}.$$

Note that in the case of the disk, we may choose all $\alpha_i = 0$ and then $\mathcal{L}_n$ is just the space of polynomials of degree at most n. Thus in that case $\mathcal{L}_n = \mathcal{P}_n$.

For the half plane though, the polynomials are less simple to recover. First suppose that all $\alpha_k = \alpha$ for some $\alpha \in \mathbb{U}$. Next we replace the basis functions $[(z - \alpha)/(z - \overline{\alpha})]^k$ for the spaces $\mathcal{L}_n$ by $[(1 - \overline{\alpha}z)/(z - \overline{\alpha})]^k$, which describes of course the same spaces. We can now let α tend to ∞ or 0 (which are both on the boundary $\overline{\mathbb{R}}$), and we find that $\mathcal{L}_n$ becomes $\mathcal{P}_n$ in the first case and the set of polynomials in $1/z$ in the second case.

The spaces $\mathcal{L}_n$ depend upon the point sets

$$A_n = \{\alpha_i : \alpha_i \in \mathbb{O}, i = 1, \ldots, n\}.$$

By $\hat{A}_n$ we shall denote the set

$$\hat{A}_n = \{\hat{\alpha}_i : \alpha_i \in A_n\}.$$

Some of the α_i can be repeated a number of times. So we could rearrange them and make the repetition explicit by setting

$$A_n^0 = \{\alpha_0\} \cup A_n = \{\underbrace{\beta_0, \ldots, \beta_0}_{\nu_0}, \underbrace{\beta_1, \ldots, \beta_1}_{\nu_1}, \ldots, \underbrace{\beta_m, \ldots, \beta_m}_{\nu_m}\}. \qquad (2.4)$$

We fix β_0 to be α_0, so that ν_i, $i = 0, \ldots, m$ are positive integers and $\sum_0^m \nu_i = n + 1$.

The basis $\{B_k : k = 0, \ldots, n\}$ is not the only possible choice to span $\mathcal{L}_n$ of course. With A_n as described in (2.4), we can use as a possible basis in the case of the disk:

$$\{w_k\}_{k=0}^n = \{1, z, \ldots, z^{\nu_0-1}, (1 - \overline{\beta}_1 z)^{-1}, \ldots, (1 - \overline{\beta}_1 z)^{-\nu_1},$$

$$\ldots, (1 - \overline{\beta}_m z)^{-1}, \ldots, (1 - \overline{\beta}_m z)^{-\nu_m}\}. \qquad (2.5)$$

For the real line, we should replace this by

$$\{w_k\}_{k=0}^n = \left\{ \frac{z + \mathbf{i}}{(z - \overline{\beta}_j)^{k_j}} : k_j = 1, \ldots, \nu_j; \, j = 0, \ldots, m \right\}$$

so that an invariant notation would be

$$\{w_k\}_{k=0}^n = \left\{ \frac{\varpi_0(z)}{\varpi_j(z)^{k_j}} : k_j = 1, \ldots, \nu_j; \, j = 0, \ldots, m \right\}, \qquad (2.6)$$

where $\varpi_i(z) = 1 - \overline{\beta}_i z$ for the disk and $\varpi_i(z) = z - \overline{\beta}_i$ for the half plane. As always $\beta_0 = \alpha_0$ is 0 for the disk and $\mathbf{i}$ for the half plane. The first ν_0 basis functions are different for both:

$$\{w_k : k = 0, \ldots, \nu_0 - 1\} = \begin{cases} \{z^{k-1} : k = 1, \ldots, \nu_0\} & \text{for } \mathbb{D}, \\ \{(z + \mathbf{i})^{1-k} : k = 1, \ldots, \nu_0\} & \text{for } \mathbb{U}. \end{cases}$$

The advantage of working with the basis $\{B_k : k = 0, \ldots, n\}$ is that repetition of points and distinction between $\alpha_i = 0$ or $\alpha_i \neq 0$ need no special notation as in some other choices such as, for instance, (2.5).

Here is yet another way to characterize the spaces $\mathcal{L}_n$. Define

$$\mathcal{M}_n = \zeta_0 B_n H_2, \qquad (2.7)$$

with B_n the finite Blaschke product associated with $\alpha_1, \ldots, \alpha_n$. Clearly, by Beurling's theorem, $\mathcal{M}_n$ is a shift invariant subspace of H_2 since $\zeta_0 B_n$ is an inner function.

The sequence $\{\mathcal{M}_n : n = 0, 1, \ldots\}$ contains shrinking subspaces, that is, $\mathcal{M}_{n+1} \subset \mathcal{M}_n \subset \cdots \subset \mathcal{M}_0 = \zeta_0 H_2 \subseteq \mathring{H}_2$. If we define

$$\mathcal{L}_n = \mathring{H}_2 \ominus \mathcal{M}_n = \mathcal{M}_n^\perp = \{f \in \mathring{H}_2, \langle f, g \rangle = 0 \text{ for all } g \in \mathcal{M}_n\}, \qquad (2.8)$$

then the sequence $\{\mathcal{L}_n : n = 0, 1, \ldots\}$ is a nested sequence of increasing subspaces: $\mathcal{L}_{n+1} \supset \mathcal{L}_n \supset \cdots \supset \mathcal{L}_0 = \mathbb{C}$. The choice of the notation $\mathcal{L}_n$ in the previous definition may seem confusing at the moment, since we reserved this notation for the spaces defined in (2.3). Our next theorem will show that the spaces of (2.3) and (2.8) are actually the same. We shall do this by proving that a basis for $\mathcal{L}_n$ is given by $\{B_k : k = 0, \ldots, n\}$.

Theorem 2.1.1. *Define the spaces $\mathcal{M}_n = \zeta_0 B_n H_2$ and $\mathcal{L}_n = \mathring{H}_2 \ominus \mathcal{M}_n = \mathcal{M}_n^\perp$, where B_n is a finite Blaschke product of degree n. Then*

$$\mathcal{L}_n = \operatorname{span}\{B_k : k = 0, \ldots, n\}.$$

Proof. We shall give the proof only for the case of the disk. The proof for the half plane is obtained by conformal mapping.

The result is obvious for $n = 0$. We shall then prove that for $n > 0$, $B_n \in \mathcal{L}_n \setminus \mathcal{L}_{n-1}$, which implies that the Blaschke products indeed form a basis. First we show that $B_n \in \mathcal{L}_n$. Choose some $f(z) = z B_n(z) g(z) \in z B_n(z) H_2$ ($g \in H_2(\mathbb{D})$). Then

$$\langle f, B_n \rangle = \int t B_n(t) g(t) B_{n*}(t) \, d\lambda(t) = \int t g(t) \, d\lambda(t) = 0$$

since $B_{n*} = 1/B_n$ and g has vanishing negative Fourier coefficients. Hence $B_n \perp z B_n H_2$ and therefore $B_n \in \mathcal{L}_n$. However, $B_n \notin \mathcal{L}_{n-1}$ since for $f \in \mathcal{M}_{n-1}$:

$$\langle f, B_n \rangle = \int t B_{n-1}(t) g(t) B_{n*}(t) \, d\lambda(t) = \int t g(t)/\zeta_n(t) \, d\lambda(t)$$

with $1/\zeta_n(z) = \alpha_n/|\alpha_n| \cdot (1 - \overline{\alpha}_n z)/(\alpha_n - z)$, which gives by Cauchy's formula

$$\langle f, B_n \rangle = -g(\alpha_n) \frac{\alpha_n}{|\alpha_n|} \alpha_n (1 - |\alpha_n|^2) \quad \text{for } \alpha_n \neq 0 \text{ or } -g(0) \quad \text{for } \alpha_n = 0,$$

which is not zero for all $g \in H_2$. Hence B_n is not orthogonal to $\mathcal{M}_{n-1}$, and thus it is not in $\mathcal{L}_{n-1}$.   □

The previous theorem shows that we can identify $\mathcal{L}_n$ as defined in (2.8) with the originally introduced space $\mathcal{L}_n$ of (2.3):

$$\mathcal{L}_n = \mathring{H}_2 \ominus \zeta_0 B_n H_2 = \mathcal{M}_n^\perp = \operatorname{span}\{B_k : k = 0, \dots, n\}.$$

The previous result says, for example, that a function in $\mathring{H}_2$ is orthogonal to $\mathcal{L}_n$ if and only if it vanishes in the point set $A_n^0 = \{\alpha_0, \alpha_1, \dots, \alpha_n\}$; thus the difference between a function $f \in \mathring{H}_2$ and its orthogonal projection onto $\mathcal{L}_n$ should vanish in A_n^0. In other words, the orthogonal projection of $f \in \mathring{H}_2$ onto $\mathcal{L}_n$ should interpolate f in the points A_n^0. We shall come back to this property in Section 7.1.

Consider the special case of the disk where we put all $\alpha_i = 0$. Then the spaces $\mathcal{L}_n$ reduce to the spaces $\mathcal{P}_n$ of polynomials. It is well known that in that case the Gram matrix of the basis $\mathbf{Z}_n^T = [1 \; z \; z^2 \; \dots \; z^n]$ in $L_2(\mu)$ is given by

$$G_n = \langle \mathbf{Z}_n, \mathbf{Z}_n^T \rangle_\mu = [\langle z^i, z^j \rangle_\mu] = [c_{j-i}],$$

which is a positive definite Toeplitz matrix containing the moments of μ. If, however, all the α_i are distinct, then the basis w_k we mentioned previously in (2.6) reduces to

$$\mathbf{W}_n^T = [w_0 \quad w_1 \quad \cdots \quad w_n] = \left[1, \frac{\varpi_0(z)}{\varpi_1(z)}, \dots, \frac{\varpi_0(z)}{\varpi_n(z)} \right].$$

Using the definition of Ω_μ, we easily obtain the Gram matrix

$$G_n = \langle \mathbf{W}_n, \mathbf{W}_n^T \rangle_{\hat{\mu}} = \frac{\varpi_0(\alpha_0)}{2} \left[\frac{\overline{\Omega_\mu(\alpha_i)} + \Omega_\mu(\alpha_j)}{\varpi_j(\alpha_i)} \right].$$

This is a so-called *Pick matrix*, named after G. Pick who used the positive definiteness of this matrix as a criterion to characterize the solvability of the Nevanlinna–Pick interpolation problem. In the more general case where some of the α_k do coincide, the Gram matrix looks more complicated and involves derivatives of Ω_μ evaluated at the α_i. To see this, we give a technical lemma first.

Lemma 2.1.2. Consider the Riesz–Herglotz kernel $D(z, w) = (z+w)(z-w)^{-1}$ for the disk. Then

$$\partial_w^k D(z, w) = 2(k!)z(z - w)^{-(k+1)}, \quad k \geq 1,$$

where ∂_w^k denotes the kth derivative with respect to w. We also have

$$\left[\partial_w^k D(z, w) \right]_* = 2(k!)z^k (1 - \overline{w}z)^{-(k+1)}, \quad k \geq 1,$$

where the substar transform is with respect to z. Furthermore, if

$$\Omega_\mu(w) = \mathbf{i}\, \mathrm{Im}\, \Omega_\mu(0) + \int D(t, w)\, d\mu(t),$$

then for $k \geq 1$

$$\Omega_\mu^{(k)}(w) = \partial_w^k \Omega_\mu(w) = \int \partial_w^k D(t, w)\, d\mu(t) = 2(k!) \int \frac{t\, d\mu(t)}{(t - w)^{k+1}}$$

and

$$\overline{\Omega_\mu^{(k)}(w)} = \int \left[\partial_w^k D(t, w) \right]_* d\mu(t)$$

(substar with respect to t). Similarly, we may consider the Nevanlinna kernel $D(z, w) = (1 + zw)/[\mathbf{i}(z - w)]$ for the half plane. We then get

$$\partial_w^k D(z, w) = -\mathbf{i}(-1)^k k!(z - w)^{-(k+1)}, \quad k \geq 1$$

and

$$\Omega_\mu^{(k)}(w) = \partial_w^k \Omega_\mu(w) = \int \partial_w^k D(t, w)\, d\mathring{\mu}(t) = -\mathbf{i}(-1)^k (k!) \int \frac{d\mathring{\mu}(t)}{(t - w)^{k+1}}.$$

Proof. This is a matter of simple algebra and we leave this to the reader. □

With this lemma, we can now prove the following theorem.

Theorem 2.1.3. *If we choose the basis (2.6) for the space $\mathcal{L}_n$, then the Gram matrix*

$$G_n = [\langle w_i, w_j \rangle_{\hat{\mu}}]$$

will only depend upon

$$\Omega_\mu^{(k)}(\beta_l), \quad k = 0, 1, \ldots, \nu_l - 1, \; l = 0, 1, \ldots, m.$$

The superscript $^{(k)}$ denotes the kth derivative and Ω_μ is the Riesz–Herglotz–Nevanlinna transform of μ.

Proof. Suppose we consider the case of the disk. One possible form of the elements in G_n involves an integral like

$$\int \frac{1}{(1 - \overline{\alpha}t)^k} \frac{t^l}{(t - \beta)^l} \, d\mu(t). \tag{2.9}$$

It should be clear from the previous lemma that if we use a partial fraction decomposition of the integrand, then it can be written as a linear combination of $\partial_w^i D(t, w)$ evaluated at $w = \beta$ for $i = 0, \ldots, l - 1$ and (substar for t) $[\partial_w^j D(t, w)]_*$ evaluated at $w = \alpha$ for $j = 0, \ldots, k - 1$. Thus (2.9) can be written as a linear combination of $\Omega_\mu^{(i)}(\beta)$ and $\overline{\Omega_\mu^{(j)}(\alpha)}$ for $i = 0, \ldots, l$ and $j = 0, \ldots, k$.

The other possibilities for w_i and w_j can be treated similarly.

For the half plane, the situation is completely similar. For example, we have to evaluate integrals of the form

$$\int \frac{1 + t^2}{(t - \overline{\alpha})^k (t - \beta)^l} \, d\mathring{\mu}(t),$$

which will, again by partial fraction decomposition and the previous lemma, only depend on the values of $\partial_z^i \Omega_\mu(z)$ evaluated at $z = \alpha$ and $z = \beta$. The remaining details are left to the reader. □

We get an immediate consequence of these expressions.

Corollary 2.1.4. *Let $\mathring{\mu}$ and $\mathring{\nu}$ be two finite, positive measures on $\overline{\partial\mathbb{O}}$. Let Ω_μ and Ω_ν be the corresponding C functions. Furthermore, let $\mathcal{L}_n$ be the spaces* *introduced before.* Then

$$\langle f, g \rangle_{\hat{\mu}} = \langle f, g \rangle_{\hat{\nu}}, \quad \forall f, g \in \mathcal{L}_n \Leftrightarrow \frac{\Omega_\mu - \Omega_\nu}{\zeta_0 B_n} \in H(\mathbb{O}).$$

The latter means that Ω_μ coincides with Ω_ν in all the points from the set $A_n^0 = \{\alpha_0, \ldots, \alpha_n\}$ (interpolation in Hermite sense).

We shall call a matrix of the form G_n as in the previous theorem a *generalized Pick matrix*. The Gram matrix for any other basis, for example, the basis $\{B_k : k = 0, \ldots, n\}$ is of course equivalent with a generalized Pick matrix. In contrast, we have

Theorem 2.1.5. *A generalized Pick matrix is equivalent with a Toeplitz matrix.*

Proof. We just note that also

$$\left\{ \ell_k^{(n)} : k = 0, \ldots, n \right\} = \left\{ \frac{z^k}{\pi_n(z)} ; k = 0, \ldots, n \right\}, \quad \pi_n(z) = \prod_{i=1}^n \varpi_i(z)$$

is a basis for $\mathcal{L}_n$. The Gram matrix for this basis is obviously a Toeplitz matrix since its entries depend only on the difference of the indices (and n):

$$\left\langle \ell_i^{(n)}, \ell_j^{(n)} \right\rangle_{\hat{\mu}} = \int \frac{t^{i-j}}{\pi_{n*}(t)\pi_n(t)} \, d\mu(t) = \langle z^i, z^j \rangle_\pi$$

with $d\pi = (\pi_{n*}\pi_n)^{-1} \, d\mathring{\mu}$.

Setting

$$W^T = [w_0, \ldots, w_n] \quad \text{and} \quad L^T = \left[\ell_0^{(n)}, \ldots, \ell_n^{(n)} \right]$$

then there must exist a regular matrix A such that $W = AL$. Hence the Pick matrix

$$P_n = \langle W, W^T \rangle_{\hat{\mu}}$$

and the Toeplitz matrix

$$T_n = \langle L, L^T \rangle_{\hat{\mu}}$$

will be related by

$$P_n = A T_n A^H. \tag{2.10}$$

$\square$

It is not difficult to find the transformation matrix explicitly, at least not in the case where all the α_i are different. This is done in the case of the disk by Delsarte, Genin, and Kamp [56]. Suppose we express the basis functions w_i of (2.5) in terms of $\ell_j^{(n)}$ as

$$w_i = \sum_{j=0}^{n} \ell_j^{(n)} a_{ij}, \quad i = 0, \ldots, n,$$

which we can also express as

$$w_i(z)\pi_n(z) = \sum_{j=0}^{n} z^j a_{ij}, \quad i = 0, \ldots, n.$$

Since this is a relation between polynomials of degree at most n, we can take the transform $p^*(z) = z^n \overline{p(1/\overline{z})}$ to get

$$\pi_{ni}(z) = \sum_{j=0}^{n} z^{n-j} \overline{a}_{ij}, \quad i = 0, \ldots, n, \tag{2.11}$$

where $\pi_{ni} = (w_i \pi_n)^*$ is a polynomial of degree at most n. So we get again a polynomial relation. We can now find the coefficients a_{ij} by requiring that the polynomial on the right interpolates the polynomial on the left in the points $\{\alpha_i : i = 0, \ldots, n\}$ where we have supposed that $\alpha_0 = 0$. Interpolation has to be understood in Hermite sense. This means that, if a point α_i appears ν_i times, then one interpolates the first $(\nu_i - 1)$ derivatives. In case of all different α_i, we get ordinary interpolation. The result then is

$$\Pi = V A^H,$$

where $A = [a_{ij}]$, V is a Vandermonde matrix $V = [\alpha_i^{n-j}]$, and Π is a diagonal matrix

$$\Pi = \operatorname{diag}[\pi_{ni}] \quad \text{with} \quad \pi_{ni} = \prod_{j\neq i, j=0}^{n} (\alpha_i - \alpha_j).$$

Thus (2.10) becomes

$$P_n = A T_n A^H = \Pi^H V^{-H} T_n V^{-1} \Pi.$$

In the case of confluent points, the matrices Π and V can still be found, only they are much more complicated to write down.

From the previous computations, it follows that another interesting basis is

$$\{n_k\}_{k=0}^{n} = \left\{ \frac{1}{\pi_n}, \frac{z}{\pi_n}, \frac{z(z-\alpha_1)}{\pi_n}, \ldots, \frac{z(z-\alpha_1)\cdots(z-\alpha_{n-1})}{\pi_n} \right\},$$

where as always $\pi_n(z) = \prod_{i=1}^{n} \varpi_i(z)$. The coefficient a_{ij} from the transformation matrix is then a jth order divided difference of the function $\pi_n w_i$ at the points $\hat{\alpha}_i = 1/\overline{\alpha}_i : i = 0, \ldots, j$. Again, here we can have confluency. See the book of Donoghue [71].

In the disk case, some other bases can be found in the literature, such as

$$\left\{ 1, \tilde{B}_0 \frac{\gamma_1}{\varpi_1}, \tilde{B}_1 \frac{\gamma_2}{\varpi_2}, \ldots, \tilde{B}_{n-1} \frac{\gamma_n}{\varpi_n} \right\}, \tag{2.12}$$

with $\gamma_k = (1 - |\alpha_k|^2)^{1/2}$ and $\tilde{B}_k = \zeta_0 B_k$. It is an orthonormal basis for $\mathcal{L}_n$ w.r.t. the Lebesgue measure. See Refs. [67] and [150] for the general case and Ref. [8, p. 138] if all the points α_i, $i = 1, 2, \ldots$ are different and nonzero. Following Walsh [200, p. 224], and more precisely in Ref. [199], it is called the Malmquist basis in Refs. [67] and [70].

For the half plane, the corresponding basis, orthonormal with respect to $d\lambda$, is

$$\left\{ 1, \tilde{B}_0(z) \frac{(z + \mathrm{i})\gamma_1}{\varpi_1(z)}, \tilde{B}_1(z) \frac{(z + \mathrm{i})\gamma_2}{\varpi_2(z)}, \ldots, \tilde{B}_{n-1}(z) \frac{(z + \mathrm{i})\gamma_n}{\varpi_n(z)} \right\},$$

with again $\tilde{B}_k = \zeta_0 B_k$ and

$$\gamma_k = \left[\frac{\alpha_k - \overline{\alpha}_k}{2\mathrm{i}} \right]^{1/2}.$$

Hence, an invariant formulation of an orthonormal basis for $\mathcal{L}_n$ in $\mathring{H}_2(\mathbb{O})$ is

$$\left\{ 1, \tilde{B}_0 \frac{\gamma_1 \varpi_0(z)}{\varpi_1(z)}, \tilde{B}_1(z) \frac{\gamma_2 \varpi_0(z)}{\varpi_2(z)}, \ldots, \tilde{B}_{n-1}(z) \frac{\gamma_n \varpi_0(z)}{\varpi_n(z)} \right\},$$

$$\gamma_k = \left[\frac{\varpi_k(\alpha_k)}{\varpi_0(\alpha_0)} \right]^{1/2}. \tag{2.13}$$

If the points α_i are renamed as in (2.5) as β_j, then, in the case of the disk, this orthogonal basis takes the form

$$\left\{ \frac{1}{1 - \overline{\beta}_i z} \left(\frac{z - \beta_i}{1 - \overline{\beta}_i z} \right)^k : k = 0, \ldots, \nu_i - 1; i = 0, \ldots, m \right\}.$$

This form has been used in, for example, Ref. [111, p. 149 ff]. See also Ref. [186, p. 27].

Let us elaborate a bit further on the orthonormal basis in $\mathring{H}_2$. Let us define (compare with (2.13))

$$v_k = B_{k-1} \frac{\gamma_k \varpi_0}{\varpi_k}; \quad k = 1, \ldots, n. \tag{2.14}$$

Then this forms, together with $v_0 = 1$, an orthonormal basis in $\mathring{H}_2$ for $\mathcal{L}_n$ if all the α_i are mutually different and different from α_0. However, it holds in general also if some points coincide or are equal to α_0, that

$$\mathcal{L}_n = \text{span}\{1, \zeta_0 v_1, \zeta_0 v_2, \ldots, \zeta_0 v_n\}.$$

The interesting thing about this basis is that we can now write

$$\mathcal{L}_n(w) = \text{span}\left\{\frac{z-w}{\varpi_0(z)}v_1(z), \ldots, \frac{z-w}{\varpi_0(z)}v_n(z)\right\} = \{f \in \mathcal{L}_n : f(w) = 0\}.$$

More precisely

$$\mathcal{L}_n(\alpha_0) = \text{span}\{\zeta_0 v_1, \ldots, \zeta_0 v_n\} \quad \text{and} \quad \mathcal{L}_n(\hat{\alpha}_0) = \text{span}\{v_1, \ldots, v_n\}.$$

Define (compare with (2.7))

$$\mathcal{M}_n(0) = B_n H_2 = \zeta_0^{-1}\mathcal{M}_n.$$

Then we can prove the following property (for the disk see also Walsh [200, p. 225]).

Theorem 2.1.6. *With the spaces as defined above we have that*

$$\mathcal{L}_n(\hat{\alpha}_0) = \{f \in \mathcal{L}_n : f(\hat{\alpha}_0) = 0\}$$

is the orthogonal complement (in $\mathring{H}_2(\mathbb{O})$) of $\mathcal{M}_n(0) = B_n H_2$. Thus

$$\mathcal{L}_n(\hat{\alpha}_0) = \mathring{H}_2 \ominus B_n H_2 = \mathcal{M}_n(0)^\perp.$$

Proof. Take a function from $B_n H_2$ of the form $B_n f$ with $f \in H_2$ and a function from $\mathcal{L}_n(\hat{\alpha}_0)$ that has the form $\varpi_0(z)p_{n-1}(z)/\pi_n(z)$, where $p_{n-1} \in \mathcal{P}_{n-1}$, a polynomial of degree at most $n-1$ and as before $\pi_n = \prod_1^n \varpi_i$. The inner product of these functions equals

$$\int B_n(t)f(t)\frac{\varpi_0(t)p_{(n-1)*}(t)}{\pi_{n*}(t)}\,d\mathring{\lambda}(t) = \eta_n \int f(t)\frac{\varpi_0^*(t)q_{n-1}(t)}{\pi_n(t)}\,d\mathring{\lambda}(t),$$

where $\eta_n = \prod_{i=1}^n z_i$, $q_{n-1}(z) \in \mathcal{P}_{n-1}$ and $\varpi_0^*(z) = z$ for the disk and $\varpi_0^*(z) = z - \mathbf{i}$ for the half plane. Since the integrand is in $\mathring{H}_2(\mathbb{O})$ and vanishes in α_0, this integral is zero by (1.8). $\qquad\square$

The next result says that we can also find a basis for $\mathcal{L}_n$ from its reproducing kernel.

Theorem 2.1.7. *Let $k_n(z, w)$ be the reproducing kernel for $\mathcal{L}_n$. Then for a set $\{\xi_0, \ldots, \xi_n\}$ of distinct points in $\mathbb{C}$, the functions $\{k_n(z, \xi_j)\}$, $j = 0, \ldots, n$ form a basis for $\mathcal{L}_n$.*

Proof. Certainly, the functions are all in $\mathcal{L}_n$. They are also linearly independent. Since

$$[k_n(z, \xi_0), \ldots, k_n(z, \xi_n)] = [\phi_0(z), \ldots, \phi_n(z)]V^H$$

with

$$V = \begin{bmatrix} \phi_0(\xi_0) & \cdots & \phi_n(\xi_0) \\ \vdots & & \vdots \\ \phi_0(\xi_n) & \cdots & \phi_n(\xi_n) \end{bmatrix}$$

and because $\{\phi_0, \ldots, \phi_n\}$ is a basis, $\{k_n(\cdot, \xi_0), \ldots, k_n(\cdot, \xi_n)\}$ will also be a basis iff V is regular. Suppose it were not, then there is a nonzero vector $A = [a_0, \ldots, a_n]^T \in \mathbb{C}^n$ such that

$$\sum_{j=0}^{n} \phi_j(\xi_k)a_j = 0, \quad k = 0, \ldots, n.$$

Thus the function $\phi = \sum_{j=0}^{n} a_j\phi_j \in \mathcal{L}_n$ is not identically zero (because $A \neq 0$), and yet it has $n + 1$ zeros $\xi_0, \ldots, \xi_n$, which is impossible. $\qquad\square$

Note that this theorem implies that the Gram matrix

$$VV^H = [k_n(\xi_i, \xi_j)] = [\langle k_n(\cdot, \xi_j), k_n(\cdot, \xi_i)\rangle_\mu]$$

is a positive definite matrix. In fact, a positive (semi-)definite (Gram) matrix is basically equivalent to a reproducing kernel. See Ref. [12, p. 344] or [71, Chap. 10].

2.2. Calculus in $\mathcal{L}_n$

Recall that we already defined the substar transform $f_*(z) = \overline{f(\bar{z})}$. Now if $f \in \mathcal{L}_n$, we shall also define the superstar transform as $f^*(z) = B_n(z)f_*(z)$, where B_n is the finite Blaschke product with zeros from $A_n = \{\alpha_i : i = 1, \ldots, n\}$. Thus, if $f = \sum a_i B_i \in \mathcal{L}_n$, then

$$\left(\sum_{i=0}^{n} a_i B_i(z)\right)^* = \sum_{i=0}^{n} \overline{a}_i B_{n\setminus i}(z) = \overline{a}_n + \overline{a}_{n-1}\zeta_n(z) + \cdots + \overline{a}_0 B_n(z),$$

where

$$B_{n\setminus i}(z) = \frac{B_n(z)}{B_i(z)} = \prod_{j=i+1}^{n} \zeta_j(z), \quad 0 \leq i \leq n.$$

Note that the definition of the superstar transform depends on n. We could have used a notation that shows this dependence explicitly, such as, for example, $f^{[n]}$, but we prefer not to for simplicity of notation and it will always be clear from the context what n we shall mean.

We call a_n the *leading coefficient* of $f \in \mathcal{L}_n$ (w.r.t. the basis $\{B_n\}$). Note that the leading coefficient of $f \in \mathcal{L}_n$ is given by $\overline{f^*(\alpha_n)}$. If the leading coefficient is 1, we say that f is *monic*.

In the case of the disk, we find the polynomials as a special case by setting $\alpha_i = 0$ for all i, so that $\mathcal{L}_n = \mathcal{P}_n$. Using the definition of superstar in this special case, it is natural to define for a polynomial $p \in \mathcal{P}_n$ the superstar as

$$p^*(z) = z^n p_*(z) \quad \text{thus} \quad \left(\sum_{i=0}^n a_k z^k \right)^* = \sum_{i=0}^n \overline{a}_i z^{n-i}$$

$$= \overline{a}_n + \overline{a}_{n-1} z + \cdots + \overline{a}_0 z^n. \quad (2.15)$$

Thus if we define $\pi_n(z) = \prod_{i=1}^n \varpi_i(z)$, then we can write $B_n(z)$ as

$$B_n(z) = \frac{\pi_n^*(z)}{\pi_n(z)} \eta_n \quad \text{with} \quad \eta_n = \prod_{i=1}^n z_i \in \mathbb{T}, \quad (2.16)$$

and

$$\left(\frac{p_n(z)}{\pi_n(z)} \right)^* = \eta_n \frac{p_n^*}{\pi_n}. \quad (2.17)$$

In the case of the half plane, however, the polynomials are *not* obtained for a special choice of the α_i. Moreover, for the half plane, the relations (2.16) and (2.17) should be replaced by

$$B_n(z) = \frac{\pi_{n*}(z)}{\pi_n(z)} \eta_n \quad \text{and} \quad \left(\frac{p_n(z)}{\pi_n(z)} \right)^* = \eta_n \frac{p_{n*}}{\pi_{n*}}.$$

If we want to keep a uniform notation as in (2.16) and (2.17) for both the disk and the half plane, then we should consider defining the superstar conjugate for polynomials in the case of the half plane not as in (2.15) but we should use $p^*(z) = p_*(z)$ instead. With this definition for the half plane and (2.15) for the disk, we can use (2.16) and (2.17) for both cases. Note that our notation $\varpi_n^*(z) = z - \alpha_n$ is conformal with this convention for the superstar for polynomials since it equals $(1 - \overline{\alpha}_n z)^*$ for the disk and it is equal to $(z - \overline{\alpha}_n)^*$ for the half plane.

The following technical properties can be trivially verified but they are very useful if one wants to do computations in $\mathcal{L}_n$.

Theorem 2.2.1. *In $\mathcal{L}_n$, the following relations hold:*

1. *If $f \in \mathcal{L}_n$, then*
   (a) $(f_*)_* = (f^*)^* = f$,
   (b) $(f_*)^* = f B_n$,
   (c) $(f^*)_* = f B_{n*} = f/B_n$.
2. *For the finite Blaschke products we have*
   (a) $B_n^* = 1$,
   (b) $B_{n*} = 1/B_n$.
3. *Define $\pi_n(z) = \prod_{i=1}^n \varpi_i(z)$, and let $f \in \mathcal{L}_n$ be given by $f = p_n/\pi_n$, with p_n a polynomial. Then*
   (a) $B_n = \eta_n \pi_n^*/\pi_n$ *with* $\eta_n = \prod_{i=1}^n z_i$,
   (b) $f_* = p_{n*}/\pi_{n*} = p_n^*/\pi_n^*$,
   (c) $f^* = \eta_n p_n^*/\pi_n$.
4. *If $f, g \in \mathcal{L}_n$, then*
   $$\langle f, g \rangle_{\hat{\mu}} = \langle g_*, f_* \rangle_{\hat{\mu}} = \langle g^*, f^* \rangle_{\hat{\mu}}$$
5. *If $\phi_n \in \mathcal{L}_n$ and $\langle \phi_n, \mathcal{L}_{n-1} \rangle_{\hat{\mu}} = 0$, i.e., $\phi_n \perp \mathcal{L}_{n-1}$, then also $\langle \phi_n^*, \zeta_n \mathcal{L}_{n-1} \rangle_{\hat{\mu}} = 0$, i.e., $\phi_n^* \perp \zeta_n \mathcal{L}_{n-1}$ or in the notation of the last section $\phi_n^* \perp \mathcal{L}_n(\alpha_n)$, where $\mathcal{L}_n(\alpha_n) = \{ f \in \mathcal{L}_n : f(\alpha_n) = 0 \}$.*
6. *If $f \in \mathcal{L}_n$ has a zero/pole β in $\mathbb{O}$ ($\mathbb{O}^e$, $\partial\mathbb{O}$), then $f_* \in \mathcal{L}_{n*}$ has a zero/pole $\hat{\beta}$ in $\mathbb{O}^e$ ($\mathbb{O}$, $\partial\mathbb{O}$) and $f^* \in \mathcal{L}_n$ has the same zeros as f_* and the same poles as f. The latter are elements from $\hat{A}_n = \{\hat{\alpha}_1, \ldots, \hat{\alpha}_n\}$.*

Proof. These results are so simple to verify that we do not give the proof explicitly.   □

Let us now introduce the orthonormal basis $\{\phi_k\}$ for the space $\mathcal{L}_n$ with respect to the inner product $\langle \cdot, \cdot \rangle_{\hat{\mu}}$. This means

$$\mathcal{L}_n = \text{span}\{\phi_k : k = 0, \ldots, n\}, \quad \phi_k \in \mathcal{L}_k \setminus \mathcal{L}_{k-1},$$

while $\langle \phi_i, \phi_j \rangle_{\hat{\mu}} = \delta_{ij}$. We can always choose the leading coefficient $\kappa_n = \overline{\phi_n^*(\alpha_n)}$ of ϕ_n to be real and positive. We shall use from now on the notation κ_n to denote this coefficient. The functions $\varphi_n = \kappa_n^{-1} \phi_n$ are then the monic orthogonal basis functions.

From a previous section we know that the reproducing kernel for $\mathcal{L}_n$ is given by

$$k_n(z, w) = \sum_{k=0}^n \phi_k(z)\overline{\phi_k(w)}.$$

We can obtain the following determinant expressions:

Theorem 2.2.2. *Let $E_n^T = [e_0, \ldots, e_n]$ be a vector of basis functions for $\mathcal{L}_n$ and let G_n denote the Gram matrix w.r.t. $\overset{\circ}{\mu}$:*

$$G_n = \left\langle E_n, E_n^T \right\rangle_{\overset{\circ}{\mu}} = [\langle e_i, e_j \rangle_{\overset{\circ}{\mu}}].$$

Then the reproducing kernel $k_n(z, w)$ for $\mathcal{L}_n$ is expressed as

$$k_n(z, w) = \frac{-1}{\det G_n} \det \begin{bmatrix} G_n & E_n(z) \\ E_n(w)^H & 0 \end{bmatrix}.$$

Proof. Suppose that by a Gram–Schmidt orthogonalization procedure, the basis E_n is transformed into an orthonormal basis $F_n = [\phi_0, \ldots, \phi_n]^T$ and that

$$\phi_k = \sum_{i=0}^{k} l_{ki} e_i.$$

Then the vector F_n can be written as $F_n = L_n E_n$, with L_n the lower triangular matrix containing the l_{ki} coefficients. Now we express that these ϕ_k form an orthonormal set. Then we get

$$I = \left\langle F_n, F_n^T \right\rangle_{\overset{\circ}{\mu}} = L_n \left\langle E_n, E_n^T \right\rangle_{\overset{\circ}{\mu}} L_n^H = L_n G_n L_n^H.$$

Hence $G_n^{-1} = L_n^H L_n$. Therefore,

$$k_n(z, w) = F_n(w)^H F_n(z) = E_n(w)^H L_n^H L_n E_n(z) = E_n(w)^H G_n^{-1} E_n(z).$$

The last line gives the result by a standard argument for determinants. □

From this fact the following simple but basic relations can be derived:

Theorem 2.2.3. *Let $k_n(z, w)$ be the reproducing kernel for $\mathcal{L}_n$ based on the points $A_n = \{\alpha_i : i = 1, \ldots, n\}$ and let $\{\phi_k\}$ be a set of orthonormal basis functions with leading coefficients $\kappa_k > 0$. Then the following relations hold:*

1. $k_n(z, w) = B_n(z)\overline{B_n(w)}k_n(\hat{w}, \hat{z})$,
2. $k_n(z, \alpha_n) = \kappa_n \phi_n^*(z)$,
3. $k_n(\alpha_n, \alpha_n) = \kappa_n^2$.

Proof.

1. If $\{e_0, \ldots, e_n\}$ is a basis for $\mathcal{L}_n$ with Gram matrix G_n, then $\{B_n e_{k*} : k = 0, \ldots, n\}$ is also a basis with Gram matrix G_n^T. Indeed,

$$\langle B_n e_{i*}, B_n e_{j*} \rangle_{\overset{\circ}{\mu}} = \langle e_j, e_i \rangle_{\overset{\circ}{\mu}}.$$

Let us denote this basis in a vector form, which, according to our earlier convention of substar transform for matrices, should be denoted as

$$B_n(z)E_{n*}(z) = B_n(z)[e_{0*}(z), \ldots, e_{n*}(z)]$$

$$\text{if } E_n^T(z) = [e_0(z), \ldots, e_n(z)].$$

Hence, when applying the determinant formula of the previous theorem for this new basis, we get

$$k_n(z, w) = \frac{-1}{\det G_n} \det \begin{bmatrix} G_n^T & B_n(z)E_{n*}(z)^T \\ \overline{B_n(w) E_{n*}(w)} & 0 \end{bmatrix}$$

$$= \frac{-B_n(z)\overline{B_n(w)}}{\det G_n} \det \begin{bmatrix} G_n & E_{n*}(w)^H \\ E_{n*}(z) & 0 \end{bmatrix}$$

$$= B_n(z)\overline{B_n(w)}k_n(\hat{w}, \hat{z}).$$

2. The first part gives

$$k_n(z, w) = \sum_{k=0}^{n} \phi_k(z)\overline{\phi_k(w)} = \sum_{k=0}^{n} \phi_k^*(z)\overline{\phi_k^*(w)} B_{n\setminus k}(z)\overline{B_{n\setminus k}(w)},$$

where $B_{n\setminus k} = B_n/B_k$. For $w = \alpha_n$, we see that $B_{n\setminus k}(\alpha_n) = \delta_{kn}$, so that

$$k_n(z, \alpha_n) = \phi_n^*(z)\overline{\phi_n^*(\alpha_n)} = \phi_n^*(z)\kappa_n.$$

3. If we also let z be equal to α_n, then we directly get the third result.   □

We can now also obtain determinant expressions for κ_n and ϕ_n.

Theorem 2.2.4. *Consider the basis of finite Blaschke products $\{B_k(z) : k = 0, \ldots, n\}$ for the space $\mathcal{L}_n$ and denote the elements from the corresponding Gram matrix as*

$$\mu_{ij} = \langle B_i, B_j \rangle_{\hat{\mu}}$$

and set $G_n = [\mu_{ij}]$. The orthonormal basis functions are denoted as ϕ_k and their leading coefficient as κ_k. Then the following equalities hold:

1. $\kappa_n^2 = \det G_{n-1} / \det G_n$,

2. $\phi_n(z) = (\det G_{n-1} \det G_n)^{-1/2} \det \begin{bmatrix} \mu_{00} & \cdots & \mu_{0,n-1} & B_0(z) \\ \vdots & & \vdots & \vdots \\ \mu_{n0} & \cdots & \mu_{n,n-1} & B_n(z) \end{bmatrix}.$

Proof. We know from the previous theorem that

$$
k_n(z, \alpha_n) = \frac{-B_n(z)}{\det G_n} \det
\begin{bmatrix}
 & & \gamma_0 \\
 & G_n & \vdots \\
 & & \gamma_n \\
B_{0*}(z)d \cdots B_{n*}(z) & & 0
\end{bmatrix},
$$

with $\gamma_k = B_{n \setminus k}(\alpha_n) = \delta_{kn}$. Hence we get

$$
k_n(z, \alpha_n) = \kappa_n \phi_n^*(z) = \frac{B_n(z)}{\det G_n} \det
\begin{bmatrix}
\mu_{00} & \cdots & \mu_{0n} \\
\vdots & & \vdots \\
\mu_{n-1,0} & \cdots & \mu_{n-1,n} \\
B_{0*}(z) & \cdots & B_{n*}(z)
\end{bmatrix},
$$

which then gives first

$$
k_n(\alpha_n, \alpha_n) = \kappa_n^2 = \frac{1}{\det G_n} \det
\begin{bmatrix}
\mu_{00} & \cdots & \mu_{0n} \\
\vdots & & \vdots \\
\mu_{n-1,0} & \cdots & \mu_{n-1,n} \\
\gamma_0 & \cdots & \gamma_n
\end{bmatrix}
$$

$$
= \frac{\det G_{n-1}}{\det G_n}
$$

because $\gamma_k = \delta_{kn}$, so that also

$$
\phi_n(z) = \kappa_n^{-1} k_n^*(z, \alpha_n)
$$

$$
= (\det G_{n-1} \det G_n)^{-1/2} \det
\begin{bmatrix}
\overline{\mu}_{00} & \cdots & \overline{\mu}_{0n} \\
\vdots & & \vdots \\
\overline{\mu}_{n-1,0} & \cdots & \overline{\mu}_{n-1,n} \\
B_0 & \cdots & B_n
\end{bmatrix}
$$

$$
= (\det G_{n-1} \det G_n)^{-1/2} \det
\begin{bmatrix}
\mu_{00} & \cdots & \mu_{0,n-1} & B_0 \\
\vdots & & \vdots & \vdots \\
\mu_{n0} & \cdots & \mu_{n,n-1} & B_n
\end{bmatrix},
$$

which concludes the proof.   □

2.3. Extremal problems in $\mathcal{L}_n$

In this section we shall review some of the extremal problems that can be solved with reproducing kernels in the special case of $\mathcal{L}_n$. From now on we shall use the notation $\| \cdot \|_{\hat{\mu}}$ to mean $\langle \cdot, \cdot \rangle_{\hat{\mu}}^{1/2}$. From the problem $P^1(a, w)$ that we considered in Section 1.4, we can now derive

Theorem 2.3.1. *All the solutions of the following optimization problem:*

$$P_n^1(1, \alpha_n): \quad \sup \{|f(\alpha_n)|^2 : \|f\|_{\hat{\mu}} = 1, f \in \mathcal{L}_n\}$$

are given by

$$f = \eta \phi_n^*, \quad |\eta| = 1,$$

where ϕ_n is the nth orthonormal basis function of $\mathcal{L}_n$ with leading coefficient $\kappa_n = \phi_n^*(\alpha_n)$. The maximum is equal to κ_n^2.

Proof. This follows immediately from the Theorem 1.4.2 and the properties given in Theorem 2.2.3.   □

Furthermore, for the second optimization problem of Theorem 1.4.2 we formulate a special case.

Theorem 2.3.2. *The optimization problem*

$$P_n^2(1, \alpha_n): \quad \inf \{\|f\|_{\hat{\mu}}^2 : f(\alpha_n) = 1, f \in \mathcal{L}_n\}$$

has a unique solution given by

$$f = \varphi_n^* = \kappa_n^{-1} \phi_n^*,$$

where φ_n is the nth monic orthogonal basis function in $\mathcal{L}_n$, and ϕ_n is the orthonormal one with $\phi_n^*(\alpha_n) = \kappa_n > 0$. The minimal value is κ_n^{-2}.

Proof. This also follows from Theorem 1.4.2 and the properties in Theorem 2.2.3.   □

Since in $\mathcal{L}_n$ it holds that

$$\|f\|_{\hat{\mu}} = \|f^*\|_{\hat{\mu}}$$

we also have solved the following problem.

Corollary 2.3.3. *The unique solution of the problem*

$$\inf \{\|f\|_{\hat{\mu}}^2 : f \in \mathcal{L}_n^M\},$$

where $\mathcal{L}_n^M$ denotes the set of all monic elements of $\mathcal{L}_n$, is given by the nth monic orthogonal basis function $\varphi_n = \kappa_n^{-1}\phi_n$ of $\mathcal{L}_n$ and the minimum is κ_n^{-2}, with $\kappa_n > 0$ the leading coefficient of the orthonormal one.

Recall the definition of $\mathcal{L}_n(w)$ given in Section 2.1:

$$\mathcal{L}_n(w) = \left\{ \frac{z-w}{\varpi_0(z)} f : f \in \mathcal{L}_n(\hat{\alpha}_0) \right\} = \{f : f \in \mathcal{L}_n : f(w) = 0\}.$$

The problem of finding the orthogonal projection of 1 onto $\mathcal{L}_n(w)$ is related to a classical Szegő problem.

Theorem 2.3.4. *Define the following problem in $\mathcal{L}_n(w)$, which is defined as above with $\mathcal{L}_n$ the rational function space based on the point set $A_n = \{\alpha_1, \ldots, \alpha_n\}$:*

$$P_n^3(w): \quad \inf\left\{\|1 - f\|_\mu^2 : f \in \mathcal{L}_n(w)\right\}.$$

Then $P_n^3(w)$ has the unique solution

$$f = 1 - \frac{k_n(z, w)}{k_n(w, w)},$$

where $k_n(z, w)$ is the reproducing kernel of $\mathcal{L}_n$. The minimum is given by $[k_n(w, w)]^{-1}$.

Proof. This problem can be reduced to problem $P_n^2(1, w)$ by noting that

$$\mathcal{L}_n(w) = \{f = 1 - g : g \in \mathcal{L}_n, g(w) = 1\}$$

and thus for $f = 1 - g$

$$\inf\left\{\|1 - f\|_\mu^2 : f \in \mathcal{L}_n(w)\right\} = \inf\left\{\|g\|_\mu^2 : g \in \mathcal{L}_n, g(w) = 1\right\}.$$

From this the result follows easily.   □

This theorem has a simple corollary.

Corollary 2.3.5. *If $k_n(z, w)$ is the reproducing kernel for $\mathcal{L}_n$ w.r.t. $\langle \cdot, \cdot \rangle_\mu$, then $k_n(w, w)$ is nondecreasing with n if $w \in \mathbb{O}$.*

Proof. Since the $\mathcal{L}_n(w)$ are nested as $\mathcal{L}_n(w) \subset \mathcal{L}_{n+1}(w)$, the minimum $[k_n(w, w)]^{-1}$ cannot increase with n.

Of course this is also obvious from $k_n(w, w) = \sum_{k=0}^n |\phi_k(w)|^2$.   □

Because there are so many optimization problems whose solutions can be expressed in terms of the reproducing kernel, one can ask whether there is an optimization problem that has this kernel for its solution. It turns out that it gives an approximation to some kernel $s_w(z)$, known as the *Szegő kernel* associated with $\mathring{\mu}$. See Ref. [102, pp. 51–52]. It is the reproducing kernel for $H_2(\mathring{\mu})$ and it is related to the spectral factor σ for the measure μ as explained in Section 1.3. More precisely, the kernel $k_n(z, w)$ approximates $s_w(z)$ in $L_2(\mathring{\mu})$-norm.

Let σ be the outer spectral factor associated with the measure μ, satisfying the Szegő condition $\log \mu' \in L_1(d\lambda)$. Then define for $w \in \mathbb{O}$ the Szegő kernel $s(z, w) = s_w(z)$ by

$$s_w(z) = \frac{1}{[1 - \zeta_0(z)\overline{\zeta_0(w)}]\sigma(z)\overline{\sigma(w)}} = \frac{1}{\varpi_0(\alpha_0)} \frac{\varpi_0(z)\overline{\varpi_0(w)}}{\varpi_w(z)\sigma(z)\overline{\sigma(w)}}. \qquad (2.18)$$

That is,

$$s_w(z) = \frac{1}{\sigma(z)(1 - \overline{w}z)\overline{\sigma(w)}} \qquad \text{for } \mathbb{D},$$

$$s_w(z) = \frac{(z + \mathbf{i})(\mathbf{i} - \overline{w})}{2\mathbf{i}\sigma(z)(z - \overline{w})\overline{\sigma(w)}} \qquad \text{for } \mathbb{U}.$$

Before we give the approximation in Theorem 2.3.8 below, we first give some properties of $s_w(z)$ in the following lemmas.

Lemma 2.3.6. The Szegő kernel is the reproducing kernel for $H_2(\mathring{\mu})$. This means that for any $f \in H_2(\mathring{\mu})$ we have

$$\langle f, s_w \rangle_{\mathring{\mu}} = f(w), \qquad w \in \mathbb{O}.$$

Proof. First note that $s_w \in H_2(\mathring{\mu})$. It is a well-known property that a function f will be in $H_2(\mathring{\mu})$ if and only if $f\sigma \in \mathring{H}_2$ (see Ref. [87, Theorem 3.4, p. 215]). We now have that the $L_2(\mathring{\lambda})$ norm of $s_w\sigma$ is

$$\begin{aligned}
\|s_w\sigma\|_2^2 &= \frac{|\varpi_0(w)|^2}{|\sigma(w)|^2|\varpi_0(\alpha_0)|^2} \int \frac{d\lambda(t)}{|\varpi_w(t)|^2} \\
&= \frac{|\varpi_0(w)|^2}{|\sigma(w)|^2 \overline{\varpi_0(\alpha_0)}\varpi_w(w)} \int P(t, w)\, d\lambda(t) \\
&= \frac{|\varpi_0(w)|^2}{|\sigma(w)|^2 \overline{\varpi_0(\alpha_0)}\varpi_w(w)} = s_w(w) < \infty
\end{aligned}$$

as long as w is in a compact subset of $\mathbb{O}$ (recall that $\overline{\varpi_0(\alpha_0)}\varpi_w(w) = 1 - |w|^2$ in the case of the disk, whereas it is equal to $4 \operatorname{Im} w$ in the case of the half plane and hence is strictly positive in both cases). This implies $s_w \in L_2(\mathring{\mu})$ and because s_w is analytic in $\mathbb{O}$, it is in $H_2(\mathring{\mu})$.

Now for any $f \in H_2(\mathring{\mu})$

$$\langle f, s_w \rangle_{\mathring{\mu}} = \int \frac{f(t)}{\varpi_0(\alpha_0)} \frac{\varpi_0(w)}{\sigma_*(t)\sigma(w)} \frac{\varpi_{0*}(t)}{\varpi_{w*}(t)} \, d\mathring{\mu}(t)$$

$$= \int \frac{f(t)}{\varpi_0(\alpha_0)} \frac{\sigma(t)}{\sigma(w)} \frac{\varpi_0(w)\varpi_{0*}(t)}{\varpi_{w*}(t)} \, d\mathring{\lambda}(t)$$

$$= \frac{\varpi_0(w)}{\sigma(w)\varpi_0(\alpha_0)} \int \frac{f(t)\sigma(t)}{\varpi_0(t)\varpi_{w*}(t)} \, d\lambda(t).$$

The second line is because $1/\sigma = 0$, $d\mathring{\mu}_s$ a.e., where $d\mathring{\mu}_s = d\mathring{\mu} - \mu' d\mathring{\lambda}$. This integral is well defined because $f \in H_2(\mathring{\mu})$, and hence $f\sigma/\varpi_0 \in H_2$. So we get

$$\langle f, s_w \rangle_{\mathring{\mu}} = \frac{\varpi_0(w)}{\sigma(w)} \int \frac{f(t)}{\varpi_0(t)} \frac{\sigma(t)t}{t-w} \, d\lambda(t) \qquad \text{for } \mathbb{D},$$

$$= \frac{\varpi_0(w)}{\sigma(w)} \int \frac{f(t)}{\varpi_0(t)} \frac{\sigma(t)}{t-w} \frac{dt}{2\pi \mathbf{i}} \qquad \text{for } \mathbb{U},$$

$$= \frac{\varpi_0(w)}{\sigma(w)} \int C(t,w) \frac{f(t)\sigma(t)}{\varpi_0(t)} \, d\lambda(t) \quad \text{for } \mathbb{O},$$

with $C(t,w)$ the Cauchy kernel. Hence, by Cauchy's integral, this equals $f(w)$.

$\square$

By taking $f = s_w$, we get the following corollary.

Corollary 2.3.7. *With s_w the Szegő kernel, we have*

$$\|s_w\|_{\mathring{\mu}}^2 = s_w(w).$$

Note that

$$s_w(w) = \frac{|\varpi_0(w)|^2}{|\sigma(w)|^2 \overline{\varpi_0(\alpha_0)}\varpi_w(w)}.$$

This means

$$s_w(w) = \frac{1}{(1-|w|^2)|\sigma(w)|^2} \qquad \text{for } \mathbb{D},$$

$$s_w(w) = \frac{|w + \mathbf{i}|^2}{4 \operatorname{Im} w |\sigma(w)|^2} \qquad \text{for } \mathbb{U}.$$

Now we are ready to state our optimization problem.

Theorem 2.3.8. *Let μ be a measure satisfying the Szegő condition $\log \mu' \in L_1(d\mathring{\lambda})$. Define the problem*

$$S_n(w): \quad \inf\left\{\|f - s_w\|_{\hat{\mu}}^2 : f \in \mathcal{L}_n\right\},$$

where s_w is the Szegő kernel. Then problem $S_n(w)$ has a solution $f(z) = k_n(z, w)$, where $k_n(z, w)$ is the reproducing kernel for $\mathcal{L}_n$ w.r.t. $\langle \cdot, \cdot \rangle_{\hat{\mu}}$ and the minimum is given by

$$s_w(w) - k_n(w, w).$$

Proof. We can also reduce this problem $S_n(w)$ to the problem $P_n^2(1, w)$ by observing that

$$\|f - s_w\|_{\hat{\mu}}^2 = \|f\|_{\hat{\mu}}^2 + \|s_w\|_{\hat{\mu}}^2 - 2\operatorname{Re}\langle f, s_w \rangle_{\hat{\mu}}.$$

The last term is equal to $-2\operatorname{Re} f(w)$ by Lemma 2.3.6. Thus we have to solve

$$\inf_a\left\{\inf_{f \in \mathcal{L}_n, f(w)=a} \|f\|_{\hat{\mu}}^2 - 2\operatorname{Re} a\right\} = \inf_a\left\{\frac{|a|^2}{k_n(w, w)} - 2\operatorname{Re} a\right\}.$$

The solution of the latter problem is easily seen to be found for $a = k_n(w, w)$. Since $\|s_w\|_{\hat{\mu}}^2 = s_w(w)$ by Corollary 2.3.7, we get the solution as given in the theorem. $\qquad\square$

This theorem can also be reformulated as follows:

Corollary 2.3.9. *Let the measure μ of the previous theorem satisfy $d\mu(t) = P(t, w)|\varpi_0(t)|^2 d\nu$ with $P(z, w)$ the Poisson kernel. Denote by σ_μ and σ_ν the spectral factors of μ and ν respectively. Then the problem*

$$\inf\left\{\|f - [\sigma_\nu \overline{\sigma_\nu(w)}]^{-1}\|_{\hat{\mu}}^2 : f \in \mathcal{L}_n\right\}, \quad w \in \mathbb{O}$$

reaches a minimum $|\sigma_\nu(w)|^{-2} - k_n(w, w)$ for $f(z) = k_n(z, w)$, where k_n is the reproducing kernel for $\mathcal{L}_n$ w.r.t. $d\mathring{\mu}$.

Proof. Note that the outer spectral factors are related by

$$\sigma_\mu = \sigma_\nu \frac{\varpi_0(z)}{\varpi_w(z)} \sqrt{\frac{\varpi_w(w)}{\varpi_0(\alpha_0)}}.$$

Fill this into the expression for s_w of the previous theorem to find that it is equal to $[\sigma_\nu(z)\overline{\sigma_\nu(w)}]^{-1}$. The result then follows easily. $\qquad\square$

3

The kernel functions

The reproducing kernels have played an important role in the theory of orthogonal polynomials. We shall study them for our spaces of rational functions. First of all we derive the Christoffel–Darboux relations, which give the generalization of the corresponding formulas for the Szegő polynomials. Next we shall derive in Section 3.2 a recurrence relation for these kernels in the style of the recurrence for the Szegő orthogonal polynomials. This is not too difficult to find once the previous relations have been found. The transition matrices that describe the recurrence are almost J-unitary matrices. We can normalize them to make them precisely J-unitary. The correspondingly normalized kernels will be produced in that case. The latter are discussed in Section 3.3.

3.1. Christoffel–Darboux relations

We now prove the Christoffel–Darboux relations. We start with some technical lemmas.

Lemma 3.1.1. Let $f \in \mathcal{L}_n$.

1. If g and h are defined by the relations $f(z) - f(w) = (z-w)g(z) = \frac{z-w}{\varpi_n(z)} h(z)$
   then
   (a) $p_1(z)g(z) \in \mathcal{L}_n$ for $p_1 \in \mathcal{P}_1$, an arbitrary polynomial of degree at most 1, especially $g(z) \in \mathcal{L}_n$.
   (b) $h \in \mathcal{L}_{n-1}$.
2. If $f(w) = 0$, then $\frac{\varpi_n(z)}{z-w} f(z) \in \mathcal{L}_{n-1}$.

Proof. Clearly $g(z)$ can be written as $p_{n-1}(z)/\pi_n(z)$ with $\pi_n(z) = \prod_{k=1}^{n} \varpi_k(z)$ and $p_{n-1}(z) \in \mathcal{P}_{n-1}$. This implies (a). Furthermore, $h(z) = \varpi_n(z)g(z)$, which gives $h(z) = p_{n-1}(z)/\pi_{n-1} \in \mathcal{L}_{n-1}$ and this is (b).

The second result is a special case of (1b) for $f(w) = 0$. □

Lemma 3.1.2. Let $\{\phi_k\}_{k=0}^n$ denote as before the orthonormal basis functions for $\mathcal{L}_n$ and ζ_k the Blaschke factor based on α_k. As functions of z, with w some parameter, we have

$$l_n^0(z, w) = \frac{\phi_{n+1}^*(z)\overline{\phi_{n+1}^*(w)} - \phi_{n+1}(z)\overline{\phi_{n+1}(w)}}{1 - \zeta_{n+1}(z)\overline{\zeta_{n+1}(w)}} \in \mathcal{L}_n$$

and for $k = 1, \ldots, n$

$$l_n^k(z, w) = \frac{\phi_n^*(z)\overline{\phi_n^*(w)} - \zeta_k(z)\overline{\zeta_k(w)}\phi_n(z)\overline{\phi_n(w)}}{1 - \zeta_k(z)\overline{\zeta_k(w)}} \in \mathcal{L}_n.$$

Proof. A straightforward computation gives

$$1 - \zeta_k(z)\overline{\zeta_k(w)} = 1 - \frac{(\alpha_k - z)(\overline{\alpha}_k - \overline{w})}{\varpi_k(z)\overline{\varpi_k(w)}} = \frac{\varpi_k(\alpha_k)\varpi_w(z)}{\varpi_k(z)\overline{\varpi_k(w)}}.$$

According to part (2) of the previous lemma, we only have to prove that the numerator of l_n^0 is zero for $z = \hat{w}$. Call this numerator $N(z, w)$. Thus we have to prove that $N(\hat{w}, w) = 0$. Now,

$$N(\hat{w}, w) = \phi_{n+1}^*(\hat{w})\overline{\phi_{n+1}^*(w)} - \phi_{n+1}(\hat{w})\overline{\phi_{n+1}(w)}$$

$$= B_{n+1}(\hat{w})\overline{\phi_{n+1}(w)}B_{n+1}(w)\phi_{n+1}(\hat{w}) - \phi_{n+1}(\hat{w})\overline{\phi_{n+1}(w)}$$

$$= 0,$$

which proves the first part.

For the second part, we have to prove, according to part (1a) of the previous lemma and as in part one of this theorem, that the numerator of $l_n^k(z, w)$ is zero for $z = \hat{w}$. Let us call this numerator again $N(z, w)$. Then

$$N(\hat{w}, w) = \phi_n^*(\hat{w})\overline{\phi_n^*(w)} - \phi_n(\hat{w})\overline{\phi_n(w)}$$

$$= B_n(\hat{w})\overline{\phi_n(w)}B_n(w)\phi_n(\hat{w}) - \phi_n(\hat{w})\overline{\phi_n(w)}$$

$$= 0,$$

and this proves the second part. □

Now we can prove the Christoffel–Darboux relations.

Theorem 3.1.3 (Christoffel–Darboux relations). *The following relations hold between reproducing kernel and orthonormal basis functions of $\mathcal{L}_n$:*

$$k_n(z, w) = \frac{\phi_{n+1}^*(z)\overline{\phi_{n+1}^*(w)} - \phi_{n+1}(z)\overline{\phi_{n+1}(w)}}{1 - \zeta_{n+1}(z)\overline{\zeta_{n+1}(w)}} \quad (= l_n^0(z, w)) \qquad (3.1)$$

and

$$k_n(z, w) = \frac{\phi_n^*(z)\overline{\phi_n^*(w)} - \zeta_n(z)\overline{\zeta_n(w)}\phi_n(z)\overline{\phi_n(w)}}{1 - \zeta_n(z)\overline{\zeta_n(w)}} \quad (= l_n^n(z, w)). \qquad (3.2)$$

Proof. The proof we give is completely analogous to the proof that Szegő gave for these relations in the polynomial case. See Szegő [196, p. 293]. We have shown in the previous lemma that the right-hand sides are elements from $\mathcal{L}_n$. We only have to show that they reproduce. So choose some $f \in \mathcal{L}_n$. Then

$$\langle f, l_n^0(\cdot, w) \rangle_{\hat{\mu}} = f(w)\langle 1, l_n^0(\cdot, w) \rangle_{\hat{\mu}} + \langle f - f(w), l_n^0(\cdot, w) \rangle_{\hat{\mu}}. \qquad (3.3)$$

By Lemma 3.1.1(1a), we find that if $f(z) - f(w) = (z - w)g(z)$, then $g \in \mathcal{L}_n$ and $p_1 g \in \mathcal{L}_n$ for all $p_1 \in \mathcal{P}_1$. Thus if we call $N(z, w)$ the numerator of l_n^0, we get

$$\langle f - f(w), l_n^0 \rangle_{\hat{\mu}} = \frac{\varpi_{n+1}(w)}{\varpi_{n+1}(\alpha_{n+1})} \left\langle (z - w)g(z), \frac{\varpi_{n+1}(z)}{\varpi_w(z)} N(z, w) \right\rangle_{\hat{\mu}}$$

$$= \frac{\varpi_{n+1}(w)}{\varpi_{n+1}(\alpha_{n+1})} \langle (z - \alpha_{n+1})g(z), N(z, w) \rangle_{\hat{\mu}}.$$

Because $(z - \alpha_{n+1})g(z) \in \mathcal{L}_n$, the inner product in the right-hand side gives

$$\langle (z - \alpha_{n+1})g(z), N(z, w) \rangle_{\hat{\mu}} = \langle (z - \alpha_{n+1})g(z), \phi_{n+1}^*(z) \rangle_{\hat{\mu}} \, \phi_{n+1}^*(w)$$

$$= \langle \zeta_{n+1}(z)h(z), \phi_{n+1}^*(z) \rangle_{\hat{\mu}} \, \phi_{n+1}^*(w),$$

with $h(z) = \bar{z}_{n+1}\varpi_{n+1}(z)g(z) \in \mathcal{L}_n$. This is zero because of Theorem 2.2.1(5). It remains to be shown that $\langle 1, l_n^0(\cdot, w) \rangle_{\hat{\mu}} = \eta(w) = 1$. Apply (3.3) to $f(\cdot) = l_n^0(\cdot, z)$. Then we get

$$\langle l_n^0(\cdot, z), l_n^0(\cdot, w) \rangle_{\hat{\mu}} = l_n^0(w, z)\eta(w).$$

If we interchange z and w, we get

$$\langle l_n^0(\cdot, w), l_n^0(\cdot, z) \rangle_{\hat{\mu}} = l_n^0(z, w)\eta(z).$$

Since the left-hand sides are each others conjugate and because also $l_n^0(z, w)$ is the complex conjugate of $l_n^0(w, z)$, we find $\eta(z) = \overline{\eta(w)}$. Thus η is a constant and this constant is 1 because for $z = w = \alpha_{n+1}$, we get

$$k_n(\alpha_{n+1}, \alpha_{n+1}) = \eta l_n^0(\alpha_{n+1}, \alpha_{n+1}) = \eta\left(\kappa_{n+1}^2 - |\phi_{n+1}(\alpha_{n+1})|^2\right)$$

$$= k_{n+1}(\alpha_{n+1}, \alpha_{n+1}) - |\phi_{n+1}(\alpha_{n+1})|^2$$

$$= \kappa_{n+1}^2 - |\phi_{n+1}(\alpha_{n+1})|^2.$$

Analogously, when $N(z, w)$ is now the numerator of l_n^n, we get

$$\left\langle f(z) - f(w), l_n^n(z, w) \right\rangle_{\hat\mu} = c \frac{\varpi_n(w)}{\varpi_n(\alpha_n)} \langle \zeta_n(z)h(z), N(z, w) \rangle_{\hat\mu},$$

with $h \in \mathcal{L}_{n-1}$ by Lemma 3.1.1. The inner product of the right-hand side is zero because $\phi_n^* \perp \zeta_n h \in \zeta_n \mathcal{L}_{n-1}$ and $\zeta_n \phi_n \perp \zeta_n h$ because $h \in \mathcal{L}_{n-1}$. The rest follows exactly as in the proof of the previous formula. □

From these relations, we find a useful corollary.

Corollary 3.1.4. *For the orthonormal functions of $\mathcal{L}_n$, it holds that for all $n \geq 0$*

1. $\phi_n^*(z) \neq 0$ *for* $z \in \mathbb{O}$ *and* $\phi_n(z) \neq 0$ *for* $z \in \mathbb{O}^e$,
2. $|\phi_{n+1}(z)/\phi_{n+1}^*(z)| < 1 (= 1, > 1)$ *for* $z \in \mathbb{O}(\partial\mathbb{O}, \mathbb{O}^e)$.

Proof. From the first Christoffel–Darboux relation (3.1), we get for $w = z$

$$(1 - |\zeta_{n+1}(z)|^2)k_n(z, z) = |\phi_{n+1}^*(z)|^2 - |\phi_{n+1}(z)|^2.$$

Because $|\zeta_{n+1}(z)| < 1$ for $z \in \mathbb{O}$, and because $k_n(z, z) > 0$, we get

$$|\phi_{n+1}^*(z)|^2 > |\phi_{n+1}(z)|^2 \geq 0 \quad \text{for } z \in \mathbb{O}.$$

Hence $\phi_{n+1}^*(z) \neq 0$ for $z \in \mathbb{O}$ and thus

$$|\phi_{n+1}(z)/\phi_{n+1}^*(z)| < 1 \quad \text{for } z \in \mathbb{O}.$$

The proof for $\partial\mathbb{O}$ and $\mathbb{O}^e$ is analogous. □

3.2. Recurrence relations for the kernels

The reproducing kernels for the spaces $\mathcal{L}_n$ satisfy some recursions that can be found from the Christoffel–Darboux relations. We shall derive them below.

Theorem 3.2.1. *Let $k_n(z, w)$ be the reproducing kernel for $\mathcal{L}_n$. Then*

$$\begin{bmatrix} k_{n+1}^*(z, w) \\ k_{n+1}(z, w) \end{bmatrix} = t_{n+1}(z, w) \begin{bmatrix} k_n^*(z, w) \\ k_n(z, w) \end{bmatrix}, \tag{3.4}$$

where the superstar conjugation is with respect to z. The matrix t_{n+1} is given by

$$t_{n+1}(z, w) = c_n \begin{bmatrix} 1 & \overline{\rho}_{n+1} \\ \rho_{n+1} & 1 \end{bmatrix} \begin{bmatrix} \zeta_{n+1}(z) & 0 \\ 0 & 1 \end{bmatrix} \begin{bmatrix} 1 & \overline{\gamma}_{n+1} \\ \gamma_{n+1} & 1 \end{bmatrix},$$

with

$$c_n = (1 - |\rho_{n+1}|^2)^{-1},$$

$$\rho_{n+1} = \rho_{n+1}(w) = \overline{\phi_{n+1}(w)}/\phi_{n+1}^*(w),$$

$$\gamma_{n+1} = \gamma_{n+1}(w) = -\zeta_{n+1}(w)\rho_{n+1}(w)$$

and where ϕ_{n+1} is the $(n + 1)$st orthonormal basis function.

Proof. Obviously,

$$k_{n+1}(z, w) = k_n(z, w) + \phi_{n+1}(z)\overline{\phi_{n+1}(w)}. \tag{3.5}$$

From the Christoffel–Darboux relation (3.1), we get

$$\phi_{n+1}(z)\overline{\phi_{n+1}(w)} = \phi_{n+1}^*(z)\overline{\phi_{n+1}^*(w)} - (1 - \zeta_{n+1}(z)\overline{\zeta_{n+1}(w)})k_n(z, w). \tag{3.6}$$

Substitute this into (3.5) to get

$$k_{n+1}(z, w) = \zeta_{n+1}(z)\overline{\zeta_{n+1}(w)}k_n(z, w) + \phi_{n+1}^*(z)\overline{\phi_{n+1}^*(w)}. \tag{3.7}$$

Now, take the superstar conjugate with respect to z of (3.5):

$$\phi_{n+1}^*(z)\overline{\phi_{n+1}^*(w)} \,\overline{\rho_{n+1}(w)} = k_{n+1}^*(z, w) - \zeta_{n+1}(z)k_n^*(z, w). \tag{3.8}$$

Substitute this into (3.7) to get

$$k_{n+1}(z, w)\overline{\rho}_{n+1} = \zeta_{n+1}(z)\overline{\zeta_{n+1}(w)}k_n(z, w)\overline{\rho}_{n+1}$$
$$+ k_{n+1}^*(z, w) - \zeta_{n+1}(z)k_n^*(z, w). \tag{3.9}$$

The superstar conjugate of (3.9) is

$$k_{n+1}^*(z, w)\rho_{n+1} = \zeta_{n+1}(w)k_n^*(z, w)\rho_{n+1} + k_{n+1}(z, w) - k_n(z, w). \tag{3.10}$$

From (3.9) and (3.10), the result follows.   □

Note that by Corollary 3.1.4, $|\rho_n(w)| < 1$ for $w \in \mathbb{O}$ and also $\gamma_n(w) = -\zeta_n(w)$
$\rho_n(w)$ is in $\mathbb{D}$ for all $w \in \mathbb{O}$. The following corollary generalizes a well-known
result from the polynomial case. See, for example, Ref. [102, p. 40].

Corollary 3.2.2. *The following statements hold.*

1. *For $n \geq 0$, define $s_n(z, w) = k_n^*(z, w)/k_n(z, w)$. Then $s_0(z, w) = 1$ and
   for $n \geq 1$ it holds that $s_n(z, w) \in \mathbb{D}$ for all $z, w \in \mathbb{O}$. In other words,
   for $n \geq 1$ and $w \in \mathbb{O}$ fixed, $s_n(z, w) \in \mathcal{B}$.*
2. *For some fixed $w \in \mathbb{O}(\mathbb{O}^e, \partial\mathbb{O})$, the reproducing kernel $k_n(z, w)$ has its zeros
   in $\mathbb{O}^e(\mathbb{O}, \partial\mathbb{O})$.*
3. *The zeros of the orthonormal basis functions ϕ_n are all in $\mathbb{O}$.*

Proof.

1. Because of the normalization of the measure $\mathring{\mu}$, which is $\langle 1, 1\rangle_{\mathring{\mu}} = \int d\mathring{\mu} = 1$,
   it follows that $\phi_0 = \kappa_0 = 1$, so that also $k_0(z, w) = 1$ and consequently
   $s_0(z, w) = k_0^*(z, w)/k_0(z, w) = 1$.

   From the recurrence relation given in the previous theorem, we can deduce
   that the function $s_{n+1}(z, w)$ is obtained from $s_n(z, w)$ by a transformation
   that we denote as τ, that is,

$$s_{n+1}(z, w) = \tau(s_n(z, w)),$$

where τ is a succession of three transformations: $\tau = \tau_1 \circ \tau_2 \circ \tau_3$ with

$$\tau_3 : t \longmapsto \frac{t + \overline{\gamma}_{n+1}}{1 + \gamma_{n+1}t},$$

$$\tau_2 : t \longmapsto \zeta_{n+1}(z)t,$$

$$\tau_1 : t \longmapsto \frac{t + \overline{\rho}_{n+1}}{1 + \rho_{n+1}t}.$$

In other words, in the notation for Möbius transformations of Section 1.2,
$\tau_3 = M_{-\overline{\gamma}_{n+1}}$ and $\tau_1 = M_{-\overline{\rho}_{n+1}}$. Because $|\gamma_{n+1}| < 1$ and $|\rho_{n+1}| < 1$ for $w \in \mathbb{O}$
and because $|\zeta_{n+1}(z)| < 1$ for $z \in \mathbb{O}$, it follows that all these transformations
are contractions; thus, all the s_n are Schur functions as this follows from
Theorem 1.2.3.
2. For this part a transcription of the proof given in Szegő [196, p. 292] can
   be made. Szegő proved this result in the polynomial case for an absolutely
   continuous measure $d\mathring{\mu} = \mu'd\mathring{\lambda}$. This absolute continuity however is not
   essential and his proof can be easily translated to a direct proof for the rational

case. If we start from the fact that the theorem is true for the polynomial case, then we can make the following observation: If

$$k_n(z, w) = p_n(z, w)/\pi_n(z), \quad p_n(z, w) \in \mathcal{P}_n,$$

where $\pi_n(z) = \prod_{k=1}^n \varpi_k(z)$, then one easily sees that $p_n(z, w)$ is a (polynomial) reproducing kernel for $(\mathcal{P}_n, \mu_n)$ with $d\mathring{\mu}_n = |\pi_n(z)|^{-2} d\mathring{\mu}$. Hence, the zeros of $k_n(z, w)$, which are the zeros of $p_n(z, w)$, are exactly as was stated.

Another direct proof can be given by using the J-unitary recursions. This kind of proof can be found in the next section.

3. If we set $w = \alpha_n$ in the previous result and use $k_n(z, \alpha_n) = \kappa_n \phi_n^*(z)$, we find that ϕ_n^* has all its zeros (and poles) in $\mathbb{O}^e$. Consequently, ϕ_n has all its zeros in $\mathbb{O}$ (and poles in $\mathbb{O}^e$). Note that this sharpens the result of Corollary 3.1.4(1). □

The recursion for the kernels can easily be inverted to give k_{n-1} and k_{n-1}^* in terms of k_n and k_n^*. Moreover, the recursion coefficients ρ_n and γ_n are completely defined in terms of k_n, since indeed $\rho_n = \overline{k_n(w, \alpha_n)^*}/k_n(w, \alpha_n)$, so that the kernel k_{n-1} is uniquely defined for given k_n. By induction, all the previous kernels k_j will also be fixed by k_n. Thus to check if in $\mathcal{L}_n$ all the kernels k_j, $j = 0, \ldots, n$ with respect to two different measures $\mathring{\mu}$ and $\mathring{\nu}$ are the same, it is sufficient to check that the last ones, k_n, are the same.

3.3. Normalized recursions for the kernels

The recursions for the reproducing kernels involved matrices that were almost J-unitary. Since J-unitary matrices have a lot of interesting properties, we want to normalize these recursions. It turns out that the recursion is given by a J-unitary matrix if we consider normalized kernels, which we shall denote with a capital:

$$K_n(z, w) = k_n(z, w)[k_n(w, w)]^{-1/2}. \tag{3.11}$$

This can be easily inverted to give $k_n(z, w) = K_n(w, w)K_n(z, w)$. Note that $K_n(z, \alpha_n) = \phi_n^*(z)$, the nth orthonormal basis function. The next theorem gives the normalized recursion.

Theorem 3.3.1. *The normalized kernels $K_n(z, w)$ defined in (3.11) satisfy the recursion*

$$\begin{bmatrix} K_n^*(z, w) \\ K_n(z, w) \end{bmatrix} = \theta_n(z, w) \begin{bmatrix} K_{n-1}^*(z, w) \\ K_{n-1}(z, w) \end{bmatrix}, \tag{3.12}$$

where the superstar conjugation is with respect to z. The matrix θ_n is given by

$$\theta_n(z, w) = c_n \begin{bmatrix} 1 & \overline{\rho}_n \\ \rho_n & 1 \end{bmatrix} \begin{bmatrix} \zeta_n(z) & 0 \\ 0 & 1 \end{bmatrix} d_n \begin{bmatrix} 1 & \overline{\gamma}_n \\ \gamma_n & 1 \end{bmatrix},$$

with

$$c_n = (1 - |\rho_n|^2)^{-1/2} \text{ and } d_n = (1 - |\gamma_n|^2)^{-1/2},$$
$$\rho_n = \rho_n(w) = \overline{\phi_n(w)}/\phi_n^*(w),$$
$$\gamma_n = \gamma_n(w) = -\zeta_n(w)\rho_n(w)$$

and where ϕ_n is the nth orthonormal basis function for $\mathcal{L}_n$.

Proof. Note that with the notation of Section 1.5, the matrix θ_n can also be written as $\theta_n = U_{-\overline{\rho}_n} B_{\alpha_n} U_{-\overline{\gamma}_n}$ with the exception of the factor $\overline{\alpha}_n/|\alpha_n|$ in ζ_n.

From the normalization, it follows that the θ_n matrix of this theorem and the t_n matrix of Theorem 3.2.1 are related by

$$t_n(z, w) = \left[\frac{k_n(w, w)}{k_{n-1}(w, w)} \right]^{1/2} \theta_n(z, w).$$

From Theorem 3.2.1 with $z = w$ and $\gamma_n = -\zeta_n(w)\rho_n$, we get

$$k_n(w, w) = (1 - |\zeta_n(w)\rho_n|^2)(1 - |\rho_n|^2)^{-1}k_{n-1}(w, w). \tag{3.13}$$

This gives the normalized recursion. □

Corollary 3.3.2. *If $\rho_k(w) = \overline{\phi_k(w)}/\phi_k^*(w)$ and $\gamma_k(w) = -\zeta_k(w)\rho_k(w)$ as in the previous theorem, we have the following expression for $k_n(w, w)$:*

$$k_n(w, w) = \prod_{k=1}^n \frac{(1 - |\gamma_k(w)|^2)}{(1 - |\rho_k(w)|^2)}. \tag{3.14}$$

Proof. This is a direct consequence of the formula (3.13), which is repeatedly applied, and the fact that $k_0(w, w) = 1$. □

Now define

$$\Theta_n = \theta_n \theta_{n-1} \cdots \theta_1, \quad n \geq 1. \tag{3.15}$$

It is not difficult to show that each $\Theta_k = [\Theta_k^{ij}]$ has the property that

$$\Theta_k^{11} = \left(\Theta_k^{22}\right)^* \text{ and } \Theta_k^{12} = \left(\Theta_k^{21}\right)^* \text{ (superstar with respect to z).} \tag{3.16}$$

Notice that $k_0 = K_0 = 1$. Let us define

$$\theta_0 = \begin{bmatrix} K_0^* & L_0^* \\ K_0 & -L_0 \end{bmatrix} = \begin{bmatrix} 1 & 1 \\ 1 & -1 \end{bmatrix}, \tag{3.17}$$

with $L_0 = K_0 = 1$. Then we immediately get

$$\Theta_n \theta_0 = \begin{bmatrix} K_n^*(z, w) & L_n^*(z, w) \\ K_n(z, w) & -L_n(z, w) \end{bmatrix}. \tag{3.18}$$

The first column is obtained as a consequence of the recurrence for the normalized kernels. The elements in the second column satisfy the same recurrence but with different initial conditions. Notice that $L_n(z, w) \in \mathcal{L}_n$ for every w. That $L_n^*(z, w)$ appears in the right top corner is a consequence of (3.16). The elements of Θ_n can be expressed in terms of these K_n and L_n by multiplying with θ_0^{-1}:

$$\Theta_n = \frac{1}{2} \begin{bmatrix} K_n^* + L_n^* & K_n^* - L_n^* \\ K_n - L_n & K_n + L_n \end{bmatrix}. \tag{3.19}$$

From property (6) of Theorem 1.5.1, we now get the next theorem.

Theorem 3.3.3. *Let $w \in \mathbb{O}$ be a given number and let K_n and L_n be as defined above by the normalized recurrence matrix. Then*

1. $\frac{1}{2}[\frac{L_n^*}{K_n^*} + \frac{L_n}{K_n}] = \frac{B_n}{K_n^* K_n} = \frac{1}{K_{n*} K_n} = \frac{1}{2}[\frac{L_{n*}}{K_{n*}} + \frac{L_n}{K_n}]$,
2. $L_n / K_n \in \mathcal{C}$,
3. $1/K_n \in \mathring{H}_2$.

Proof. Take $a = \theta_{11} - \theta_{12} = L_n^*$, $b = \theta_{11} + \theta_{12} = K_n^*$, and use $(f^*)_* = B_n^{-1} f$ from Theorem 2.2.1. Theorems 1.5.1 and 1.5.3 then lead to the result. □

The following is a useful observation.

Theorem 3.3.4. *If K_n^* and K_n are generated from (3.12) with some coefficients ρ_n and γ_n, then L_n^* and L_n as defined in (3.18) are generated by exactly the same relation (3.12), where we have to replace ρ_n by $-\rho_n$ and γ_n by $-\gamma_n$.*

Proof. We can remove the minus sign in the defining relation (3.18) by writing it as

$$\begin{bmatrix} L_n^* \\ L_n \end{bmatrix} = J\theta_n J \begin{bmatrix} L_{n-1}^* \\ L_{n-1} \end{bmatrix},$$

with

$$J\theta_n J = c_n \begin{bmatrix} 1 & -\overline{\rho}_n \\ -\rho_n & 1 \end{bmatrix} \begin{bmatrix} \zeta_n & 0 \\ 0 & 1 \end{bmatrix} d_n \begin{bmatrix} 1 & -\overline{\gamma}_n \\ -\gamma_n & 1 \end{bmatrix},$$

where the coefficients have the same meaning as in Theorem 3.3.1. □

4

Recurrence and second kind functions

One might expect that, if there exist Szegő style recurrence relations for the reproducing kernels, then it should be possible to derive recurrence relations in the Szegő style for the orthogonal functions themselves. However, deriving these is not as simple as for the reproducing kernels, in the sense that the transition matrices that give the recurrence are not precisely J-unitary, but it is still possible to get some recurrences that coincide with the Szegő recurrence in the polynomial case. This will be done in the first section of this chapter.

Related to this recurrence are the so-called functions of the second kind. They are also solutions of the same recurrence but with different initial conditions. Because they will be important in obtaining several interpolation properties, we shall study them in some detail in Section 4.2.

General solutions of the recurrence, which are linear combinations of first and second kind functions, are then treated in Section 4.3 and we include there an analog of Green's formula. The latter will be used in Chapter 10 on moment problems.

Since the convergents of a continued fraction are linked by a three-term recurrence relation, there is a natural link between three-term recurrence relations and continued fractions. This is explained in Section 4.4.

Finally, Section 4.5 gives some remarks about the situation when not all the points α_k are in $\mathbb{O}$, but when they are arbitrarily distributed in $\mathbb{C} \setminus \partial\mathbb{O}$. The case where they are all on $\partial\mathbb{O}$ is discussed in Chapter 11.

4.1. Recurrence for the orthogonal functions

Because of the importance of the recurrence relations for the orthogonal polynomials as studied by Szegő, it is a natural question to ask whether it is possible to find such a recurrence relation also for the rational case. These turn out to be a bit more complicated. In view of the J-unitary recursions we saw in the previous chapter for the reproducing kernels, the recursions for the orthonormal basis seems not to be so nice. Recall that for the disk $\varpi_i(z) = 1 - \bar{\alpha}_i z$ and $z_i = -\bar{\alpha}_i/|\alpha_i|$ while $\alpha_0 = 0$. For the half plane, $\varpi_i(z) = z - \bar{\alpha}_i$ and $z_i = |\alpha_i^2 + 1|/(\alpha_i^2 + 1)$ while $\alpha_0 = \mathbf{i}$.

Theorem 4.1.1. *For the orthonormal basis functions in $\mathcal{L}_n$, a recursion of the following form exists:*

$$\begin{bmatrix} \phi_n(z) \\ \phi_n^*(z) \end{bmatrix} = N_n \frac{\varpi_{n-1}(z)}{\varpi_n(z)} \begin{bmatrix} 1 & \bar{\lambda}_n \\ \lambda_n & 1 \end{bmatrix} \begin{bmatrix} \zeta_{n-1}(z) & 0 \\ 0 & 1 \end{bmatrix} \begin{bmatrix} \phi_{n-1}(z) \\ \phi_{n-1}^*(z) \end{bmatrix}, \qquad (4.1)$$

where the matrix N_n is a positive constant e_n times a unitary matrix

$$N_n = e_n \begin{bmatrix} \eta_n^1 & 0 \\ 0 & \eta_n^2 \end{bmatrix},$$

with η_n^1 and $\eta_n^2 \in \mathbb{T}$. The constant η_n^1 is chosen such that our normalization condition $\phi_n^*(\alpha_n) > 0$ is maintained. The other constant η_n^2 is related to η_n^1 by

$$\eta_n^2 = \overline{\eta_n^1 \bar{z}_{n-1} z_n}.$$

The parameter λ_n is given by

$$\lambda_n = \eta \frac{\overline{\phi_n(\alpha_{n-1})}}{\phi_n^*(\alpha_{n-1})} = \eta \rho_n(\alpha_{n-1}) \quad \text{with } \eta = \frac{\varpi_{n-1}(\alpha_n)}{\varpi_n(\alpha_{n-1})} z_n \bar{z}_{n-1} \in \mathbb{T}$$

and $\rho_n(w)$ as defined in Theorem 3.2.1, that is, $\rho_n(w) = \overline{\phi_n(w)}/\phi_n^*(w)$.

Proof. First we prove the existence of constants c_n and d_n such that

$$\frac{\varpi_n(z)}{z - \alpha_{n-1}} \phi_n - d_n \phi_{n-1} - c_n \frac{\varpi_{n-1}(z)}{z - \alpha_{n-1}} \phi_{n-1}^* \in \mathcal{L}_{n-2}. \qquad (4.2)$$

Let us define as before $\pi_k(z) = \prod_{i=1}^{k} \varpi_i(z)$, and the polynomials p_k are defined by $\phi_k = p_k/\pi_k$. Note that we can use Theorem 2.2.1 to rewrite this as

$$\frac{p_n - d_n(z - \alpha_{n-1})p_{n-1} - c_n \varpi_{n-1}(z)p_{n-1}^* \eta_{n-1}}{(z - \alpha_{n-1})\pi_{n-1}(z)} = \frac{N(z)}{D(z)},$$

where we have used $\eta_k \in \mathbb{T}$ as defined in Theorem 2.2.1. If this has to be in $\mathcal{L}_{n-2}$, then we should require that $N(\alpha_{n-1}) = N(\hat{\alpha}_{n-1}) = 0$ or, which is the same, $N(\alpha_{n-1}) = N^*(\alpha_{n-1}) = 0$. The first condition gives

$$c_n = \frac{\overline{\eta}_{n-1}}{\varpi_{n-1}(\alpha_{n-1})} \frac{p_n(\alpha_{n-1})}{p^*_{n-1}(\alpha_{n-1})}.$$

The second condition defines d_n:

$$\overline{d}_n = \frac{1}{\varpi_{n-1}(\alpha_{n-1})} \frac{p^*_n(\alpha_{n-1})}{p^*_{n-1}(\alpha_{n-1})}.$$

Note that

$$p^*_{n-1}(\alpha_{n-1}) = \phi^*_{n-1}(\alpha_{n-1})\pi_{n-1}(\alpha_{n-1})\overline{\eta}_{n-1}$$
$$= \kappa_{n-1}\pi_{n-1}(\alpha_{n-1})\overline{\eta}_{n-1} \neq 0.$$

We can therefore also write

$$c_n = \frac{\varpi_n(\alpha_{n-1})}{\varpi_{n-1}(\alpha_{n-1})} \frac{\phi_n(\alpha_{n-1})}{\kappa_{n-1}}$$

and

$$\overline{d}_n = \overline{z}_n \frac{\varpi_n(\alpha_{n-1})}{\varpi_{n-1}(\alpha_{n-1})} \frac{\phi^*_n(\alpha_{n-1})}{\kappa_{n-1}}. \tag{4.3}$$

Thus we have proved that with the previous choices of c_n and d_n, the expression in (4.2) is in $\mathcal{L}_{n-2}$. However, at the same time it is orthogonal to $\mathcal{L}_{n-2}$. To check this, we note that for every $k \leq n-2$, ϕ_k is orthogonal to the first term in (4.2) because

$$\left\langle \frac{\varpi_n(z)}{z - \alpha_{n-1}}\phi_n, \phi_k \right\rangle_{\hat{\mu}} = \left\langle \phi_n, \frac{z - \alpha_n}{\varpi_{n-1}(z)}\phi_k \right\rangle_{\hat{\mu}}$$

and this is zero because the right factor is in $\mathcal{L}_{n-1}$. The function ϕ_k is trivially orthogonal to the second term in (4.2) because $k \leq n-2$. Finally, it is also orthogonal to the third term since

$$\left\langle \frac{\varpi_{n-1}(z)}{z - \alpha_{n-1}}\phi^*_{n-1}, \phi_k \right\rangle_{\hat{\mu}} = \left\langle \phi^*_{n-1}, \frac{z - \alpha_{n-1}}{\varpi_{n-1}(z)}\phi_k \right\rangle_{\hat{\mu}}$$

and this is zero by Theorem 2.2.1. We may thus conclude that the expression in (4.2) is zero. Hence

$$\phi_n = d_n \frac{z - \alpha_{n-1}}{\varpi_n(z)}\phi_{n-1} + c_n \frac{\varpi_{n-1}(z)}{\varpi_n(z)}\phi^*_{n-1}$$

$$= d_n\overline{z}_{n-1}\frac{\varpi_{n-1}(z)}{\varpi_n(z)}[\zeta_{n-1}(z)\phi_{n-1} + \overline{\lambda}_n\phi^*_{n-1}], \tag{4.4}$$

with

$$\lambda_n = \bar{z}_{n-1}\bar{c}_n/\bar{d}_n. \tag{4.5}$$

Note that we can write λ_n as

$$\lambda_n = \eta \frac{\overline{\phi_n(\alpha_{n-1})}}{\phi_n^*(\alpha_{n-1})} = \eta\rho_n(\alpha_{n-1}), \quad \text{with } \eta = z_n\bar{z}_{n-1}\frac{\overline{\varpi_{n-1}(\alpha_n)}}{\varpi_n(\alpha_{n-1})} \in \mathbb{T}$$

and $\rho_n(w)$ as defined in Theorem 3.2.1. We then know from Corollary 3.1.4 that $\rho_n \in \mathbb{D}$ and thus also $\lambda_n \in \mathbb{D}$. Recall that $z_0 = 1$. Taking the superstar conjugate, we can find the recurrence as claimed. One can choose, for example, $e_n = |d_n| \in \mathbb{R}$. The values of η_n^1 and η_n^2 can readily be computed to be

$$\eta_n^1 = \bar{z}_{n-1}\frac{d_n}{|d_n|} \quad \text{and} \quad \eta_n^2 = z_n\frac{\bar{d}_n}{|d_n|}. \tag{4.6}$$

It remains to check the initial conditions of the recurrence, that is, for $n = 1$. Now, since $\phi_0 = \phi_0^* = 1$, we can always put

$$\phi_1(z) = e_1\eta_1^1\frac{\varpi_0(z)}{\varpi_1(z)}[\zeta_0(z)\phi_0 + \bar{\lambda}_1\phi_0^*], \tag{4.7}$$

where $e_1 \in \mathbb{R}$ and $\eta_1^1 \in \mathbb{T}$. Hence the constants η_1^1 and λ_1 should satisfy

$$\phi_1(\alpha_0) = e_1\eta_1^1\frac{\varpi_0(\alpha_0)}{\varpi_1(\alpha_0)}\bar{\lambda}_1 \quad \text{and} \quad \phi_1^*(\alpha_0) = e_1\bar{\eta}_1^1\frac{\varpi_0(\alpha_0)}{\varpi_1(\alpha_0)}z_1.$$

These can be solved for η_1^1 and λ_1, and as one can easily check, the result corresponds to the general formula if one takes $\alpha_0 = 0$ and uses $z_0 = 1$ in the disk case and $\alpha_0 = \mathbf{i}$ and $z_0 = 1$ for the half plane case.

This gives the first of the two coupled recursions of (4.1). The other recurrence is found by taking the superstar conjugate of the first one. They are equivalent to each other. □

Sometimes, it is simpler to rewrite the previous recurrence relation in a somewhat different format. We can easily see that

$$\phi_n(z) = \frac{\kappa_n}{\kappa_{n-1}}\left[\varepsilon_n\frac{\varpi_{n-1}^*(z)}{\varpi_n(z)}\phi_{n-1}(z) + \delta_n\frac{\varpi_{n-1}(z)}{\varpi_n(z)}\phi_{n-1}^*(z)\right] \tag{4.8}$$

and its superstar conjugate is

$$\phi_n^*(z) = z_n\frac{\kappa_n}{\kappa_{n-1}}\left[\bar{\delta}_n\frac{\varpi_{n-1}^*(z)}{\varpi_n(z)}\phi_{n-1}(z) + \bar{\varepsilon}_n\frac{\varpi_{n-1}(z)}{\varpi_n(z)}\phi_{n-1}^*(z)\right]. \tag{4.9}$$

The parameters ε_n and δ_n are given by

$$\varepsilon_n = z_n \frac{\varpi_{n-1}(\alpha_n)}{\varpi_{n-1}(\alpha_{n-1})} \frac{\overline{\phi_n^*(\alpha_{n-1})}}{\kappa_n}, \tag{4.10}$$

$$\delta_n = \frac{\varpi_n(\alpha_{n-1})}{\varpi_{n-1}(\alpha_{n-1})} \frac{\phi_n(\alpha_{n-1})}{\kappa_n}. \tag{4.11}$$

Comparison of (4.8)–(4.9) with the recurrence (4.1) shows that the relation between both requires that

$$\eta_n^1 e_n z_{n-1} = \varepsilon_n \frac{\kappa_n}{\kappa_{n-1}},$$

$$\eta_n^1 e_n \overline{\lambda}_n = \delta_n \frac{\kappa_n}{\kappa_{n-1}}.$$

Taking their ratio gives

$$\overline{\lambda}_n = z_{n-1} \frac{\delta_n}{\varepsilon_n}.$$

Using the above expressions for ε_n and δ_n shows that we get indeed the same λ_n as we had before.

Furthermore, we note that it follows from the Christoffel–Darboux relation that

$$|\varepsilon_n|^2 - |\delta_n|^2 = \frac{\kappa_{n-1}^2}{\kappa_n^2} \frac{\varpi_n(\alpha_n)}{\varpi_{n-1}(\alpha_{n-1})} > 0. \tag{4.12}$$

Thus $|\varepsilon_n| > |\delta_n|$ or $|\lambda_n| < 1$, as we have seen before.

We also note that instead of giving a direct proof, the recurrence relation can also be obtained from the Christoffel–Darboux relation. Indeed, take the superstar conjugate of that relation with respect to w and one will get

$$\frac{\phi_n^*(z)\phi_n(w) - \phi_n(z)\phi_n^*(w)}{\zeta_n(w) - \zeta_n(z)} = \sum_{k=0}^{n-1} \phi_k(z) B_{n-1 \setminus k}(w)\phi_k^*(w).$$

Another way of obtaining this is as a special case of the Liouville–Ostrogradskii relation (see later in Section 4.3). A similar relation can be obtained from Christoffel–Darboux by taking the superstar conjugate for both z and w. Putting $w = \alpha_{n-1}$ then yields the following two relations:

$$\frac{\phi_n^*(z)\phi_n(\alpha_{n-1}) - \phi_n(z)\phi_n^*(\alpha_{n-1})}{\zeta_n(\alpha_{n-1}) - \zeta_n(z)} = \phi_{n-1}(z)\kappa_{n-1}$$

and

$$\frac{\phi_n^*(z)\overline{\phi_n^*(\alpha_{n-1})} - \phi_n(z)\overline{\phi_n(\alpha_{n-1})}}{1 - \zeta_n(z)\overline{\zeta_n(\alpha_{n-1})}} = \phi_{n-1}^*(z)\kappa_{n-1}.$$

The first recurrence relation then follows from these two by elimination of $\phi_n^*(z)$. The second one follows by taking its superstar conjugate.

The previous expressions for the recursion coefficients λ_n are not very practical, since they use function values of ϕ_n and ϕ_n^* to compute these. The following theorem gives, at least in principle, more reasonable expressions.

Theorem 4.1.2. *The recursion coefficient λ_n from the previous theorem can also be expressed as*

$$\lambda_n = -\overline{z}_{n-1} \frac{\left\langle \phi_k, \frac{\varpi_{n-1}^*}{\varpi_n} \phi_{n-1} \right\rangle_{\hat{\mu}}}{\left\langle \phi_k, \frac{\varpi_{n-1}}{\varpi_n} \phi_{n-1}^* \right\rangle_{\hat{\mu}}}, \quad k \in \{0, 1, \dots, n-1\},$$

and the value of $e_n > 0$ can be obtained as the positive square root of

$$e_n^2 = \frac{\varpi_n(\alpha_n)}{\varpi_{n-1}(\alpha_{n-1})} \frac{1}{1 - |\lambda_n|^2}.$$

Proof. Use the relation (4.4) for ϕ_n and express that it is orthogonal to ϕ_k. Then one gets

$$\left\langle \phi_k, \frac{z - \alpha_{n-1}}{\varpi_n(z)} \phi_{n-1} \right\rangle_{\hat{\mu}} \overline{d}_n + \left\langle \phi_k, \frac{\varpi_{n-1}(z)}{\varpi_n(z)} \phi_{n-1}^* \right\rangle_{\hat{\mu}} \overline{c}_n = 0.$$

Then use the defining relation of (4.5) and the expression for the ratio of $\overline{c}_n/\overline{d}_n$ that one can get from the previous relation. Then one will find the expression for λ_n.

To find the expression for e_n^2, we should prove that

$$e_n^2(1 - |\lambda_n|^2) = \frac{\varpi_n(\alpha_n)}{\varpi_{n-1}(\alpha_{n-1})}. \tag{4.13}$$

Fill in $e_n^2 = |d_n|^2$ with d_n given by (4.3) and the expression for λ_n to find

$$e_n^2(1 - |\lambda_n|^2) = \frac{|\varpi_n(\alpha_{n-1})|^2}{|\varpi_{n-1}(\alpha_{n-1})|^2} \frac{|\phi_n^*(\alpha_{n-1})|^2}{|\phi_{n-1}^*(\alpha_{n-1})|^2} \left(1 - \frac{|\phi_n(\alpha_{n-1})|^2}{|\phi_n^*(\alpha_{n-1})|^2} \right)$$

$$= \frac{|\varpi_n(\alpha_{n-1})|^2}{|\varpi_{n-1}(\alpha_{n-1})|^2} \frac{|\phi_n^*(\alpha_{n-1})|^2 - |\phi_n(\alpha_{n-1})|^2}{|\phi_{n-1}^*(\alpha_{n-1})|^2}$$

$$= \frac{|\varpi_n(\alpha_{n-1})|^2}{|\varpi_{n-1}(\alpha_{n-1})|^2} (1 - |\zeta_n(\alpha_{n-1})|^2),$$

where for the third line we used the Christoffel–Darboux relation. It is just a matter of writing $\zeta_n(\alpha_{n-1})$ explicitly and simplifying to find that one gets indeed the right-hand side of (4.13). $\qquad\square$

The presence of the η_n^1 and η_n^2 in the recurrence relation of Theorem 4.1.1 are a bit cumbersome to deal with in certain circumstances. They are needed because of our choice of the orthonormal functions that had to satisfy $\phi_n^*(\alpha_n) = \kappa_n > 0$. It is possible to get rid of the ηs by rotating the orthonormal functions. That is, we multiply them by some number $\epsilon_n \in \mathbb{T}$ (note the difference with the parameter ε_n from the recurrence relation). This number can be chosen to avoid the rotations needed in the recurrence (4.1). Therefore we define

$$\epsilon_0 = 1 \quad \text{and} \quad \epsilon_n = \epsilon_{n-1} z_n \frac{\overline{d_n}}{|d_n|} = \epsilon_{n-1}\eta_n^2 \quad \text{for } n \geq 1, \qquad (4.14)$$

where d_n is as in (4.3) and η_n^2 is as in (4.6). We use this ϵ_n to rotate ϕ_n. The rotated orthonormal functions, which are still orthonormal, will be denoted by $\Phi_n = \epsilon_n\phi_n$. These basis functions now satisfy a recurrence relation as given in the next theorem.

Theorem 4.1.3. *Let ϕ_n be the orthonormal functions satisfying the recurrence relation of Theorem 4.1.1 and denote by Φ_n the rotated orthonormal functions $\Phi_n = \epsilon_n\phi_n$ as introduced above. Then these satisfy the recurrence relation*

$$\begin{bmatrix} \Phi_n(z) \\ \Phi_n^*(z) \end{bmatrix} = e_n \frac{\varpi_{n-1}(z)}{\varpi_n(z)} \begin{bmatrix} 1 & \overline{\Lambda_n} \\ \Lambda_n & 1 \end{bmatrix} \begin{bmatrix} Z_{n-1}(z) & 0 \\ 0 & 1 \end{bmatrix} \begin{bmatrix} \Phi_{n-1}(z) \\ \Phi_{n-1}^*(z) \end{bmatrix}, \qquad (4.15)$$

where

$$\Lambda_n = \overline{\epsilon}_{n-1}^2 \overline{Z}_n z_{n-1} \lambda_n$$
$$= \overline{\epsilon}_{n-1}^2 \frac{\varpi_{n-1}(\alpha_n)}{\varpi_n(\alpha_{n-1})} \frac{\overline{\Phi_n(\alpha_{n-1})}}{\Phi_n^*(\alpha_{n-1})}$$

and

$$Z_{n-1} = z_n \overline{z}_{n-1} \zeta_{n-1}.$$

Proof. We can start with the recurrence (4.1) and express the ϕ_n in terms of the Φ_n, which results in the relation

$$\begin{bmatrix} \Phi_n(z) \\ \Phi_n^*(z) \end{bmatrix} = e_n \frac{\varpi_{n-1}(z)}{\varpi_n(z)} M_n \begin{bmatrix} \zeta_{n-1}(z) & 0 \\ 0 & 1 \end{bmatrix} \begin{bmatrix} \Phi_{n-1}(z) \\ \Phi_{n-1}^*(z) \end{bmatrix},$$

with the matrix M_n defined by

$$M_n = \begin{bmatrix} \epsilon_n \eta_n^1 \overline{\epsilon}_{n-1} & \epsilon_n \eta_n^1 \epsilon_{n-1} \overline{\lambda}_n \\ \overline{\epsilon}_n \eta_n^2 \overline{\epsilon}_{n-1} \lambda_n & \overline{\epsilon}_n \eta_n^2 \epsilon_{n-1} \end{bmatrix}.$$

Use in this matrix the definitions of ϵ_n, of η_n^i and Λ_n and some algebra to get the result. □

In the next sections, we continue to develop the results for the ϕ_n, but virtually the same results hold true for the rotated functions Φ_n. Occasionally we shall state the result for Φ_n in a remark. The rotated functions are, however, important for the interpolation algorithm to be given later in Section 6.5.

It is also possible to get relations between successive orthogonal functions from the recursions for the (normalized) kernels $K_n(z, w)$ in the previous section. We give the result in a slightly more general form.

Theorem 4.1.4. *Let the J-unitary contractive matrices $\Theta_n(z, w)$ be as defined in (3.15). Then*

$$\begin{bmatrix} \phi_n(z) \\ \phi_n^*(z) \end{bmatrix} = \Theta_n(z, \alpha_n) \Theta_{n-k}^{-1}(z, \alpha_{n-k}) \begin{bmatrix} \phi_{n-k}(z) \\ \phi_{n-k}^*(z) \end{bmatrix}. \tag{4.16}$$

Proof. We recall that $K_n(z, \alpha_n) = \phi_n^*(z)$. Hence it follows that

$$\begin{bmatrix} \phi_n(z) \\ \phi_n^*(z) \end{bmatrix} = \Theta_n(z, \alpha_n) \begin{bmatrix} \phi_0(z) \\ \phi_0^*(z) \end{bmatrix}.$$

Because Θ_n is J-unitary and therefore invertible (see Theorem 1.5.1) the result easily follows. □

Recall the definition of the $L_n(z, w)$ from (3.19) and suppose we set by definition $L_n(z, \alpha_n) = \chi_n^*(z) \in \mathcal{L}_n$. As a special case of (3.1.9), we thus get

$$\Theta_n(z, \alpha_n) = \frac{1}{2} \begin{bmatrix} \phi_n + \chi_n & \phi_n - \chi_n \\ \phi_n^* - \chi_n^* & \phi_n^* + \chi_n^* \end{bmatrix}. \tag{4.17}$$

We can now formulate a special case of Theorem 3.3.3.

Theorem 4.1.5. *Let ϕ_n and χ_n be as defined above. Then*

1. $\frac{1}{2} \left[\frac{\chi_n}{\phi_n} + \frac{\chi_n^*}{\phi_n^*} \right] = \frac{B_n}{\phi_n \phi_n^*} = \frac{1}{\phi_n \phi_{n*}},$

2. $\chi_n^*/\phi_n^* = \chi_{n*}/\phi_{n*} \in \mathcal{C}$,
3. $1/\phi_n^*$ *and hence also* $1/\phi_{n*} \in \mathring{H}_2$,
4. $\phi_n/\phi_n^* \in \mathcal{B}$.

Proof. Use Theorems 3.3.3 and 3.2.2 and some properties from Section 2.2.
$\square$

It will be useful to write an inverse form of the recursion formulas as in the next theorem.

Theorem 4.1.6. *Given the orthonormal function* ϕ_n *with* $\phi_n^*(\alpha_n) = \kappa_n > 0$, *all the previous orthonormal functions* ϕ_k, $k < n$ *are uniquely defined if they are normalized by* $\phi_k^*(\alpha_k) = \kappa_k > 0$. *They can be found with the recursions*

$$\begin{bmatrix} \phi_{n-1}(z) \\ \phi_{n-1}^*(z) \end{bmatrix} = V_n(z) \begin{bmatrix} \phi_n(z) \\ \phi_n^*(z) \end{bmatrix}, \qquad (4.18)$$

with

$$V_n(z) = \frac{1}{1 - |\lambda_n|^2} \frac{\varpi_n(z)}{\varpi_{n-1}(z)} \begin{bmatrix} 1/\zeta_{n-1}(z) & 0 \\ 0 & 1 \end{bmatrix} \begin{bmatrix} 1 & -\bar{\lambda}_n \\ -\lambda_n & 1 \end{bmatrix} N_n^{-1}$$

and with all the quantities appearing in this formula as in Theorem 4.1.1.

Proof. The formula (4.18) is evidently the inverse of the recurrence formula (4.1). Since the coefficients λ_n and the matrix N_n are completely defined in terms of ϕ_n, the ϕ_{n-1} are uniquely defined. By induction, all the previous ϕ_k are uniquely defined.
$\square$

In fact this is a simple consequence of the note given at the end of Section 3.2. The kernels are uniquely defined in terms of the last one. The orthonormal functions will also be unique if they have the normalization mentioned.

4.2. Functions of the second kind

In this section we shall define some functions ψ_k that are the rational analogues of the polynomials of the second kind that appear in the Szegő theory. We shall call them *functions of the second kind*. They are defined first in terms of the orthogonal functions ϕ_n. We then show that they satisfy the same recurrence relation as the orthogonal functions and that they can be used to get rational approximants for the positive real function Ω_μ. Later, in Section 6.2, it will be shown that these functions of the second kind are also orthogonal rational functions with respect to a measure that is related to the given measure μ.

Define the following kernel:

$$E(t, z) = D(t, z) + 1,$$

with $D(t, z)$ the Riesz–Herglotz–Nevanlinna kernel. It is easily checked that

$$E(t, z) = 2 \frac{\varpi_0^*(t)\varpi_0(z)}{\varpi_0(\alpha_0)(t - z)} = \begin{cases} \dfrac{2t}{t - z} & \text{for } \mathbb{D}, \\[2ex] \dfrac{(t - \mathbf{i})(z + \mathbf{i})}{\mathbf{i}(t - z)} & \text{for } \mathbb{U}. \end{cases}$$

Note that taking the substar w.r.t. z implies for $t \in \partial\mathbb{O}$ that $D(t, z)_* = -D(t, z) = D(z, t)$. Consequently it also is true that

$$E(z, t) = E(t, z)_* = 1 - D(t, z).$$

Here are some equivalent definitions:

$$\psi_n(z) = \int [E(t, z)\phi_n(t) - D(t, z)\phi_n(z)] \, d\mathring{\mu}(t) \tag{4.19}$$

$$= \int D(t, z)[\phi_n(t) - \phi_n(z)] \, d\mathring{\mu}(t) + \int \phi_n(t) \, d\mathring{\mu}(t) \tag{4.20}$$

$$= \begin{cases} 1 & \text{if } n = 0, \\ \int D(t, z)[\phi_n(t) - \phi_n(z)] \, d\mathring{\mu}(t) & \text{if } n \geq 1. \end{cases} \tag{4.21}$$

The last equality follows from the fact that $\langle 1, \phi_n \rangle_{\mathring{\mu}} = \delta_{0n}$.

We shall first show that these are functions from $\mathcal{L}_n$.

Lemma 4.2.1. The functions ψ_n of the second kind belong to $\mathcal{L}_n$.

Proof. This is trivially true for $n = 0$. For $n \geq 1$, note that the integrand in (4.21) has the form

$$[\phi_n(t) - \phi_n(z)]D(t, z).$$

The term in square brackets vanishes for $t = z$, so that the integral can be written as

$$\psi_n(z) = \int \frac{(t - z)\sum_{k=0}^n a_n(t)z^k}{(t - z)\pi_n(z)} \, d\mathring{\mu}(t) = \frac{\sum_{k=0}^n [\int a_n(t) \, d\mathring{\mu}(t)]z^k}{\pi_n(z)}$$

and this is clearly an element in $\mathcal{L}_n$. $\qquad\qquad\square$

We can obtain more general expressions for these functions of the second kind as shown below.

Lemma 4.2.2. To define the functions of the second kind for $n > 0$, we may replace (4.21) by

$$\psi_n(z)f(z) = \int D(t,z)[\phi_n(t)f(t) - \phi_n(z)f(z)]\,d\mathring{\mu}(t)$$

$$= \int [E(t,z)\phi_n(t)f(t) - D(t,z)\phi_n(z)f(z)]\,d\mathring{\mu}(t) \quad (4.22)$$

for any function $f \in \mathcal{L}_{(n-1)*}$. The second formula holds also for $n = 0$, if we then take $f = 1$.

In particular, we could take for f any of the inverse Blaschke products $1/B_k$ with $0 \le k < n$.

Proof. The case $n = 0$ is trivial. We only consider the case $n > 0$. To prove the first or the second formula, we only have to check that

$$\int D(t,z)\left[\frac{f(t)}{f(z)} - 1\right]\phi_n(t)\,d\mathring{\mu}(t) = 0$$

or

$$\int E(t,z)\left[\frac{f(t)}{f(z)} - 1\right]\phi_n(t)\,d\mathring{\mu}(t) = 0$$

depending on the case. The proof is the same for both of them. Since the term in square brackets vanishes for $z = t$, it follows that we can write the integral as

$$\int \frac{p(t)}{\pi_k^*(t)}\phi_n(t)\,d\mathring{\mu}(t),$$

with p a polynomial of degree at most $n - 1$. The latter integral is of the form $\langle \phi_n, f \rangle_\mu$, with $f \in \mathcal{L}_{n-1}$. Since $\phi_n \perp \mathcal{L}_{n-1}$, this is zero.   □

We show next an expression for ψ_n^*.

Lemma 4.2.3. The superstar conjugates of the functions of the second kind satisfy

$$\psi_n^*(z)g(z) = \int D(z,t)[\phi_n^*(t)g(t) - \phi_n^*(z)g(z)]\,d\mathring{\mu}(t)$$

$$= \int [E(z,t)\phi_n^*(t)g(t) - D(z,t)\phi_n^*(z)g(z)]\,d\mathring{\mu}(t) \quad (4.23)$$

for any function $g \in \mathcal{L}_{n*}(\hat{\alpha}_n)$. The second relation also holds for $n = 0$ if we take $g = 1$.

As a special case one could take for $g(z)$ any of the functions $1/B_{n\setminus k} = B_k/B_n$, where $0 \le k < n$ with the convention that it equals 1 for $n = 0$.

Note: We used the notation $\mathcal{L}_{n*}(\hat{\alpha}_n)$ to mean $\mathcal{L}_{n*}(\hat{\alpha}_n) = \{f \in \mathcal{L}_{n*} : f(\hat{\alpha}_n) = 0\}$. This is equivalent with $\mathcal{L}_{n*}(\hat{\alpha}_n) = B_{n*}\mathcal{L}_{n-1} = \zeta_{n*}\mathcal{L}_{(n-1)*}$.

Proof. Note that the second expression implies that $\psi_0^* = 1$, since we get

$$\psi_0^*(z) = \int [E(z,t) - D(z,t)]\, d\mathring{\mu}(t) = \int d\mathring{\mu}(t) = 1.$$

So, suppose that $n > 0$. The relations (4.23) then follow immediately from the corresponding ones in (4.22) by taking the superstar conjugate. In fact $g(z) = f_*(z)/B_n(z) \in \mathcal{L}_{n*}(\hat{\alpha}_n)$. This proves the lemma. $\qquad\square$

Note that as in (4.20), we can give an equivalent form of (4.23) as follows:

$$-\psi_n^*(z)g(z) = \int D(t,z)[\phi_n^*(t)g(t) - \phi_n^*(z)g(z)]\, d\mathring{\mu}(t) - \int \phi_n^*(t)g(t)\, d\mathring{\mu}(t),$$

where, as we know, the last term is δ_{0n}. For $g = 1/B_n$, this takes the even simpler form

$$-\psi_{n*}(z) = \int D(t,z)[\phi_{n*}(t) - \phi_{n*}(z)]\, d\mathring{\mu}(t) - \delta_{0n}.$$

As in the polynomial case, these functions of the second kind satisfy the same recurrence relations as the orthogonal functions but with opposite signs for the parameters λ_k. Taking this sign outside the transition matrix of the recurrence gives the formula (4.24) as proved in the next theorem.

Theorem 4.2.4. *For the functions of the second kind a recursion of the following form exists:*

$$\begin{bmatrix} \psi_n(z) \\ -\psi_n^*(z) \end{bmatrix} = N_n \frac{\varpi_{n-1}(z)}{\varpi_n(z)} \begin{bmatrix} 1 & \bar{\lambda}_n \\ \lambda_n & 1 \end{bmatrix} \begin{bmatrix} \zeta_{n-1}(z) & 0 \\ 0 & 1 \end{bmatrix} \begin{bmatrix} \psi_{n-1}(z) \\ -\psi_{n-1}^*(z) \end{bmatrix}, \qquad (4.24)$$

where the recurrence matrix is exactly as in Theorem 4.1.1.

Proof. As in the case of Theorem 4.1.1, it is sufficient to prove only one of the two associated recursions. The other one follows by applying the superstar conjugate. We shall prove the second one. First note that by our previous lemmas we can write for $n > 1$

$$
\begin{bmatrix} \psi_{n-1}(z) \\ -\psi_{n-1}^*(z) \end{bmatrix} = -\Omega_\mu(z) \begin{bmatrix} \phi_{n-1}(z) \\ \phi_{n-1}^*(z) \end{bmatrix} + \int D(t, z) \begin{bmatrix} \phi_{n-1}(t) \\ \frac{\zeta_{n-1}(z)}{\zeta_{n-1}(t)} \phi_{n-1}^*(t) \end{bmatrix} d\mathring{\mu}(t).
$$

Multiply from the left with

$$
e_n \eta_n^2 \frac{\varpi_{n-1}(z)}{\varpi_n(z)} [\lambda_n \zeta_{n-1}(z) \quad 1].
$$

Then the right-hand side becomes

$$
-\Omega_\mu(z)\phi_n^*(z) + e_n \eta_n^2 \int D(t, z) f(t, z) \, d\mathring{\mu}(t), \qquad (4.25)
$$

with

$$
f(t, z) = \frac{\varpi_{n-1}(z)}{\varpi_n(z)} \left[\lambda_n \zeta_{n-1}(z)\phi_{n-1}(t) + \frac{\zeta_{n-1}(z)}{\zeta_{n-1}(t)} \phi_{n-1}^*(t) \right]
$$

$$
= \frac{\zeta_{n-1}(z)}{\zeta_{n-1}(t)} \frac{\varpi_{n-1}(z)}{\varpi_n(z)} [\lambda_n \zeta_{n-1}(t)\phi_{n-1}(t) + \phi_{n-1}^*(t)].
$$

Using the recursion for ϕ_n^*, we thus get that (4.25) can be replaced by

$$
-\Omega_\mu(z)\phi_n^*(z) + \int D(t, z) \frac{\alpha_{n-1} - z}{\alpha_{n-1} - t} \frac{\varpi_n(t)}{\varpi_n(z)} \phi_n^*(t) \, d\mathring{\mu}(t).
$$

This will equal $-\psi_n^*(z)$ if we may replace the latter integral by

$$
\int D(t, z) \frac{\alpha_n - z}{\alpha_n - t} \frac{\varpi_n(t)}{\varpi_n(z)} \phi_n^*(t) \, d\mathring{\mu}(t).
$$

This can indeed be done, since the difference equals

$$
\int D(t, z) \left[\frac{\alpha_{n-1} - z}{\alpha_{n-1} - t} - \frac{\alpha_n - z}{\alpha_n - t} \right] \frac{\varpi_n(t)}{\varpi_n(z)} \phi_n^*(t) \, d\mathring{\mu}(t)
$$

$$
= \int f_*(t)\phi_n^*(t) \, d\mathring{\mu}(t) = \langle \phi_n^*, f \rangle_\mu,
$$

with $f \in \zeta_n \mathcal{L}_{n-1}$. This gives zero because of the orthogonality. This proves the theorem for $n > 1$.

We check the case $n = 1$ only for the case of the disk. The proof for the half plane is similar.

We have to show that $\psi_1(z) = e_1\eta_1^!(z-\bar{\lambda}_1)/(1-\bar{\alpha}_1 z)$. From the definition, we get

$$\psi_1(z) = \int D(t,z)[\phi_1(t) - \phi_1(z)]\,d\mathring{\mu}(t).$$

Now we replace ϕ_1 by its expression from the recurrence relation, which is $\phi_1(z) = e_1\eta_1^!(z+\bar{\lambda}_1)/(1-\bar{\alpha}_1 z)$. Compare with (4.7). After some computations (see (4.7)) this results in

$$\psi_1(z) = \frac{e_1\eta_1^!}{1-\bar{\alpha}_1 z}\int \frac{t+z}{t-z}\left[\frac{(t+\bar{\lambda}_1)(1-\bar{\alpha}_1 z)}{1-\bar{\alpha}_1 t} - (z+\bar{\lambda}_1)\right]d\mathring{\mu}(t)$$

$$= \frac{e_1\eta_1^!(1+\bar{\alpha}_1\bar{\lambda}_1)}{1-\bar{\alpha}_1 z}\int \frac{t+z}{1-\bar{\alpha}_1 t}\,d\mathring{\mu}(t).$$

Now we use the expression we get from Theorem 4.1.1 for ϕ_1 in terms of λ_1 and make the orthogonality relation $\langle\phi_1, 1\rangle_{\mathring{\mu}} = 0$ explicit to find

$$\int \frac{t}{1-\bar{\alpha}_1 t}\,d\mathring{\mu}(t) = -\bar{\lambda}_1\int \frac{1}{1-\bar{\alpha}_1 t}\,d\mathring{\mu}(t).$$

Filling this into the last expression we get

$$\psi_1(z) = e_1\eta_1^!\frac{z-\bar{\lambda}_1}{1-\bar{\alpha}_1 z}(1+\bar{\alpha}_1\bar{\lambda}_1)\int \frac{1}{1-\bar{\alpha}_1 t}\,d\mathring{\mu}(t). \qquad (4.26)$$

We have to find an expression for the remaining integral. Therefore we again use the expression for λ_1 from Theorem 4.1.2 to get

$$\int \frac{d\mathring{\mu}(t)}{1-\bar{\alpha}_1 t} = 1 + \bar{\alpha}_1\int \frac{t\,d\mathring{\mu}(t)}{1-\bar{\alpha}_1 t}$$

$$= 1 - \bar{\alpha}_1\bar{\lambda}_1\int \frac{d\mathring{\mu}(t)}{1-\bar{\alpha}_1 t}.$$

From this relation we finally get

$$(1+\bar{\alpha}_1\bar{\lambda}_1)\int \frac{d\mathring{\mu}(t)}{1-\bar{\alpha}_1 t} = 1.$$

The recursion for ψ_1 is proved and this concludes the proof of the theorem. $\square$

We shall now derive some determinant formulas and some other properties of these functions as we did for the kernels and for the orthogonal functions at the end of the previous section. Therefore we need a J-unitary matrix. This is obtained in the next lemma.

Lemma 4.2.5. Let t_n denote the recursion matrix

$$t_n = N_n \frac{\varpi_{n-1}(z)}{\varpi_n(z)} \begin{bmatrix} 1 & \bar{\lambda}_n \\ \lambda_n & 1 \end{bmatrix} \begin{bmatrix} \zeta_{n-1}(z) & 0 \\ 0 & 1 \end{bmatrix},$$

with all the parameters as defined in Theorems 4.1.1 and 4.2.4. Set $T_n = t_n t_{n-1} \cdots t_1$ (recall $\alpha_0 = 0$). Then

$$T_n = \frac{1}{2} \begin{bmatrix} \phi_n + \psi_n & \phi_n - \psi_n \\ \phi_n^* - \psi_n^* & \phi_n^* + \psi_n^* \end{bmatrix}. \tag{4.27}$$

There exists a positive constant c_n such that

$$\Theta_n = \frac{\varpi_n(z)}{c_n \varpi_0(z)} T_n$$

is a J-unitary matrix that is J-contractive in $\mathbb{O}$.

Proof. The first relation follows easily from

$$\begin{bmatrix} \psi_n & \phi_n \\ -\psi_n^* & \phi_n^* \end{bmatrix} = T_n \begin{bmatrix} 1 & 1 \\ -1 & 1 \end{bmatrix} \tag{4.28}$$

by inverting the right-most matrix.

Now note that t_k can be written as

$$|d_k|(1 - |\lambda_k|^2)^{1/2} \frac{\varpi_{k-1}(z)}{\varpi_k(z)} \theta_k$$

with

$$\theta_k = \begin{bmatrix} \eta_k^1 & 0 \\ 0 & \eta_k^2 \end{bmatrix} (1 - |\lambda_k|^2)^{-1/2} \begin{bmatrix} 1 & \bar{\lambda}_k \\ \lambda_k & 1 \end{bmatrix} \begin{bmatrix} \zeta_{k-1} & 0 \\ 0 & 1 \end{bmatrix},$$

a J-unitary matrix, which is also J-contractive in $\mathbb{O}$ since

$$|\lambda_k| = |\rho_k(\alpha_{k-1})| < 1.$$

Multiply this out to find $\Theta_n = \theta_n \theta_{n-1} \cdots \theta_1$ and

$$c_n = \prod_{k=1}^{n} |d_k|(1 - |\lambda_k|^2)^{1/2}.$$

This completes the proof.   □

With the previous result, we can now prove the following theorem.

Theorem 4.2.6. *With the notation introduced in the previous lemma, we have:*

1. The determinant formula

$$\frac{1}{2}[\psi_n(z)\phi_{n*}(z) + \psi_{n*}(z)\phi_n(z)] = \varpi_0(z)\varpi_{0*}(z)P(z, \alpha_n),$$

whence

$$\frac{1}{2}\left[\frac{\psi_n(z)}{\phi_n(z)} + \frac{\psi_n^*(z)}{\phi_n^*(z)}\right] = \varpi_0(z)\varpi_{0*}(z)\frac{P(z, \alpha_n)}{\phi_n\phi_{n*}},$$

where $P(z, w)$ is the Poisson kernel.
2. $\psi_n^*/\phi_n^* = \psi_{n*}/\phi_{n*} \in \mathcal{C}$. The Riesz–Herglotz–Nevanlinna measure for this positive real function is given in

$$\frac{\psi_n^*(z)}{\phi_n^*(z)} = \int D(t, z)\, d\mathring{\mu}_n(t) \quad with \quad d\mathring{\mu}_n(t) = \frac{P(t, \alpha_n)}{|\phi_n^*|^2}\, d\lambda(t).$$

Proof. The first determinant relation follows by taking the determinant of (4.28), giving

$$\frac{1}{2}[\psi_n(z)\phi_n^*(z) + \psi_n^*(z)\phi_n(z)] = \frac{c_n^2\varpi_0(z)^2}{\varpi_n(z)^2}\det\Theta_n$$

$$= \zeta_0(z)\frac{\varpi_0(z)^2}{\varpi_n(z)^2}B_{n-1}(z)\prod_{k=1}^{n}\frac{\varpi_k(\alpha_k)}{\varpi_{k-1}(\alpha_{k-1})}\bar{z}_{k-1}z_k$$

$$= \frac{\zeta_0(z)}{\varpi_n(z)(z-\alpha_n)}B_n(z)\frac{\varpi_n(\alpha_n)\varpi_0(z)^2}{\varpi_0(\alpha_0)}$$

$$= \varpi_0(z)\varpi_{0*}(z)B_n(z)P(z, \alpha_n).$$

The second relation is a direct consequence.

That $\psi_n^*/\phi_n^* \in \mathcal{C}$ is because Θ_n is J-contractive and the factor $c_n\varpi_0(z)/\varpi_n(z)$ relating Θ_n and T_n drops out of the ratio.

If $\Omega_n = \psi_n^*/\phi_n^* \in \mathcal{C}$, then Re $\Omega_n(t) = |\varpi_0(t)|^2 P(t, \alpha_n)/|\phi_n(t)|^2$, $t \in \partial\mathbb{O}$, as we have just seen. Hence the Riesz–Herglotz–Nevanlinna representation has the form

$$\Omega_n(z) = \int D(t, z)\text{Re }\Omega_n(t)\, d\mathring{\lambda}(t) = \int D(t, z)\, d\mathring{\mu}(t).$$

The theorem is completely proved. □

As with the functions ϕ_n, we could rotate the functions of the second kind to give $\Psi_n = \epsilon_n\psi_n$, where the ϵ_n are as defined in (4.14). For these rotated

Ψ_n a recurrence as in Theorem 4.1.3 exists. Most of the properties of ψ_n are transferred to Ψ_n. For further reference we note that in analogy with (4.27) we have

$$T_n = \frac{1}{2} \begin{bmatrix} \Phi_n + \Psi_n & \Phi_n - \Psi_n \\ \Phi_n^* - \Psi_n^* & \Phi_n^* + \Psi_n^* \end{bmatrix} \tag{4.29}$$

when T_n is the product $T_n = t_n t_{n-1} \cdots t_1$ with the elementary matrices t_n as used in the recurrence (4.15).

4.3. General solutions

We have seen that (ϕ_n, ϕ_n^*) and $(\psi_n, -\psi_n^*)$ are independent solutions of the recurrence relation

$$\begin{bmatrix} X_n(z) \\ X_n^+(z) \end{bmatrix} = t_n(z) \begin{bmatrix} X_{n-1}(z) \\ X_{n-1}^+(z) \end{bmatrix}, \tag{4.30}$$

with

$$
\begin{aligned}
t_n(z) &= N_n \frac{\varpi_{n-1}(z)}{\varpi_n(z)} \begin{bmatrix} 1 & \overline{\lambda}_n \\ \lambda_n & 1 \end{bmatrix} \begin{bmatrix} \zeta_{n-1}(z) & 0 \\ 0 & 1 \end{bmatrix} \\[2mm]
&= \frac{\kappa_n}{\kappa_{n-1}} \begin{bmatrix} \varepsilon_n \dfrac{\varpi_{n-1}^*(z)}{\varpi_n(z)} & \delta_n \dfrac{\varpi_{n-1}(z)}{\varpi_n(z)} \\[3mm] -z_n \overline{\delta}_n \dfrac{\varpi_{n-1}^*(z)}{\varpi_n(z)} & -z_n \overline{\varepsilon}_n \dfrac{\varpi_{n-1}(z)}{\varpi_n(z)} \end{bmatrix} \\[2mm]
&= \frac{\kappa_n}{\kappa_{n-1}} \begin{bmatrix} 1 & 0 \\ 0 & -z_n \end{bmatrix} \begin{bmatrix} \varepsilon_n & \delta_n \\ \overline{\delta}_n & \overline{\varepsilon}_n \end{bmatrix} \begin{bmatrix} a_n(z) & 0 \\ 0 & b_n(z) \end{bmatrix}
\end{aligned} \tag{4.31}
$$

and with

$$a_n(z) = \frac{\varpi_{n-1}^*(z)}{\varpi_n(z)} \quad \text{and} \quad b_n(z) = \frac{\varpi_{n-1}(z)}{\varpi_n(z)}.$$

They form a basis for the solution space. We want to give more general relations for arbitrary solutions of this recurrence relation that will generalize formulas such as the Christoffel–Darboux relation, the determinant formula, and others. In Akhiezer's book [2], such formulas are referred to as the Liouville–Ostrogradskii formula (determinant formula) and the Green formula. We shall derive similar relations for the rational case. We start with a general summation formula.

Theorem 4.3.1. *Consider two solutions* $(x_n(z), x_n^+(z))$ *and* $(y_n(z), y_n^+(z))$ *of the recurrence relation (4.30). Denoting the Blaschke products* $B_0 = 1$, $B_n = \zeta_1 \cdots \zeta_n$, $n \geq 1$, *and* $B_{n \setminus k} = B_n/B_k$, *we have for*

$$F_n(z, w) = x_n^+(z)y_n(w) - x_n(z)y_n^+(w), \quad n = 0, 1, \ldots$$

the following summation formula:

$$\frac{F_n(z, w)}{1 - \zeta_n(z)/\zeta_n(w)} - B_n(w)\frac{F_0(z, w)}{1 - \zeta_0(z)/\zeta_0(w)} = -\sum_{k=0}^{n-1} x_k(z) B_{n \setminus k}(w) y_k^+(w).$$
(4.32)

Proof. We fill the recurrence into the definition of $F_n(z, w)$, which gives

$$\frac{\kappa_{n-1}^2}{\kappa_n^2} F_n(z, w) = z_n[|\delta_n|^2 - |\varepsilon_n|^2]a_n(z)b_n(w)x_{n-1}(z)y_{n-1}^+(w)$$

$$+ z_n[|\varepsilon_n|^2 - |\delta_n|^2]a_n(w)b_n(z)x_{n-1}^+(z)y_{n-1}(w)$$

$$= z_n[|\varepsilon_n|^2 - |\delta_n|^2]\{a_n(w)b_n(z)x_{n-1}^+(z)y_{n-1}(w)$$

$$- a_n(z)b_n(w)x_{n-1}(z)y_{n-1}^+(w)\}$$

$$= z_n[|\varepsilon_n|^2 - |\delta_n|^2]a_n(w)b_n(z)$$

$$\cdot \{F_{n-1}(z, w) + M_n(z, w)x_{n-1}(z)y_{n-1}^+(w)\},$$

with

$$M_n(z, w) = 1 - \frac{a_n(z)b_n(w)}{a_n(w)b_n(z)} = 1 - \frac{\varpi_{n-1}^*(z)\varpi_{n-1}(w)}{\varpi_{n-1}^*(w)\varpi_{n-1}(z)}$$

$$= 1 - \frac{\zeta_{n-1}(z)}{\zeta_{n-1}(w)} = \frac{(w - z)\varpi_{n-1}(\alpha_{n-1})}{\varpi_{n-1}^*(w)\varpi_{n-1}(z)}.$$

Let us define as auxiliary quantities

$$c_0 = 1, \quad \frac{c_{n-1}}{c_n} = z_n[|\varepsilon_n|^2 - |\delta_n|^2]a_n(w)b_n(z), \quad n = 1, 2, \ldots.$$

Then it is easily obtained by multiplication that

$$\frac{1}{c_n} = \prod_{k=1}^n [|\varepsilon_k|^2 - |\delta_k|^2] \cdot \prod_{k=1}^n \zeta_k(w) \cdot \prod_{k=1}^n \frac{\varpi_{k-1}^*(w)}{\varpi_k^*(w)} \cdot \prod_{k=1}^n \frac{\varpi_{k-1}(z)}{\varpi_k(z)}.$$

Now using (4.12), that is,

$$|\varepsilon_n|^2 - |\delta_n|^2 = \frac{\kappa_{n-1}^2}{\kappa_n^2} \frac{\varpi_n(\alpha_n)}{\varpi_{n-1}(\alpha_{n-1})},$$

we find that

$$c_n = \frac{\kappa_n^2}{\kappa_0^2} \frac{\varpi_0(\alpha_0)\varpi_n(z)\varpi_n^*(w)}{\varpi_n(\alpha_n)\varpi_0(z)\varpi_0^*(w)} \frac{1}{B_n(w)}$$

and also

$$c_{n-1}\left[1 - \frac{a_n(z)b_n(w)}{a_n(w)b_n(z)}\right] = -\frac{\kappa_{n-1}^2}{\kappa_0^2} \frac{(z-w)\varpi_0(\alpha_0)}{\varpi_0(z)\varpi_0^*(w)B_{n-1}(w)}.$$

Hence with $d_n = c_n/\kappa_n^2$,

$$d_n F_n(z,w) - d_{n-1}F_{n-1}(z,w) = \frac{-d_0(z-w)\varpi_0(\alpha_0)}{\varpi_0(z)\varpi_0^*(w)B_{n-1}(w)}x_{n-1}(z)y_{n-1}^+(w).$$

Summation now gives

$$\frac{\kappa_0^2}{\kappa_n^2}c_n F_n(z,w) - F_0(z,w) = -\frac{(z-w)\varpi_0(\alpha_0)}{\varpi_0(z)\varpi_0^*(w)}\sum_{k=1}^{n}\frac{x_{k-1}(z)y_{k-1}^+(w)}{B_{k-1}(w)}.$$

Now we can use the expression for c_n and we get

$$\frac{\varpi_n(z)\varpi_n^*(w)}{\varpi_n(\alpha_n)(z-w)}\frac{F_n(z,w)}{B_n(w)} - \frac{\varpi_0(z)\varpi_0^*(w)F_0(z,w)}{(z-w)\varpi_0(\alpha_0)} = -\sum_{k=0}^{n-1}\frac{x_k(z)y_k^+(w)}{B_k(w)}.$$

Multiplication with $B_n(w)$ gives the result.   □

We can formulate some special cases as a corollary

Corollary 4.3.2. *We use the notation of the previous theorem. Furthermore, let $\{\phi_n\}$ be the orthonormal functions and $\{\psi_n\}$ the corresponding functions of the second kind and let $P(z,w)$ be the general Poisson kernel (1.33). Then we have:*

1. *The following determinant formula corresponds to the Liouville–Ostrogradskii formula:*

$$F_n(z,z) = x_n^+(z)y_n(z) - x_n(z)y_n^+(z)$$
$$= P(z,\alpha_n)\varpi_0(z)\varpi_{0*}(z)B_n(z)F_0(z,z).$$

2. *We also find the determinant formula of Theorem 4.2.6 as*

$$\phi_n^*(z)\psi_n(z) + \phi_n(z)\psi_n^*(z) = 2P(z,\alpha_n)\varpi_0(z)\varpi_{0*}(z)B_n(z).$$

3. *The superstar conjugate with respect to w of the Christoffel–Darboux formula is*

$$\frac{\phi_n^*(w)\phi_n(z) - \phi_n(z)\phi_n^*(w)}{\zeta_n(w) - \zeta_n(z)} = \sum_{k=0}^{n-1} \phi_k(z) B_{n-1\backslash k}(w)\phi_k^*(w).$$

Proof.

1. First multiply by the denominator of $F_n(z, w)$ in the general formula (4.32) and then take $z = w$.
2. Use $x_n = \phi_n$, $x_n^+ = \phi_n^*$, $y_n = \psi_n$, and $y_n^+ = -\psi_n^*$ in the previous formula.
3. Choose $x_n = \phi_n$, $x_n^+ = \phi_n^*$, $y_n = \phi_n$, and $y_n^+ = \phi_n^*$ in the general formula.

$\square$

Similarly, we can derive an analog of Green's formula, which has the following formulation.

Theorem 4.3.3 (Green's formula). *Consider two arbitrary solutions $(x_n(z),$ $x_n^+(z))$ and $(y_n(z), y_n^+(z))$ of the recurrence relation (4.30). Set*

$$G_n(z, w) = x_n^+(z)\overline{y_n^+(w)} - x_n(z)\overline{y_n(w)}, \quad n = 0, 1, \ldots.$$

Then we have the following summation formula:

$$\frac{G_n(z, w)}{1 - \zeta_n(z)\overline{\zeta_n(w)}} - \frac{G_0(z, w)}{1 - \zeta_0(z)\overline{\zeta_0(w)}} = \sum_{k=0}^{n-1} x_k(z)\overline{y_k(w)}. \tag{4.33}$$

Proof. We start by filling the recurrence relation into the defining expression for G_n and we get

$$\frac{\kappa_{n-1}^2}{\kappa_n^2} G_n(z, w) = [|\varepsilon_n|^2 - |\delta_n|^2]b_n(z)\overline{b_n(z)}$$

$$\cdot [G_{n-1}(z, w) + M_n(z, w)x_{n-1}(z)\overline{y_{n-1}(w)}],$$

where

$$M_n(z, w) = 1 - \frac{a_n(z)\overline{a_n(w)}}{b_n(z)\overline{b_n(w)}} = 1 - \zeta_{n-1}(z)\overline{\zeta_{n-1}(w)}.$$

Introducing the auxiliary quantities

$$c_0 = 1, \quad \frac{c_{n-1}}{c_n} = [|\varepsilon_n|^2 - |\delta_n|^2]b_n(z)\overline{b_n(w)}, \quad n = 1, 2, \ldots$$

we get after multiplication

$$c_n = \frac{\kappa_n^2}{\kappa_0^2} \frac{\varpi_n(z)\overline{\varpi_n(w)}\,\overline{\varpi_0(\alpha_0)}}{\varpi_n(\alpha_n)\varpi_0(z)\overline{\varpi_0(w)}}$$

$$= \frac{\kappa_n^2}{\kappa_0^2} \frac{\overline{\varpi_z(w)}}{1 - \zeta_n(z)\overline{\zeta_n(w)}} \frac{\overline{\varpi_0(\alpha_0)}}{\varpi_0(z)\overline{\varpi_0(w)}}$$

$$= \frac{\kappa_n^2}{\kappa_0^2} \frac{1 - \zeta_0(z)\overline{\zeta_0(w)}}{1 - \zeta_n(z)\overline{\zeta_n(w)}}.$$

Furthermore

$$c_{n-1} M_n(z, w) = \frac{\kappa_{n-1}^2}{\kappa_0^2} (1 - \zeta_0(z)\overline{\zeta_0(w)}),$$

so that we obtain

$$\frac{c_n}{\kappa_n^2} G_n(z, w) - \frac{c_{n-1}}{\kappa_{n-1}^2} G_{n-1}(z, w) = \frac{1}{\kappa_0^2}(1 - \zeta_0(z)\overline{\zeta_0(w)})x_{n-1}(z)\overline{y_{n-1}(w)},$$

which, after summation, leads to

$$\frac{c_n}{\kappa_n^2} G_n(z, w) - \frac{1}{\kappa_0^2} G_0(z, w) = \frac{1}{\kappa_0^2}(1 - \zeta_0(z)\overline{\zeta_0(w)}) \sum_{k=0}^{n-1} x_k(z)\overline{y_k(w)}.$$

With the expression for c_n we finally get the desired formula.   □

We can easily formulate some special cases of this general formula.

Corollary 4.3.4. *We use the notation of the previous theorem. Furthermore, let $\{\phi_n\}$ be the orthonormal functions and $\{\psi_n\}$ the corresponding functions of the second kind. Then we have:*

1. $\frac{\phi_n^*(z)\overline{\psi_n^*(w)} + \phi_n(z)\overline{\psi_n(w)}}{1 - \zeta_n(z)\overline{\zeta_n(w)}} - \frac{2}{1 - \zeta_0(z)\overline{\zeta_0(w)}} = -\sum_{k=0}^{n-1} \phi_k(z)\overline{\psi_k(w)},$

2. $\frac{\psi_n^*(z)\overline{\psi_n^*(w)} - \psi_n(z)\overline{\psi_n(w)}}{1 - \zeta_n(z)\overline{\zeta_n(w)}} = \sum_{k=0}^{n-1} \psi_k(z)\overline{\psi_k(w)},$

3. $\frac{|\psi_n^*(z)|^2 - |\psi_n(z)|^2}{1 - |\zeta_n(z)|^2} = \sum_{k=0}^{n-1} |\psi_k(z)|^2,$

4. $\frac{|\psi_n^*(z)|^2 - |\psi_n(z)|^2}{1 - |\zeta_n(z)|^2} = \sum_{k=0}^{n-1} |B_{n-1 \setminus k}(z)|^2 |\psi_k(z)|^2.$

Proof.

1. Take $x_n = \phi_n$, $x_n^+ = \phi_n^*$, $y_n = \psi_n$, and $y_n^+ = -\psi_n^*$ in the general formula.
2. Take $x_n = \psi_n$, $x_n^+ = -\psi_n^*$, $y_n = \psi_n$, and $y_n^+ = -\psi_n^*$ in the general formula.

3. Take $w = z$ in the previous formula.
4. Take the superstar conjugate in the previous formula.   □

4.4. Continued fractions and three-term recurrence

A continued fraction can be defined as the sequence of convergents or approximants

$$\frac{A_n}{B_n} = (\tau_0 \circ \tau_1 \circ \cdots \circ \tau_n)(0), \qquad \tau_k(w) = \frac{a_k}{b_k + w}, \quad k \geq 0,$$

where the (a_k, b_k) are complex numbers with all a_k nonzero. This continued fraction is also denoted as

$$\sum_{k=0}^{\infty} \frac{a_k \mid}{\mid b_k} = \frac{a_0 \mid}{\mid b_0} + \frac{a_1 \mid}{\mid b_1} + \cdots$$

with convergents

$$\frac{A_n}{B_n} = \sum_{k=0}^{n} \frac{a_k \mid}{\mid b_k} = \cfrac{a_0}{b_0 + \cfrac{a_1}{b_1 + \cfrac{}{\ddots \\ b_{n-1} + \cfrac{a_n}{b_n}}}}.$$

The numerators A_n and B_n both solve the second-order linear difference equation

$$f_n = f_{n-2} a_n + f_{n-1} b_n, \quad n = 0, 1, \ldots,$$

where the initial conditions $(f_{-2}, f_{-1}) = (1, 0)$ give the numerators A_n, whereas the initial conditions $(f_{-2}, f_{-1}) = (0, 1)$ result in the denominators B_n. We can also write this as

$$\begin{bmatrix} A_{n-1} & B_{n-1} \\ A_n & B_n \end{bmatrix} = \begin{bmatrix} 0 & 1 \\ a_n & b_n \end{bmatrix} \begin{bmatrix} A_{n-2} & B_{n-2} \\ A_{n-1} & B_{n-1} \end{bmatrix}, \qquad \begin{bmatrix} A_{-2} & B_{-2} \\ A_{-1} & B_{-1} \end{bmatrix} = \begin{bmatrix} 1 & 0 \\ 0 & 1 \end{bmatrix}.$$

Thus continued fractions are inherently related to three-term recurrence relations. Note that each step in the continued fraction, that is, each transformation τ_k, is represented in the above relation as a multiplication with a 2×2 matrix whose first row is just $[0 \ 1]$.

However, the type of recurrences we have seen before are related to more general rational transformations than the simple τ_k. This corresponds to multiplying with a more general 2×2 matrix whose first row need not be just $[0 \ 1]$. Nonetheless, such recurrences can be transformed into transformations of the continued fraction type; hence the recurrences can be translated into three-term recurrence relations. This basically means that we eliminate one superstar conjugated function from the coupled recurrence relation.

We shall now give some continued fractions, hence three-term recurrences, that can be associated with the recurrence relations we have introduced so far. Therefore, we borrow a result from Ref. [29, pp. 19–21].

Lemma 4.4.1. Let us define

$$t_k = \begin{bmatrix} c_k & d_k \\ a_k & b_k \end{bmatrix}, \quad k \geq 0 \quad \text{and} \quad T_n = \begin{bmatrix} C_n & D_n \\ A_n & B_n \end{bmatrix},$$

with $T_0 = t_0$ and $T_n = t_n T_{n-1}$ for $n > 0$. Suppose $d_k c_k \neq 0$ and $a_k d_k - b_k c_k \neq 0$ for all $k > 0$. Furthermore, define $R_n = F B_n - A_n$ and $S_n = F D_n - C_n$. Then the formal continued fraction expansion

$$F = \frac{c_0}{d_0} + \frac{(a_0 d_0 - b_0 c_0) d_0^{-2}}{\left| \begin{array}{c} b_0 d_0^{-1} \end{array} \right.} + \frac{c_1}{\left| \begin{array}{c} d_1 \end{array} \right.} + \frac{(b_1 c_1 - a_1 d_1) c_1^{-1}}{\left| \begin{array}{c} a_1 c_1^{-1} \end{array} \right.}$$

$$+ \cdots + \frac{c_n}{\left| \begin{array}{c} d_n \end{array} \right.} + \frac{(b_n c_n - a_n d_n) c_n^{-1}}{\left| \begin{array}{c} a_n c_n^{-1} \end{array} \right.} + \begin{cases} \dfrac{c_{n+1}}{d_{n+1}} - \dfrac{S_{n+1}}{R_n} \\[2mm] -\dfrac{R_n}{S_n} \end{cases} \quad (4.34)$$

holds and the successive convergents are

$$\frac{C_0}{D_0}, \frac{A_0}{B_0}, \frac{C_1}{D_1}, \frac{A_1}{B_1}, \ldots, \frac{C_n}{D_n}, \frac{A_n}{B_n}, \ldots.$$

Proof. This is a matter of simple algebra. See, for example, Ref. [29, Property 2.9 and Property 2.5]. ☐

This theorem can now be applied to the situation of a recursion like the one for the orthonormal functions. Note that t_n there looks like

$$t_n = \begin{bmatrix} c_k & d_k \\ a_k & b_k \end{bmatrix} = e_k \frac{\varpi_{k-1}(z)}{\varpi_k(z)} \begin{bmatrix} \eta_k^1 \zeta_{k-1} & \eta_k^1 \bar{\lambda}_k \\ \eta_k^2 \lambda_k \zeta_{k-1} & \eta_k^2 \end{bmatrix}.$$

A general term, therefore, is of the form

$$\frac{c_k}{\lceil d_k} + \frac{(b_k c_k - a_k d_k)c_k^{-1}}{\lceil a_k c_k^{-1}} = \frac{\zeta_{k-1}}{\lceil \bar{\lambda}_k} + \frac{\bar{\eta}_k^1 \eta_k^2 (1 - |\lambda_k|^2)}{\lceil \bar{\eta}_k^1 \eta_k^2 \lambda_k}.$$

Recall that $\bar{\eta}_k^1 \eta_k^2 = \bar{z}_{k-1} z_k \in \mathbb{T}$ and $|\lambda_k| < 1$.

We can now try several initial conditions and get different convergents. For example, to get the convergents

$$\frac{\psi_0}{\phi_0}, -\frac{\psi_0^*}{\phi_0^*}, \ldots, \frac{\psi_n}{\phi_n}, -\frac{\psi_n^*}{\phi_n^*}, \ldots \tag{4.35}$$

one needs the initial conditions

$$\begin{bmatrix} c_0 & d_0 \\ a_0 & b_0 \end{bmatrix} = \begin{bmatrix} 1 & 1 \\ -1 & 1 \end{bmatrix}.$$

This example illustrates how the ψ_k and ϕ_k appear in the convergents of a continued fraction and thus how they appear as solutions of a three-term recurrence relation. The three-term recurrence is not exactly what we would expect since it relates, three successive elements in the sequence

$$\phi_0, \phi_0^*, \phi_1, \phi_1^*, \ldots, \phi_k, \phi_k^*, \ldots,$$

whereas we would rather have a relation between three successive elements in the sequence

$$\phi_0, \phi_1, \phi_2, \ldots.$$

This is a matter of eliminating one out of two convergents, which corresponds to a contraction of the continued fraction. We shall come back to this at the end of this section.

First, we want to complete our description of the present continued fraction. For example, we did not identify the functions R_n and S_n in the tails of the continued fraction (4.34), or, equivalently, we did not say what function F we are (formally) expanding. This can be derived from the interpolating properties of the approximants. These interpolating properties are given in Chapter 6, which deals with interpolation. More precisely, the interpolating properties we refer to are given in Section 6.5. In fact the interpolation algorithms given there are performing successive linear fractional transforms and are thus strongly related to continued fractions. In order to complete our story about continued fractions and three-term recurrence relations at this place, we need a forward reference to what will be obtained there.

As we shall see in Section 6.5, we have the following interpolation properties:

$$\phi_n(z)\Omega_\mu(z) + \psi_n(z) = 0 \quad \text{for } z = \alpha_0, \alpha_1, \ldots, \alpha_{n-1},$$

$$\phi_n^*(z)\Omega_\mu(z) - \psi_n^*(z) = 0 \quad \text{for } z = \alpha_0, \alpha_1, \ldots, \alpha_n,$$

where Ω_μ is the positive real function $\Omega_\mu(z) = \int D(t, z)\, d\mathring{\mu}(t)$. Thus we may define the remainders $r_{ni}(z)$, $i = 1, 2$ as

$$r_{n1}(z) = [\phi_n(z)\Omega_\mu(z) + \psi_n(z)]/\tilde{B}_{n-1}(z),$$

$$r_{n2}(z) = [\phi_n^*(z)\Omega_\mu(z) - \psi_n^*(z)]/\tilde{B}_n(z),$$

with $\tilde{B}_n(z) = \zeta_0(z)B_n(z)$. Comparing this with (6.23), which can be written as

$$\begin{bmatrix} -\bar{\epsilon}_n \tilde{B}_{n-1} R_{n1}(z) \\ -\epsilon_n \tilde{B}_n R_{n2}(z) \end{bmatrix} = \begin{bmatrix} \phi_n(z) \\ \phi_n^*(z) \end{bmatrix} (-\Omega_\mu(z)) - \begin{bmatrix} \psi_n(z) \\ -\psi_n^*(z) \end{bmatrix},$$

we see that the r_{ni} defined above and the R_{ni} defined in Section 6.5 are related by $r_{n1} = \bar{\epsilon}_n R_{n1}$ and $r_{n2} = \epsilon_n R_{n2}$. Thus we can identify the tail functions S_n and R_n that appear in (4.34) as

$$S_n(z) = -\tilde{B}_{n-1}(z)r_{n1}(z) = -\bar{\epsilon}_n \tilde{B}_{n-1}(z)R_{n1}(z),$$

$$R_n(z) = -\tilde{B}_n(z)r_{n2}(z) = -\epsilon_n \tilde{B}_n(z)R_{n2}(z).$$

Putting all our findings together gives the complete description of the continued fraction whose convergents are given by (4.35), namely

$$-\Omega_\mu = 1 - \frac{2|}{|1} + \sum_{k=1}^{n}\left(\frac{\zeta_{k-1}|}{|\bar{\lambda}_k} + \frac{\bar{\eta}_k^1\eta_k^2(1 - |\lambda_k|^2)|}{|\bar{\eta}_k^1\eta_k^2\lambda_k}\right)$$

$$+ \begin{cases} \dfrac{\zeta_n|}{|\bar{\lambda}_{n+1}} - \dfrac{\bar{\epsilon}_{n+1}\varpi_{n+1}R_{n+1,1}|}{|e_{n+1}\epsilon_n\varpi_n\eta_{n+1}^1 R_{n,2}} \\ -\dfrac{\epsilon_n\zeta_n R_{n2}|}{|\bar{\epsilon}_n R_{n1}} \end{cases} . \qquad (4.36)$$

The R_{ni}, $i = 1, 2$ are as defined above. From this, we can derive that

$$\frac{\phi_n - \psi_n}{\phi_n + \psi_n}, \quad \frac{\phi_n^* - \psi_n^*}{\phi_n^* + \psi_n^*} \quad \text{for } n \geq 1$$

are the successive convergents of the previous continued fraction without the two initial terms, that is,

$$\sum_{k=1}^{\infty}\left(\frac{\zeta_{k-1}|}{|\bar{\lambda}_k} + \frac{\bar{\eta}_k^1\eta_k^2(1 - |\lambda_k|^2)|}{|\bar{\eta}_k^1\eta_k^2\lambda_k}\right).$$

By interchanging rows and columns in the matrices of the recurrence, we get the same convergents in the other order. For example, the convergents of the continued fraction expansion

$$\Omega_\mu = 1 - \frac{2|}{|1} + \sum_{k=1}^{n} \left(\frac{1|}{|\zeta_{k-1}\lambda_k} + \frac{\overline{\eta}_k^2 \eta_k^1 (1 - |\lambda_k|^2)\zeta_{k-1}|}{|\overline{\eta}_k^2 \eta_k^1 \lambda_k} \right)$$

$$+ \begin{cases} \dfrac{1|}{|\zeta_n \lambda_{n+1}} - \dfrac{\epsilon_{n+1} B_{n+1} \varpi_{n+1} R_{n+1,2}|}{|e_{n+1}\eta_{n+1}^2 \overline{\epsilon}_n B_{n-1} \varpi_n R_{n1}} \\[2ex] - \dfrac{\overline{\epsilon}_n B_{n-1} R_{n1}|}{|\epsilon_n B_n R_{n2}} \end{cases} \tag{4.37}$$

are now

$$\frac{\psi_0^*}{\phi_0^*}, \ -\frac{\psi_0}{\phi_0}, \ \ldots, \ \frac{\psi_n^*}{\phi_n^*}, \ -\frac{\psi_n}{\phi_n}, \ \ldots .$$

Taking contractions of these continued fractions will give us the even or odd parts and these give genuine three-term recurrence relations. By definition, the even contraction of the continued fraction

$$\frac{a_0}{b_0} + \frac{a_1|}{|b_1} + \frac{a_2|}{|b_2} + \cdots \tag{4.38}$$

is the continued fraction whose convergents are the even convergents of the contracted one. We have the following expression for an even contraction.

Lemma 4.4.2. If $b_{2k} \neq 0$ for all k, then the even contraction of the continued fraction (4.38) is given by

$$\frac{a_0}{b_0} + \frac{a_1 b_2|}{|a_2 + b_1 b_2} + \sum_{k=1}^{\infty} \frac{-a_{2k} b_{2k}^{-1} a_{2k+1} b_{2k+2}|}{|a_{2k+2} + \left(b_{2k+1} + b_{2k}^{-1} a_{2k+1}\right) b_{2k+2}} .$$

Proof. This can be found, for example, in Ref. [29, Property 2.8]. □

For example, if all $\lambda_k \neq 0$ and $|\lambda_k| \neq 1$, then the even convergents of (4.36) that is, ψ_k/ϕ_k for $k = 0, 1, \ldots$, are the successive convergents of the following formal continued fraction expansion:

$$-\Omega_\mu \sim 1 - \frac{2\overline{\lambda}_1|}{|z + \overline{\lambda}_1} + \sum_{k=1}^{\infty} \frac{-\zeta_{k-1}\eta_k^2 \overline{\eta}_k^1 (1 - |\lambda_k|^2)\dfrac{\overline{\lambda}_{k+1}}{\overline{\lambda}_k}|}{|\zeta_k + \eta_k^2 \overline{\eta}_k^1 \dfrac{\overline{\lambda}_{k+1}}{\overline{\lambda}_k}} .$$

It follows that the orthonormal functions as well as the functions of the second kind satisfy the three-term recurrence relation

$$\bar{\lambda}_k f_{k+1}(z) = \left(\zeta_k(z)\bar{\lambda}_k + \eta_k^2\bar{\eta}_k^1\bar{\lambda}_{k+1}\right) f_k(z)$$
$$- \left(\zeta_{k-1}(z)\eta_k^2\bar{\eta}_k^1(1 - |\lambda_k|^2)\bar{\lambda}_{k+1}\right) f_{k-1}(z). \qquad (4.39)$$

With the initial conditions

$$f_0 = 1, \qquad f_1 = e_1\eta_1^1\varpi_0(z)\frac{\zeta_0(z) + \bar{\lambda}_1}{\varpi_1(z)}$$

one generates $f_k = \phi_k$, and with initial conditions

$$f_0 = 1, \qquad f_1 = e_1\eta_1^1\varpi_0(z)\frac{\zeta_0(z) - \bar{\lambda}_1}{\varpi_1(z)}$$

one generates $f_k = \psi_k$. For the even contraction of (4.37), we find

$$\Omega_\mu \sim 1 - \left|\frac{2\zeta_0\lambda_1}{1 + \zeta_0\lambda_1}\right| + \sum_{k=1}^{\infty} \left|\frac{-\zeta_k\eta_k^1\bar{\eta}_k^2(1 - |\lambda_k|^2)\frac{\lambda_{k+1}}{\lambda_k}}{1 + \zeta_k\eta_k^1\bar{\eta}_k^2\frac{\lambda_{k+1}}{\lambda_k}}\right|$$

if all $\lambda_k \neq 0$ and $|\lambda_k| \neq 1$ with convergents ψ_k^*/ϕ_k^* for $k = 0, 1, \ldots$.

We can also use the rotated functions and thus avoid the ηs. The reader can easily translate the results. The rule is to multiply ϕ_n, ψ_n, and R_{n1} by ϵ_n. For ϕ_n and ψ_n this means replace them by Φ_n and Ψ_n, etc. In short, here are the translation rules:

$$\phi \to \Phi, \qquad \psi \to \Psi, \qquad \lambda \to \Lambda, \qquad \zeta \to Z, \qquad \eta \to 1, \qquad \epsilon \to 1.$$

As an example, the translation of (4.37) reads like

$$\Omega_\mu = 1 - \left|\frac{2}{1}\right| + \sum_{k=1}^{n}\left(\left|\frac{1}{Z_{k-1}\Lambda_k}\right| + \left|\frac{(1 - |\Lambda_k|^2)Z_{k-1}}{\Lambda_k}\right|\right)$$
$$+ \begin{cases} \left|\dfrac{1}{Z_n\Lambda_{n+1}}\right| - \left|\dfrac{B_{n+1}\varpi_{n+1}R_{n+1,2}}{e_{n+1}B_{n-1}\varpi_n R_{n1}}\right| \\ - \left|\dfrac{B_{n-1}R_{n1}}{B_n R_{n2}}\right| \end{cases}.$$

The convergents are

$$\frac{\Psi_0^*}{\Phi_0^*}, \quad -\frac{\Psi_0}{\Phi_0}, \quad \ldots, \quad \frac{\Psi_n^*}{\Phi_n^*}, \quad -\frac{\Psi_n}{\Phi_n}, \quad \ldots.$$

The three-term recurrence for the rotated functions is

$$\overline{\Lambda}_k F_{k+1}(z) = (Z_k(z)\overline{\Lambda}_k + \overline{\Lambda}_{k+1}) F_k(z) - (Z_{k-1}(z)(1 - |\Lambda_k|^2) \overline{\Lambda}_{k+1}) F_{k-1}(z). \tag{4.40}$$

It will be clear from our discussion in this section that it is also possible to get continued fractions whose convergents are the ratios of kernels $L_k(z, w)/K_k(z, w)$, etc. Likewise, we can obtain three-term recursions for these kernels. We leave this to the reader, for it is a matter of simple algebra.

4.5. Points not on the boundary

In this section we briefly deteriorate to the more general situation where not all the interpolation points α_k need be in $\mathbb{O}$. There are two different situations: The one that does not deviate very much from the previous situation is when the points are in $\mathbb{O} \cup \mathbb{O}^e$, and the second one is when all the interpolation points are on the boundary $\partial \mathbb{O}$, which needs a special treatment. The latter situation will be referred to as the *boundary situation* and is postponed until Chapter 11.

We consider here the situation where the interpolation points α_k are all in $\mathbb{O} \cup \mathbb{O}^e$. This situation is not much different from the situation we had previously. At least, when we do not go into convergence results, the algebraic aspects are nearly the same. Therefore we shall not go far into the development of this situation. This section is given as an example to show that the recurrence relations can be easily obtained in fairly the same way as before.

We suppose that the spaces $\mathcal{L}_n$ are as before and that the ϕ_n are the corresponding orthogonal rational functions, whenever they exist. In this more general situation, there may be some trouble spots though. We give these a name. We say that ϕ_n is *degenerate* if $\phi_n^*(\alpha_{n-1}) = 0$, and nondegenerate otherwise and we say that ϕ_n is *exceptional* if $\phi_n(\alpha_{n-1}) = 0$ and nonexceptional otherwise.

When all the points α_k coincide in $\alpha \in \mathbb{O} \cup \mathbb{O}^e$, then, because $\phi_n^*(\alpha) = \kappa_n$ is the leading coefficient of ϕ_n, it can not be zero if ϕ_n exists, so that the degenerate case will not occur in that situation. For example, when $\mathbb{O} = \mathbb{D}$ and all $\alpha_k = 0$, we have the case of the Szegő polynomials, and we do not have degeneracy there.

This section mainly serves to show the role of degeneracy and exceptionality.

It is easily seen that the proof of the Christoffel–Darboux relation that we derived before does not depend on the points α_k being all in $\mathbb{O}$. Thus, in the present situation, it still holds that

$$\phi_n^*(z)\overline{\phi_n^*(w)} - \phi_n(z)\overline{\phi_n(w)} = [1 - \zeta_n(z)\overline{\zeta_n(w)}] \sum_{k=0}^{n-1} \phi_k(z)\overline{\phi_k(w)}. \tag{4.41}$$

Theorem 4.5.1. *When the functional M is positive definite and all α_k are in $\mathbb{O}$ or all α_k are in $\mathbb{O}^e$, then ϕ_n is nondegenerate for all n.*

Proof. This follows from setting $z = w$ in the previous Christoffel–Darboux relation, since

$$[1 - |\zeta(z)|^2]^{-1}[|\phi_n^*(z)|^2 - |\phi_n(z)|^2] > 0. \qquad (4.42)$$

Thus $\phi_n^*(z) \neq 0$ for $z \in \mathbb{O}$ if all $\alpha_k \in \mathbb{O}$ and $\phi_n^*(z) \neq 0$ for $z \in \mathbb{O}^e$ if all $\alpha_k \in \mathbb{O}^e$. In particular $\phi_n^*(\alpha_{n-1}) \neq 0$ in both situations. □

We turn next to the recurrence relations. It was pointed out that the recurrence relations discussed in Chapter 4 can be derived from the Christoffel–Darboux relations so that they do not rely upon the fact that the α_k are all in $\mathbb{O}$. Thus, we may state without further proof that we have,

Theorem 4.5.2. *The orthonormal rational functions with leading coefficient $\kappa_n = \phi_n^*(\alpha_n) > 0$ satisfy the recurrence relations*

$$\phi_n(z) = \frac{\kappa_n}{\kappa_{n-1}}\left[\varepsilon_n \frac{\varpi_{n-1}^*(z)}{\varpi_n(z)}\phi_{n-1}(z) + \delta_n \frac{\varpi_{n-1}(z)}{\varpi_n(z)}\phi_{n-1}^*(z)\right], \qquad (4.43)$$

$$\phi_n^*(z) = -z_n \frac{\kappa_n}{\kappa_{n-1}}\left[\bar{\delta}_n \frac{\varpi_{n-1}^*(z)}{\varpi_n(z)}\phi_{n-1}(z) + \bar{\varepsilon}_n \frac{\varpi_{n-1}(z)}{\varpi_n(z)}\phi_{n-1}^*(z)\right], \qquad (4.44)$$

valid for n = 1, 2, . . . , where

$$\varepsilon_n = z_n \frac{\varpi_{n-1}(\alpha_n)}{\varpi_{n-1}(\alpha_{n-1})} \frac{\overline{\phi_n^*(\alpha_{n-1})}}{\kappa_n}, \qquad (4.45)$$

$$\delta_n = \frac{\varpi_n(\alpha_{n-1})}{\varpi_{n-1}(\alpha_{n-1})} \frac{\phi_n(\alpha_{n-1})}{\kappa_n}. \qquad (4.46)$$

We immediately get the following corollary.

Corollary 4.5.3. *The function ϕ_n can not be both degenerate and exceptional at the same time.*

Proof. This follows from the previous theorem, since degenerate and exceptional at the same time would result in $\phi_n \equiv 0$, which is absurd. □

The boundedness of the reflection coefficients also follows easily if all α_k are on one side of the boundary.

Corollary 4.5.4. *If M is positive definite and all α_k are in $\mathbb{O}$ or all α_k are in $\mathbb{O}^e$, then $|\delta_n| < |\varepsilon_n|$.*

Proof. We already know that, in this situation, we do not have degeneracy; hence $\varepsilon_n \neq 0$. On the other hand, the expressions for ε_n and δ_n yield

$$\frac{|\delta_n|}{|\varepsilon_n|} = \frac{|\phi_n(\alpha_{n-1})|}{|\phi_n^*(\alpha_{n-1})|}.$$

Since (4.42) implies that $|\phi_n^*(\alpha_m)| > |\phi_n(\alpha_m)|$ for any n, m, the result follows. □

We now give some other recurrence relations that hold under certain conditions.

Theorem 4.5.5. *Assume that ϕ_n is nondegenerate. Then the following recurrence is satisfied for $n = 1, 2, \ldots$:*

$$\phi_n(z) = \frac{1}{\bar{\varepsilon}_n} \left[\bar{z}_n \delta_n \phi_n^*(z) + (|\varepsilon_n|^2 - |\delta_n|^2) \frac{\kappa_{n-1}}{\kappa_n} \frac{\varpi_n(\alpha_n)\varpi_{n-1}(z)}{\varpi_{n-1}(\alpha_{n-1})\varpi_n(z)} \phi_{n-1}(z) \right].$$

(4.47)

Proof. This formula is obtained by taking ϕ_{n-1}^* from (4.46) and substituting it into (4.45). Formula (4.12) then gives the result. □

In the disk situation, with all $\alpha_k = 0$, we get the Szegő polynomials and then this formula is well known. See, for example, Ref. [120].

A combination of (4.46) and (4.47) leads in the nondegenerate case to a three-term recurrence relation of the sequence $\{\phi_0, \phi_1^*, \phi_1, \phi_2^*, \phi_2, \ldots\}$, hence to a continued fraction, as the one studied in Section 4.4. For the general situation we need that the ϕ_n are nondegenerate and that $\delta_n| \neq |\varepsilon_n|$. This is automatically satisfied when all the α_k are in $\mathbb{O}$ or when they are all in $\mathbb{O}^e$.

Also, the contracted form holds, that is, a three-term recurrence between the elements of the sequence $\{\phi_0, \phi_1, \phi_2, \ldots\}$, but since we should be able to divide by λ_n (see (4.39)), we now have to require that the δ_n are nonzero (i.e., that the sequence ϕ_n is nonexceptional).

Thus without further proof, we may state

Theorem 4.5.6. *Assume that the sequence ϕ_n is nondegenerate and $|\delta_n| \neq |\varepsilon_n|$. Define the sequence*

$$f_{2k} = \phi_k^*, \quad k = 1, 2, \ldots,$$

$$f_{2k+1} = \phi_k, \quad k = 0, 1, \ldots.$$

Then the subsequent f_ks are the denominators of the subsequent convergents of the continued fraction

$$\sum_{k=1}^{\infty} \frac{a_k(z)}{\left| b_k(z) \right.},$$

where for $k = 1, 2, \ldots$

$$a_{2k}(z) = z_k \bar{\varepsilon}_k \frac{\kappa_k \varpi_{k-1}(z)}{\kappa_{k-1} \varpi_k(z)}, \quad a_{2k+1}(z) = \frac{|\varepsilon_k|^2 - |\delta_k|^2}{\bar{\varepsilon}_k} \frac{\kappa_{k-1}}{\kappa_k} \frac{\varpi_k(\alpha_k)\varpi_{k-1}^*(z)}{\varpi_{k-1}(\alpha_{k-1})\varpi_k(z)},$$

$$b_{2k}(z) = z_k \bar{\delta}_k \frac{\kappa_k \varpi_{k-1}^*(z)}{\kappa_{k-1} \varpi_k(z)}, \quad b_{2k+1}(z) = \bar{z}_k \frac{\delta_k}{\bar{\varepsilon}_k}.$$

The recurrence relations for numerators and denominators are

$$f_k(z) = b_k(z) f_{k-1}(z) + a_k(z) f_{k-2}(z).$$

The initial conditions $f_0 = f_1 = \kappa_0 = 1$ give the denominators; $f_0 = -\kappa_0^{-1} = -1$ and $f_1 = \kappa_0^{-1} = 1$ give the corresponding numerators (which are the associated functions of the second kind).

In the disk case, with all $\alpha_k = 0$, these continued fractions are known as Perron–Carathéodory fractions (or PC-fractions for short); see Refs. [120, 122]. Since the previous continued fractions are related to the Nevanlinna–Pick generalization of the Carathéodory coefficient problem, it is appropriate to call them Nevanlinna–Pick fractions (or NP-fractions for short).

Contracted versions of the previous relations give the following result:

Theorem 4.5.7. *Assume that ϕ_{n-1} is nonexceptional; then the following relation holds:*

$$\phi_n(z) = b_n(z)\phi_{n-1}(z) + a_n(z)\phi_{n-2}(z),$$

where

$$a_n(z) = -z_{n-1} \frac{\kappa_n \delta_n}{\kappa_{n-2} \delta_{n-1}} \frac{\varpi_{n-2}^*(z)}{\varpi_n(z)} (|\varepsilon_{n-1}|^2 - |\delta_{n-1}|^2),$$

$$b_n(z) = \frac{\kappa_n}{\kappa_{n-1}} \left[\frac{\varepsilon_n \varpi_{n-1}^*(z)}{\varpi_n(z)} + z_{n-1} \frac{\bar{\varepsilon}_{n-1} \delta_n}{\delta_{n-1}} \frac{\varpi_{n-1}(z)}{\varpi_n(z)} \right].$$

while

$$\phi_n^*(z) = \zeta_n(z)b_{n*}(z)\phi_{n-1}^*(z) + \zeta_n(z)\zeta_{n-1}(z)a_{n*}(z)\phi_{n-2}^*(z).$$

The initial conditions are

$$\phi_0 = \kappa_0, \qquad \phi_1(z) = \frac{\kappa_1}{\kappa_0}\frac{\delta_1 + \varepsilon_1 z}{\varpi_1(z)}.$$

If the sequence ϕ_n is nonexceptional and $|\varepsilon_n| \neq |\delta_n|$ for all n, then the ϕ_n are the denominators of the convergents of the continued fraction

$$\sum_{k=1}^{\infty} \frac{a_k(z)}{\left| b_k(z) \right.}.$$

The numerators are the corresponding functions of the second kind.

Taking once more the special case of the disk with all $\alpha_k = 0$, we get the recurrences

$$\phi_n(z) = \left(z + \frac{\delta_n}{\delta_{n-1}}\right)\frac{\kappa_n}{\kappa_{n-1}}\phi_{n-1}(z) - (1 - |\delta_{n-1}|^2)\frac{\delta_n\kappa_n}{\delta_{n-1}\kappa_{n-2}}z\phi_{n-2}(z),$$

$$\phi_n^*(z) = \left(1 + \frac{\overline{\delta}_n}{\overline{\delta}_{n-1}}z\right)\frac{\kappa_n}{\kappa_{n-1}}\phi_{n-1}^*(z) - (1 - |\delta_{n-1}|^2)\frac{\overline{\delta}_n\kappa_n}{\overline{\delta}_{n-1}\kappa_{n-2}}z\phi_{n-2}^*(z).$$

The continued fractions corresponding to the latter recurrences were called M-fractions and T-fractions respectively (see Ref. [123]). The more general fractions of the previous theorem are called Multipoint-Padé fractions (or MP-fractions).

5

Para-orthogonality and quadrature

In this chapter we consider quadrature formulas that can be associated with rational functions. For the case of an interval, it is well known that the abscissas of the Gaussian quadrature formulas are zeros of orthogonal polynomials. For quadrature over the unit circle, the zeros of orthogonal polynomials are inside the unit disk. One can obtain abscissas on the unit circle itself by introducing para-orthogonal polynomials whose zeros are simple and guaranteed to lie on the unit circle. In this chapter, we shall generalize such results to the rational case.

5.1. Interpolatory quadrature

Consider the space of rational functions

$$\mathcal{L}_{p,q} = \mathcal{L}_q \cdot \mathcal{L}_{p*} = \{fg : f \in \mathcal{L}_q, g \in \mathcal{L}_{p*}\}, \quad p, q \geq 0,$$

where $\mathcal{L}_{p*} = \{f : f_* \in \mathcal{L}_p\}$. The space $\mathcal{L}_{p,q}$ can also be characterized as

$$\mathcal{L}_{p,q} = \left\{ \frac{M(z)}{\pi_q(z)\pi_p^*(z)} : M \in \mathcal{P}_{p+q} \right\},$$

where $\pi_n(z) = \prod_{k=0}^{n} \varpi_k(z)$.

Define the point sets $A_n = \{\alpha_1, \ldots, \alpha_n\} \subset \mathbb{C}$ and $\hat{A}_n = \{\hat{\alpha}_1, \ldots, \hat{\alpha}_n\}$. Select a set of n mutually distinct points

$$N_n = \{\xi_1, \ldots, \xi_n\} \subset \mathbb{C} \setminus \{A_n \cup \hat{A}_n\}.$$

We want to construct an element $R_n \in \mathcal{L}_{p,q}$, where $p + q = n - 1$, that interpolates some function f in the points N_n. In order to give a Lagrange formula

106

for the solution, we define the nodal polynomial

$$x_n(z) = \prod_{k=1}^{n} (z - \xi_k).$$

Then (the prime denotes derivative)

$$\ell_{ni}(z) = \prod_{\substack{k=1 \\ k \neq i}}^{n} \frac{z - \xi_k}{\xi_i - \xi_k} = \frac{x_n(z)}{(z - \xi_i)x_n'(\xi_i)}, \quad i = 1, \ldots, n$$

are the well-known Lagrange polynomials, satisfying $\ell_{ni}(\xi_k) = \delta_{ik}$, $1 \le i$, $k \le n$. Thus the polynomial $P_n(z) = \sum_{k=1}^{n} f(\xi_k)\ell_{nk}(z)$ is the interpolating polynomial of degree $n - 1$ for f in the interpolation points N_n. Let L_{ni} be defined by

$$L_{ni}(z) = \ell_{ni}(z) \frac{\pi_q(\xi_i)\pi_p^*(\xi_i)}{\pi_q(z)\pi_p^*(z)} \in \mathcal{L}_{p,q}.$$

Note that we still have $L_{nk}(\xi_i) = \delta_{ki}$, so that $\sum_1^n f(\xi_k)L_{nk}(z) \in \mathcal{L}_{p,q}$ is an interpolant for $f(z)$ in the points N_n.

Defining

$$X_n(z) = \frac{x_n(z)}{\pi_{q+1}(z)\pi_p^*(z)} \in \mathcal{L}_{p,q+1},$$

it can be easily shown that

$$L_{ni}(z) = \frac{\varpi_{q+1}(z)}{\varpi_{q+1}(\xi_i)} \frac{X_n(z)}{(z - \xi_i)X_n'(\xi_i)}, \tag{5.1}$$

where the prime means derivative.

As a special case, one can take $p = 0$, $q = n - 1$, so that $\mathcal{L}_{p,q} = \mathcal{L}_q = \mathcal{L}_{n-1}$.

We are now interested in quadrature formulas $I_n\{f\}$ that approximate the integral $I_{\mathring{\mu}}\{f\}$:

$$I_n\{f\} = \sum_{k=1}^{n} \lambda_{nk} f(\xi_k) \approx I_{\mathring{\mu}}\{f\} = \int f(t) \, d\mathring{\mu}(t).$$

The space $\mathcal{L}_{p,q}$ is called a *domain of validity* for the quadrature formula $I_n\{f\}$ if the quadrature is exact for all $f \in \mathcal{L}_{p,q}$. It is called a *maximal domain of validity* when neither $\mathcal{L}_{p+1,q}$ nor $\mathcal{L}_{p,q+1}$ is a domain of validity.

We can immediately construct an *interpolatory quadrature formula* as being the integral of the interpolating function from $\mathcal{L}_{p,q}$, that is,

$$I_{\mathring{\mu}}\{f\} \approx I_n\{f\} = \sum_{k=1}^{n} \lambda_{nk} f(\xi_k), \quad \text{with} \quad \lambda_{nk} = \int L_{nk}(t)\, d\mathring{\mu}(t).$$

It can be shown that an n-point quadrature formula with distinct nodes ξ_k, $k = 1, \ldots, n$ is of interpolatory type in $\mathcal{L}_{p,q}$, $p+q = n-1$ iff $\mathcal{L}_{p,q}$ is a domain of validity for that formula.

Concerning the maximality of the domain of validity we have the following theorem.

Theorem 5.1.1. *Suppose the quadrature formula $I_n\{f\}$ has n nodes on $\partial\mathbb{O}$ and suppose that $\mathcal{R}_{n-1} = \mathcal{L}_{n-1,n-1}$ is a domain of validity; then it is a maximal domain of validity.*

Proof. We have to show that neither $\mathcal{L}_{n-1,n}$ nor $\mathcal{L}_{n,n-1}$ is a domain of validity. Suppose that $I_n\{f\}$ is exact in $\mathcal{L}_{n-1,n}$. Define Q_n as the unique $Q_n \in \mathcal{L}_n$ with zeros ξ_k, $k = 1, \ldots, n$, of unit norm: $\langle Q_n, Q_n \rangle_{\mathring{\mu}} = 1$ and with positive highest degree coefficient. Because $Q_n B_{k*} \in \mathcal{L}_{n-1,n}$ for all $0 \leq k < n$, we have

$$\langle Q_n, B_k \rangle_{\mathring{\mu}} = I_n\{Q_n B_{k*}\} = \sum_{j=1}^{n} \lambda_{nj} \frac{Q_n(\xi_j)}{B_k(\xi_j)} = 0.$$

Thus $Q_n \in \mathcal{L}_n$ is orthogonal to $\mathcal{L}_{n-1}$, and since it satisfies the precise normalizations, it has got to be the orthonormal function ϕ_n. However, we have shown before that all the zeros of ϕ_n are in $\mathbb{O}$ and thus we get a contradiction.

The proof for $\mathcal{L}_{n,n-1}$ is similar. $\square$

The next question to be solved is: How can we construct quadrature formulas with maximal domains of validity? The answer will be that they are of interpolatory type and use as nodes the zeros of a para-orthogonal function. This concept will be considered in the next section.

5.2. Para-orthogonal functions

In this section we introduce para-orthogonal rational functions. The zeros of such para-orthogonal functions are simple and are all on the boundary $\overline{\partial\mathbb{O}}$. They will play an important role as abscissas of quadrature formulas.

We follow the development and terminology used in Ref. [122].

A sequence of functions $f_n \in \mathcal{L}_n$ is called para-orthogonal when $f_n \perp \mathcal{L}_{n-1}(\alpha_n)$ while $\langle f_n, 1 \rangle \neq 0 \neq \langle f_n, B_n \rangle$.

Note that

$$
\begin{aligned}
\mathcal{L}_{n-1} \cap \zeta_n \mathcal{L}_{n-1} &= \{ g \in \mathcal{L}_n : g(\alpha_n) = g^*(\alpha_n) = 0 \} \\
&= \left\{ \frac{(z - \alpha_n) p_{n-2}(z)}{\pi_{n-1}(z)} : p_{n-2} \in \mathcal{P}_{n-2} \right\} \\
&= \mathcal{L}_{n-1}(\alpha_n)
\end{aligned}
$$

(see Theorem 2.2.1).

The sequence of orthogonal rational functions ϕ_n is not para-orthogonal since $\langle \phi_n, 1 \rangle = 0$ and neither is the sequence ϕ_n^* because $\langle \phi_n^*, B_n \rangle = 0$. However, define the following functions for u and v nonzero complex numbers:

$$
Q_n(z, u, v) = u\phi_n(z) + v\phi_n^*(z) \in \mathcal{L}_n.
$$

Obviously these functions are para-orthogonal. It is not difficult to show (see Ref. [34]) that all functions of the form $c_n Q_n(z, u_n, v_n)$ are para-orthogonal when c_n, u_n, v_n are nonzero and any para-orthogonal function can be written in this form.

Next we observe that since neither u nor v are allowed to be zero, there is no loss of generality if we choose $(u, v) = (1, \tau)$. The normalizing constant in front does not really matter. Thus we consider in what follows only

$$
Q_n(z) = Q_n(z, \tau) = \phi_n(z) + \tau \phi_n^*(z) \in \mathcal{L}_n, \qquad 0 \neq \tau \in \mathbb{C}. \tag{5.2}
$$

Another concept used in this context is k-invariance. A function $f_n \in \mathcal{L}_n$ is called k-*invariant* iff $f_n^* = k f_n$. Again, neither ϕ_n nor ϕ_n^* can be k-invariant for any k, but Q_n is $\bar{\tau}$-invariant if $\tau \in \mathbb{T}$. It is possible to prove [34] that any function of the form $c_n Q_n(z, \tau_n)$, $|\tau_n| = 1$, $c_n \neq 0$ is k_n-invariant with $k_n = \bar{c}_n \bar{\tau}_n / c_n$ and conversely that any para-orthogonal f_n that is k_n-invariant is of that form.

Thus a para-orthogonal function with $|\tau| = 1$ is k-invariant and conversely. In view of the previous remarks, we may only discuss the para-orthogonal functions Q_n with $|\tau| = 1$. These functions have simple zeros that all lie in $\partial \mathbb{O}$ as we shall presently show.

Theorem 5.2.1. *Let $\tau \in \mathbb{T}$ be given and define the para-orthogonal functions $Q_n(z) = Q_n(z, \tau)$ by (5.2). Then all the zeros of $Q_n(z)$ are on $\overline{\partial \mathbb{O}}$ and they are simple.*

Proof. We shall give this proof only for the case of $\partial \mathbb{O} = \mathbb{T}$. The case $\partial \mathbb{O} = \mathbb{R}$ is even simpler and we leave it as an exercise.

Since ϕ_n^* does not vanish in $\mathbb{D}$, the ratio ϕ_n/ϕ_n^* is well defined in $\mathbb{D}$ and it equals p_n/p_n^* if the polynomial $p_n \in \mathcal{P}_n$ is defined by $\phi_n = p_n/\pi_n$ with as always $\pi_n(z) = \prod_{i=1}^{n}(1 - \overline{\alpha}_i z)$. Since $b_n = p_n/p_n^*$ is a finite Blaschke product, with all its poles in $\mathbb{E}$, we have $|b_n(z)| < 1$ in $\mathbb{D}$. Suppose that α is a zero of Q_n in $\mathbb{D}$; then $Q_n(\alpha) = 0$ and this implies $p_n(\alpha)/p_n^*(\alpha) = b_n(\alpha) = -\tau$. Since $\tau \in \mathbb{T}$ and $|b_n(\alpha)| < 1$, we get a contradiction. Hence Q_n has no zeros in $\mathbb{D}$. Because, however, $Q_n^*(z, \tau) = \overline{\tau} Q_n(z, \tau)$, it follows that if α is a zero of $Q_n(z)$, then $Q_n(1/\overline{\alpha}) = \overline{B_{n*}(\alpha)} \tau \overline{Q_n(\alpha)} = 0$. Thus zeros appear in pairs $(\alpha, 1/\overline{\alpha})$. Because we showed that there are no zeros in $\mathbb{D}$, there can also be no zeros in $\mathbb{E}$ by this duality property. We may therefore conclude that all the zeros are on $\mathbb{T}$. Taking derivatives, it also follows that the multiplicity of a zero α is equal to the multiplicity of the zero $1/\overline{\alpha}$.

Now we prove that they are simple zeros. Suppose there are only $s \le n - 2$ zeros with an odd multiplicity. Call them $\xi_1, \dots, \xi_s$, possibly repeated if they are multiple. They are all in $\mathbb{T}$ and depend on τ. The remaining zeros are all of even multiplicity and hence we can arrange them in doublets (ξ_i, ξ_i) for $i = s + 1, \dots, s + r$. Again, we assume that a doublet (ξ_i, ξ_i) is repeated if the multiplicity of ξ_i is more than 2, whence $n = s + 2r$ with $r \ge 1$. Now we note that $(z - \xi_i)^2 = (z - \xi_i)(z - 1/\overline{\xi}_i) = c_i z(z - \xi_i)(z - \xi_i)_*$ with $c_i = -\xi_i$. Hence, $Q_n(z) = N/\pi_n$ with N of the form

$$N(z) = c \prod_{i=1}^{s}(z - \xi_i) z^r \prod_{i=s+1}^{s+r} (z - \xi_i)(z - \xi_i)_* \quad \text{with} \quad n - s = 2r, \quad r \ge 1$$

for some constant c. Consider now the function $T(z) = M(z)/\pi_{n-1}(z)$ with $M(z)$ of the form $c \prod_{i=1}^{s}(z - \xi_i) z^{r-1}(z - \alpha_n)$. Clearly $T \in \mathcal{L}_{n-1}$, and hence it is orthogonal to ϕ_n, but also $T(z) \in \mathcal{L}_{n-1}(\alpha_n)$. This means that it is of the form $(z - \alpha_n)p_{n-2}(z)/\pi_{n-1}$ with $p_{n-2} \in \mathcal{P}_{n-2}$. Thus it is also in $\zeta_n \mathcal{L}_{n-1}$ and therefore $\langle T, \phi_n^* \rangle_{\mathring{\mu}} = 0$. Consequently, T is orthogonal to Q_n. In contrast, if we write explicitly $\langle Q_n, T \rangle_{\mathring{\mu}}$, we get

$$|c|^2 \int \frac{\prod_{i=1}^{s}|t - \xi_i|^2}{|\pi_{n-1}(t)|^2} \prod_{i=s+1}^{s+r} |t - \xi_i|^2 d\mathring{\mu}(t) = \|S\|_{\mathring{\mu}}^2 > 0$$

since $S \not\equiv 0$. This is a contradiction, so that $s \ge n - 1$. This means that all the zeros of Q_n should be simple and on $\mathbb{T}$.   □

Note that in the case of the real line, there can be zeros at infinity. For example, in the case of the Lebesgue measure $\mu = \lambda$, then $\phi_0 = 1$ and $\phi_1(z) = (z - i)/(z + i)$ are the first two orthonormal functions and $Q_1(z) = [(1 + \tau)z +$

$i(\tau - 1)]/(z + i)$, which is zero at ∞ for $\tau = -1$. Of course it is always possible to choose $\tau = \tau_n$ such that there is no zero at infinity.

With the para-orthogonal functions Q_n, we associate corresponding functions of the second kind as follows:

$$P_n(z) = P_n(z, \tau) = \psi_n(z) - \tau\psi_n^*(z), \qquad (5.3)$$

where the ψ_n are the usual functions of the second kind. The following is true:

Lemma 5.2.2. Let $Q_n(z) = Q_n(z, \tau)$ be the para-orthogonal functions and let $P_n(z) = P_n(z, \tau)$ be the associated functions of the second kind as defined by (5.3); then for $n \geq 2$, we have

$$P_n(z)f(z) = \int D(t, z)[Q_n(t)f(t) - Q_n(z)f(z)]\,d\mathring{\mu}(t)$$

for an arbitrary function $f \in \mathcal{L}_{(n-1)*}(\hat{\alpha}_n) = \{f : f_* \in \mathcal{L}_{(n-1)} : f(\hat{\alpha}_n) = 0\}$.

Proof. From our results in Lemmas 4.2.2 and 4.2.3, it follows that for $n \geq 2$ and $f \in \mathcal{L}_{(n-1)*} \cap \zeta_{n*}\mathcal{L}_{(n-1)*} = \mathcal{L}_{(n-1)*}(\hat{\alpha}_n)$

$$\begin{bmatrix} \psi_n(z) \\ -\psi_n^*(z) \end{bmatrix} = \int D(t, z)\left\{\begin{bmatrix} \phi_n(t) \\ \phi_n^*(t) \end{bmatrix}\frac{f(t)}{f(z)} - \begin{bmatrix} \phi_n(z) \\ \phi_n^*(z) \end{bmatrix}\right\}\,d\mathring{\mu}(t).$$

Multiplying this from the left by $[1 \ \tau]$ gives the result. $\qquad\square$

The previous lemma says nothing about the case $n = 1$. The following result will be useful in that case.

Lemma 5.2.3. Let Q_n denote the para-orthogonal functions and suppose Q_1 has a zero $\xi \in \partial\mathbb{O}$. Let P_1 be the associated second kind function. Then

$$P_1(z) = -D(\xi, z)Q_1(z),$$

with $D(t, z)$ the Riesz–Herglotz–Nevanlinna kernel.

More explicitly, this means that if η is the zero of P_1, then η and ξ are related by $\eta = -\xi$ for $\partial\mathbb{O} = \mathbb{T}$ and $\eta = -1/\xi$ for $\partial\mathbb{O} = \mathbb{R}$.

Proof. The proof requires some calculations, but one uses essentially the first step of the recurrence relation to find explicit expressions for $Q_1 = \phi_1 + \tau\phi_1^*$ and $P_1 = \psi_1 - \tau\psi_1^*$. Since for $\partial\mathbb{O} = \mathbb{T}$, $\zeta_0(z) = z$, one sees that the only difference then gives a sign change, which accounts for $\eta = -\xi$. In the case $\partial\mathbb{O} = \mathbb{R}$, this essentially corresponds to a Cayley transform of this result. $\qquad\square$

5.3. Quadrature

Now we consider the space of rational functions

$$\mathcal{R}_n = \mathcal{L}_n \cdot \mathcal{L}_{n*} = \mathcal{L}_{n,n},$$

where $\mathcal{L}_{n*} = \{f : f_* \in \mathcal{L}_n\}$. The space $\mathcal{R}_n$ can also be characterized as

$$\mathcal{R}_n = \left\{ \frac{p_n(z)}{\pi_n(z)\pi_n^*(z)} : p_n \in \mathcal{P}_{2n} \right\},$$

where $\pi_n(z) = \prod_{k=0}^n \varpi_k(z)$. Let $\ell_{ni}(z) = \prod_{k \neq i}(z - \xi_k)/(\xi_i - \xi_k)$ denote the Lagrange polynomial for the interpolation points $\xi_1, \xi_2, \ldots, \xi_n$, so that $\sum_{k=1}^n R(\xi_k)\ell_{nk}(z)$ is the interpolating polynomial of degree $n-1$ for R, which we take from $\mathcal{R}_{n-1}$. Let L_{ni} be defined by (compare with (5.1))

$$L_{ni}(z) = \ell_{ni}(z)\frac{\pi_{n-1}(\xi_i)}{\pi_{n-1}(z)} = \frac{\varpi_n(z)}{\varpi_n(\xi_i)}\frac{Q_n(z)}{(z - \xi_i)Q_n'(\xi_i)} \in \mathcal{L}_{n-1},$$

where the prime means derivative and $\xi_k, k = 1, \ldots, n$ are the zeros of $Q_n(z) = \phi_n(z) + \tau\phi_n^*(z)$, which are all on $\partial\mathbb{O}$. Note that $L_{nk}(\xi_i) = \delta_{ki}$, so that $\sum_1^n R(\xi_k)L_{nk}(z)$ is an interpolant for $R(z)$ from $\mathcal{L}_{n-1}$ in the points $\xi_1, \ldots, \xi_n$. Consider the interpolation error

$$E(z) = R(z) - \sum_{k=1}^n R(\xi_k)L_{nk}(z)$$

$$= \frac{t_{n-1}(z) - \sum_{k=1}^n R(\xi_k)\ell_{nk}(z)\pi_{n-1}(\xi_k)\pi_{(n-1)*}(z)}{\pi_{n-1}(z)\pi_{(n-1)*}(z)}.$$

Here $t_{n-1}(z) = z^{-(n-1)}p_{n-1}(z)$ for the disk and $t_{n-1}(z) = p_{n-1}(z)$ for the half plane, with $p_{n-1} \in \mathcal{P}_{2n-2}$ in both cases. From the interpolating property, we find

$$E(z) = \frac{\prod_{k=1}^n(z - \xi_k)q(z)}{\pi_{n-1}(z)\pi_{(n-1)*}(z)},$$

with $q(z)$ of the form $q(z) = q_{n-2}(z)z^{-(n-1)}$ for the disk and $q(z) = q_{n-2}(z)$ for the half plane, where $q_{n-2} \in \mathcal{P}_{n-2}$ in both cases. Now, because $\xi_i, i = 1, \ldots, n$ are the zeros of $Q_n(z)$, we can also write this as

$$E(z) = Q_n(z)\frac{\varpi_n(z)q(z)}{\pi_{(n-1)*}(z)}. \tag{5.4}$$

The second factor can be written as S_* with S defined by

$$S(z) = \frac{(z - \alpha_n)q_{n-2}^*(z)}{\pi_{n-1}(z)}, \quad q_{n-2} \in \mathcal{P}_{n-2}.$$

Observe that $S \in \mathcal{L}_{n-1}$ and also $S \in \zeta_n \mathcal{L}_{n-1}$, so that S is orthogonal to ϕ_n and to ϕ_n^* and hence is orthogonal to Q_n. Thus $\langle Q_n, S \rangle_{\mathring{\mu}} = \int E \, d\mathring{\mu} = 0$. In other words, if $R \in \mathcal{R}_{n-1}$ and ξ_k, $k = 1, \ldots, n$ are the zeros of $Q_n = \phi_n + \tau \phi_n^*$, with $|\tau| = 1$, then

$$\int R \, d\mathring{\mu} = \sum_{i=1}^{n} R(\xi_i) \int L_{ni} \, d\mathring{\mu} = \sum_{i=1}^{n} R(\xi_i)\lambda_{ni},$$

which is a quadrature on the boundary $\partial \mathbb{O}$. We shall refer to this as a Rational Szegő quadrature formula or an R-Szegő formula for short.

Since $L_{ni} L_{ni*} \in \mathcal{R}_{n-1}$ and also $L_{ni} L_{ni*} - L_{ni} \in \mathcal{R}_{n-1}$, we get

$$\int (L_{ni} L_{ni*} - L_{ni}) \, d\mathring{\mu} = \sum \lambda_{ni} 0 = 0.$$

Thus

$$\lambda_{ni} = \int L_{ni} \, d\mathring{\mu} = \int |L_{ni}|^2 \, d\mathring{\mu} > 0.$$

Thus we have proved the following theorem.

Theorem 5.3.1. *Let $\{\phi_k\}$ be an orthonormal system for $\mathcal{L}_n$, $\tau \in \mathbb{T}$ and $Q_n = \phi_n + \tau \phi_n^*$. Then Q_n has n simple zeros on $\partial \mathbb{O}$: $\xi_1, \ldots, \xi_n$. Let ℓ_{ni} denote the Lagrange polynomials for the interpolation points $\xi_1, \ldots, \xi_n$. Then the R-Szegő quadrature formula*

$$\sum_{i=1}^{n} R(\xi_i)\lambda_{ni} \quad \text{with} \quad \lambda_{ni} = \int \frac{\ell_{ni}(z)}{\pi_{n-1}(z)} \pi_{n-1}(\xi_i) \, d\mathring{\mu} > 0 \qquad (5.5)$$

is exact for all $R \in \mathcal{R}_{n-1} = \mathcal{L}_{n-1} \cdot \mathcal{L}_{(n-1)*}$.

Consider the discrete measure $\mathring{\mu}_n$ for this quadrature formula. This means

$$\mathring{\mu}_n(t) = \sum_{j=1}^{n} \lambda_{nj} \delta_{\xi_j}(t), \qquad (5.6)$$

where $\delta_{\xi_j}(t)$ is the Dirac delta function at $t = \xi_j$.

The previous theorem now says that

Corollary 5.3.2. *With the notation of the previous theorem and the discrete measure $\mathring{\mu}_n$ as just introduced, it holds that*

$$\int R(t)\,d\mathring{\mu}(t) = \int R(t)\,d\mathring{\mu}_n(t) \quad \textit{for all } R \in \mathcal{R}_{n-1},$$

or, equivalently,

$$\langle f, g \rangle_{\mathring{\mu}} = \langle f, g \rangle_{\mathring{\mu}_n} \quad \textit{for all } f, g \in \mathcal{L}_{n-1}.$$

Proof. The first observation is a restatement of the previous problem. The second one is a consequence of the fact that $R \in \mathcal{R}_{n-1}$ iff $R = f \cdot g_*$ for $f, g \in \mathcal{L}_{n-1}$. □

Clearly, using the Riesz–Herglotz–Nevanlinna kernel $D(t, z)$,

$$R_n(z) = \int D(t, z)\,d\mathring{\mu}_n(t) = \sum_{j=1}^{n} \lambda_{nj} D(\xi_j, z)$$

can be written as

$$R_n(z) = -\frac{S_n(z)}{Q_n(z)}, \tag{5.7}$$

where Q_n is the para-orthogonal function and S_n is some element from $\mathcal{L}_n$. We shall currently show that $S_n = P_n = \psi_n - \tau \psi_n^*$ is the associated function of the second kind.

Theorem 5.3.3. *Let Q_n be the para-orthogonal function $Q_n = \phi_n + \tau \phi_n^*$ with $|\tau| = 1$. Let $\xi_1, \ldots, \xi_n \in \partial \mathbb{O}$ be its zeros and $\mathring{\mu}_n$ the discrete measure (5.6) for the associated quadrature formula. Then, for $n \geq 1$,*

$$R_n(z) = \int D(t, z)\,d\mathring{\mu}_n(t) = -\frac{P_n(z)}{Q_n(z)},$$

where $P_n = \psi_n - \tau \psi_n^*$ is the function of the second kind associated with Q_n.

Proof. To avoid notational confusion, suppose S_n is defined by (5.7), that is,

$$S_n(z) = -Q_n(z) \int D(t, z)\,d\mathring{\mu}_n(t)$$

while $P_n = \psi_n - \tau \psi_n^*$ is the second kind function. We have to prove that $S_n = P_n$.

We first consider the case $n \geq 2$. We know by Lemma 5.2.2 that then

$$P_n(z) = \int D(t, z) \left[Q_n(t) \frac{f(t)}{f(z)} - Q_n(z) \right] d\mathring{\mu}(t),$$

where $f \in \mathcal{L}_{(n-1)*} \cap \zeta_{n*} \mathcal{L}_{(n-1)*}$. For such functions f, it can be verified that

$$D(t, z) \left[Q_n(t) \frac{f(t)}{f(z)} - Q_n(z) \right] \in \mathcal{R}_{n-1} = \mathcal{L}_{n-1} \cdot \mathcal{L}_{(n-1)*}.$$

By the previous theorem, we may then replace the integral by the quadrature formula (i.e., we may replace $\mathring{\mu}$ by $\mathring{\mu}_n$). Thus (recall that ξ_j are zeros of Q_n)

$$P_n(z) = -Q_n(z) \sum_{j=1}^{n} \lambda_{nj} D(\xi_j, z) = -Q_n(z) \int D(t, z) \, d\mathring{\mu}_n(t) = S_n(z).$$

Hence the theorem is proved for $n \geq 2$.

For the case $n = 1$, we use Lemma 5.2.3. Note that $\lambda_{1,1} = \int d\mathring{\mu}(t) = 1$, and thus

$$S_1(z) = -Q_1(z) R_1(z) = -Q_1(z) \int D(t, z) \, d\mathring{\mu}_1(t)$$

$$= -Q_1(z) D(\xi_1, z) = P_1(z).$$

The proof of the theorem is now complete. $\qquad\square$

We have shown in this section how to construct a quadrature formula with maximal domain of validity $\mathcal{R}_{n-1}$. It was of interpolatory type and the nodes were the zeros of a para-orthogonal function. We can also prove that these are essentially all the possible quadrature formulas with this domain of validity.

Theorem 5.3.4. *Let $I_n\{f\}$ be an n-point quadrature formula with distinct nodes ξ_k, $k = 1, \ldots, n$, all on $\partial\mathbb{O}$. Then it has maximal domain of validity $\mathcal{R}_{n-1}$ iff it is of interpolatory type in $\mathcal{L}_{p,q}$ with $p + q = n - 1$ and the nodes are the zeros of a para-orthogonal function in $\mathcal{L}_n$ with $|\tau| = 1$.*

Proof. $\Rightarrow$: Let p and q be nonnegative integers with $p + q = n - 1$; then $\mathcal{L}_{p,q} \subset \mathcal{R}_{n-1}$. Hence $I_n\{f\}$ is exact in $\mathcal{L}_{p,q}$.

Let $x_n(z) = \prod_{k=1}^{n}(z - \xi_k)$ be the nodal polynomial and define $X_n(z) = x_n(z)/\pi_n(z)$, $\pi_n(z) = \prod_{k=1}^{n} \varpi_k(z)$; then

$$X_n^*(z) = B_n(z) \prod_{k=1}^{n} \frac{z - \xi_k}{\varpi_{k*}(z)} = k_n X_n(z),$$

with $k_n = \prod_{k=1}^{n} z_k \bar{\xi}_k$ for $\mathbb{D}$ and $k_n = \prod_{k=1}^{n} z_k$ for $\mathbb{U}$. Hence X_n is k_n-invariant and thus, by the remarks preceding Theorem 5.2.1, it is para-orthogonal since $|\tau| = 1$.

$\Leftarrow$: Conversely, for nonnegative integers p and q with $p + q = n - 1$, define $L_k^{(p)} = L_k \in \mathcal{L}_{p,q}$ by $L_k(\xi_j) = \delta_{kj}$ and set $\lambda_k^{(p)} = \lambda_k = I_{\mathring{\mu}}\{L_k\}$, $k = 1, \ldots, n$. We prove that $I_n\{f\} = \sum_{k=1}^{n} \lambda_k f(\xi_k)$ is exact in $\mathcal{R}_{n-1}$. So let $f \in \mathcal{R}_{n-1}$ and consider the function $E(z) = f(z) - \sum_{k=1}^{n} L_k(z) f(\xi_k) \in \mathcal{R}_{n-1}$. Since $E(\xi_k) = 0$ for $k = 1, \ldots, n$, we can now show as in (5.4) that it is of the form

$$E(z) = X_n(z) g_{n*}(z),$$

where $X_n \in \mathcal{L}_n$ is para-orthogonal and $g(z) = (z - \alpha_n) q_{n-2}^*(z) / \pi_{n-1}(z)$ with $q_{n-2} \in \mathcal{P}_{n-2}$, and thus with $g \in \mathcal{L}_{n-1}(\alpha_n)$. Consequently,

$$I_{\mathring{\mu}}\{E\} = \langle X_n, g \rangle_{\mathring{\mu}} = 0.$$

This means that the error is zero for any $f \in \mathcal{R}_{n-1}$.

We conclude by showing that the weights are independent of p (i.e., that $\lambda_k^{(p)} = \lambda_k^{(\tilde{p})}$). It suffices to prove this for $\tilde{p} = p + 1$. Define $X_n^{(p)}(z)$ as $x_n(z)/[\pi_p^*(z)\pi_{q+1}(z)]$; then we know that (prime means derivative)

$$L_k^{(p)}(z) = \frac{\varpi_{q+1}(z)}{\varpi_{q+1}(\xi_k)} \frac{X_n^{(p)}(z)}{(z - \xi_k) X_n^{(p)'}(\xi_k)}.$$

Now

$$X_n^{(p+1)}(z) = \frac{\varpi_{q+1}(z)}{z - \alpha_{p+1}} X_n^{(p)}(z) \quad \text{and} \quad X_n^{(p+1)'}(\xi_k) = \frac{\varpi_{q+1}(\xi_k)}{\xi_k - \alpha_{p+1}} X_n^{(p)'}(\xi_k).$$

Hence

$$L_k^{(p+1)}(z) = \frac{\varpi_q(z)(\xi_k - \alpha_{p+1})}{\varpi_q(\xi_k)(z - \alpha_{p+1})} L_k^{(p)}(z),$$

or

$$L_k^{(p+1)}(z) = L_k^{(p)}(z) + \frac{(z - \xi_k)(\alpha_p - \bar{\alpha}_{p+1})}{(z - \alpha_{p+1})\varpi_q(\xi_k)} L_k^{(p)}(z).$$

Thus we have to prove that

$$\int \frac{t - \xi_k}{t - \alpha_{p+1}} L_k^{(p)}(t) \, d\mathring{\mu}(t) = 0,$$

that is,

$$\int \frac{\varpi_{q+1}(t)}{t - \alpha_{p+1}} X_n^{(p)}(t) \, d\mathring{\mu}(t) = 0.$$

The latter is equal to

$$\int \frac{x_n(t)}{\pi_q(t)\pi_{p+1}^*(t)} \, d\mathring{\mu}(t) = \int X_n^{(p)}(t) \frac{\prod_{k=q+1}^n \varpi_k(t)}{\pi_{p+1}^*(t)} \, d\mathring{\mu}(t)$$

$$= \int X_n^{(p)}(t) h_*(t) \, d\mathring{\mu}(t),$$

with $h \in \mathcal{L}_{n-1}(\alpha_n)$, and this is zero by para-orthogonality of $X_n^{(p)}$. □

This theorem shows that there is a one-parameter family (depending on $\tau \in \mathbb{T}$) of quadrature formulas that have maximal domain of validity $\mathcal{R}_{n-1}$. It has the following properties:

1. Its nodes $\xi_k, k = 1, \ldots, n$ are the distinct zeros of the para-orthogonal function $Q_n = \phi_n + \tau\phi_n^*, \tau \in \mathbb{T}$, which are all on $\overline{\partial\mathbb{O}}$,
2. The positive weights are given as $\lambda_{nk} = \int L_{nk}(t) \, d\mathring{\mu}(t)$, where the $L_{nk} \in \mathcal{L}_{n-1}$ are defined by $L_{nk}(\xi_j) = \delta_{kj}, k, j = 1, \ldots, n$.

This characterizes the set of all n-point rational Szegő quadrature formulas.

5.4. The weights

We shall derive in this section some alternative expressions for the weights of the quadrature formula as given in Theorem 5.3.1.

A first result expresses them in terms of the para-orthogonal functions and the associated ones of the second kind.

Theorem 5.4.1. *The weights λ_{nk} of the quadrature formula of Theorem 5.3.1 are given by*

$$\lambda_{nk} = \frac{1}{2} \frac{\varpi_0(\alpha_0)}{\varpi_0(\xi_k)\varpi_0^*(\xi_k)} \frac{P_n(\xi_k)}{Q_n'(\xi_k)}, \quad k = 1, \ldots, n,$$

where the prime means differentiation with respect to z, Q_n are the para-orthogonal functions $Q_n = \phi_n + \tau\phi_n^*, |\tau| = 1$ with zeros $\xi_1, \ldots, \xi_n \in \overline{\partial\mathbb{O}}$, and $P_n = \psi_n - \tau\psi_n^*$ are the associated functions of the second kind.

Note that for $\partial\mathbb{O} = \mathbb{T}$ the middle factor is just $1/\xi_k$, and for $\partial\mathbb{O} = \mathbb{R}$, it is $2\mathrm{i}/(1 + \xi_k^2)$.

Proof. From the partial fraction decomposition

$$R_n(z) = -\frac{P_n(z)}{Q_n(z)} = \sum_{j=1}^n \lambda_{nj} D(\xi_j, z),$$

it follows that

$$(z - \xi_k) R_n(z) = \sum_{j \neq k} \lambda_{nj} D(\xi_j, z)(z - \xi_k) + \lambda_{nk} D(\xi_k, z)(z - \xi_k).$$

Taking the limit for $z \to \xi_k$ cancels all the terms in the sum and we get

$$-\frac{P_n(\xi_k)}{Q_n'(\xi_k)} = \lambda_{nk} \lim_{z \to \xi_k} D(\xi_k, z)(z - \xi_k).$$

This gives the result.   □

Another expression for the weights can be obtained in terms of the Christoffel function. This is given in the next theorem.

Theorem 5.4.2. *With the same notation as in the previous theorem, it holds that*

$$\lambda_{nk} = \left[\sum_{j=0}^{n-1} |\phi_j(\xi_k)|^2 \right]^{-1}.$$

Proof. We note that $\xi_k \in \partial\mathbb{O}$, so that $\overline{\zeta_n(\xi_k)} = 1/\zeta_n(\xi_k)$. Therefore, we can use the Christoffel–Darboux relation with $w = \xi_j$ and $z = t$ to get

$$\zeta_n(\xi_k)\overline{\phi_n(\xi_k)}\frac{Q_n(\xi_k) - Q_n(t)}{\xi_k - t} = \frac{\zeta_n(\xi_k) - \zeta_n(t)}{\xi_k - t} \sum_{j=0}^{n-1} \overline{\phi_j(\xi_k)}\phi_j(t).$$

Using

$$\frac{\zeta_n'(\xi_k)}{\zeta_n(\xi_k)} = \frac{\varpi_n(\alpha_n)}{\varpi_n(\xi_k)\varpi_n^*(\xi_k)},$$

the limit for $t \to \xi_k$ yields

$$Q_n'(\xi_k) = \frac{1}{\overline{\phi_n(\xi_k)}} \frac{\varpi_n(\alpha_n)}{\varpi_n(\xi_k)\varpi_n^*(\xi_k)} \sum_{j=0}^{n-1} |\phi_j(\xi_k)|^2.$$

The determinant formula also gives

$$P_n(\xi_k) = 2\frac{B_n(\xi_k)}{\phi_n^*(\xi_k)} \frac{\varpi_n(\alpha_n)}{\varpi_n(\xi_k)\varpi_{n*}(\xi_k)} \frac{\varpi_0(\xi_k)\varpi_{0*}(\xi_k)}{\varpi_0(\alpha_0)}.$$

Using the formula of the previous theorem then yields

$$\lambda_{nk} = \frac{B_n(\xi_k)\overline{\phi_n(\xi_k)}}{\phi_n^*(\xi_k)} \left[\sum_{j=0}^{n-1} |\phi_j(\xi_k)|^2 \right]^{-1}.$$

Since $\xi_k \in \overline{\partial \mathbb{O}}$, and hence

$$\phi_n^*(\xi_k) = \overline{\phi_n(\xi_k)} B_n(\xi_k),$$

this proves the theorem.   □

5.5. An alternative approach

An R-Szegő formula can also be characterized in an alternative way, by using Hermite interpolation. We shall presently explain this. It is similar to the approach given by Markov for classical Gauss formulas. Using the same notation as in the rest of this chapter, we consider a set $N_n = \{\xi_1, \ldots, \xi_n\} \subset \mathbb{C} \backslash \{A_n \cup \hat{A}_n\}$ of distinct nodes. Since $\mathcal{L}_{p,q}$ is a Haar subspace (i.e., it is spanned by a Chebyshev system) there exists a unique $R \in \mathcal{L}_{p,q}$, $(p + q = 2(n - 1))$ that is a Hermite interpolant for a given function f, in the sense that

$$R(\xi_k) = f(\xi_k), \quad k = 1, \ldots, n \quad \text{and} \quad R'(\xi_k) = f'(\xi_k), \quad k = 1, \ldots, n - 1. \tag{5.8}$$

We can represent R as given in the following lemma, which can be found in Ref. [39].

Lemma 5.5.1. With the notation introduced above, the Hermite interpolant R can be given as

$$R(z) = \sum_{j=1}^{n} H_{j0}(z) f(\xi_j) + \sum_{j=1}^{n-1} H_{j1}(z) f'(\xi_j),$$

where $H_{j0}, H_{j1} \in \mathcal{R}_{n-1}$ can be characterized by

$$\begin{aligned}
H_{i0}(\xi_j) &= \delta_{ij}, \quad 1 \le i, j \le n; \\
H_{i0}'(\xi_j) &= 0, \quad 1 \le i \le n, \, 1 \le j \le n - 1; \\
H_{i1}(\xi_j) &= 0, \quad 1 \le i \le n - 1, \, 1 \le j \le n; \\
H_{i0}'(\xi_j) &= 0, \quad 1 \le i, j \le n.
\end{aligned}$$

The functions H_{i1} are of the form

$$H_{i1}(z) = c_{in} \frac{x_n^2(z)}{\pi_{n-1}(z) \pi_{n-1}^*(z)(z - \xi_i)(z - \xi_n)},$$

where $x_n(z)$ is the nodal polynomial and c_{in} is a constant.

Obviously, such an interpolant gives rise to a quadrature formula

$$I_n\{f\} = \sum_{j=1}^{n} \lambda_{nj} f(\xi_j) + \sum_{j=1}^{n-1} \tilde{\lambda}_{nj} f'(\xi_j), \qquad (5.9)$$

with

$$\lambda_{nj} = \int H_{i0}(t)\,d\mathring{\mu}(t), \ 1 \le j \le n, \quad \tilde{\lambda}_{nj} = \int H_{i1}(t)\,d\mathring{\mu}(t), \ 1 \le j \le n-1.$$

It can now be shown that this is in fact the same formula we had before.

Theorem 5.5.2. *Consider the quadrature formula (5.9) where the nodes ξ_k are supposed to be the zeros of the para-orthogonal function $Q_n = \phi_n + \tau \phi_n^*$, $|\tau| = 1$; then it is an R-Szegő formula, that is, it is identical to the n-point quadrature formula (5.5).*

Proof. We have to show that the weights $\tilde{\lambda}_{ni}$ vanish for $1 \le i \le n - 1$. Using the expression for H_{i1} from the previous lemma, we get, up to a constant factor, that the weights are given by

$$\int \frac{x_n(t)}{\pi_n(t)} \cdot \frac{\varpi_n(t)x_{n-1}(t)}{(t - \xi_i)\pi_{n-1}^*(t)}\,d\mathring{\mu}(t) = \int Q_n(t)h_*(t)\,d\mathring{\mu}(t) = \langle Q_n, h \rangle_{\mathring{\mu}},$$

where $h \in \mathcal{L}_{n-1}(\alpha_n)$, and therefore the latter inner product vanishes. This proves the theorem. $\qquad \square$

6

Interpolation

In this chapter we discuss several aspects related to interpolation. In the first section, we derive some simple interpolation properties that can be easily obtained from the properties of the functions of the second kind that were studied earlier. It also turns out that interpolation of the positive real function Ω_μ, whose Riesz–Herglotz–Nevanlinna measure μ is the measure that we used for the inner product, will imply that in $\mathcal{L}_n$ the measure can be replaced by the rational Riesz–Herglotz–Nevanlinna measure for the interpolant without changing the inner product. Some general theorems in this connection will be proved in Section 6.2. This will be important for the constructive proof of the Favard theorems to be discussed in Chapter 8. We then resume the interpolation results that can be obtained with the reproducing kernels and some functions that are in a sense reproducing kernels of the second kind.

We then show the connection with the algorithm of Nevanlinna–Pick in Section 6.4. This algorithm provides an alternative way to find the coefficients for the recurrence of the reproducing kernels that we gave in Section 3.2, without explicitly generating the kernels themselves. If all the interpolation points are at the origin, then the algorithm reduces to the Schur algorithm. It was designed originally to check whether a given function is in the Schur class. It basically generates a sequence of Schur functions by Möbius transforms and extractions of zeros. Section 6.5 gives a similar algorithm that works for the orthogonal functions rather than the reproducing kernels.

6.1. Interpolation properties for orthogonal functions

We shall give some interpolation properties that are easily derived from the properties given in Section 4.2 in connection with the functions of the second kind.

Since by definition

$$\int D(t,z)\phi_n(z)\,d\mathring{\mu}(t) = \phi_n(z)\int D(t,z)\,d\mathring{\mu}(t) = \phi_n(z)\Omega_\mu(z),$$

we can derive the following interpolation properties.

Theorem 6.1.1. *Let Ω_μ be the positive real function with Riesz–Herglotz–Nevanlinna measure μ. We introduce the Blaschke products $\tilde{B}_{-1} = 1$ and $\tilde{B}_n = \zeta_n \tilde{B}_{n-1} = \zeta_0 B_n$ for $n \geq 0$. Then for the orthonormal functions and the functions of the second kind, it holds that*

$$h = \frac{\phi_n \Omega_\mu + \psi_n}{\tilde{B}_{n-1}} \in H(\mathbb{O}), \quad n \geq 0. \tag{6.1}$$

For their superstar conjugates, we find

$$g = \frac{\phi_n^* \Omega_\mu - \psi_n^*}{\tilde{B}_n} \in H(\mathbb{O}), \quad n \geq 0. \tag{6.2}$$

Proof. For $n = 0$, the relation (6.1) is obvious knowing that $\phi_0 = \psi_0 = 1$ and that $\Omega_\mu \in H(\mathbb{O})$.

Use (4.22) for $f = 1/B_{n-1}$ and $n > 0$ to get

$$\frac{\phi_n \Omega_\mu + \psi_n}{B_{n-1}} = \int D(t,z)\frac{\phi_n(t)}{B_{n-1}(t)}\,d\mathring{\mu}(t). \tag{6.3}$$

For $z = \alpha_0$, the integral equals

$$\int \frac{\phi_n}{B_{n-1}}\,d\mathring{\mu} = \langle \phi_n, B_{n-1}\rangle_{\mathring{\mu}} = 0.$$

Moreover, Equation (6.3) is analytic in $\mathbb{O}$ as a Cauchy–Stieltjes integral.

For the relation (6.2), one can similarly check the case $n = 0$ (use $\Omega_\mu(\alpha_0) = 1$), and for $n > 0$, use (4.23) with $g = 1/B_n$ to see that

$$\frac{\phi_n^* \Omega_\mu - \psi_n^*}{B_n} = \int D(t,z)\frac{\phi_n^*(t)}{B_n(t)}\,d\mathring{\mu}(t), \tag{6.4}$$

which for $z = \alpha_0$ equals zero because $D(t,\alpha_0) = 1$ and $\langle \phi_n^*, B_n\rangle_{\mathring{\mu}} = 0$. Also, Equation (6.4) is analytic in $\mathbb{O}$ as a Cauchy–Stieltjes integral.

This proves the theorem. $\square$

We have a simple consequence for the para-orthogonal functions.

Corollary 6.1.2. *With the same notation as in the previous theorem and with the para-orthogonal functions $Q_n(z) = \phi_n(z) + \tau \phi_n^*(z)$, $\tau \in \mathbb{T}$, and the associated functions of the second kind $P_n(z) = \psi_n(z) - \tau \psi_n^*(z)$, it holds that*

$$\frac{Q_n \Omega_\mu + P_n}{\tilde{B}_{n-1}} \in H(\mathbb{O}), \quad n \geq 0.$$

Recall $\tilde{B}_{n-1} = \zeta_0 B_{n-1} = \zeta_0 \cdots \zeta_{n-1}$, for $n \geq 1$ and $B_{-1} = 1$.

Proof. This function is of the form $f(z) + \tau(z - \alpha_n)g(z)$ with $f, g \in H(\mathbb{O})$, which gives the result. $\qquad\qquad\qquad\qquad\qquad\qquad\qquad\qquad\qquad\qquad$ $\square$

More general interpolation results can be obtained as follows.

Theorem 6.1.3. *Let Ω_μ be the Riesz–Herglotz–Nevanlinna transform of the measure μ. Let $Q_n = \phi_n + \tau \phi_n^*$ be a para-orthogonal function for this measure with $\tau \in \mathbb{T}$ and $P_n = \psi_n - \tau \psi_n^*$ the associated function of the second kind. Then with $R_n(z) = -P_n(z)/Q_n(z)$, it holds that*

$$[B_{n-1}(z)][\Omega_\mu(z) - R_n(z)] = \hat{h}(z) \in H(\mathbb{O}^e), \quad n \geq 1,$$

with $\hat{h}(\hat{\alpha}_0) = 0$, and

$$[B_{n-1}(z)]^{-1}[\Omega_\mu(z) - R_n(z)] = h(z) \in H(\mathbb{O}), \quad n \geq 1,$$

with $h(\alpha_0) = 0$.

Proof. We use the fact that

$$\int g(t) d(\mathring{\mu} - \mathring{\mu}_n)(t) = 0, \quad \forall g \in \mathcal{R}_{n-1}$$

when

$$\mathring{\mu}_n(t) = \sum_{k=1}^{n} \lambda_{n,k} \delta_{\xi_k}(t)$$

is the discrete measure associated with the quadrature formula of Theorem 5.3.1. If $g \in \mathcal{R}_{n-1}$, then obviously

$$D(t, z)[g(t) - g(z)] \in \mathcal{R}_{n-1}$$

and thus

$$\int D(t, z)[g(t) - g(z)] d(\mathring{\mu} - \mathring{\mu}_n)(t) = 0.$$

Because

$$\Omega_\mu(z) - R_n(z) = \int D(t,z)d(\mathring\mu - \mathring\mu_n)(t),$$

we get for any $g \in \mathcal{R}_{n-1}$

$$h(z) = \int D(t,z)g(t)d(\mathring\mu - \mathring\mu_n)(t) = g(z)\int D(t,z)d(\mathring\mu - \mathring\mu_n)(t)$$
$$= g(z)[\Omega_\mu(z) - R_n(z)].$$

Clearly, $h \in H(\mathbb{O} \cup \mathbb{O}^e)$ and $h(\alpha_0) = 0$ and $h(\mathring\alpha_0) = 0$. By choosing $g(z) = B_{n-1}(z)$, we get the first result and by choosing $g(z) = 1/B_{n-1}(z)$ we get the second one. $\qquad\square$

For the orthonormal functions, we similarly obtain extra interpolating properties in $\mathbb{O}^e$.

Theorem 6.1.4. *Let ϕ_n be the orthonormal functions and ψ_n the functions of the second kind. Then*

$$[B_n(z)]\left[\Omega_\mu(z) + \frac{\psi_n(z)}{\phi_n(z)}\right] = \hat h(z) \in H(\mathbb{O}^e) \quad \text{and} \quad \hat h(\mathring\alpha_0) = 0, \qquad n \geq 0,$$

while

$$[B_{n-1}(z)][\Omega_\mu(z)\phi_{n*}(z) - \psi_{n*}(z)] = \hat g(z) \in H(\mathbb{O}^e) \quad \text{and} \quad \hat g(\mathring\alpha_0) = 0,$$
$$n \geq 1.$$

Also,

$$[B_{n-1}(z)]^{-1}[\Omega_\mu(z)\phi_n(z) + \psi_n(z)] = h(z) \in H(\mathbb{O}) \quad \text{and} \quad h(\alpha_0) = 0,$$
$$n \geq 1,$$

and

$$[B_n(z)]^{-1}\left[\Omega_\mu(z) - \frac{\psi_n^*(z)}{\phi_n^*(z)}\right] = g(z) \in H(\mathbb{O}) \quad \text{and} \quad g(\alpha_0) = 0, \quad n \geq 0.$$

Proof. Note that this does not follow from the previous results by simply setting $\tau = 0$ or $\tau = \infty$ since the previous result relied on the fact that $\tau \in \mathbb{T}$. We shall use a direct proof instead.

For the first result, we should prove that

$$\frac{B_n(z)}{\phi_n(z)}[\Omega_\mu(z)\phi_n(z) + \psi_n(z)]$$

is analytic in $\mathbb{O}^e$ and vanishes at $\hat{\alpha}_0$. Clearly $B_n(z)/\phi_n(z)$ is analytic in $\mathbb{O}^e$ because $\phi_n(z)$ has all its zeros in $\mathbb{O}$. We prove that the other factor is also analytic. Use (4.22) with $f(z) = 1$ to get

$$\Omega_\mu(z)\phi_n(z) + \psi_n(z) = \int D(t,z)[\phi_n(z) + \phi_n(t) - \phi_n(z)]\,d\mathring{\mu}(t)$$

$$= \int D(t,z)\phi_n(t)\,d\mathring{\mu}(t);$$

this is analytic in $\mathbb{O}^e$ as a Cauchy–Stieltjes integral. Moreover, because $D(t, \hat{\alpha}_0) = -1$, we find for $z = \hat{\alpha}_0$ and $n > 0$ that the latter integral equals $-\langle \phi_n, 1\rangle_{\mathring{\mu}} = 0$. For $n = 0$, we note that $D(t, \hat{\alpha}_0) = -1$, so that $\Omega_\mu(\hat{\alpha}_0) = -\int d\mathring{\mu} = -1$. Hence the first relation also holds for $n = 0$ because $\phi_0 = \psi_0 = 1$.

For the second one, we observe that we should show that for $n \geq 1$

$$\frac{1}{\zeta_n(z)}[\Omega_\mu(z)\phi_n^*(z) - \psi_n^*(z)]$$

is analytic in $\mathbb{O}^e$. We use (4.23) with $f = 1/\zeta_n$ to get

$$\frac{\Omega_\mu(z)\phi_n^*(z) - \psi_n^*(z)}{\zeta_n(z)} = \int D(t,z)\frac{\phi_n^*(t)}{\zeta_n(t)}\,d\mathring{\mu}(t),$$

and this is again analytic in $\mathbb{O}^e$ as a Cauchy–Stieltjes integral. Moreover, for $z = \hat{\alpha}_0$, this is equal to $\langle \phi_n^*, \zeta_n\rangle_{\mathring{\mu}} = 0$.

The third result is equivalent with (6.1) and the fourth is a simple consequence of (6.2) since ϕ_n^* has no zeros in $\mathbb{O}$.   □

We can give expressions for the errors of the interpolants $R_n(z)$ of Theorem 6.1.3. We first need the following lemma, which is easily proved by induction.

Lemma 6.1.5. If $D(t, z)$ is the Riesz–Herglotz–Nevanlinna kernel, then the Newton interpolating polynomial for $D(t, z)$ in the points $A_m^0 = \{\alpha_0, \ldots, \alpha_m\}$ plus the error term are given by

$$D(t,z) = 1 + 2\sum_{k=1}^m a_k(t)(z - \alpha_0)\pi_{k-1}^*(z) + 2\frac{a_m(t)}{t-z}(z - \alpha_0)\pi_m^*(z), \quad (6.5)$$

where

$$a_k(t) = \frac{\varpi_0(t)}{\varpi_0(\alpha_0)\pi_k^*(t)}, \quad k = 1, 2, \ldots.$$

We prove next

Lemma 6.1.6. *If $R_n(z) = -P_n(z)/Q_n(z)$ is the rational interpolant to Ω_μ of Theorem 6.1.3, then for all $z \in \mathbb{O} \cup \mathbb{O}^e$, the approximation error $E_n = \Omega_\mu - R_n$ is given by*

$$E_n(z) = \frac{2\varpi_0^*(z)\pi_{n-1}^*(z)\pi_{n-1}(z)}{\varpi_0(\alpha_0)\chi_n(z)} \int \frac{\chi_n(t)\varpi_0(t)}{\pi_{n-1}(t)\pi_{n-1}^*(t)} \frac{d\mathring{\mu}(t)}{t-z}, \quad n \geq 2,$$

where $\chi_n(z) = Q_n(z)\pi_n(z)$.

Proof. By taking $f(z) = \varpi_n(z)/\varpi_{n-1}^*(z) \in \mathcal{L}_{(n-1)*}(\hat{\alpha}_n)$ in Lemma 5.2.2, we find

$$E_n(z) = \frac{\varpi_{n-1}^*(z)}{\varpi_n(z)Q_n(z)} \int D(t,z)Q_n(t)\frac{\varpi_n(t)}{\varpi_{n-1}^*(t)} d\mathring{\mu}(t). \qquad (6.6)$$

We replace $D(t,z)$ by (6.5) with $m = n - 2$, and use orthogonality properties of $Q_n(z)$ to get

$$E_n(z) = \frac{2\varpi_0^*(z)\varpi_{n-1}^*(z)\pi_{n-2}^*(z)}{\varpi_0(\alpha_0)\varpi_n(z)Q_n(z)} \int \frac{\varpi_0(t)\varpi_n(t)}{\varpi_{n-1}^*(t)\pi_{n-2}^*(t)} \frac{Q_n(t)}{t-z} d\mathring{\mu}(t).$$

Since $Q_n(z) = \chi_n(z)/\pi_n(z)$, the proof follows.   □

Note that in the case of $\tau = 0$, and thus $R_n = -\psi_n/\phi_n$, we can use Lemma 4.2.2 instead of Lemma 5.2.2 and choose $f = 1/\varpi_{n-1}^*$ in the first line of the previous proof. We then obtain

Corollary 6.1.7. *If $R_n(z) = -\psi_n(z)/\phi_n(z)$ is the rational interpolant to Ω_μ of Theorem 6.1.1, then for all $z \in \mathbb{O} \cup \mathbb{O}^e$, the approximation error $E_n = \Omega_\mu - R_n$ is given by*

$$E_n(z) = \frac{2\varpi_0^*(z)\pi_{n-1}^*(z)\pi_n(z)}{\varpi_0(\alpha_0)\chi_n(z)} \int \frac{\chi_n(t)\varpi_0(t)}{\pi_n(t)\pi_{n-1}^*(t)} \frac{d\mathring{\mu}(t)}{t-z}, \quad n \geq 1,$$

where $\chi_n(z) = \phi_n(z)\pi_n(z)$.

We transform the error expression in Lemma 6.1.6 somewhat further.

Lemma 6.1.8. For $n \geq 2$, the error $E_n(z)$ of Lemma 6.1.6 can also be expressed as

$$E_n(z) = \frac{2\varpi_0^*(z)\varpi_0(z)\pi_{n-1}^*(z)\pi_{n-1}(z)}{\varpi_0(\alpha_0)\chi_n^2(z)}\left[\int \frac{\chi_n^2(t)\,d\mathring{\mu}(t)}{\pi_{n-1}(t)\pi_{n-1}^*(t)(t-z)} + \delta_n\right],$$

$$z \in \mathbb{O} \cup \mathbb{O}^e, \quad (6.7)$$

where $\delta_n = c_n \int Q_n(t)\varpi_n(t)\,d\mathring{\mu}(t)$, with $Q_n(z) = \chi_n(z)/\pi_n(z)$ and $\chi_n(z) = c_n z^n + \cdots$.

Proof. First we show that in the integral of Lemma 6.1.6 we can bring the factor $\varpi_0(t)$ outside. This is obvious for $\mathbb{O} = \mathbb{D}$ because then $\varpi_0(t) = 1$. For $\mathbb{O} = \mathbb{U}$, we have $\varpi_0(t) = t + \mathbf{i}$. Now

$$\int \frac{(t+\mathbf{i})\chi_n(t)\,d\mathring{\mu}(t)}{\pi_{n-1}^*(t)\pi_{n-1}(t)(t-z)} = \int \frac{(t+\mathbf{i})(t-\bar{\alpha}_n)Q_n(t)\,d\mathring{\mu}(t)}{\pi_{n-1}^*(t)(t-z)}$$

$$= \int \frac{(t-\bar{\alpha}_n)Q_n(t)\,d\mathring{\mu}(t)}{\pi_{n-1}^*(t)}$$

$$+ (z+\mathbf{i})\int \frac{(t-\bar{\alpha}_n)Q_n(t)\,d\mathring{\mu}(t)}{\pi_{n-1}^*(t)(t-z)}.$$

The first integral is zero by the para-orthogonality of Q_n. Thus

$$\int \frac{\varpi_0(t)\chi_n(t)\,d\mathring{\mu}(t)}{\pi_{n-1}^*(t)\pi_{n-1}(t)(t-z)} = \varpi_0(z)\int \frac{\chi_n(t)\,d\mathring{\mu}(t)}{\pi_{n-1}^*(t)\pi_{n-1}(t)(t-z)}.$$

It remains to transform the integral in the right-hand side. Since $\chi_n(z) = c_n z^n + \cdots$,

$$\frac{\chi_n(t) - \chi_n(z)}{t-z} = c_n t^{n-1} + P(t), \quad P \in \mathcal{P}_{n-2},$$

so that

$$\frac{\chi_n(t) - \chi_n(z)}{\pi_{n-1}^*(t)\pi_{n-1}(t)(t-z)} = \frac{c_n t^{n-1} + P(t)}{\pi_{n-1}(t)\pi_{n-1}^*(t)} = \frac{\varpi_n(t)[c_n t^{n-1} + P(t)]}{\pi_n(t)\pi_{n-1}^*(t)}.$$

Therefore,

$$\chi_n(t)\left[\frac{\chi_n(t) - \chi_n(z)}{\pi_{n-1}(t)\pi_{n-1}^*(t)(t-z)}\right] = Q_n(t)\frac{\varpi_n(t)[c_n t^{n-1} + P(t)]}{\pi_{n-1}^*(t)}$$

and also

$$\int \chi_n(t) \left[\frac{\chi_n(t) - \chi_n(z)}{\pi_{n-1}(t)\pi_{n-1}^*(t)(t-z)} \right] d\mathring{\mu}(t) = \int \frac{Q_n(t)c_n t^{n-1} \varpi_n(t)}{\pi_{n-1}^*(t)} d\mathring{\mu}(t)$$
$$+ \int Q_n(t) \frac{P(t)\varpi_n(t)}{\pi_{n-1}^*(t)} d\mathring{\mu}(t).$$

The second term in the right-hand side is zero by the para-orthogonality of Q_n. Since $t^{n-1} = \pi_{n-1}^*(t) + \sum_{j=0}^{n-2} \gamma_j \pi_j^*(t)$, we find, again by the para-orthogonality of Q_n, that the first term is

$$\int \frac{Q_n(t)c_n t^{n-1} \varpi_n(t)}{\pi_{n-1}^*(t)} d\mathring{\mu}(t) = c_n \int Q_n(t)\varpi_n(t) d\mathring{\mu}(t).$$

If we call this δ_n, then we have by Lemma 6.1.6 the desired form for the error. □

Remark. The presence of the "strange" term δ_n in (6.7) is due to the deficiencies in the orthogonality properties of Q_n. For a Stieltjes function with a measure on the real line, true orthogonal instead of para-orthogonal functions are used and this term will not appear. Compare with, for example, Ref. [193].

As an introduction to the next section, we derive the following result.

The interpolation properties imply that the following theorem holds true. (Note that the μ_n in this theorem is different from the discrete measure defined in Theorem 5.3.1.)

Theorem 6.1.9. *Let ϕ_n be the orthonormal basis functions of $\mathcal{L}_n$ with respect to the measure $\mathring{\mu}$. Define the absolutely continuous measure $\mathring{\mu}_n$ by*

$$d\mathring{\mu}_n(t) = \frac{|\varpi_0(t)|^2 P(t, \alpha_n)}{|\phi_n(t)|^2} d\mathring{\lambda}(t) = \frac{P(t, \alpha_n)}{|\phi_n(t)|^2} d\lambda(t),$$

where P is the Poisson kernel. Then on $\mathcal{L}_n$, the inner product with respect to $\mathring{\mu}_n$ and $\mathring{\mu}$ is the same: $\langle \cdot, \cdot \rangle_{\mathring{\mu}} = \langle \cdot, \cdot \rangle_{\mathring{\mu}_n}$.

Proof. We prove first that the norm of ϕ_n is the same. This is obvious since

$$\|\phi_n\|_{\mathring{\mu}_n}^2 = \int \frac{P(t, \alpha_n)|\phi_n|^2}{|\phi_n|^2} d\lambda = 1 = \|\phi_n\|_{\mathring{\mu}}^2.$$

Next we show that $\langle \phi_n, \phi_k \rangle_{\mathring{\mu}}$ and $\langle \phi_n, \phi_k \rangle_{\mathring{\mu}_n}$ is the same for $k < n$. They are both zero.

$$\langle \phi_n, \phi_k \rangle_{\mathring{\mu}_n} = \int \frac{\phi_{k*}(t)}{\phi_{n*}(t)} P(t, \alpha_n) \, d\lambda(t)$$

$$= \int \frac{\phi_k^*(t) B_{n \backslash k}(t)}{\phi_n^*(t)} P(t, \alpha_n) \, d\lambda(t).$$

Since ϕ_n^* has its zeros in $\mathbb{O}^e$, we know that $B_{n \backslash k} \phi_k^* / \phi_n^*$ is analytic in the closed domain $\mathbb{O} \cup \partial \mathbb{O}$ and then we may apply Poisson's formula, which gives zero because $B_{n \backslash k}(\alpha_n) = 0$. Of course also $\langle \phi_n, \phi_k \rangle_{\mathring{\mu}} = 0$. Hence ϕ_n is a function of norm 1 and orthogonal to $\mathcal{L}_{n-1}$ both with respect to $\mathring{\mu}_n$ and with respect to $\mathring{\mu}$. By Theorem 4.1.6 this ϕ_n will uniquely define all the previous ϕ_k, provided they are normalized properly with $\phi_k^*(\alpha_k) = \kappa_k > 0$. Thus the orthonormal system in $\mathcal{L}_n$ for $\mathring{\mu}_n$ and for $\mathring{\mu}$ is the same: $\langle \phi_k, \phi_i \rangle_{\mathring{\mu}} = \langle \phi_k, \phi_i \rangle_{\mathring{\mu}_n} = \delta_{ki}$. Since every element from $\mathcal{L}_n$ can be expressed as a linear combination of the ϕ_k, it also holds that $\langle f, g \rangle_{\mathring{\mu}} = \langle f, g \rangle_{\mathring{\mu}_n}$ for every f and $g \in \mathcal{L}_n$. $\quad\square$

The previous theorem was proved for orthogonal polynomials in Ref. [87, p. 199].

6.2. Measures and interpolation

In this section, we want to show how interpolation properties of positive real functions are practically equivalent with the equality of inner products in some $\mathcal{L}_n$ spaces. In Section 4.2 we already saw that $\Omega_n = \psi_n^* / \phi_n^*$ was in $\mathcal{C}$ and interpolated $\Omega_\mu \in \mathcal{C}$ in the points of $A_n^0 = \{\alpha_0, \alpha_1, \ldots, \alpha_n\}$ in Hermite sense. By the latter we mean that if in A_n^0 some of the α_k are repeated, then the interpolation conditions satisfied refer to function values, first derivatives, etc. In the same section, it was shown that $d\mathring{\mu}_n = |\varpi_0|^2 P(\cdot, \alpha_n) / |\phi_n|^2 d\mathring{\lambda} = \mathrm{Re}\, \Omega_n d\mathring{\lambda}$ and $d\mathring{\mu}$ are two measures defining the same inner product in $\mathcal{L}_n$. That this is not a coincidence will be shown in this section. See also Section 2.1 for results giving a relation between interpolation and equality of the inner product.

We can already conclude from Theorem 2.1.3 that the following holds:

Theorem 6.2.1. *Let $\mathring{\mu}$ and $\mathring{\nu}$ be two normalized measures on $\overline{\partial \mathbb{O}}$ with associated positive real functions Ω_μ and Ω_ν respectively:*

$$\Omega_\mu(z) = \int D(t, z) \, d\mathring{\mu}(t) \quad and \quad \Omega_\nu(z) = \int D(t, z) \, d\mathring{\nu}(t).$$

Then the equality of the inner products $\langle \cdot, \cdot \rangle_{\mathring{\mu}} = \langle \cdot, \cdot \rangle_{\mathring{\nu}}$ holds in $\mathcal{L}_n$ if and only of Ω_μ and Ω_ν mutually interpolate (in Hermite sense) in the point set

$A_n^0 = \{\alpha_0, \dots, \alpha_n\}$. *This means that*

$$\frac{\Omega_\mu(z) - \Omega_\nu(z)}{\pi_n^*(z)} = g(z) \in H(\mathbb{O}) \quad and \quad g(\alpha_0) = 0,$$

where $\pi_n^*(z) = \prod_{k=1}^n (z - \alpha_k)$.

Proof. Indeed, when the functions in $\mathcal{L}_n$ are written in the appropriate basis $\{w_k : k = 0, \dots, n\}$ (see Section 2.1), then the metric or Gram matrix G_n of the space $\mathcal{L}_n$ is a matrix that is completely defined in terms of $\Omega^{(k)}(\beta_l)$, and by the previous interpolation conditions, these are precisely the values that for Ω_μ and Ω_ν coincide.   □

We want to give a more general result where we want to replace α_0 by an arbitrary point $w \in \mathbb{O}$.

We shall therefore first prove a simple lemma.

Lemma 6.2.2. Let $\hat{\mu}$ be a normalized positive measure on $\mathbb{T}$ and let the positive real function Ω_μ be associated with it by (1.27) with $c = 0$. This means that $\Omega_\mu(0) > 0$. Define also the positive real function $\Omega_\mu(z, w)$, with $w \in \mathbb{D}$ some parameter, by

$$\Omega_\mu(z, w) = \int \frac{D(t, z)}{P(t, w)} d\mu(t) + c,$$

where $P(t, w)$ is the Poisson kernel, and the constant c is given by

$$c = -\mathbf{i} \int \frac{\operatorname{Re} D(t, w)}{\operatorname{Im} D(t, w)} = c_w[\overline{w}\mu_{-1} - w\mu_1] \in \mathbf{i}\mathbb{R},$$

where $c_w = 1/(1 - |w|^2)$, and $\mu_k = \overline{\mu}_{-k} = \int t^{-k} d\mu(t)$ are the moments of μ. Then the relation between $\Omega_\mu(z)$ and $\Omega_\mu(z, w)$ is given by

$$\Omega_\mu(z, w) = c_w[(z - w)(z^{-1} - \overline{w})\Omega_\mu(z) + (z^{-1}w - \overline{w}z)\mu_0]$$

$$= \frac{\Omega_\mu(z)}{P(z, w)} + c_w(z^{-1}w - \overline{w}z)\mu_0$$

for $z \neq 0$. Moreover, $\Omega_\mu(z, 0) = \Omega_\mu(z)$ and $\mu_0 = \Omega_\mu(0) = \Omega_\mu(w, w)$, for any $w \in \mathbb{D}$, also for $w = 0$.

Proof. First, we recall that the real part of the Riesz–Herglotz–Nevanlinna kernel is given by the Poisson kernel, that is, for $t \in \mathbb{T}$

$$\operatorname{Re} D(t, w) = \frac{1 - |w|^2}{(t - w)(t^{-1} - \overline{w})},$$

whereas it can be computed that

$$\text{Im } D(t, w) = -\mathbf{i}\frac{wt^{-1} - \overline{w}t}{(t - w)(t^{-1} - \overline{w})}.$$

Hence

$$-\mathbf{i}\int\frac{\text{Im } D(t, w)}{\text{Re } D(t, w)}\, d\mu(t) = c_w\int(\overline{w}t - wt^{-1})\, d\mu(t) = c_w(\overline{w}\mu_{-1} - w\mu_1),$$

which is indeed in $\mathbf{i}\mathbb{R}$. Next we compute

$$\frac{D(t, z)}{P(t, w)} - \frac{D(t, z)}{P(z, w)} = \frac{t + z}{t - z}c_w[(t - w)(t^{-1} - \overline{w}) - (z - w)(z^{-1} - \overline{w})]$$

$$= c_w(t + z)\left[\frac{w}{tz} - \overline{w}\right]$$

$$= c_w[(wz^{-1} - \overline{w}z) + (wt^{-1} - \overline{w}t)].$$

After integration, this results in

$$\Omega_\mu(z, w) = \frac{\Omega_\mu(z)}{P(z, w)} + c_w(wz^{-1} - \overline{w}z)\mu_0,$$

and this proves the lemma for $z \neq 0$.

Since for $w = 0$, we have $P(t, 0) = 1$ and $D(t, 0) = 1$, so that $c = 0$, then we get $\Omega_\mu(z, 0) = \Omega_\mu(z)$. However, $1/P(w, w) = 0$, so that $\Omega_\mu(w, w)$ is indeed equal to $\mu_0 = 1 = \Omega_\mu(0)$. □

For the case of the half plane, one can formulate a similar result.

Lemma 6.2.3. Let $\mathring{\mu}$ be a normalized positive measure on $\overline{\mathbb{R}}$ and let the positive real function Ω_μ be associated with it by the Nevanlinna transform $\Omega(z) = \int D(t, z)\, d\mathring{\mu}(t)$. This means that $\Omega_\mu(\mathbf{i}) = 1 > 0$. Define also the positive real function $\Omega_\mu(z, w)$ with $w \in \mathbb{U}$ some parameter, by

$$\Omega_\mu(z, w) = \int\frac{D(t, z)}{P(t, w)}\frac{d\mathring{\mu}(t)}{(1 + t^2)} + c,$$

with $P(t, w)$ the Poisson kernel and

$$c = -\mathbf{i}\int\frac{\text{Im } D(t, w)}{\text{Re } D(t, w)}\frac{d\mathring{\mu}(t)}{(1 + t^2)} \in \mathbf{i}\mathbb{R}.$$

If $\int d\mathring{\mu} = c_0 (=1)$, then the following relation between $\Omega_\mu(z)$ and $\Omega_\mu(z, w)$ holds for $z \neq \mathbf{i}$:

$$\Omega_\mu(z, w) = \frac{\Omega_\mu(z)}{P(z, w)(1 + z^2)} + \frac{c_w[(z^2 - 1)\,\mathrm{Re}\,w + z(1 - |w|^2)]}{1 + z^2}\mu_0,$$

with $c_w = -\mathbf{i}/\mathrm{Im}\,w$. Furthermore, $\Omega_\mu(z, \mathbf{i}) = \Omega_\mu(z)$ and $\mu_0 = 1 = \Omega_\mu(\mathbf{i}) = \Omega_\mu(w, w)$ for all $w \in \mathbb{U}$, also for $w = \mathbf{i}$.

Proof. The calculations are long and technical, but they are along the lines of the previous proof. We only give a sketch.

We first note that

$$\frac{D(t, z)}{P(t, w)}\frac{1}{1 + t^2} = \frac{D(t, z)}{P(z, w)}\frac{1}{1 + z^2} + c_w A(t, z, w),$$

with

$$A(t, z, w) = \frac{1 + tz}{(1 + t^2)(1 + z^2)}[(tz - 1)(w + \overline{w}) + (t + z)(1 - |w|^2)].$$

Integration gives

$$\Omega_\mu(z, w) = \frac{\Omega_\mu(z)}{P(z, w)(1 + z^2)} + I(z, w) + c,$$

where

$$I(z, w) = c_w \int A(t, z, w)\,d\mathring{\mu}(t).$$

A long computation then yields

$$I(z, w) + c = \frac{c_w}{1 + z^2}[(z^2 - 1)\,\mathrm{Re}\,w + z(1 - |w|^2)]\mu_0,$$

from which the theorem follows. Note that the latter formula is just μ_0 if $z = w$.

For $w = \mathbf{i}$ we have $P(t, \mathbf{i}) = (1 + t^2)^{-1}$ and $D(t, \mathbf{i}) = 1$ so that $c = 0$ and thus $\Omega_\mu(z, \mathbf{i}) = \Omega_\mu(z)$. Hence we get for all $w \in \mathbb{U}$ that $\mu_0 = 1 = \Omega_\mu(w, w) = \Omega_\mu(\mathbf{i})$.   □

We note that in the two previous lemmas, the constant c depends on w, and it is chosen such that $\Omega_\mu(w, w) = \mu_0 = 1$ in both cases. Also, it is interesting to focus attention on the fact that the expression

$$\Omega_\mu(z, w) - \frac{\Omega_\mu(z)}{P(z, w)}$$

depends on z and w, but the measure enters only via the moment μ_0. Thus, it is independent of the actual measure in the sense that it depends only on its normalization.

The previous lemmas give the following corollary.

Corollary 6.2.4. *Suppose $\mathring{\mu}$ and $\mathring{\nu}$ are two normalized measures on $\overline{\partial \mathbb{O}}$ to which we associate the positive real functions*

$$\Omega_\mu(z) = \int D(t, z) \, d\mathring{\mu}(t) \quad and \quad \Omega_\nu(z) = \int D(t, z) \, d\mathring{\nu}(t)$$

and also

$$\Omega_\mu(z, w) = \int \frac{D(t, z)}{P(t, w)} \frac{d\mathring{\mu}(t)}{|\varpi_0(t)|^2} + c_\mu$$

and

$$\Omega_\nu(z, w) = \int \frac{D(t, z)}{P(t, w)} \frac{d\mathring{\nu}(t)}{|\varpi_0(t)|^2} + c_\nu,$$

with proper normalizing constants c_μ and c_ν to make $\Omega_\mu(w, w) = 1 = \Omega_\nu(w, w)$ and where $P(z, w)$ is the Poisson kernel. Then

$$\frac{\Omega_\mu(z) - \Omega_\nu(z)}{\pi_n^*(z)} = g(z) \in H(\mathbb{O}) \quad and \quad g(\alpha_0) = 0 \tag{6.8}$$

if and only if

$$\frac{\Omega_\mu(z, w) - \Omega_\nu(z, w)}{\pi_n^*(z)} = g_w(z) \in H(\mathbb{O}) \quad and \quad g_w(w) = 0. \tag{6.9}$$

In other words, $\Omega_\mu(z)$ is a Hermite interpolant for $\Omega_\nu(z)$ in the point set $A_n^0 = \{\alpha_0, \alpha_1, \dots, \alpha_n\}$ if and only if $\Omega_\mu(z, w)$ is a Hermite interpolant for $\Omega_\nu(z, w)$ in the point set $A_n^w = \{w, \alpha_1, \dots, \alpha_n\}$. By Theorem 6.2.1, this happens if and only if equality of the inner products $\langle \cdot, \cdot \rangle_{\mathring{\mu}} = \langle \cdot, \cdot \rangle_{\mathring{\nu}}$ holds in $\mathcal{L}_n$.

Proof. It follows from the previous lemmas that

$$\Omega_\mu(z, w) - \Omega_\nu(z, w) = \frac{\Omega_\mu(z) - \Omega_\nu(z)}{P(z, w)\varpi_0(z)\varpi_{0*}(z)}.$$

Thus, if (6.8) holds, then

$$\frac{\Omega_\mu(z, w) - \Omega_\nu(z, w)}{\pi_n^*(z)} = g_w(z),$$

within the case of the disk

$$g_w(z) = \frac{(z-w)(1-\overline{w}z)}{1-|w|^2} \frac{g(z)}{z}.$$

Clearly, $g_w \in H(\mathbb{D})$ because $g(0) = 0$. For the half plane,

$$g_w(z) = \frac{(z-w)(z-\overline{w})}{\mathrm{Im}\, w} \frac{g(z)}{(z-\mathbf{i})(z+\mathbf{i})}$$

and also here $g_w \in H(\mathbb{U})$ because $g(\mathbf{i}) = 0$. In both cases, the factor $z - w$ also gives $g_w(w) = 0$.

The converse is also true and the result follows.   □

With the previous results on interpolation, it is now easy to prove that the functions of the second kind ψ_n associated with the orthonormal rational functions ϕ_n with respect to the measure μ are also orthogonal rational functions in $\mathcal{L}_n$ with respect to some associated measure ν.

First note that since $\Omega_\mu \in C$ has a positive real part in $\mathbb{O}$, and hence no zeros, its inverse is also in C. This means that there is a uniquely defined measure ν such that

$$\frac{1}{\Omega_\mu(z)} = \Omega_\nu(z) = \int D(t,z)\, d\mathring{\nu}(t).$$

By the normalization $\Omega_\mu(\alpha_0) = 1$, hence $\Omega_\nu(\alpha_0) = 1$, we have $d\mathring{\nu}(t) = 1$. The positive real function $\psi_n^*(z)/\phi_n^*(z)$ is also invertible in $\mathbb{O}$, and by Theorem 6.1.4, which says

$$[B_n(z)]^{-1}\left[\Omega_\mu(z) - \frac{\psi_n^*(z)}{\phi_n^*(z)}\right] = g(z) \in H(\mathbb{O}) \quad \text{and} \quad g(\alpha_0) = 0,$$

it follows that

$$[B_n(z)]^{-1}\left[\Omega_\nu(z) - \frac{\phi_n^*(z)}{\psi_n^*(z)}\right] = \tilde{g}(z) \in H(\mathbb{O}) \quad \text{and} \quad \tilde{g}(\alpha_0) = 0,$$

where $\tilde{g}(z) = -g(z)\phi_n^*(z)/[\Omega_\mu(z)\psi_n^*(z)]$. Thus the positive real function $\phi_n^*(z)/\psi_n^*(z)$ interpolates $\Omega_\nu(z)$ in the points $A_n^0 = \{\alpha_0, \ldots, \alpha_n\}$. By Theorem 6.2.1 and the determinant formula, it then follows that in $\mathcal{L}_n$ the inner product with respect to $\mathring{\nu}$ and with respect to the measure $\mathring{\nu}$ defined by

$$d\mathring{\nu}(t) = \frac{P(t, \alpha_n)}{|\psi_n^*(t)|^2}\, d\lambda(t)$$

are the same. We can repeat the proof of Theorem 6.1.9 with ϕ replaced by ψ and find

Theorem 6.2.5. *Let* $\Omega_\mu(z) = \int D(t, z) \, d\mathring{\mu}(t)$ *and define* $\Omega_\nu(z) = 1/\Omega_\mu(z)$ *with uniquely defined Riesz–Herglotz–Nevanlinna measure* ν. *Then the functions of the second kind* ψ_n *with respect to the measure* μ *are orthogonal rational functions for* $\mathcal{L}_n$ *with respect to the measure* ν.

6.3. Interpolation properties for the kernels

In Section 6.1, we discussed rational interpolants for Ω_μ, the positive real function associated with the measure μ. These rational interpolants were constructed from the orthogonal or para-orthogonal functions and the corresponding functions of the second kind.

In this section we shall give rational interpolants for $\Omega = \Omega_\mu$, which are based on the reproducing kernels for $\mathcal{L}_n$. By the results of the previous section, we also find approximants of the measure μ, and hence also of the spectral factor σ defined in (1.43). Obviously, if we make use of the spectral factor, then we assume it exists and thus that Szegő's condition $\log \mu' \in L_1(\mathring{\lambda})$ is satisfied.

The following theorem is such an interpolation result for the spectral factor. It is a consequence of Theorems 2.3.8 and 2.1.6. It says that the normalized kernel for $\mathcal{L}_n$ interpolates $1/\sigma$ in the points A_n^0.

Theorem 6.3.1. *Let* $K_n(z, w)$ *denote the normalized reproducing kernel for* $\mathcal{L}_n$ *w.r.t. the measure* $\mathring{\mu}$. *Suppose that* $\log \mu' \in L_1(d\mathring{\lambda})$ *and let* $\sigma(z)$ *be the outer spectral factor of* μ *such that* $\sigma(\alpha_0) > 0$. *Then*

$$\frac{K_n(z, \alpha_0) - 1/\sigma(z)}{\tilde{B}_n(z)} \in H(\mathbb{O}),$$

where $\tilde{B}_n = \zeta_0 B_n$, $n \geq 0$, *with* B_n *the finite Blaschke products.*

Proof. By Theorem 2.3.8, we know that $k_n(z, w)$ is the projection in $\mathring{H}_2$ of the Szegő kernel s_w (see (2.18)). Theorem 2.1.6 then says that $k_n(z, w) = s_w(z)$ for $z \in A_n^0 = \{\alpha_0, \ldots, \alpha_n\}$ (interpolation in Hermite sense). By setting $z = w = \alpha_0$, we obtain

$$k_n(\alpha_0, \alpha_0) = K_n(\alpha_0, \alpha_0)^2 = |\sigma(\alpha_0)|^{-2}.$$

Hence, because $k_n(z, w) = K_n(z, w) K_n(w, w)$, the result follows. $\square$

We next prove the following theorem, which can be found in Ref. [61, p. 48] or [62, p. 654] in the case of the disk.

Theorem 6.3.2. *Let $K_n(z, w)$ be the normalized reproducing kernel of $\mathcal{L}_n$ w.r.t.*
$\mathring{\mu}$ and define the absolutely continuous measure $\mathring{\mu}$ on the boundary $\overline{\partial\mathbb{O}}$ by
$d\mathring{\mu}_n(t) = |K_n(t, \alpha_0)|^{-2} d\mathring{\lambda}(t)$, $t \in \overline{\partial\mathbb{O}}$. Then in the space $\mathcal{L}_n$ we have equality
of the inner products $\langle \cdot, \cdot \rangle_{\mathring{\mu}_n} = \langle \cdot, \cdot \rangle_{\mathring{\mu}}$

Proof. We only have to show that for all $f \in \mathcal{L}_n$ and arbitrary $w \in \mathbb{D}$

$$\langle f, K_n(\cdot, w)\rangle_{\mathring{\mu}_n} = \langle f, K_n(\cdot, w)\rangle_{\mathring{\mu}} = f(w)/K_n(w, w), \qquad (6.10)$$

because the $\{K_n(\cdot, w_i)\}$, with w_i, $i = 0, \ldots, n$ some set of distinct points in $\mathbb{O}$,
form a basis for $\mathcal{L}_n$.

For the proof we shall need the spaces $\mathcal{L}_n$ where we let all the points α_k
coincide at α_0. For the circle, this corresponds to the polynomials (= rationals
with all poles at $\hat{\alpha}_0 = \infty$), but for the real line, these are rational functions
with all poles in $\hat{\alpha}_0 = -\mathbf{i}$. Thus we consider the spaces ($\varpi_0 = 1$ for $\mathbb{T}$ and
$\varpi_0(z) = z + \mathbf{i}$ for $\mathbb{R}$)

$$\mathcal{L}_0^n = \{p_n(z)/[\varpi_0(z)]^n : p_n \in \mathcal{P}_n\}.$$

Note that any $f = p/\pi_n \in \mathcal{L}_n$ can be written as

$$f(z) = \frac{p(z)}{\varpi_0(z)^n} \cdot \frac{\varpi_0(z)^n}{\pi_n(z)}.$$

The first factor is in $\mathcal{L}_0^n$ and the second factor is incorporated in the measure $\mathring{\pi}$
defined by

$$d\mathring{\pi}(t) = \frac{|\varpi_0(t)|^{2n}}{|\pi_n(t)|^2} d\mathring{\mu}(t).$$

Let $\{p_k\}$ be a system of orthonormal functions for $\mathcal{L}_0^n$ with respect to the measure
$\mathring{\pi}$; then

$$k_\pi(z, w) = \sum_{k=0}^{n} p_k(z)\overline{p_k(w)}$$

is a reproducing kernel for the space $\mathcal{L}_0^n$ with measure $\mathring{\pi}$. Thus for any function
$q \in \mathcal{L}_0^n$, it holds that

$$\langle q, k_\pi(\cdot, w)\rangle_{\mathring{\pi}} = q(w).$$

Hence, multiplying by $\varpi_0(w)^n/\pi_n(w)$ and extracting again $d\mathring{\mu}$ from $d\mathring{\pi}$, we
get

$$\left\langle \frac{q(\cdot)[\varpi_0(\cdot)]^n}{\pi_n(\cdot)}, \frac{k_\pi(\cdot, w)[\varpi_0(\cdot)\overline{\varpi_0(w)}]^n}{\pi_n(\cdot)\overline{\pi_n(w)}} \right\rangle_{\mathring{\mu}} = \frac{q(w)[\varpi_0(w)]^n}{\pi_n(w)},$$

which means that

$$\frac{[\varpi_0(z)\overline{\varpi_0(w)}]^n}{\pi_n(z)\overline{\pi_n(w)}}k_\pi(z, w) = k_n(z, w) \tag{6.11}$$

is the reproducing kernel for $\mathcal{L}_n$ w.r.t. $\mathring{\mu}$.

However, if the p_n are chosen such that $p_n^*(\alpha_0)\varpi_0(\alpha_0)^n/\pi_n(\alpha_0) > 0$, then

$$K_n(z, \alpha_0) = p_n^*(z)[\varpi_0(z)]^n/\pi_n(z). \tag{6.12}$$

To see this, note that by the Christoffel–Darboux relation obtained from Theorem 3.1.3 by setting all $\alpha_i = \alpha_0$ we have

$$k_\pi(z, w) = \frac{p_n^*(z)\overline{p_n^*(w)} - \zeta_0(z)\overline{\zeta_0(w)}p_n(z)\overline{p_n(w)}}{1 - \zeta_0(z)\overline{\zeta_0(w)}},$$

so that $k_\pi(z, \alpha_0) = p_n^*(z)\overline{p_n^*(\alpha_0)}$. Using the relation (6.11) between k_π and k_n, we get after normalization $K_n(z, w) = k_n(z, w)/\sqrt{k_n(w, w)}$ the expression (6.12).

With these tools, we can now see that (6.10) will be proved if we can show that

$$\left\langle \frac{p(\cdot)\varpi_0(\cdot)^n}{\pi_n(\cdot)}, k_n(\cdot, w) \right\rangle_{\mathring{\mu}_n} \frac{\pi_n(w)}{\varpi_0(w)^n} = p(w)$$

for arbitrary $p \in \mathcal{L}_0^n$ and $d\mathring{\mu}(t) = |K_n(t, \alpha_0)|^{-2}d\mathring{\lambda}(t)$. The left-hand side can be written explicitly as

$$\int \frac{p(t)}{|p_n^*(t)|^2}\left[\frac{\zeta_0(t)\overline{p_n^*(t)}p_n^*(w) - \zeta_0(w)\overline{p_n(t)}p_n(w)}{\zeta_0(t) - \zeta_0(w)}\right]d\mathring{\lambda}(t) = I_1 - I_2,$$

where

$$I_1 = p_n^*(w)\int \frac{p(t)}{p_n^*(t)}\frac{\zeta_0(t)}{\zeta_0(t) - \zeta_0(w)}\,d\mathring{\lambda}(t),$$

$$I_2 = \zeta_0(w)p_n(w)\int \frac{p(t)}{p_n(t)}\frac{1}{\zeta_0(t) - \zeta_0(w)}\,d\mathring{\lambda}(t)$$

since $|p_n^*(t)|^2 = |p_n(t)|^2$ for $t \in \mathbb{T}$. Now we can use in the case of the disk that

$$\frac{\zeta_0(t)\,d\mathring{\lambda}(t)}{\zeta_0(t) - \zeta_0(w)} = \frac{t\,d\lambda(t)}{t - w} = C(t, w)\,d\lambda(t),$$

where $C(t, w) = t/(t - w)$ is the Cauchy kernel for the disk. Similarly, we find for the case of the half plane that

$$\frac{\zeta_0(t)\, d\mathring{\lambda}(t)}{\zeta_0(t) - \zeta_0(w)} = C(t, w)\frac{w + \mathbf{i}}{t + \mathbf{i}}\, d\lambda(t),$$

where $C(t, w) = [2\mathbf{i}(t - w)]^{-1}$ is the Cauchy kernel for the half plane. Thus we find that in both cases $I_1 = p(w)$ and $I_2 = 0$ by the Cauchy formula and the fact that $p_n^*(z) \neq 0$ in $\overline{\mathbb{O}}$ while $p_n(z) \neq 0$ in $\overline{\mathbb{O}^e}$.   □

Note that for the polynomial case on the circle (i.e., when all $\alpha_i = 0$), then $d\mathring{\mu}_n = |\phi_n^*(z)|^{-2}d\lambda = |\phi_n(z)|^{-2}d\lambda$; a theorem in this style can be found in Ref. [87, p. 198].

From the previous result and the theorems from Section 6.2 we can now find the interpolation property.

Theorem 6.3.3. *Let us define*

$$\Omega_\mu(z) = \int D(t, z)\, d\mu(t) \quad \text{and} \quad \Omega_n(z) = \int D(t, z)\, d\mathring{\mu}_n(t),$$

with $D(t, z)$ the Riesz–Herglotz–Nevanlinna kernel and with $d\mathring{\mu}_n(z)] = |K_n(z, \alpha_0)|^{-2}d\mathring{\lambda}(z)$ as defined in Theorem 6.3.2. Then

$$\Omega_n(z) = \frac{L_n(z, \alpha_0)}{K_n(z, \alpha_0)}, \tag{6.13}$$

with $K_n(t, z)$ the normalized kernel and $L_n(t, z)$ the associated function as defined in (3.18). Furthermore (recall $\mu' = d\mu/d\lambda = d\mathring{\mu}/d\mathring{\lambda}$)

$$\mu'_n = \frac{1}{2}[\Omega_n + \Omega_{n*}] = \frac{1}{K_n(z, \alpha_0)K_{n*}(z, \alpha_0)} \tag{6.14}$$

(which is obviously positive on $\partial\mathbb{O}$) has a spectral factor given by

$$\sigma_n(z) = 1/K_n(z, \alpha_0). \tag{6.15}$$

Moreover, the function g defined by

$$\frac{\Omega_\mu - \Omega_n}{\tilde{B}_n} = g, \qquad \tilde{B}_n = \zeta_0 B_n \tag{6.16}$$

is analytic in $\mathbb{O}$.

Proof. By Theorem 3.3.3, we get for $t \in \partial \mathbb{O}$,

$$\frac{1}{|K_n(t, \alpha_0)|^2} = \text{Re} \left[\frac{L_n(t, \alpha_0)}{K_n(t, \alpha_0)} \right] = \text{Re}\, \Omega_n(t)$$

while $\Omega_n \in \mathcal{C}$. Thus Ω_n of (6.13) is the class $\mathcal{C}$ function associated with the measure μ_n as follows from (1.36). Hence $1/K_n(z, \alpha_0)$ is outer in H_2, so that also (6.15) follows and (6.16) is a consequence of the previous theorem. □

The interpolation result of the last theorem states that Ω_n is a *partial multipoint Padé approximant* of Ω_μ in the points of $A_n^0 = \{\alpha_0, \alpha_1, \ldots, \alpha_n\}$. It is only a *partial* or multipoint Padé-*type* interpolant since it is of degree type (n/n) whereas only $n + 1$ interpolation conditions are satisfied. Because $\Omega_\mu - \Omega_n = \zeta_0 B_n g$, with g analytic in $\mathbb{O}$, we find by taking the substar conjugate that $\Omega_{\mu*} - \Omega_{n*} = \zeta_0^{-1} B_{n*} g_*$, with g_* analytic in $\mathbb{O}^e$. Summing up gives for $\mu_n' = [K_n(z, \alpha_0) K_{n*}(z, \alpha_0)]^{-1}$

$$\mu' - \mu_n' = \zeta_0 B_n g + \zeta_0^{-1} B_{n*} g_*.$$

In the case of the disk, this generalizes the notion of Laurent–Padé approximant [29], since in the case where all $\alpha_i = 0$, μ_n' is the inverse of a Laurent polynomial of degree n, which fits the expansion of μ' from $-n$ till $+n$. In the present case, μ_n' takes the form

$$\mu_n' = \frac{1}{K_n(z, \alpha_0) K_{n*}(z, \alpha_0)} = \frac{\pi_n(z) \pi_{n*}(z)}{p_n(z) p_{n*}(z)},$$

where $p_n(z) = K_n(z, \alpha_0) \pi_n(z) \in \mathcal{P}_n$ is a polynomial. Thus μ_n' is the ratio of two Laurent polynomials of degree n and fits only $2n + 2$ interpolation conditions. It is a *partial* Laurent–Padé approximant since by fixing the interpolation points α_i, one fixes the zeros and hence the numerator of the approximant.

We could repeat here the same kind of arguments as we used at the end of Section 6.2 concerning the orthogonality of the functions of the second kind with respect to the associated measure ν defined through

$$1/\Omega_\mu(z) = \int D(t, z)\, d\mathring{\nu}(t).$$

We would find that the kernels of the second kind $L_n(z, w)$ are normalized reproducing kernels for $\mathcal{L}_n$ with respect to the measure ν. Thus the kernels $l_n(z, w) = L_n(z, w) L_n(w, w)$ are the reproducing kernels, and since the ψ_n are the orthonormal functions, we have

$$l_n(z, w) = \sum_{k=0}^{n} \psi_k(z) \overline{\psi_k(w)}.$$

6.4. The interpolation algorithm of Nevanlinna–Pick

In this section we describe the algorithm of Nevanlinna–Pick for interpolation of class $\mathcal{C}$ functions or, equivalently, class $\mathcal{B}$ functions. The recursions can be described by J-unitary matrices. This approach gives an alternative way of computing the recurrence coefficients ρ_k and γ_k, exactly like the duality of the two algorithms considered in Ref. [29].

Suppose we start with some function $S_0 \in \mathcal{B}$, which is zero at $w \in \mathbb{O}$. Since it depends on the parameter w, we write it as a function in z but include the dependence on w explicitly when appropriate. We now transform this S_0 into some other $S_1 \in \mathcal{B}$ in three steps. Let α_1 in $\mathbb{O}$ be given. Then

$$S_1(z, w) = \tau_{31} \circ \tau_{21} \circ \tau_{11}(S_0(z, w)) = \tau_1(S_0(z, w)),$$

where

$$\tau_{11} : S_0 \mapsto S_1' = \frac{S_0 - \gamma_1}{1 - \overline{\gamma}_1 S_0}, \quad \gamma_1 = \gamma_1(w) = S_0(\alpha_1, w);$$

$$\tau_{21} : S_1' \mapsto S_1'' = S_1'/\zeta_1, \quad \zeta_1(z) = z_1 \frac{z - \alpha_1}{\varpi_1(z)};$$

$$\tau_{31} : S_1'' \mapsto S_1 = \frac{S_1'' - \rho_1}{1 - \overline{\rho}_1 S_1''}, \quad \rho_1 = \rho_1(w) = S_1''(w, w).$$

Clearly, S_1 is again a function in $\mathcal{B}$ and it will be zero at w. This follows from Theorem 1.2.3. What was done in the previous transformations is the following. First, S_0 is transformed into S_1' to make it zero in α_1. This zero can be taken out by dividing by ζ_1. The last step will normalize S_1 by making it zero at w just as S_0 was. We are now in a position like the one we started with and we can repeat the same procedure with some point α_2 from $\mathbb{O}$ to produce S_2. Note that if S_0 is not a rational function, this procedure can go on indefinitely as long as the points α_i are chosen in $\mathbb{O}$ since both γ_i and ρ_i will be in $\mathbb{D}$ as evaluations of functions in $\mathcal{B}$. Note the following relation between γ_k and ρ_k. Since $S_{k-1}(w, w) = 0$, it follows that

$$S_k'(w, w) = \frac{S_{k-1}(w, w) - \gamma_k}{1 - \overline{\gamma}_k S_{k-1}(w, w)} = -\gamma_k,$$

whereas

$$S_k'(w, w) = \zeta_k(w) S_k''(w, w) = \zeta_k(w)\rho_k,$$

so that for all $k > 0 : \gamma_k = -\zeta_k(w)\rho_k$. Note that if S_0 were a rational function, then each step will decrease the degree of the function by 1, so that one will end up with a constant unimodular function and then the algorithm will break down. Everything we said however still holds up to the step where the algorithm breaks down.

We can also invert the previous procedure. Suppose the coefficients ρ_k and γ_k for $k = 1, \ldots, n$ are produced by the previous algorithm starting from some S_0. Now choose some $\Gamma_0 \in \mathcal{B}$ such that $\Gamma_0(w) = 0$. Then generate the sequence $\Gamma_{k+1} = \tau_{n-k}^{-1}(\Gamma_k)$ for $k = 0, \ldots, n-1$. It turns out that Γ_n shall interpolate S_0 in the points $A_n^w = \{w, \alpha_1, \ldots, \alpha_n\}$. Indeed, if we denote $n-k$ as j, then it holds in general that Γ_j will interpolate S_k in the points $\{w, \alpha_n, \ldots, \alpha_{k+1}\}$. We show this for the case where all α_k are different. If some of them are confluent, the proof becomes messy by technicalities. For that case, we refer to the homogeneous formulation to be given later in this section. If the interpolation points are all different, then the result follows easily by induction. For $j = 0$, we have interpolation at w since both Γ_0 and S_n are zero in that point. The induction step can be proved by noting that $\Gamma_{j+1} = \tau_k^{-1}(\Gamma_j)$ and $S_{k-1} = \tau_k^{-1}(S_k)$ so that the interpolation from the previous step is inherited. There is one extra interpolation condition satisfied, namely, in the point α_k since $\Gamma_{j+1}(\alpha_k) = \gamma_k = S_{k-1}(\alpha_k, w)$.

We have now (almost) proved the following theorem.

Theorem 6.4.1. *Let* $S_0(z, w) \in \mathcal{B}$ *be a Schur function that is zero at* $z = w$ *and that is not a rational function. Construct iteratively for a sequence of points* $\{\alpha_k : k > 0\} \subset \mathbb{O}$ *the functions* S_k *by* $S_k(z, w) = \tau_k(S_{k-1}(z, w))$, *where* $\tau_k = \tau_{1k} \circ \tau_{2k} \circ \tau_{3k}$,

$$\tau_{1k} : S_{k-1} \mapsto S_k' = \frac{S_{k-1} - \gamma_k}{1 - \overline{\gamma}_k S_{k-1}}, \quad \gamma_k = S_{k-1}(\alpha_k, w), \qquad (6.17)$$

$$\tau_{2k} : S_k' \mapsto S_k'' = S_k'/\zeta_k, \qquad \zeta_k(z) = z_k \frac{z - \alpha_k}{\varpi_k(z)}, \qquad (6.18)$$

$$\tau_{3k} : S_k'' \mapsto S_k = \frac{S_k'' - \rho_k}{1 - \overline{\rho}_k S_k''}, \qquad \rho_k = \rho_k(w) = S_k''(w, w). \quad (6.19)$$

Then all the S_k *are in* $\mathcal{B}$ *and* $S_k(w, w) = 0$. *All the* $\gamma_k(w)$ *and* $\rho_k(w)$ *are in* $\mathbb{D}$ *and* $\gamma_k(w) = -\zeta_k(w)\rho_k(w)$.

Conversely, by choosing an arbitrary $\Gamma_0(z, w) \in \mathcal{B}$ *that vanishes for* $z = w$, *we can construct, using the previous* γ_k *and* ρ_k, *the functions* Γ_k *by* $\Gamma_{k+1} = \tau_{n-k}^{-1}(\Gamma_k)$. *All these* Γ_k *are in* $\mathcal{B}$ *and* Γ_k *will interpolate* S_{n-k} *in the points* $\{w, \alpha_n, \ldots, \alpha_{n-k+1}\}$. *Specifically,*

$$\frac{\Gamma_n - S_0}{B_n(z)} = h(z), \quad h \in H(\mathbb{O}) \quad and \quad h(w) = 0.$$

If we start from a rational function $S_0(z, w) \in \overline{\mathcal{B}}$, then it will happen that at some stage we get $\gamma_n \in \mathbb{T}$. The algorithm then stops because obviously

$S_{n-1}(z, w) = \gamma_n(w)$ for all $z \in \overline{\mathbb{D}}$. Thus $S'_n \equiv 0$ and all the remaining γs and ρs are zero. In this case, S_0 was a Blaschke product of degree n.

The previous theorem is an elaborated version of the Schur lemma (see, for example, Refs. [2], [200], and others), which says

Theorem 6.4.2 (Schur lemma). *We have* $S(z) \in \overline{B}$ *iff*

$$\gamma = S(\alpha) \in \overline{\mathbb{D}} \quad and \quad S'(z) = \frac{S(z) - \gamma}{1 - \overline{\gamma} S(z)} \in \overline{B}$$

for some $\alpha \in \mathbb{D}$.

If $\gamma \in \mathbb{T}$, then by the maximum modulus principle, $S(z) = \gamma$ for all $z \in \overline{\mathbb{D}}$. Since we are here mainly interested in the case where all the orthogonal functions ϕ_n exist for $n = 0, 1, 2, \ldots$, we shall concentrate on the case where $S_0(z, w) \in B$ is not rational, so that the algorithm never breaks down.

We shall now give an equivalent homogeneous formulation of the previous algorithm. Suppose that a Schur function $S \in B$ is described as the ratio of two functions $S = \Delta_1/\Delta_2$ that are both holomorphic in $\mathbb{O}$ and where Δ_2 is zero-free in $\mathbb{O}$. If $S(w) = 0$, then of course $\Delta_1(w) = 0$ too. We place these two functions in a vector

$$\Delta = [\Delta_1 \quad \Delta_2],$$

which can be considered as a set of homogeneous coordinates for S. Following Dewilde–Dym [60], we shall call the set of such Δ-matrices *admissible* and denote it as

$$\mathcal{A} = \{\Delta = [\Delta_1 \quad \Delta_2] : \Delta_1, \Delta_2 \in H(\mathbb{O}), \Delta_2(z) \neq 0, z \in \mathbb{O}, \Delta_1/\Delta_2 \in B\}. \tag{6.20}$$

We replace $\mathcal{A}$ by $\overline{\mathcal{A}}$ if B is replaced by $\overline{B}$ in the previous definition. Note that $\Delta_1/\Delta_2 \in B$ can also be written as $\Delta J \Delta^H < 1$, where $J = 1 \oplus -1$ is our usual signature matrix.

We can now describe the Nevanlinna–Pick algorithm of the previous theorem in terms of J-unitary matrix multiplications applied to admissible matrix functions. Let $\Delta_n = [\Delta_{n1} \quad \Delta_{n2}]$ be the admissible matrix containing the homogeneous coordinates for S_n. Then the inverse transform $S_{n-1} = \tau_n^{-1}(S_n)$ can be written as

$$\Delta_{n-1} = \Delta_n \theta_n,$$

with the matrix θ_n given by

$$\theta_n(z, w) = c_n \begin{bmatrix} 1 & \bar{\rho}_n \\ \rho_n & 1 \end{bmatrix} \begin{bmatrix} \zeta_n(z) & 0 \\ 0 & 1 \end{bmatrix} d_n \begin{bmatrix} 1 & \bar{\gamma}_n \\ \gamma_n & 1 \end{bmatrix}$$

with

$$c_n = (1 - |\rho_n|^2)^{-1/2} \quad \text{and} \quad d_n = (1 - |\gamma_n|^2)^{-1/2},$$

$$\gamma_n = \gamma_n(w) = \Delta_{n-1,1}(\alpha_n, w)/\Delta_{n-1,2}(\alpha_n, w),$$

$$\rho_n = \rho_n(w) = \begin{cases} \gamma_n(w)/\zeta_n(w) & \text{if } w \neq \alpha_n, \\ \partial_z \left(\Delta_{n-1,1}(z, w)/\Delta_{n-1,2}(z, w)\right)\big|_{z=w} & \text{if } w = \alpha_n \end{cases}$$

(∂_z means derivative w.r.t. z). Let us define the J-unitary matrix Θ_n as $\Theta_n = \theta_n \cdots \theta_1$ for $n \geq 1$. Because this matrix is formally the same as the Θ_n matrix of Section 3.3, there must exist some K_n and L_n, both functions in $\mathcal{L}_n$, parametrized in w, such that a relation like (3.19) holds. The interpolation property given in Theorem 6.4.1 implies that if we choose $\Delta_n = [0 \quad 1] \in \mathcal{A}$, then $\Delta_n \Theta_n = [K_n - L_n \quad K_n + L_n] \in \mathcal{A}$ has the property that $(K_n - L_n)/(K_n + L_n)$ interpolates S_0 in the points of the set $A_n^w = \{w, \alpha_1, \ldots, \alpha_n\}$ if they are all different. We can now easily give the proof for confluent points too. If we define Θ_n^* as $B_n \Theta_{n*}$ we get the form

$$\Theta_n^* = \frac{1}{2} \begin{bmatrix} K_n + L_n & K_n^* - L_n^* \\ K_n - L_n & K_n^* + L_n^* \end{bmatrix}.$$

Because we know that for a J-unitary matrix $\Theta_n^{-1} = J\Theta_{n*}J = B_{n*}J\Theta_n^*J$, we get

$$\Delta_0 J\Theta_n^* J = B_n \Delta_n.$$

Furthermore, since $S_0 = \Delta_{01}/\Delta_{02} \in \mathcal{B}$ and vanishes at $z = w$, the positive real function $\Omega(z, w) = (1 - S_0(z, w))/(1 + S_0(z, w)) \in \mathcal{C}$ will be 1 for $z = w$: $\Omega(w, w) = 1$. We can write $\Delta_0 = [1 - \Omega(z, w) \quad 1 + \Omega(z, w)]$. If we multiply this from the right with $J\Theta_n^*J$, then we get

$$\frac{1}{2}[1 - \Omega \quad 1 + \Omega] \begin{bmatrix} K_n + L_n & -K_n^* + L_n^* \\ -K_n + L_n & K_n^* + L_n^* \end{bmatrix} = B_n[\Delta_{n1} \quad \Delta_{n2}]. \quad (6.21)$$

Thus

$$[L_n - K_n\Omega \quad L_n^* + K_n^*\Omega] = B_n[\Delta_{n1} \quad \Delta_{n2}].$$

The first of these relations shows that

$$L_n(z, w) - K_n(z, w)\Omega(z, w) = B_n(z)\Delta_{n1}(z, w),$$

with

$$\Delta_{n1} \in H(\mathbb{O}) \quad \text{and} \quad \Delta_{n1}(w, w) = 0.$$

Because K_n does not vanish in $\mathbb{O}$ and $L_n/K_n \in C$ by a property of J-unitary matrices, we can also say that the positive real function Ω is approximated by the positive real function $\Omega_n = L_n/K_n$ such that

$$\frac{\Omega - \Omega_n}{B_n} = h \in H(\mathbb{O}) \quad \text{with} \quad h(w) = 0.$$

Taking a Cayley transform results in the interpolation property of the Schur function S_0 by the Schur function $\Gamma_n = (K_n - L_n)/(K_n + L_n) \in \mathcal{B}$:

$$\frac{S_0 - \Gamma_n}{B_n} = g \in H(\mathbb{O}) \quad \text{with} \quad g(w) = 0.$$

Suppose $\mathring{\mu}$ is some positive measure on $\partial\mathbb{O}$ and that we associate with it the positive real function $\Omega_\mu(z, w)$ as in Lemmas 6.2.2 and 6.2.3. Suppose we start the Nevanlinna–Pick algorithm as described above with Ω equal to this $\Omega_\mu(z, w)$. Since we just showed that then Ω_n will interpolate this starting Ω_μ at the point set A_n^w, it follows from Corollary 6.2.4 that the inner product on $\mathcal{L}_n$ is the same for the measure $\mathring{\mu}(t)$ and for the measure $\mathring{\mu}_n(t, w)$ defined by

$$d\mathring{\mu}_n(t, w) = \frac{|\varpi_0(t)|^2 P(t, w)}{|K_n(t, w)|^2} d\mathring{\lambda}(t) = \frac{P(t, w)}{|K_n(t, w)|^2} d\lambda(t).$$

Thus, because $\mathring{\mu}(t)$ does not depend on w, it means that as far as the inner product in $\mathcal{L}_n$ is concerned, $\mathring{\mu}(t, w)$ does not depend on w. Therefore, we may replace w, for example, by α_0 and we thus have

$$\int f(t)g_*(t)d\mathring{\mu}(t) = \int f(t)g_*(t)\, d\mathring{\mu}_n(t, w)$$

$$= \int f(t)g_*(t)\frac{P(t, w)}{|K_n(t, w)|^2}\, d\lambda(t)$$

$$= \int f(t)g_*(t)\frac{P(t, \alpha_0)}{|K_n(t, \alpha_0)|^2}\, d\lambda(t).$$

This has the important consequence that $K_n(z, w)$ as generated by the Nevanlinna–Pick algorithm applied to the positive real $\Omega_\mu(z, w)$ is the normalized reproducing kernel for $\mathcal{L}_n$ w.r.t. the measure $\mathring{\mu}$. Therefore we have to show that, up to normalization, K_n reproduces every $f \in \mathcal{L}_n$; and it does indeed since

$$
\begin{aligned}
\langle f, K_n(\cdot, z) \rangle_{\mathring{\mu}} &= \langle f, K_n(\cdot, z) \rangle_{\mathring{\mu}_n} \\
&= \int f(t) K_{n*}(t, z) \frac{P(t, z)}{K_{n*}(t, z) K_n(t, z)} \, d\lambda(t) \\
&= \int \frac{f(t)}{K_n(t, z)} P(t, z) \, d\lambda(t) = \frac{f(z)}{K_n(z, z)}
\end{aligned}
$$

because f / K_n is analytic in $\mathbb{O}$ so that the Poisson formula holds. Finally, we note that the real part of $\Omega_\mu(z, w)$ has a radial limit satisfying a.e. Re $\Omega_\mu(t, w) = \mu'(t)/P(t, w)$, $t \in \partial\mathbb{O}$, $w \in \mathbb{O}$.

For further reference, we state the next theorem, which follows from our previous remarks.

Theorem 6.4.3. *Let $d\mathring{\mu}$ be a normalized measure ($\int d\mathring{\mu} = 1$) and define for $w \in \mathbb{O}$*

$$
\Omega_\mu(z, w) = \mathcal{S}_w(\mu)(z) = \int \frac{D(t, z)}{P(t, w)} \frac{d\mathring{\mu}(t)}{|\varpi_0(t)|^2} + \mathrm{i}c, \quad c \in \mathbb{R}, \qquad (6.22)
$$

where c is chosen such that $\Omega_\mu(w, w) > 0$. (D is the Riesz–Herglotz–Nevanlinna kernel as always and P is the Poisson kernel.) Furthermore, let $K_n(z, w)$ be the kernels produced by the Nevanlinna–Pick algorithm starting from $\Delta_0 = [1 - \Omega_\mu(z, w) \quad 1 + \Omega_\mu(z, w)]$. Define the absolutely continuous measure

$$
d\mathring{\mu}_n(t) = \frac{|\varpi_0(t)|^2 P(t, w)}{|K_n(t, w)|^2} \, d\mathring{\lambda}(t) = \frac{P(t, w)}{|K_n(t, w)|^2} \, d\lambda(t).
$$

Then the inner product $\langle \cdot, \cdot \rangle_{\mathring{\mu}_n}$ will not depend on w for functions in $\mathcal{L}_n$ and it is the same as the inner product $\langle \cdot, \cdot \rangle_{\mathring{\mu}}$ on $\mathcal{L}_n$. Moreover, $K_n(z, w)$ is the normalized reproducing kernel for $\mathcal{L}_n$ with respect to the measure $\mathring{\mu}_n$ and hence with respect to the measure $\mathring{\mu}$.

6.5. Interpolation algorithm for the orthonormal functions

We shall in this section give an algorithm in the style of the Nevanlinna–Pick algorithm, which, based on the idea of successive interpolation, will generate the recursion for the orthonormal functions ϕ_n and the functions of the second kind ψ_n. As a matter of fact, it is difficult to do this for these functions because of the rotating factors η_n^1 and η_n^2 in the recurrence relation. These rotations depend on the angle that $\phi_n(\alpha_{n-1})$ forms with the real axis and this is difficult to find without evaluating $\phi_n(\alpha_{n-1})$. However, the rotated functions Φ_n and Ψ_n satisfied a recurrence that got rid of these ηs and it will be possible to find an interpolation algorithm for these rotated functions. That is what we shall currently do.

Recall that $\Phi_n = \epsilon_n \phi_n$ and $\Psi_n = \epsilon_n \psi_n$, with $\epsilon_n \in \mathbb{T}$ as defined by (4.14). We shall use once more the following notation:

$$\tilde{B}_{-1} = 1, \qquad \tilde{B}_n(z) = \tilde{B}_{n-1}(z)\zeta_n(z) = \zeta_0(z)B_n(z), \quad n = 0, 1, \ldots.$$

We now define R_{n1} and R_{n2} by

$$\begin{bmatrix} \tilde{B}_{n-1} R_{n1}(z) \\ \tilde{B}_n R_{n2}(z) \end{bmatrix} = \begin{bmatrix} \Phi_n(z) \\ \Phi_n^*(z) \end{bmatrix} \Omega(z) + \begin{bmatrix} \Psi_n(z) \\ -\Psi_n^*(z) \end{bmatrix}, \tag{6.23}$$

where $\Omega = \Omega_\mu = \int D(t, \cdot) \, d\mathring{\mu}(t)$ is the positive real function associated with the measure μ for which the orthogonality holds. Note that the functions in the left-hand side are in fact rotated versions of the functions g and h as defined in (6.1) and (6.2) respectively. These are indeed the remainders in the linearized interpolation properties of the rotated functions. We shall call the functions R_{n1} and R_{n2} the *remainder functions*. Since $\tilde{B}_{-1} = 1$ it follows that $R_{01} = \Omega + 1$ and $R_{02} = \Omega - 1$. Note also that for $n > 0$, both R_{n1} and R_{n2} are zero in α_0.

The right-hand side in the defining relation (6.23) of the remainder functions satisfies the recurrence for the rotated functions as in Theorem 4.1.3. Hence, the left-hand side shall satisfy

$$\begin{bmatrix} \tilde{B}_{n-1} R_{n1}(z) \\ \tilde{B}_n R_{n2}(z) \end{bmatrix} = e_n \frac{\varpi_{n-1}(z)}{\varpi_n(z)} \begin{bmatrix} 1 & \overline{\Lambda}_n \\ \Lambda_n & 1 \end{bmatrix} \begin{bmatrix} Z_{n-1}(z) & 0 \\ 0 & 1 \end{bmatrix} \begin{bmatrix} \tilde{B}_{n-2} R_{n-1,1}(z) \\ \tilde{B}_{n-1} R_{n-1,2}(z) \end{bmatrix}. \tag{6.24}$$

This can be rewritten as given in the next theorem.

Theorem 6.5.1. *The remainder functions as defined above satisfy the following recursion:*

$$\varpi_n(z) \begin{bmatrix} R_{n1}(z) \\ R_{n2}(z) \end{bmatrix} = e_n \begin{bmatrix} 1 & 0 \\ 0 & 1/\zeta_n(z) \end{bmatrix} \begin{bmatrix} 1 & \overline{\Lambda}_n \\ \Lambda_n & 1 \end{bmatrix} \varpi_{n-1}(z) \begin{bmatrix} \eta_{n-1} R_{n-1,1}(z) \\ R_{n-1,2}(z) \end{bmatrix}, \tag{6.25}$$

with $e_n > 0$ and

$$\Lambda_n = -\overline{\eta}_{n-1} \lim_{z \to \alpha_n} \frac{R_{n-1,2}(z)}{R_{n-1,1}(z)}, \quad \eta_{n-1} = z_n \overline{z}_{n-1}, \quad and$$

$$e_n^2 = \frac{\varpi_n(\alpha_n)}{\varpi_{n-1}(\alpha_{n-1})} \frac{1}{1 - |\Lambda_n|^2}. \tag{6.26}$$

The Λ_n in the previous expression are the same as the Λ_n of Theorem 4.1.3.

We can make the recursion even simpler and avoid the explicit use of the η_{n-1} by introducing

$$r_{n1}(z) = -z_n R_{n1}(z) \quad \text{and} \quad r_{n2}(z) = R_{n2}(z). \tag{6.27}$$

With this notation, the recursion (6.25) becomes

$$\varpi_n(z) \begin{bmatrix} r_{n1}(z) \\ r_{n2}(z) \end{bmatrix} = e_n \begin{bmatrix} 1 & 0 \\ 0 & 1/\zeta_n(z) \end{bmatrix} \begin{bmatrix} 1 & \overline{L}_n \\ L_n & 1 \end{bmatrix} \varpi_{n-1}(z) \begin{bmatrix} r_{n-1,1}(z) \\ r_{n-1,2}(z) \end{bmatrix},$$

$$\tag{6.28}$$

with

$$L_n = -\overline{z}_n \Lambda_n = -\lim_{z \to \alpha_n} \frac{r_{n-1,2}(z)}{r_{n-1,1}(z)}, \qquad e_n = \left[\frac{\varpi_n(\alpha_n)}{\varpi_{n-a}(\alpha_{n-1})} \frac{1}{1 - |L_n|^2} \right]^{1/2}.$$

$$\tag{6.29}$$

Proof. We shall only prove (6.26), because (6.29) is a direct consequence. We can start from the relation (6.24) and use $\zeta_{n-1} = \eta_{n-1} Z_{n-1}$ to get

$$\tilde{B}_{n-1}(z) \begin{bmatrix} R_{n1}(z) \\ \zeta_n R_{n2}(z) \end{bmatrix} = e_n \frac{\varpi_{n-1}(z)}{\varpi_n(z)} \begin{bmatrix} 1 & \overline{\Lambda}_n \\ \Lambda_n & 1 \end{bmatrix} \tilde{B}_{n-1}(z) \begin{bmatrix} \eta_{n-1} R_{n-1,1}(z) \\ R_{n-1,2}(z) \end{bmatrix},$$

$$\tag{6.30}$$

which now easily gives (6.25). To find the expression for Λ_n, one can use the last line of (6.30) for $z = \alpha_n$, which gives

$$0 = \Lambda_n \eta_{n-1} R_{n-1,1}(\alpha_n) + R_{n-1,2}(\alpha_n),$$

from which the expression for Λ_n follows. The expression for e_n was shown in Theorem 4.1.2. □

The previous theorem has the following consequence.

Corollary 6.5.2. *Define the function $\Gamma_n(z)$ in terms of the remainder functions by*

$$\Gamma_n(z) = -\overline{z}_n \frac{R_{n2}(z)}{R_{n1}(z)} = \frac{r_{n2}(z)}{r_{n1}(z)}, \quad n = 0, 1, \dots, \tag{6.31}$$

especially

$$\Gamma_0(z) = \frac{1}{\zeta_0(z)} \frac{1 - \Omega(z)}{1 + \Omega(z)}.$$

Then for all $k \geq 0$: $\Gamma_k \in \mathcal{B}$ and they satisfy the recurrence

$$\Gamma_n = \frac{1}{\zeta_n} \left(\frac{L_n + \Gamma_{n-1}}{1 + \overline{L}_n \Gamma_{n-1}} \right),$$

with $L_n = -\Gamma_{n-1}(\alpha_n)$.

Proof. The proof follows immediately from the previous theorem. All the Γ_k are in $\mathcal{B}$ because Γ_0 is, whereas the Möbius transforms are done with $L_k \in \mathbb{D}$. Moreover, the division by ζ_n respects the analyticity because the function between brackets was made zero in $z = \alpha_n$ by the choice of L_n. $\square$

7

Density of the rational functions

As a step toward some convergence results to be considered in Chapter 9, we consider here some asymptotic phenomena. For instance, what happens with the spaces $\mathcal{L}_n$ and $\mathcal{R}_n = \mathcal{L}_n \cdot \mathcal{L}_{n*}$ when n tends to infinity? In general, are these spaces dense in H_p or L_p as they are in the polynomial case? This is important if we want to know how good functions in H_p or L_p can be approximated within the space of rationals that we considered. These results will of course depend on the spaces, that is, on the placement of the poles α_k. More precisely, they will depend on the convergence or divergence of the Blaschke products for these numbers.

7.1. Density in L_p and H_p

The functions $t^k, k \in \mathbb{Z}$ are known to be complete in L_p. Similarly, it is possible to prove that the basis functions of finite Blaschke products and their inverses are complete in the space L_p if and only if $\sum (1 - |\alpha_k|) = \infty$. In analogy with the powers of z we can define the finite Blaschke products B_n for $n = 0, 1, \ldots$ as before and we set by definition $B_{-n} = B_{n*} = 1/B_n$ for $n = 1, 2, \ldots$. Hence, the $\{B_k\}_{|k| \leq n}$ span the spaces $\mathcal{R}_n = \mathcal{L}_n \cdot \mathcal{L}_{n*}, n = 0, 1, \ldots$. Given the spaces $\mathcal{L}_n$ and $\mathcal{R}_n$, we define

$$\mathcal{L}_\infty = \bigcup_{n=0}^{\infty} \mathcal{L}_n \quad \text{and} \quad \mathcal{R}_\infty = \bigcup_{n=0}^{\infty} \mathcal{R}_n$$

just as $\mathcal{P}_\infty$ is the set of all polynomials. The notation $\mathcal{P}, \mathcal{L}$, and $\mathcal{R}$ will be used for the closures of $\mathcal{P}_\infty, \mathcal{L}_\infty$, and $\mathcal{R}_\infty$ respectively in some topological space (for example L_p). The completeness of the functions $\{B_n\}_{n \in \mathbb{Z}}$ in L_p is by definition the same as the density of $\mathcal{R}_\infty$ in L_p. If $\mathcal{R}$ denotes the closure in L_p of $\mathcal{R}_\infty$, then if $\mathcal{R}_\infty$ is dense in L_p, $\mathcal{R}$ should coincide with L_p.

149

In order to prove this density property for the disk, we start with a lemma that can be found in Akhiezer [1, p. 243].

Lemma 7.1.1. Let $z_1, z_2, \ldots, z_n$ be some fixed points in $\mathbb{C}$. We have the following optimization result in $L_p(\mathbb{T})$, $0 < p \leq \infty$:

$$\min_q \left\{ \|f\|_p : f(z) = \frac{q(z)}{(z - z_1) \cdots (z - z_n)}, q \in \mathcal{P}_N^M, N \geq n \right\} = \prod_{k=1}^{n} \frac{1}{|z_k|_+},$$
(7.1)

where q ranges over $\mathcal{P}_N^M$, the set of all monic polynomials of degree N and $|z|_+ = \max\{|z|, 1\}$. The unique solution is obtained for $q = Q$ with Q as in (7.2) below.

Proof. Suppose without loss of generality that the points are ordered such that $z_1, \ldots, z_d$ are all in $\mathbb{E}$ while $z_{d+1}, \ldots, z_n$ are all in $\overline{\mathbb{D}}$. It is clear that the right-hand side value can be reached. It is obtained for $q = Q$ with

$$Q(z) = \frac{z^{N-n}(z\overline{z}_1 - 1) \cdots (z\overline{z}_d - 1)(z - z_{d+1}) \cdots (z - z_n)}{\overline{z}_1 \cdots \overline{z}_d}.$$
(7.2)

To prove that this is the unique solution, we have to show that for any other monic polynomial $P \in \mathcal{P}_N^M$

$$\left\| \frac{P(z)}{\pi(z)} \right\|_p > \prod_{k=1}^{d} \frac{1}{|z_k|} = \left\| \frac{Q(z)}{\pi(z)} \right\|_p,$$
(7.3)

where $\pi(z) = (z - z_1) \cdots (z - z_n)$. Since $\|f\|_\infty \geq \|f\|_p$, it is sufficient to prove that (7.3) is true for $0 < p < \infty$. Suppose that $q_1, \ldots, q_j$ are all the zeros of P in the open unit disk. Thus $P(z) = (z - q_1) \cdots (z - q_j)R(z)$ with $|R(0)| \geq 1$ because $R(z) = \prod_k (z - d_k)$ with all $|d_k| \geq 1$. The function

$$g(z) = \frac{(1 - \overline{q}_1 z) \cdots (1 - \overline{q}_j z)R(z)}{(z - z_1) \cdots (z - z_d)(1 - \overline{z}_{d+1} z) \cdots (1 - \overline{z}_n z)}$$
(7.4)

is analytic and nonzero in $\mathbb{D}$. It is even analytic in $\overline{\mathbb{D}}$ since we can assume that the z_k that are on the boundary $\mathbb{T}$ are canceled by zeros of P, otherwise the inequality (7.3) would be trivial. Therefore,

$$\int \left| \frac{P}{\pi} \right|^p d\lambda = \int |g|^p d\lambda \geq \left| \int g(t)^p d\lambda(t) \right| = |g(0)|^p = \frac{|R(0)|^p}{|z_1 \cdots z_d|^p}$$

$$\geq \frac{1}{|z_1 \cdots z_d|^p},$$

with inequality if $P \neq Q$. $\square$

With the previous lemma, we can now prove the following completeness theorem. It says that $\mathcal{R}_\infty$ is dense in the spaces $L_p(\mathbb{T})$ for $1 \le p \le \infty$ and in the space of continuous 2π-periodic functions if and only if the Blaschke product diverges. It is a slight generalization of a result in Akhiezer [1, p. 244] where the poles were supposed to be simple. However, Akhiezer's result is more general since poles need not appear in reflection pairs $(\alpha_k, 1/\overline{\alpha}_k)$ as we suppose here.

Theorem 7.1.2. *For given $\alpha_1, \alpha_2, \ldots$ all in $\mathbb{D}$, let the finite Blaschke products B_n be defined as before for $n \ge 0$ and $B_{-n} = 1/B_n$ for $n = 1, 2, \ldots$. Then the system $\{B_n\}_{n\in\mathbb{Z}}$ is complete in any $L_p(\mathbb{T})$ space $(1 \le p \le \infty)$ as well as in the class $C(\mathbb{T})$ of continuous 2π-periodic functions (with respect to the Chebyshev norm) if and only if $\sum(1 - |\alpha_k|) = \infty$.*

Proof. First note that the case where infinitely many α_k are equal to zero can immediately be discarded because in that case the system contains the trigonometric polynomials and these are known to be complete while the sum certainly diverges. So we suppose that only finitely many (say q) of the α_k are zero. Without loss of generality we suppose $\alpha_1 = \cdots = \alpha_q = 0$. In this case the system contains the trigonometric polynomials of degree at most q. Suppose we set

$$\mathcal{R}_n = \mathrm{span}\{B_k : k = -n, \ldots, n\} = \mathcal{L}_n \cdot \mathcal{L}_{n*}$$

$$= \left\{ \frac{p_{2n}(z)}{D(z)} : p_{2n} \in \mathcal{P}_{2n}, D(z) = z^q \prod_{k=q+1}^{n} (z - 1/\overline{\alpha}_k)(z - \alpha_k) \right\}.$$

In view of the inclusions $C(\mathbb{T}) \subset L_\infty(\mathbb{T}) \subset \cdots \subset L_1(\mathbb{T})$, it is sufficient to prove that the divergence of the sum implies completeness in $C(\mathbb{T})$ and conversely that completeness in $L_1(\mathbb{T})$ implies divergence of the sum.

Completeness in $L_1(\mathbb{T}) \Rightarrow$ divergence of the sum. By the previous lemma we have for $p = 1$ that

$$\inf_{f\in\mathcal{R}_n} \|z^{q+1} - f(z)\|_1 = \inf_{P\in\mathcal{P}_{2n}} \left\| z^{q+1} - \frac{P(z)}{D(z)} \right\|_1$$

$$= \inf_{Q\in\mathcal{P}_{2n+1}^M} \left\| \frac{Q(z)}{D(z)} \right\|_1 = \prod_{k=q+1}^{n} |\alpha_k|.$$

Since the system is supposed to be complete in $L_1(\mathbb{T})$, the previous expression should go to zero as $n \to \infty$. Whence $\sum(1 - |\alpha_k|) = \infty$.

Divergence of the sum $\Rightarrow$ completeness in $C(\mathbb{T})$. We already know that all the powers z^k, $k = -q, \ldots, q$ are in the system, so we should prove that all z^k for $|k| > q$ can be approximated arbitrarily close in $C(\mathbb{T})$ (the norm is the

Chebyshev norm) by elements from $\mathcal{R}_n$ for n sufficiently large. This follows from the following observations:

$$\inf_{f\in\mathcal{R}_n, a_i} \|z^{m+q} + a_1 z^{m+q-1} + \cdots + a_{m-1}z^{q+1} + f(z)\|_\infty$$

$$= \inf_{p\in\mathcal{P}_{2n+m-1}} \left\| \frac{z^{2n+m} + p(z)}{D(z)} \right\|_\infty = \prod_{k=q+1}^{n} |\alpha_k|$$

for $m = 1, 2, \ldots$ and

$$\inf_{f\in\mathcal{R}_n, a_i} \|z^{-m-q} + a_1 z^{-m-q+1} + \cdots + a_{m-1}z^{-q-1} + f(z)\|_\infty$$

$$= \inf_{f_*\in\mathcal{R}_n, a_i} \|z^{m+q} + \bar{a}_1 z^{m+q-1} + \cdots + \bar{a}_{m-1}z^{q+1} + f_*(z)\|_\infty$$

$$= \prod_{k=q+1}^{n} |\alpha_k|$$

for $m = 1, 2, \ldots$. Since the sum diverges to infinity, we must have that the right-hand side $\prod_{q+1}^{n} |\alpha_k| \to 0$. By an induction argument, it then follows that $z^{\pm(q+m)}$ for $m = 1, 2, \ldots$ can be approximated arbitrary close in $C(\mathbb{T})$ by the system $\{B_n\}$. This means that the system $\{B_n\}$ is complete in $C(\mathbb{T})$. □

The same proof, with some simplifications can be used to prove that $\mathcal{L}_\infty$ is dense in $H_p(\mathbb{D})$.

Corollary 7.1.3. *The system* $\{B_n : n = 0, 1, \ldots\}$ *of Blaschke products, with zeros* $\alpha_1, \alpha_2, \ldots$ *all in* $\mathbb{D}$, *is complete in the spaces* $H_p(\mathbb{D})$, $1 \le p \le \infty$ *if and only if* $\sum(1 - |\alpha_k|) = \infty$.

Along the same lines, one can adapt the completeness condition given by Akhiezer [1, pp. 246–249] for the real line. We leave it to the reader to check the details. The result is

Theorem 7.1.4. *For given* $\alpha_1, \alpha_2, \ldots$ *all in* $\mathbb{U}$, *let the finite Blaschke products* B_n *be defined as before for* $n \ge 0$ *and* $B_{-n} = 1/B_n$ *for* $n = 1, 2, \ldots$. *Then the system* $\{B_n : n \in \mathbb{Z}\}$ *is complete in any* $L_p(\mathbb{R})$ *space* $(1 < p \le \infty)$ *as well as in the class* $C(\bar{\mathbb{R}})$ *of continuous functions in* $\bar{\mathbb{R}}$ *(with respect to the Chebyshev norm) if and only if* $\sum \text{Im}\,\alpha_k/(1 + |\alpha_k|^2) = \infty$.

Note that continuous in $\bar{\mathbb{R}}$ means continuous in $\mathbb{R}$ and that the limits $\lim_{x\to+\infty} f(x)$ and $\lim_{x\to-\infty} f(x)$ both exist and are equal to each other.

Thus in the previous results we had a density result iff the Blaschke product diverges, that is, iff

$$\sum(1 - |\alpha_k|) = \infty \quad \text{for } \mathbb{D} \quad \text{and} \quad \sum \frac{\text{Im } \alpha_k}{1 + |\alpha_k|^2} = \infty \quad \text{for } \mathbb{U}. \quad (7.5)$$

We now have a look at the case $p = 2$ and consider the density of $\mathcal{L}_\infty$ in $\mathring{H}_2(\mathbb{O})$. It can already be expected from Theorem 2.1.1, which characterized $\mathcal{L}_n$ as $\mathring{H}_2 \ominus \zeta_0 B_n H_2$, that if the Blaschke product B_n diverges to 0 in $\mathbb{O}$, then $\mathcal{L}_\infty$ will be dense in $\mathring{H}_2$.

In fact the previous characterization links interpolation with least squares approximation as we can find in the next theorem. It is in fact a restatement of Theorem 2.1.1. For the disk case, this theorem can be found in the book of Walsh [200, p. 224].

Theorem 7.1.5. *Let $f \in \mathring{H}_2$ be given. Then the following least squares approximation problem in $\mathring{H}_2$ norm,*

$$\inf_{f_n}\{\|f - f_n\|_2 : f_n \in \mathcal{L}_n\},$$

has a unique solution, which is the function $f_n \in \mathcal{L}_n$ that interpolates f in the point set $A_n^0 = \{\alpha_0, \alpha_1, \dots, \alpha_n\}$.

Proof. Obviously, f_n is the projection in $\mathring{H}_2$ of f onto $\mathcal{L}_n$. Therefore, the residual $f - f_n$ must be orthogonal to $\mathcal{L}_n = \mathring{H}_2 \ominus \mathcal{M}_n$ with $\mathcal{M}_n = \zeta_0 B_n H_2$ and thus it is an element from $\mathcal{M}_n$. This means that $f(z) = f_n(z)$ for all the points $z \in A_n^0$ in Hermite sense. This identifies the interpolant as the least squares approximant. $\square$

In fact Walsh observes that we may replace in the previous theorem $f \in H_2(\mathbb{D})$ by $f_1 \in L_2(\mathbb{T})$. The best least squares approximant is the interpolant for its Cauchy integral

$$f(z) = \int \frac{f_1(t)}{t - z} t \, d\lambda(t)$$

in the point set A_n^0. See Walsh [200, p. 225].

Completely analogous is the following result, which uses Theorem 2.1.6.

Theorem 7.1.6. *Let $f \in \mathring{H}_2$ be given and recall (Theorem 2.1.6) $\mathcal{L}_n(\mathring{\alpha}_0) = \{f_n \in \mathcal{L}_n : f_n(\mathring{\alpha}_0) = 0\} = \mathring{H}_2 \ominus B_n H_2$. The unique solution of the least* *squares approximation problem*

$$\inf_{f_n}\{\|f - f_n\|_2 : f_n \in \mathcal{L}_n(\hat{\alpha}_0)\}$$

in $\mathring{H}_2$ is the function $f_n \in \mathcal{L}_n(\hat{\alpha}_0)$ that interpolates f in the point set $A_n = \{\alpha_1, \ldots, \alpha_n\}$.

Let us now return to our density problem. The idea that the interpolation error, which now turns out to be also related to the L_2 approximation error, is proportional to a Blaschke product forms the backbone of the following convergence result that gives a local uniform convergence result in $\mathbb{O}$, which means uniform convergence on compact subsets of $\mathbb{O}$. For the disk, this theorem can be found again in Walsh [200, pp. 305–306].

Theorem 7.1.7. *Let $f \in H_2(\mathbb{D})$ and suppose that $\sum(1 - |\alpha_k|)$ diverges. Let $f_n \in \mathcal{L}_n$ be the function that interpolates f in the point set $A_n^0 = \{0, \alpha_1, \ldots, \alpha_n\}$. (The zero in this theorem is not essential. It could be replaced by any other $\alpha_0 \in \mathbb{D}$.) Then f_n converges uniformly on compact subsets of $\mathbb{D}$ to $f(z)$.*

If $f \in H_2(\mathbb{D})$ is also analytic on $\mathbb{T}$, then the convergence is also uniform on $\mathbb{T}$.

Proof. The details of the proof can be found in Walsh's book. The basic idea is to use the error formula

$$f(z) - f_n(z) = \int B_n(z) B_{n*}(t) \frac{f(t)t}{t - z} d\lambda(t).$$

For $z \in \mathbb{D}$, use the fact that $B_n(z)$ converges to zero while the modulus of $B_{n*}(t)$ is 1.

For $z \in \mathbb{T}$, take the integral over a circle slightly larger than the unit circle. Then $|B_n(z)| = 1$ while ($|t| > 1$)

$$B_{n*}(t) = \prod \frac{\alpha_i}{|\alpha_i|} \frac{1 - \overline{\alpha}_i t}{\alpha_i - t} = \prod \frac{\alpha_i}{|\alpha_i|} \frac{\overline{\alpha}_i - 1/t}{1 - \alpha_i/t} \to 0,$$

from which the theorem follows.   □

The second part of the theorem implies that the set $\{B_n\}_{n \geq 0}$ is complete in H_2. The result is stronger. It says that if the sum diverges, it is possible for an arbitrary polynomial p and any $\epsilon > 0$ to find n sufficiently large such that there is an $f_n \in \mathcal{L}_n$ with $|p - f_n| < \epsilon$ uniformly on $\overline{\mathbb{D}}$.

7.2. Density in $L_2(\mu)$ and $H_2(\mu)$

For a more general positive measure μ, it is well known that [1, p. 261] the system $\{t^k\}_{k\geq0}$ is complete in $L_p(\mu, \mathbb{T})$, $p \geq 1$ iff $\int |\log \mu'(t)| d\lambda(t) = \infty$, and for the real line, that the system $\{e^{iax}\}_{a\geq0}$, or equivalently the system $\{(t + i)^{-k}\}_{k\geq0}$, is complete in $L_p(\mathring{\mu}, \mathbb{R})$, $p \geq 1$ iff $\int |\log \mu'(t)| d\mathring{\lambda}(t) = \infty$ [1, pp. 263–266]. We call

$$\log \mu'(t) \in L_1(\mathring{\lambda}) \tag{7.6}$$

the Szegő condition. Thus from classical polynomial theory [2, p. 186; 116, p. 50; 92, p. 144] we have.

Theorem 7.2.1. *Let* $p \geq 1$ *be an integer. The set of polynomials is dense in* $L_p(\mu, \mathbb{T})$ *if and only if* $\log \mu' \notin L_1(\lambda, \mathbb{T})$. *The set of rational functions* $\{(z + i)^{-k}\}_{k\geq0}$ *is dense in* $L_p(\mathring{\mu}, \mathbb{R})$ *if and only if* $\log \mu' \notin L_1(\mathring{\lambda}, \mathbb{R})$.

We want to investigate the density of $\mathcal{L}_\infty$. From the comments at the end of the previous section, we may immediately conclude the second part of the following theorem.

Theorem 7.2.2. *Let* $1 \leq p \leq \infty$ *and* $\mathring{\mu}$ *be a finite positive measure on* $\overline{\partial\mathbb{O}}$. *Define* $\mathcal{L}$ *as the* $L_p(\mathring{\mu})$-*closure of* $\mathcal{L}_\infty$. *We can then give the following statements:*

1. *It is always true that* $\mathcal{P}_\infty$ *is uniformly dense in* $\mathcal{L}_\infty$, *that is, dense w.r.t.* $\|\cdot\|_\infty$ *norm. Thus also* $\mathcal{L} \subseteq H_p(\mathring{\mu})$ *for any* p.
2. *If the Blaschke product diverges, that is, if (7.5) is satisfied, then* $\mathcal{L}_\infty$ *is uniformly dense in* $\mathcal{P}_\infty$. *We then have* $\mathcal{L} = H_p(\mathring{\mu})$ *for any* p.

Proof. Recall that $H_p(\mathring{\mu})$ is the $L_p(\mathring{\mu})$-closure of $\mathcal{P}_\infty$.
For part 1, we note that since every element from some $\mathcal{L}_n$ is meromorphic in $\mathbb{C}$ and analytic in $\mathbb{O} \cup \partial\mathbb{O}$ and since for $1 \leq p \leq \infty$, $L_p(\mathring{\mu})$ is a complete metric space [187, p. 69], all $\mathcal{L}_n \subseteq H_p(\mathring{\mu})$. Hence $\mathcal{L}_\infty \subseteq \mathcal{L} \subseteq H_p(\mathring{\mu})$.
For part 2, it follows from the remarks given above that $H_p(\mathring{\mu}) \subseteq \mathcal{L}$, hence, in combination with part 1 we get equality. $\qquad\square$

If $\mathcal{L}$ is the $L_2(\mathring{\mu})$-closure of $\mathcal{L}_\infty$, then we know from Theorem 7.2.2 that $\mathcal{L} \subseteq H_2(\mathring{\mu})$ and if the Blaschke product diverges, we have equality. Thus the divergence of the Blaschke product is a sufficient condition for completeness. It is, however, not necessary (see Theorem 10.3.4). In the case $p = 2$, we have the following property concerning the density of $\mathcal{L}_\infty$ in $H_2(\mathring{\mu})$, which includes a partial converse.

Theorem 7.2.3. *Let $\mathring{\mu}$ be a positive measure on $\partial\mathbb{O}$. Then $\mathcal{L}_\infty$ is dense in $H_2(\mathring{\mu})$ if the Blaschke product diverges, that is, if (7.5) holds. Conversely, let $\log\mu' \in L_1(\mathring{\lambda})$, then the density of $\mathcal{L}_\infty$ in $H_2(\mathring{\mu})$ implies the divergence of the the Blaschke product.*

Proof. The first part is the second part of the previous theorem for $p = 2$.

For the second part, we can use a similar construct as in Ref. [60]. If (7.5) is *not* satisfied, then it is known that $B_n(z)$ converges to a Blaschke product $B(z)$, which is an inner function, that is, with modulus bounded by 1 in $\mathbb{O}$, while its radial limit has modulus 1 λ a.e. It has zeros in $z = \alpha_1, \alpha_2, \ldots$. We have to show that there exists a nonzero function in $H_2(\mathring{\mu})$ that is not in $\mathcal{L}$. Let us take $f(z) = \zeta_0(z)B(z)\sigma(z)/\sigma_*(z)$, with $\sigma(z)$ the outer spectral factor of μ. Since $|f(t)| = 1$ a.e. on $\partial\mathbb{O}$ we have $f \in L_2(\mathring{\mu})$. Let g be the orthogonal projection of f onto $\mathring{H}_2$ and define $h = g/\sigma$. Clearly, h is not identically zero and $h \in H_2(\mathring{\mu})$ because $h\sigma = g \in \mathring{H}_2$ (see Ref. [87, Theorem 3.4, p. 215]). However, it is orthogonal to $\mathcal{L}_\infty$ and thus the inclusion $\mathcal{L}_\infty \subset H_2(\mathring{\mu})$ is proper. That h is orthogonal to $\mathcal{L}_\infty$ is because the orthogonality $h \perp_{\mathring{\mu}} \mathcal{L}_\infty$ does not depend on the singular part $\mathring{\mu}_s$ of the measure $\mathring{\mu}$ because

$$\left|\int \frac{g(t)}{\sigma(t)B_k(t)}d\mathring{\mu}_s(t)\right|^2 = |\langle h, B_k\rangle_{\mathring{\mu}_s}|^2 \le \left\|\frac{g(t)}{B_k(t)}\right\|_{\mathring{\mu}_s}^2 \left\|\frac{1}{\sigma(t)}\right\|_{\mathring{\mu}_s}^2$$

$$= \int\left|\frac{g(t)}{B_k(t)}\right|^2 d\mathring{\mu}_s \cdot \int\left|\frac{1}{\sigma(t)}\right|^2 d\mathring{\mu}_s.$$

The last factor is zero because $1/\sigma$ vanishes $d\mathring{\mu}_s$ a.e., and the other factor is finite because

$$\int\left|\frac{g}{B_k}\right|^2 d\mathring{\mu}_s = \int |g|^2 d\mathring{\mu}_s \le \int |g|^2 d\mathring{\mu} = \|g\|_{2,\mathring{\mu}}^2 \le \|f\|_{2,\mathring{\mu}}^2 < \infty.$$

Hence $\langle h, B_k\rangle_{\mathring{\mu}_s} = 0$ for all $k = 0, 1, \ldots$. Thus the situation is exactly as in the case of an absolutely continuous measure. $\square$

We can combine Theorem 7.2.1 with the previous theorem to get the following corollary.

Corollary 7.2.4. *The following holds:*

1. *If the Blaschke product diverges (i.e., (7.5) holds), then $\log\mu' \notin L_1(\mathring{\lambda})$ iff $\mathcal{L}_\infty$ is dense in $L_2(\mathring{\mu})$.*

2. If $\log \mu' \in L_1(\overset{\circ}{\lambda})$, then the Blaschke product diverges iff $\mathcal{L}_\infty$ is dense in $H_2(\overset{\circ}{\mu})$.

Proof.

1. If the Blaschke product diverges then $\mathcal{L}$, the $L_2(\overset{\circ}{\mu})$-closure of $\mathcal{L}_\infty$, is equal to $H_2(\overset{\circ}{\mu})$ by Corollary 7.2.2. Moreover, by Theorem 7.2.1, $\log \mu' \notin L_1(\overset{\circ}{\lambda})$ iff $H_2(\overset{\circ}{\mu}) = L_2(\overset{\circ}{\mu})$. Thus if the Blaschke product diverges $\log \mu' \notin L_1(\overset{\circ}{\lambda})$ iff $\mathcal{L} = L_2(\overset{\circ}{\mu})$, which proves the first part.
2. It is always true that the divergence of the Blaschke product implies the density of $\mathcal{L}_\infty$ in $H_2(\overset{\circ}{\mu})$. However, if $\log \mu' \in L_1(\overset{\circ}{\lambda})$, then the density of $\mathcal{L}_\infty$ in $H_2(\overset{\circ}{\mu})$ implies the divergence of the Blaschke product by the second part of Theorem 7.2.3. This proves the second part. $\square$

Later, in Section 9.4, we shall give other equivalent conditions for density, convergence, and boundedness under the more restrictive condition that the $|\alpha_n|$ are bounded away from the boundary. If course if the $|\alpha_n|$ are bounded away from the boundary, then the Blaschke product will automatically diverge.

We can also prove that when the Blaschke product diverges then $\mathcal{R}_\infty$ is dense in $L_2(\mu)$.

Corollary 7.2.5. *Suppose the Blaschke product diverges to zero and that $\overset{\circ}{\mu}$ is a positive measure on $\overline{\partial \mathbb{O}}$. Then $\mathcal{R}_\infty = span\{B_n\}_{n\in\mathbb{Z}}$ is dense in $L_2(\overset{\circ}{\mu})$.*

Proof. To prove the density, we have to show that if $f \in L_2(\overset{\circ}{\mu})$ is orthogonal to $\mathcal{R}_\infty$, then it is zero. Consider first the case of the disk, so that $\mathbb{O} = \mathbb{D}$ and $d\overset{\circ}{\mu} = d\mu$. By the previous theorem $\mathcal{L}_\infty$ is dense in $H_2(\mu)$, so that if $\int f(t)B_k(t)d\mu(t) = 0$ for all $k = 0, 1, \ldots$, then $\int f(t)t^{-k}d\mu(t) = 0$ for all $k = 0, 1, \ldots$. Now consider the function

$$F(z) = \int C(t, z) f_*(t) d\mu(t),$$

where $C(t, z)$ is the Cauchy kernel. This F is analytic in $\mathbb{D}$ because it is the Cauchy–Stieltjes integral of the complex measure $d\nu = f_* d\mu$. Moreover $F(z)$ is of bounded variation and belongs to H_p for any $p < 1$ [76, Theorem 3.5, p. 39]. By our assumption, $\langle f, B_k \rangle = 0$ for $k \in \mathbb{Z}$, and in particular $\int f_* B_k d\mu = 0, k = 0, -1, -2, \ldots$. This implies that the α_k are zeros of F since $C(t, \alpha_k) \in$ span $\{B_0, B_{-1}, \ldots\}$ and hence

$$F(\alpha_k) = \int C(t, \alpha_k) f_*(t) d\mu(t) = 0 \quad \text{for all } k = 0, 1, \ldots.$$

Suppose for simplicity that there are infinitely many different α_k. If not one has to show first that the α_k that is infinitely many times repeated is also a zero of F of infinite order. Because F vanishes in the zeros α_k that are the zeros of a divergent Blaschke product, and because F belongs to the Nevanlinna class $N \supset H_p$, it follows from its inner–outer decomposition that it vanishes identically in $\mathbb{D}$. This then implies that

$$\int f_*(t)t^{-k}d\mu(t) = \int f(t)t^k d\mu(t) = 0 \quad \text{for all } k = 0, 1, \ldots$$

(compare with Ref. [92, p. 62]). Thus f is orthogonal to all the elements in $\{t^n\}_{n\in\mathbb{Z}}$, which is complete in $L_2(\mu)$. Thus f is at the same time in $L_2(\mu)$ and orthogonal to it. Hence it is zero μ a.e.

The case of the half plane is treated similarly. We give a brief sketch. First it is observed that $f \perp_{\mathring{\mu}} H_2(\mathring{\mu})$; hence

$$\int f(t)e^{-\mathrm{i}xt}d\mathring{\mu}(t) = 0 \quad \text{for } x \geq 0.$$

Furthermore, we note that $C(t, z)$ can be written as an integral (compare with Ref. [92, p. 62]), $C(t, z) = \int_0^\infty \exp\{\mathrm{i}(z - t)x\}d\lambda(x)$, so that in that case $\int f_*(t)C(t, \alpha_k)\, d\mathring{\mu}(t) = 0$ implies

$$\int f_*(t)e^{-\mathrm{i}xt}d\mathring{\mu}(t) = 0 \quad \text{for all } x \geq 0$$

and thus also

$$\int f(t)e^{\mathrm{i}xt}d\mathring{\mu}(t) = 0 \quad \text{for all } x \geq 0.$$

Thus $f \in L_2(\mathring{\mu})$, while at the same time, its Fourier transform vanishes identically. Thus $f = 0$ $\mathring{\mu}$ a.e.   □

Another kind of density result relates to the representation of positive functions in L_1. It is for instance well known that $f \in L_1$, $f \geq 0$ a.e. on $\mathbb{T}$ and $\log f \in L_1$ if and only if $f = |g|^2$ with $g \in H_2$. In fact, any positive trigonometric polynomial can be written as the square modulus of a polynomial of the same degree. For a more general function f, we can take g to be the outer spectral factor of f. One finds this result in any standard work, for example, in Grenander and Szegő [102, pp. 23–26]. The density of the trigonometric polynomials then implies that any positive function from L_1 with integrable logarithm can be approximated arbitrarily well by a positive trigonometric polynomial, and hence by the square modulus of an outer polynomial. This result can be generalized as follows.

Theorem 7.2.6. *Let the Blaschke product diverge to 0 and take* $f \in L_1(\mathring{\lambda})$. *Suppose* $f \geq 0$ *a.e. on* $\overline{\partial \mathbb{O}}$ *and* $\log f \in L_1(\mathring{\lambda})$. *Then for every* $\epsilon > 0$, *there is some* $f_n \in \mathcal{L}_n$ *for n sufficiently large such that* $\|f - |f_n|^2\|_1 < \epsilon$ *(norm in* $L_1(\mathring{\lambda}t))$.

Proof. By the above-mentioned property, we may replace f by $|g|^2$ with $g \in \mathring{H}_2$ and outer. Thus we have to prove that there exists an f_n such that $\||g|^2 - |f_n|^2\|_1 < \epsilon$. By a property that can be found in the book by Rudin [187, p. 78], we may use for $p = 2$

$$\||g|^p - |h|^p\|_1 \leq 2pR^{p-1}\|g - h\|_p, \quad \text{with} \quad \max\{\|g\|_p, \|h\|_p\} \leq R.$$

Thus

$$\||g|^2 - |f_n|^2\|_1 \leq 4R \left[\int |g - f_n|^2 d\lambda \right]^{1/2}.$$

We can always find an interpolant f_n that makes $|g - f_n|$ arbitrary small and hence also $|f_n| < |g| + \epsilon_1$ with $\epsilon_1 > 0$ arbitrary small. Thus, because $g \in \mathring{H}_2$, f_n will also be in $\mathring{H}_2$, so that R is bounded, whereas $\|g - f_n\|_2$ can be made as small as we want. This proves the theorem. $\qquad\square$

The space $\mathcal{L}_\infty$ will be dense in $L_2(\mathring{\mu})$ whenever the Parseval equality holds for all functions (or for a set of basis functions) in $L_2(\mathring{\mu})$. This result was stated for the polynomial case in Ref. [2, Theorem 2.2.3]. The same argument can be used in the rational case. Consider a function $f \in L_2(\mathring{\mu})$. Suppose it has the formal Fourier series expansion

$$f(t) \sim \sum_{k=0}^{\infty} c_k \phi_k(t), \quad c_k = \langle f, \phi_k \rangle_{\mathring{\mu}}.$$

It is well known that

$$\min \left\{ \|f - f_n\|_{\mathring{\mu}}^2 : f_n \in \mathcal{L}_n \right\}$$

is obtained for

$$f_n(t) = \sum_{k=0}^{n} c_k \phi_k(t)$$

and the minimum is

$$\|f\|_{\mathring{\mu}}^2 - \sum_{k=0}^{n} |c_k|^2.$$

Obviously, this implies the Bessel inequality

$$\sum_{k=0}^{\infty} |c_k|^2 \le \|f\|_{\mathring{\mu}}^2 = \int |f(t)|^2 d\mathring{\mu}(t).$$

If, however, equality holds, then we get Parseval's equality and this is equivalent with the fact that the function f can be approximated arbitrary close in $L_2(\mathring{\mu})$ by functions from $\mathcal{L}_n$ if n is sufficiently large.

Basically, the previous argument is used to obtain another density result in Chapter 10; see Theorem 10.3.4. A solution of the moment problem considered in Chapter 10 is called N-extremal if the Parseval equality holds for a certain function. If the Blaschke product diverges, then the moment problem considered in Chapter 10 will have a unique solution, which will be N-extremal and Theorem 10.3.4 then says that $\mathcal{L}_\infty$ is dense in $L_2(\mathring{\mu})$. Thus it is sufficient that the Blaschke product diverges for $\mathcal{L}_\infty$ to be dense in $L_2(\mathring{\mu})$. It is, however, not a necessary condition because the same theorem says that if $\mathring{\mu}$ is an N-extremal solution of the moment problem, which may well exist even if the Blaschke product does not diverge, then also $\mathcal{L}_\infty$ will be dense in $L_2(\mathring{\mu})$.

8

Favard theorems

In this chapter we shall prove two Favard type theorems, which say that if we have a sequence of functions generated by a recurrence of the type we studied in Chapter 4, then there will be a measure with respect to which these are orthogonal. We not only prove the existence of the measure, but really give a constructive proof. Since in our development, we have derived the recurrence relation for the orthogonal functions from the recurrence relation for the reproducing kernels, it seems a natural question to ask whether a Favard theorem exists for reproducing kernels as well. Thus, if we have a recurrence relation for some kernel functions as we gave them in Chapter 3, are these then reproducing kernels for the rational spaces with respect to some positive measure? We were not completely successful in this respect, but we include the results obtained so far.

8.1. Orthogonal functions

In Sections 3.3 and 4.1, we have seen how the kernels, as well as the orthogonal functions, satisfied certain recurrence relations that generalize the Szegő recurrence relations. It is thus true that all rational functions that are orthogonal (or reproducing kernels) with respect to a certain positive measure on the boundary $\partial \mathbb{O}$ will satisfy such a recurrence. The converse of this theorem is known as a Favard theorem, named after Favard's paper [79]. Such a Favard theorem states that if functions satisfy recurrence relations as we have given in Section 4.1, then they will give orthogonal rational functions with respect to some positive measure on the boundary $\partial \mathbb{O}$. A simple proof for the Szegő polynomials was given in Ref. [78]. There, not only the existence of the measure is proved, but the measure is actually constructed. We shall follow in this section a similar approach for the rational case. Related results in a somewhat different setting were obtained in Refs. [112] and [31].

161

We shall give a sequence of lemmas that will eventually lead to the proof of the Favard Theorem.

Lemma 8.1.1. Suppose we are given two sequences of numbers $\alpha_k \in \mathbb{O}$ and $\lambda_k \in \mathbb{D}$ for $k = 1, 2, \ldots$. Define the numbers $e_k > 0$ by their squares

$$e_k^2 = \frac{\varpi_k(\alpha_k)}{\varpi_{k-1}(\alpha_{k-1})} \frac{1}{1 - |\lambda_k|^2} \quad \text{for } k = 1, 2, \ldots. \tag{8.1}$$

Finally, define the functions ϕ_k by

$$\phi_0 = 1, \quad \phi_k(z) = e_k \eta_k^1 \frac{\varpi_{k-1}(z)}{\varpi_k(z)} [\zeta_{k-1}(z)\phi_{k-1}(z) + \bar{\lambda}_k \phi_{k-1}^*(z)], \quad k = 1, 2, \ldots, \tag{8.2}$$

where the numbers $\eta_k^1 \in \mathbb{T}$ are chosen such that $\phi_k^*(\alpha_k) > 0$.
Then the functions ϕ_k^* satisfy the following recurrence:

$$\phi_0^* = 1, \quad \phi_k^*(z) = e_k \eta_k^2 \frac{\varpi_{k-1}(z)}{\varpi_k(z)} [\zeta_{k-1}(z)\lambda_k \phi_{k-1}(z) + \phi_{k-1}^*(z)], \quad k = 1, 2, \ldots, \tag{8.3}$$

with

$$\eta_k^2 = \bar{\eta}_k^1 \bar{z}_{k-1} z_k, \tag{8.4}$$

with z_k as in (2.1). Moreover, $1/\phi_n^* \in \mathring{H}_2$.

Proof. We can leave the proof of the first part to the reader because it comes down to simple calculus. For the second part, note that we can couple the two recurrences (8.2) and (8.3) into the form (4.1). We can use this form recursively and end up with (recall $\alpha_0 = 0$)

$$\begin{bmatrix} \phi_n \\ \phi_n^* \end{bmatrix} = \frac{c_n}{\varpi_n(z)} \Theta_n \begin{bmatrix} 1 \\ 1 \end{bmatrix},$$

where $c_n = \prod_{k=1}^n e_k$ and where Θ_n is given by $\Theta_n = \theta_n \cdots \theta_0$ with

$$\theta_k = \begin{bmatrix} \eta_k^1 & 0 \\ 0 & \eta_k^2 \end{bmatrix} \begin{bmatrix} 1 & \bar{\lambda}_k \\ \lambda_k & 1 \end{bmatrix} \begin{bmatrix} \zeta_{k-1}(z) & 0 \\ 0 & 1 \end{bmatrix},$$

which is a J-contractive matrix. Hence, because

$$\phi_n^* = \frac{c_n}{\varpi_n(z)} [(\Theta_n)_{21} + (\Theta_n)_{22}]$$

it follows by Theorem 1.5.3(5) that

$$1/\phi_n^* = c_n^{-1}(\varpi_n(z))[(\Theta_n)_{21} + (\Theta_n)_{22}]^{-1} \in \mathring{H}_2,$$

which concludes the proof. $\square$

Lemma 8.1.2. Under the conditions of Lemma 8.1.1, it holds that

$$\lambda_k = \eta_k \frac{\overline{\phi_k(\alpha_{k-1})}}{\phi_k^*(\alpha_{k-1})} \quad \text{with} \quad \eta_k = z_k \bar{z}_{k-1} \frac{\varpi_{k-1}(\alpha_k)}{\varpi_k(\alpha_{k-1})} \in \mathbb{T}. \tag{8.5}$$

Proof. Using

$$\overline{\left(\frac{\varpi_{k-1}(\alpha_{k-1})}{\varpi_k(\alpha_{k-1})}\right)} = \frac{\varpi_{k-1}(\alpha_{k-1})}{\varpi_{k-1}(\alpha_k)}$$

and $\phi_{k-1}^*(\alpha_{k-1}) > 0$, we find from the recurrences for ϕ_k and ϕ_k^* that

$$\overline{\phi_k(\alpha_{k-1})} = e_k \bar{\eta}_k^1 \frac{\varpi_{k-1}(\alpha_{k-1})}{\varpi_{k-1}(\alpha_k)}[0 + \lambda_k \phi_{k-1}^*(\alpha_{k-1})] \tag{8.6}$$

and

$$\phi_k^*(\alpha_{k-1}) = e_k \eta_k^2 \frac{\varpi_{k-1}(\alpha_{k-1})}{\varpi_k(\alpha_{k-1})}[0 + \phi_{k-1}^*(\alpha_{k-1})]. \tag{8.7}$$

Dividing (8.6) by (8.7) gives

$$\frac{\overline{\phi_k(\alpha_{k-1})}}{\phi_k^*(\alpha_{k-1})} = \frac{\bar{\eta}_k^1}{\eta_k^2} \frac{\varpi_k(\alpha_{k-1})}{\varpi_{k-1}(\alpha_k)}\lambda_k = z_{k-1}\bar{z}_k \frac{\varpi_k(\alpha_{k-1})}{\varpi_{k-1}(\alpha_k)}\lambda_k = \bar{\eta}_k \lambda_k.$$

The result now follows. $\square$

Lemma 8.1.3. Let the functions ϕ_n be as constructed in Lemma 8.1.1. Define the positive measure μ_n on $\overline{\partial\mathbb{O}}$ by

$$d\mathring{\mu}_n(t) = \frac{|\varpi_0(t)|^2 P(t, \alpha_n)}{|\phi_n(t)|^2} d\mathring{\lambda}(t) = \frac{P(t, \alpha_n)}{|\phi_n(t)|^2} d\lambda(t), \tag{8.8}$$

with $P(z, w)$ the Poisson kernel. Then the functions ϕ_k satisfy the orthogonality relations

$$\langle \phi_k, \phi_\ell \rangle_{\mathring{\mu}_n} = \delta_{k\ell} \quad \text{for } 0 \le k, \ell \le n. \tag{8.9}$$

Proof. We shall first prove that ϕ_n is orthonormal with respect to all its predecessors:

$$\langle \phi_n, \phi_m \rangle_{\tilde{\mu}_n} = \delta_{nm} \quad \text{for } 0 \leq m \leq n. \tag{8.10}$$

This is shown as follows:

$$
\begin{aligned}
\langle \phi_n, \phi_m \rangle_{\tilde{\mu}_n} &= \int \frac{\phi_n(t)\phi_{m*}(t)}{\phi_n(t)\phi_{n*}(t)} P(t, \alpha_n)\, d\lambda(t) \\
&= \int \frac{\phi_{m*}(t)}{\phi_{n*}(t)} P(t, \alpha_n)\, d\lambda(t) \\
&= \int B_{n \setminus m}(t) \frac{\phi_m^*(t)}{\phi_n^*(t)} P(t, \alpha_n)\, d\lambda(t) \\
&= B_{n \setminus m}(z) \frac{\phi_m^*(z)}{\phi_n^*(z)}\Big|_{z=\alpha_n} = \delta_{nm}.
\end{aligned}
$$

The general orthogonality follows from Theorem 4.1.6, which says that all the previous orthogonal functions are defined in terms of ϕ_n, orthonormal to $\mathcal{L}_{n-1}$ by the inverse recurrence, and we have just proved by the previous Lemma 8.1.2 that the recurrence in the Lemma 8.1.1 is the same as the one from Theorem 4.1.1. □

We are now ready to prove the following Favard type theorem.

Theorem 8.1.4. *There exists a Borel measure on $\overline{\partial \mathbb{O}}$ for which the ϕ_n as constructed in Lemma 8.1.1 are the orthonormal functions.*

The measure is unique when the Blaschke product with zeros α_k diverges (to zero).

Proof. For notational reasons, we give the proof for the case of the disk. It can be easily adapted for the half plane. Define the linear functional M on $\mathcal{R}_\infty = \mathcal{L}_\infty \cdot \mathcal{L}_{\infty*}$ by $M(\phi_k \phi_{l*}) = \delta_{kl}, k, l = 0, 1, \ldots$ Obviously the ϕ_k are orthogonal with respect to this functional. We prove that this functional can be represented as an integral such that

$$M(f) = \int f\, d\mu \quad \text{for all } f \in C(\mathbb{T}).$$

Define the measure μ_n as in (8.8). Let us temporarily switch back from our notation for the unit circle to the notation for the interval $(0, 2\pi]$. To avoid confusion, we set $\tilde{\mu}_n(\theta)\,d\theta = d\mu_n(e^{i\theta})$. Thus for $0 < t \le 2\pi$

$$\tilde{\mu}_n(t) = \int_0^t \frac{P(e^{i\theta}, \alpha_n)}{|\phi_n(e^{i\theta})|^2}\,d\lambda(\theta) = \int_0^t d\tilde{\mu}_n(\theta).$$

These are all increasing functions and uniformly bounded ($\int d\tilde{\mu}_n = 1$). Hence, there exists a subsequence $\{\tilde{\mu}_{n_k}\}$ and a distribution function $\tilde{\mu}$ such that

$$\lim_{k\to\infty} \tilde{\mu}_{n_k}(\theta) = \tilde{\mu}(\theta) = \mu(e^{i\theta})$$

and

$$\lim_{k\to\infty} \int f(e^{i\theta})\,d\tilde{\mu}_{n_k}(\theta) = \int f(e^{i\theta})\,d\tilde{\mu}(\theta) = \int f(t)\,d\mu(t)$$

for all functions f continuous on $\mathbb{T}$. Thus the $\{\phi_n\}$ are an orthonormal system with respect to this measure $\mu(e^{it}) = \tilde{\mu}(t)$. Hence, the linear functional M defined on $\mathcal{R}_\infty = \mathcal{L}_\infty \cdot \mathcal{L}_{\infty*}$ by $M(\phi_k \phi_{l*}) = \delta_{kl}$, $k, l = 0, 1, \ldots$ can be extended to a bounded functional on $C(\mathbb{T})$.

The uniqueness follows from the representation of bounded linear functionals on $C(\partial\mathbb{O})$ (F. Riesz) and the density property of the previous chapter. $\square$

8.2. Kernels

We shall in this section try to formulate a Favard theorem for the reproducing kernels. We were not completely successful, but we give the results obtained so far. Even in the polynomial case, this problem has not been considered before.

Lemma 8.2.1. Recall the definition $\alpha_0 = 0$ for $\mathbb{O} = \mathbb{D}$ and $\alpha_0 = i$ for $\mathbb{O} = \mathbb{U}$. Furthermore, let $\alpha_1, \alpha_2, \ldots$ be complex numbers in $\mathbb{O}$. Let $\rho_k, k = 1, 2, \ldots$ be given functions in $\mathcal{B}$ (i.e., $|\rho_k(w)| < 1$ for $w \in \mathbb{O}$). Define $\gamma_k(w) = -\zeta_k(w)\rho_k(w) \in \mathcal{B}$ for $k = 1, 2, \ldots$. Let the functions $K_n(z, w)$ be generated by (3.12), that is,

$$\begin{bmatrix} K_n^*(z, w) \\ K_n(z, w) \end{bmatrix} = \theta_n(z, w) \begin{bmatrix} K_{n-1}^*(z, w) \\ K_{n-1}(z, w) \end{bmatrix}; \qquad \begin{bmatrix} K_0^*(z, w) \\ K_0(z, w) \end{bmatrix} = \begin{bmatrix} 1 \\ 1 \end{bmatrix}, \qquad (8.11)$$

with the matrix θ_n given by

$$\theta_n(z, w) = \frac{1}{\sqrt{1 - |\rho_n|^2}} \begin{bmatrix} 1 & \overline{\rho}_n \\ \rho_n & 1 \end{bmatrix} \begin{bmatrix} \zeta_n(z) & 0 \\ 0 & 1 \end{bmatrix} \frac{1}{\sqrt{1 - |\gamma_n|^2}} \begin{bmatrix} 1 & \overline{\gamma}_n \\ \gamma_n & 1 \end{bmatrix}.$$

$$(8.12)$$

Then the following properties hold:

$$1/K_n(z, w) \in \mathring{H}_2 \quad \text{for } w \in \mathbb{O} \text{ fixed.} \tag{8.13}$$

$$K_n(w, w) = K_{n-1}(w, w)\sqrt{\frac{1 - |\gamma_n(w)|^2}{1 - |\rho_n(w)|^2}} = \prod_{k=1}^{n} \left(\frac{1 - |\gamma_n(w)|^2}{1 - |\rho_n(w)|^2}\right)^{1/2} > 0. \tag{8.14}$$

$$\overline{\rho_n(w)} = \frac{K_n^*(\alpha_n, w)}{K_n(\alpha_n, w)}. \tag{8.15}$$

Proof. Again, (8.13) follows from Theorem 3.3.3 as for the orthogonal functions because the recurrence is J-unitary by definition.

The proof of (8.14) is pure calculus: Just replace z by w and play a bit with the formulas.

Equality (8.15) is also immediate, since

$$\frac{K_n^*(\alpha_n, w)}{K_n(\alpha_n, w)} = \frac{\overline{\rho}_n[K_{n-1}(\alpha_n, w) + \gamma_n(w)K_{n-1}^*(\alpha_n, w)]}{[K_{n-1}(\alpha_n, w) + \gamma_n(w)K_{n-1}^*(\alpha_n, w)]} = \overline{\rho}_n. \qquad \square$$

Note that as in the case of the reproducing kernels, the previous result implies that K_n defines all the previous K_j, $j = n - 1, n - 2, \ldots, 0$ uniquely. Thus if $K_n(z, w)$ is a normalized reproducing kernel for $\mathcal{L}_n$ with respect to some measure μ, then the $K_j(z, w)$ will be normalized reproducing kernels for $\mathcal{L}_j$, $0 \le j \le n$.

Concerning the reproducing property of $K_n(z, w)$ we have the following lemma, which is somewhat deceiving as we shall discuss after its proof.

Lemma 8.2.2. Let the K_n be generated as in the previous Lemma 8.2.1 and define

$$k_m(z, w) = K_m(z, w)K_m(w, w) \quad \text{for } m = 0, 1, \ldots. \tag{8.16}$$

Define the measure μ_n by

$$d\mathring{\mu}_n(t) = \frac{|\varpi_0(t)|^2 P(t, w)}{|K_n(t, w)|^2} d\mathring{\lambda}(t) = \frac{P(t, w)}{|K_n(t, w)|^2} d\lambda(t), \tag{8.17}$$

where $P(z, w)$ is the Poisson kernel. Then

$$\langle f(t), k_m(t, w)\rangle_{\mathring{\mu}_n} = f(w), \quad \forall f \in \mathcal{L}_m, \ 0 \le m \le n. \tag{8.18}$$

Proof. We first prove this for $m = n$. This is easily seen as follows. It holds for all $f \in \mathcal{L}_n$ that

$$\langle f, k_n(\cdot, w) \rangle_{\hat{\mu}_n} = \int f(t) k_{n*}(t, w) \frac{|\varpi_0(t)|^2 P(t, w) \, d\mathring{\lambda}(t)}{K_n(t, w) K_{n*}(t, w)}$$

$$= K_n(w, w) \int f(t) \frac{k_{n*}(t, w)}{k_{n*}(t, w)} \frac{P(t, w)}{K_n(t, w)} \, d\lambda(z)$$

$$= f(w) \frac{K_n(w, w)}{K_n(w, w)} = f(w).$$

We show next the induction step, which starts from the fact that (8.18) is true for a certain m, and we shall prove that it will be also true for $m - 1$. That is, we have to prove

$$\langle f, k_m(\cdot, w) \rangle_{\hat{\mu}_n} = f(w), \quad \forall f \in \mathcal{L}_m \Rightarrow \langle f, k_{m-1}(\cdot, w) \rangle_{\hat{\mu}_n} = f(w),$$

$$\forall f \in \mathcal{L}_{m-1}. \quad (8.19)$$

We first derive the following expression from the recurrence, which can be obtained with some tedious calculations:

$$k_m(z, w)[\zeta_m(z)\overline{\gamma}_m + \overline{\rho}_m]$$
$$= -[\zeta_m(z)(1 - |\gamma_m|^2)]k_{m-1}^*(z, w) + [\zeta_m(z)\overline{\gamma}_m \rho_m + 1]k_m^*(z, w).$$

After dividing by B_m this becomes

$$\frac{k_m(z, w)}{B_m(z)}[\zeta_m(z)\overline{\gamma}_m + \overline{\rho}_m]$$
$$= -(1 - |\gamma_m|^2)k_{m-1*}(z, w) + [\zeta_m(z)\overline{\gamma}_m \rho_m + 1]k_{m*}(z, w).$$

Now we use this in the right-hand side of (8.19) to get

$$\langle f, k_{m-1}(\cdot, w) \rangle_{\hat{\mu}_n} = \frac{1}{1 - |\gamma_m|^2} \langle f, [(\zeta_m \overline{\gamma}_m \rho_m + 1)k_{m*}]_* \rangle_{\hat{\mu}_n}$$

$$- \frac{1}{1 - |\gamma_m|^2} \left\langle f, \left[\frac{k_m}{B_m}(\zeta_m \overline{\gamma}_m + \overline{\rho}_m) \right]_* \right\rangle_{\hat{\mu}_n}$$

$$= \frac{1}{1 - |\gamma_m|^2} \langle (\zeta_m \overline{\gamma}_m \rho_m + 1)f, k_m \rangle_{\hat{\mu}_n}$$

$$- \frac{1}{1 - |\gamma_m|^2} \langle (\zeta_m \overline{\gamma}_m + \overline{\rho}_m)f, k_m^* \rangle_{\hat{\mu}_n}.$$

The first term can be simplified because $(\zeta_m \bar{\gamma}_m \rho_m + 1) f \in \mathcal{L}_m$ and k_m reproduces f in w. Therefore, the inner product in the first term equals

$$f(w)(\zeta_m(w)\bar{\gamma}_m(w)\rho_m(w) + 1) = -(1 - |\gamma_m(w)|^2).$$

The inner product in the second term equals

$$\langle (\zeta_m \bar{\gamma}_m + \bar{\rho}_m) f, k_m^* \rangle_{\mathring{\mu}_n} = \langle k_m, f^*(\gamma_m + \zeta_m \rho_m) \rangle_{\mathring{\mu}_n}.$$

The latter has a factor $(\gamma_m + \zeta_m \rho_m) f^*$, which is again in $\mathcal{L}_m$ and we can again use the reproducing property of k_m to find that inner product equals the complex conjugate of

$$f^*(w)(\gamma_m(w) + \zeta_m(w)\rho_m(w)),$$

which is zero by the definition of γ_m. Thus, after filling in the last two results for the inner products, we get

$$\langle f, k_{m-1}(\cdot, w) \rangle_{\mathring{\mu}_n} = f(w) + 0 = f(w),$$

which proves the induction step.   □

Note that the previous result does not imply that k_m is a reproducing kernel. Indeed, it reproduces f in w but only with respect to a measure $\mathring{\mu}_n$ that depends upon the specific w in which the function is reproduced. It will only be a reproducing kernel if there exists a measure that is independent of w.

This is about as far as we can get without further specification of how ρ_n depends upon w. For an arbitrary sequence of numbers $\rho_k(w)$, depending on w and satisfying $|\rho_k(w)| < 1$, one may not expect that the corresponding Θ_n matrix given by $\Theta_n = \theta_n \theta_{n-1} \cdots \theta_1$ contains (normalized) reproducing kernels for $\mathcal{L}_n$ with respect to any measure whatsoever. The situation is different from the Nevanlinna–Pick algorithm of Section 6.4 where the Θ_n matrix did give these kernels, but there the $\rho_k(w)$ were special since they were generated starting from a measure μ that did not depend on w. The w there was introduced during the definition of the starting values for the Nevanlinna–Pick algorithm.

For arbitrary $\rho_k(w)$, $k = 1, \ldots, n$, one can build Θ_n as the product of the θ_k given by Lemma 8.2.1, from which we can extract $K_n = (\Theta_n)_{21} + (\Theta_n)_{22}$ and the corresponding measure μ_n as in Lemma 8.2.2. We then do have (8.18), which does not identify $k_n(z, w)$ as a reproducing kernel as we said before.

If $\{\phi_0, \ldots, \phi_n\}$ is an orthonormal basis for $\mathcal{L}_n$, then the kernels are

$$k_n(z, w) = \sum_{k=0}^{n} \phi_k(z)\overline{\phi_k(w)},$$

and this reflects a specific symmetry in z and w. It implies, for example, that as a function of w, $\overline{k_n(z, w)}$ should be in $\mathcal{L}_n$. In general, a reproducing kernel should be sesqui-analytic, that is, $\overline{k_n(z, w)} = k_n(w, z)$ and, more specifically, in $\mathcal{L}_n$ all the relations given in Theorem 2.2.3 should hold. This means that the way in which $k_n(z, w)$ depends upon w is very special, and one should not expect that the choice of arbitrary $\rho_k(w)$, which depend in some exotic way on w, will provide this. One can easily check this by considering the simple case of $n = 1$ for example.

So we shall have to introduce the notion of a sequence $\rho_k(w)$ having the property that the corresponding k_n are indeed reproducing kernels. We shall say that such a sequence $\rho_k(w)$ has the RK (reproducing kernel) property.

Since the $k_n(z, w)$ as they were generated in the previous lemmas depend upon w via $\rho_i(w)$ in a very complex way, it is not easy to find conditions on how the coefficients $\rho_i(w)$ should depend upon w to ensure that $\overline{k_n(z, w)}$, as a function of w, is in $\mathcal{L}_n$. The reader is invited to try and check this for the simplest possible case $n = 1$.

It remains an open problem to find a direct and simple characterization of the $\rho_i(w)$ having the RK property. For the moment we content ourselves with a characterization that is in the line of how these $\rho_i(w)$ are produced by the Nevanlinna–Pick algorithm and we shall formulate some equivalent conditions. Unfortunately, none of these will give a direct characterization of how the coefficients ρ_k should depend on w. If such a characterization exists, it is still to be found.

As explained in the previous lemmas, the coefficients $\{\rho_i(w) : i = 1, \ldots, n\}$ define uniquely the normalized kernels $\{K_i(z, w) : i = 1, \ldots, n\}$ and thus also the kernels $\{k_i(z, w) : i = 1, \ldots, n\}$, as well as the J-inner factors $\{\theta_i(z, w) : i = 1, \ldots, n\}$ and thus also the products $\{\Theta_i(z, w) = \theta_i \cdots \theta_1 : i = 1, \ldots, n\}$. Conversely, the $K_i(z, w) = (\Theta_i)_{21} + (\Theta_i)_{22}$ (and similarly the $L_i(z, w)$) can be recovered from the Θ_i. These $K_i(z, w)$ define uniquely the $k_i(z, w)$ and also the $\rho_i(w)$ and thus the θ_i as well. In other words, there is a one-to-one correspondence between all these sets of quantities.

We shall say that one of these (and therefore also all the others) has the RK property, if on $\mathcal{L}_n$, the inner product $\langle \cdot, \cdot \rangle_{\tilde{\mu}_n}$ is independent of w, where μ_n is the measure defined in terms of the $K_n(z, w)$ by an expression like (8.17).

It is an immediate consequence of Theorem 6.4.3 that the $\rho_i(w)$ will have the RK property if they can be generated by the Pick–Nevanlinna algorithm applied to some $\Delta_0 \in \mathcal{A}$, $\Delta_{01}(w) = 0$ of the form

$$\Delta_0 = [1 - \mathcal{S}_w(\tilde{\mu}) \quad 1 + \mathcal{S}_w(\tilde{\mu})] \tag{8.20}$$

for some measure $\tilde{\mu}$ satisfying $\int d\tilde{\mu} = 1$ and independent of w. Recall that S_w is as defined in (6.22). Thus we have by Lemma 6.2.2 for the disk case

$$\Omega_{\tilde{\mu}}(z, w) = S_w(\tilde{\mu})(z) = \frac{\Omega_{\tilde{\mu}}(z)}{P(z, w)} + \frac{1}{1 - |w|^2}\left(\frac{w}{z} - z\overline{w}\right), \qquad (8.21)$$

and for the half plane we have by Lemma 6.2.3

$$\Omega_{\tilde{\mu}}(z, w) = S_w(\tilde{\mu})(z) = \frac{\Omega_{\tilde{\mu}}(z)}{P(z, w)(1 + z^2)} + \frac{(z^2 - 1)\operatorname{Re} w + z(1 - |w|^2)}{(i\operatorname{Im} w)(1 + z^2)},$$

$$\qquad (8.22)$$

where in both cases $\Omega_{\tilde{\mu}}(z) = S_0(\tilde{\mu})(z)$, with S_0 the usual Riesz–Herglotz–Nevanlinna transform, which is a special case of S_w obtained by setting $w = \alpha_0$.

Since $\Omega_n(z, w) = L_n(z, w)/K_n(z, w)$ as generated by the Pick–Nevanlinna algorithm shall interpolate this $S_w(\tilde{\mu})$ in $A_n^w = \{w, \alpha_1, \ldots, \alpha_n\}$, also $L_n(z, \alpha_0)/K_n(z, \alpha_0)$ will interpolate $\Omega_{\tilde{\mu}} = S_0(\tilde{\mu})$ in $A_n^0 = \{\alpha_0, \alpha_1, \ldots, \alpha_n\}$. Thus in the case of the disk, for example, $\Theta_n(z, w)$ will have the RK property if

$$\frac{L_n(z, w)}{K_n(z, w)} = \frac{L_n(z, 0)/K_n(z, 0)}{P(z, w)} + \frac{1}{1 - |w|^2}\left(\frac{w}{z} - z\overline{w}\right) \quad \text{for } z \in A_n^w.$$

The construction of these interpolants Ω_n can be done by applying the Nevanlinna–Pick algorithm as explained in Section 6.4 since this algorithm constructs Θ_n and the K_n and L_n can be obtained from the latter. However, they can also be obtained by running the algorithm backwards. Indeed, setting $\Delta_n = [0\ 2]$, and forming the array $\Delta_0 = \Delta_n\Theta_n$ where Θ_n is the J-unitary matrix generated by the Nevanlinna–Pick algorithm, we shall get

$$\Omega_n(z, w) = \frac{L_n(z, w)}{K_n(z, w)} = \frac{\Delta_{02}(z, w) - \Delta_{01}(z, w)}{\Delta_{02}(z, w) + \Delta_{01}(z, w)}.$$

More generally, one can choose any $\tilde{\Delta}_n \in \mathcal{A}$ with $\tilde{\Delta}_{n1}(w) = 0$ and generate $\tilde{\Delta}_0 = \tilde{\Delta}_n\Theta_n$, which gives

$$\tilde{\Omega}_n(z, w) = \frac{\tilde{\Delta}_{02}(z, w) - \tilde{\Delta}_{01}(z, w)}{\tilde{\Delta}_{02}(z, w) + \tilde{\Delta}_{01}(z, w)},$$

which will also interpolate $\Omega_{\tilde{\mu}}(z, w)$ for $z \in A_n^w$. If this $\Omega(z, w)$ equals $S_w(\tilde{\mu})(z)$ for some measure $\tilde{\mu}$ that does not depend on w, then in view of Theorem 6.4.3, the inner product with respect to $\mathring{\mu}_n$ will in $\mathcal{L}_n$ not depend on w since it is there equal to the inner product with respect to $\tilde{\mu}$. Thus the $\Theta_i(z, w)$ will have the RK property if there exists some $\tilde{\Delta}_n \in \mathcal{A}$ with $\tilde{\Delta}_{n1}(w) = 0$ and $\tilde{\Delta}_0 = \tilde{\Delta}_n\Theta_n$ of the form (8.20)–(8.21).

We can now use $\tilde{\Delta}_n(z, w) = [\tilde{S}_n(z, w) \; 1]$, with $\tilde{S}_n(z, w) \in \mathcal{B}$ and $\tilde{S}_n(w, w) = 0$, to get

$$\tilde{\Delta}_0 = \tilde{\Delta}_n \Theta_n$$

$$= \frac{1}{2}[\tilde{S}_n \quad 1] \begin{bmatrix} K_n^* + L_n^* & K_n^* - L_n^* \\ K_n - L_n & K_n + L_n \end{bmatrix}$$

$$= \frac{1}{2}[\tilde{S}_n(K_n^* + L_n^*) + (K_n - L_n) \quad \tilde{S}_n(K_n^* - L_n^*) + (K_n + L_n)].$$

If this has to be of the form (8.20), then

$$S_w(\tilde{\mu}) = \frac{\tilde{\Delta}_{02} - \tilde{\Delta}_{01}}{\tilde{\Delta}_{02} + \tilde{\Delta}_{01}} = \frac{L_n - \tilde{S}_n L_n^*}{K_n + \tilde{S}_n K_n^*}.$$

We may thus conclude that Θ_i, $i = 1, \ldots, n$ will have the RK property if there exists some function $\tilde{S}_n(z, w) \in \mathcal{B}$, which may depend upon a parameter w and which satisfies $\tilde{S}_n(w, w) = 0$, such that the function $\Omega_n(z)$, defined by (case of the disk)

$$\tilde{\Omega}_n(z) = \left[\frac{L_n(z, w) - \tilde{S}_n(z, w) L_n^*(z, w)}{K_n(z, w) - \tilde{S}_n(z, w) K_n^*(z, w)} - \frac{z^{-1}w - z\overline{w}}{1 - |w|^2} \right] P(z, w),$$

belongs to $\mathcal{C}$ and is independent of w.

Similar observations can be made for the half plane.

We now have a Favard type theorem.

Theorem 8.2.3. *Let the $k_n(z, w)$ be generated as in the previous lemmas and let $K_n(z, w) = k_n(z, w)/\sqrt{k_n(w, w)}$ be their normalized versions. Suppose the $\rho_n(w)$ form a sequence with the RK property. Then there exists a Borel measure on $\partial \mathbb{O}$ such that for $n = 0, 1, 2, \ldots$ the function $k_n(z, w)$ is a reproducing kernel for $\mathcal{L}_n$. Thus there is a measure μ such that for $n = 0, 1, 2, \ldots$*

$$\langle f(z), k_n(z, w) \rangle_{\tilde{\mu}} = f(w), \quad \forall f \in \mathcal{L}_n, \; \forall w \in \mathbb{D}.$$

If the rational functions $\cup_{n=0}^{\infty} \mathcal{R}_n$, where $\mathcal{R}_n = \mathcal{L}_n \cdot \mathcal{L}_{n*}$ and $\mathcal{L}_{n*} = \{f_* : f \in \mathcal{L}_n\}$, are dense in the space of continuous functions $C(\partial \mathbb{O})$, then the measure μ is unique.

Proof. If the ρ_n have the RK property, then $\mathring{\mu}_n^w(t) = \mathring{\mu}_n(t, w)$ as defined in (8.17) will define an inner product $\langle \cdot, \cdot \rangle_{\tilde{\mu}_n}$ that on $\mathcal{L}_n$ will be independent of w, which implies as in Theorem 6.4.3 that the $k_n(z, w)$ is a reproducing kernel for $\mathcal{L}_n$ with respect to $\mathring{\mu}_n(t) = \mathring{\mu}_n^0(t) = \mathring{\mu}_n(t, \alpha_0)$. Because the kernel k_n uniquely defines all the previous ones, we shall also have that $k_j(z, w)$ is a reproducing kernel for $\mathcal{L}_j$ with respect to the measure $\mu_n(t)$ for $j = n - 1, n - 2, \ldots$.

We can now use the same reasoning as in the case of the Favard theorem for the orthogonal functions given before. We shall again restrict the formulation to the case of the disk.

Since the distribution functions (recall our convention that $\tilde{\mu}_n(\theta)\, d\theta = d\mu(e^{i\theta})$)

$$\tilde{\mu}_n(t) = \int_0^t \frac{P(e^{i\theta}, 0)}{|K_n(e^{i\theta}, 0)|^2}\, d\lambda(\theta) = \int_0^t \frac{d\lambda(\theta)}{|K_n(e^{i\theta}, 0)|^2}$$

are increasing functions and uniformly bounded ($\int d\mu_n = 1$, because $\mathcal{S}_0(\mu_n) = \Omega_n(z) = L_n(z, 0)/K_n(z, 0)$ and $\Omega_n(0) = 1$ and $\int d\mu = c_0 = \Omega_n(0)$), there exists a subsequence such that

$$\lim_{k\to\infty} \tilde{\mu}_{n_k}(\theta) = \tilde{\mu}(\theta) \quad \text{and} \quad \lim_{k\to\infty} \int_0^{2\pi} f(e^{i\theta})\, d\tilde{\mu}_{n_k}(\theta) = \int f(t)\, d\mu(t)$$

for all continuous f with $d\mu(e^{i\theta}) = \tilde{\mu}(\theta)\, d\theta$. Thus, for $n = 0, 1, \ldots$, the kernels $k_n(z, w)$ are all reproducing in $\mathcal{L}_n$ with respect to this measure μ.

To prove the uniqueness, we note that, because these k_n are reproducing kernels, there exists a sequence of complex numbers $w_n, n = 0, 1, \ldots$ such that the sequence of functions $\{k_n(z, w_n), n = 0, 1, \ldots\}$ forms a basis for $\mathcal{L}_\infty$. Thus we may define a linear bounded functional M on $\mathcal{R}_\infty$ (and, because of the denseness, also in the space of continuous functions) by means of

$$M(k_i(z, w_i)k_{j*}(z, w_j)) = \int k_i(z, w_i)\overline{k_j(z, w_j)}\, d\mathring{\mu} = k_m(w_j, w_i),$$

where $m = \min\{i, j\}$. By the Riesz representation theorem of bounded linear functionals, it follows that μ is unique.   □

9

Convergence

This rather long chapter contains many different convergence results. We shall start by recalling the background of the classical Szegő problem of weighted least squares approximation, which is related to the construction of the Szegő kernel. Traditionally, finite-dimensional approximants are taken as reproducing kernels in the set of polynomials $\mathcal{P}_n$. We give its generalization in the case where these approximants are reproducing kernels for the rational spaces $\mathcal{L}_n$. In Section 9.2 we give some further preliminary convergence results related to rational interpolants for the Riesz–Herglotz–Nevanlinna transform of the measure. Such results hold locally uniformly, that is, uniformly on compact subsets of the unit disk or the half plane. In Section 9.3 we study some preliminary convergence results that hold for the reciprocal orthonormal functions ϕ_n^*. These are called preliminary, because some rather strong conditions are imposed on the location of the points α_k. They are supposed to stay away from the boundary, so that they are all in a compact subset of $\mathbb{O}$. When this is assumed, we obtain in the course of this section several other results, and many equivalent formulations of the Szegő condition $\log \mu' \in L_1(\mathring{\lambda})$ are used. These equivalences are collected in Section 9.4. These more restrictive conditions on the α_k can be deleted if we apply results from orthogonal rational functions with respect to varying measures. Such orthogonal polynomials can be found in the work of Lopéz, which is reviewed in Section 9.5. These are used in the subsequent Section 9.6 to obtain stronger results for the convergence of the orthogonal rational functions and the reproducing kernels. Some theorems about convergence in the weak star topology are given in Section 9.7. Ratio asymptotics are described in Section 9.8. These ratio asymptotics are typical when the measure is assumed to satisfy the Erdős–Turán condition $\mu' > 0$ a.e., which is weaker than the Szegő condition. Root asymptotics, as given in Section 9.9, involve the convergence of expressions such as $|\phi_n|^{1/n}$. These typically involve some potential theory and

173

are related to estimates for the rates of convergence. Such estimates for the rates of convergence of the rational interpolants for the Riesz–Herglotz–Nevanlinna transform of the measure and for R-Szegő quadrature formulas are given in the final Section 9.10.

As suggested by this survey, there is a hierarchy in the different kinds of asymptotics. In power asymptotics, one typically considers limits of the form $\lim_{n\to\infty} \phi_n(z)/p_n(z)$, with p_n a polynomial. In ratio asymptotics, one considers limits of the form $\lim_{n\to\infty} \phi_{n+1}(z)/\phi_n(z)$, and in root asymptotics, one considers $\lim_{n\to\infty} \phi_n(z)^{1/n}$. The first type of convergence implies the second kind, which implies in turn convergence of the third kind. Thus naturally one can prove convergence of power, ratio, or root type under conditions that are decreasingly restrictive. We do not claim to have all possible results, proved under the weakest possible conditions, but we give examples of each, showing what kind of generalizations and complications the rational case provides in comparison to the polynomial case.

9.1. Generalization of the Szegő problem

For the first convergence results, we look at the classical Szegő problem, which is to construct the Szegő kernel. We start with a resume of the treatment given by Grenander and Szegő [102, Chap. 3] for the polynomial case.

The original Szegő problem involved the case of the disk. So in these introductory lines, we shall only refer to the disk case. Szegő's problem was to solve the extremal problem, which in the notation of Section 2.3, can be denoted as $P_\infty^2(1, 0)$. That is: Find the solution of

$$17 P_\infty^2(1, 0) : \quad \inf \left\{ \|f\|_\mu^2 : f(0) = 1, f \in \mathcal{P}_\infty \right\},$$

where $\mathcal{P}_\infty$ represents the set of polynomials. Since the previous problem is set in $L_2(\mu)$, we could as well replace $\mathcal{P}_\infty$ by $\mathcal{P}$, which is the $L_2(\mu)$ closure of the polynomials. In other words, $\mathcal{P}$ is just another notation for $H_2(\mu)$. The problem was solved by successively considering $f \in \mathcal{P}_n$ for $n = 0, 1, 2, \ldots$. An immediate generalization to the rational case forces us to consider the problem $P_n^2(1, \alpha_n)$, treated in Theorem 2.3.2, which was to find $\inf \|f\|_\mu^2$ when f is in $\mathcal{L}_n$ and $f(\alpha_n) = 1$. The minimum is reached for $f = \kappa_n^{-1}\phi_n^*$ and the minimum is κ_n^{-2}. One recovers the polynomial results by setting all α_k equal to zero. The κ_n are then the positive leading coefficients in the orthonormal Szegő polynomials $\phi_n(z)$. If we take this process to the limit when $n \to \infty$, we shall have solved the Szegő problem for a general $f \in L_2(\mu)$ if $\mathcal{P}_\infty$ is dense in $L_2(\mu)$. So the intimately related problem is to find conditions for the latter to happen. The answer is that $\mathcal{P}_\infty$ is dense in $L_2(\mu)$ if and only if the infimum $\kappa^{-2} = 0$, which happens if and only if $\int \log \mu' \, d\lambda = -\infty$ [102, pp. 49–50]. In general, the limiting value of κ_n^{-2} as $n \to \infty$ is given by $|\sigma(0)|^2 = \exp\{\int \log \mu' \, d\lambda\}$. This is equal to the geometric mean of μ'. If $\log \mu' \notin L_1$, then this has to be replaced by 0. This was proved, for example, in Ref. [102, p. 44]. See also Ref. [87, p. 200 ff]. As a generalization, Grenander and Szegő consider next the problem $P_n^2(1, w)$ with $w \in \mathbb{D}$ again for the polynomial case. We know that the infimum is reached for $f = k_n(z, w)[k_n(w, w)]^{-1}$ and that the minimum is found to be $[k_n(w, w)]^{-1}$. The latter function is known as the *Christoffel function*. For $w = 0$ we rediscover the previous result in the case of polynomials. In Ref. [102, Section 3.2] it is shown that this minimum converges to $1/s_w(w) = (1 - |w|^2)|\sigma(w)|^2$, where $s_w(z)$ is the Szegő kernel. We shall generalize the latter approach to the rational case. However, in the rational case, it requires putting $w = \alpha_n$ to rediscover the former problem. This is an extra complication since this w is supposed to be a constant in $\mathbb{D}$, whereas when replacing it by α_n, it depends upon n and when n tends to ∞ this may converge to a point on the unit circle $\mathbb{T}$ or to a point in the disk $\mathbb{D}$ or it may not converge at all. We shall also have to consider $\mathcal{L}_\infty$ and $\mathcal{L}$. In analogy with the polynomials, the latter is again supposed to denote the closure in $L_2(\mu)$ of $\mathcal{L}_\infty$. We shall have solved the same problem as Szegő did if $\mathcal{L} = \mathcal{P} \equiv H_2(\mu)$. We have seen in Section 7.1 that this happens when the sum $\sum(1 - |\alpha_i|)$ diverges.

In Dewilde and Dym [60] we find many of the results of this section proved for an absolutely continuous measure μ, say $d\mu = \mu' \, d\lambda$ on $\mathbb{T}$. We now generalize this to the case of a general finite positive measure on $\partial\mathbb{O}$. We follow closely the development in Ref. [60]. Assume that $\log \mu' \in L_1(\mathring{\lambda})$, so that the outer spectral factor σ exists. In Ref. [102, p. 51] we find that for polynomials the limit function $\lim_{n\to\infty} k_n(z, w)$ equals the *Szegő kernel*,

$$s_w(z) = s(z, w) = [(1 - \overline{w}z)\sigma(z)\overline{\sigma(w)}]^{-1}. \tag{9.1}$$

We also introduce its normalized form given by

$$S_w(z) = s_w(z)/\sqrt{s_w(w)}.$$

Thus

$$s_w(z) = S_w(z)S_w(w).$$

This is the function appearing in Theorem 2.3.8. Note that it satisfies $s_w(w) > 0$. Theorem 2.3.8 then stated that

$$\|k_n(z, w) - s_w(z)\|_{\hat{\mu}}^2 = s_w(w) - k_n(w, w). \tag{9.2}$$

We now turn to the invariant formulation including the disk as well as the half plane case. Thus we replace (9.1) by (2.18):

$$s_w(z) = \frac{1}{[1 - \zeta_0(z)\overline{\zeta_0(w)}]\sigma(z)\overline{\sigma(w)}} = \frac{1}{\varpi_0(\alpha_0)} \frac{\varpi_0(z)\overline{\varpi_0(w)}}{\varpi_w(z)\sigma(z)\overline{\sigma(w)}}. \qquad (9.3)$$

We shall denote by Π_n the orthogonal projection operator in $L_2(\mathring{\mu})$ onto $\mathcal{L}_n$. Theorem 2.3.8 then says that $\Pi_n[s_w(z)] = S_w(w)\Pi_n[S_w(z)] = k_n(z, w)$ and that the squared norm of the error is given by the expression (9.2).

Recall that if $\log \mu' \in L_1(\mathring{\lambda})$, then by Theorem 7.2.4 the divergence of the Blaschke product, that is, (7.5), is equivalent with the density of $\mathcal{L}_\infty$ in $H_2(\mathring{\mu})$. Now we formulate our first convergence result:

Theorem 9.1.1. *Let $k_n(z, w)$ be the reproducing kernel for $\mathcal{L}_n$, assume $\log \mu' \in L_1(\mathring{\lambda})$, and let $s_w(z)$ be the Szegő kernel as defined in (9.3). If the Blaschke product $B_\infty(z)$ diverges, that is, if (7.5) is satisfied, then*

$$\lim_{n \to \infty} \|k_n(z, w) - s_w(z)\|_{\mathring{\mu}}^2 = 0 \qquad (9.4)$$

and

$$\lim_{n \to \infty} k_n(w, w) = s_w(w) \quad and \quad \lim_{n \to \infty} \phi_n(w) = 0, \qquad (9.5)$$

pointwise in $\mathbb{O}$.

Proof. Note that $s_w \in H_2(\mathring{\mu})$ (it is the reproducing kernel for $H_2(\mathring{\mu})$). Since the Blaschke product diverges, we know by Theorem 7.2.3 that $\mathcal{L}_\infty$ is dense in $H_2(\mathring{\mu})$. Therefore $\Pi_n s_w$, the projection of s_w onto $\mathcal{L}_n$, converges to s_w in $L_2(\mathring{\mu})$, which is the first result.

Thus the squared norm of the error, which by (9.2) equals $s_w(w) - k_n(w, w)$, has to go to zero, which is Parseval's equality.

Since $k_n(w, w) = \sum_{k=0}^n |\phi_k(w)|^2$ converges, the terms $\phi_n(w)$ should go to zero, and this concludes the proof.   □

Note that (9.5) implies in fact a solution of the generalized Szegő problem in the rational case. It says

Theorem 9.1.2. *Let $\log \mu' \in L_1(\mathring{\lambda})$, let s_w denote the Szegő kernel (9.3), and assume that the Blaschke product diverges. Then*

$$\inf\{\|f\|_{\mathring{\mu}} : f \in \mathring{H}_2, f(w) = 1\} = \frac{1}{s_w(w)},$$

that is, the infimum equals

$$(1 - |w|^2)|\sigma(w)|^2 \quad for \ \mathbb{D} \quad and \quad \frac{4 \operatorname{Im} w |\sigma(w)|^2}{|w + \mathbf{i}|^2} \quad for \ \mathbb{U}.$$

Proof. Indeed, Theorem 1.4.2 says that in $\mathcal{L}_n$ the optimization problem

$$\inf\left\{\|f\|_{\mathring{\mu}}^2 : f(w) = 1\right\}$$

has the solution $f_n(z) = k_n(z, w)/k_n(w, w)$ and that the minimum is given by $1/k_n(w, w)$. Since the Blaschke product diverges, the space $\mathcal{L}_\infty$ is dense in $H_2(\mathring{\mu})$ and by the previous theorem $k_n(w, w)$ converges to $s_w(w)$, so that the infimum of $\|f\|_{\mathring{\mu}}^2$ in $H_2(\mathring{\mu})$ is equal to $1/s_w(w)$. $\qquad\square$

Theorem 9.1.1 gives $L_2(\mathring{\mu})$ convergence and pointwise convergence of the reproducing kernels for $\mathcal{L}_n$ to the Szegő kernel. In Section 9.3 we shall prove under somewhat more restrictive conditions on the α_i that $k_n(z, w)$ converges locally uniformly to $s_w(z)$ in $\mathbb{O}$. See also Theorem 9.6.7.

Sometimes it is more interesting to work with the normalized kernels, which were defined as $K_n(z, w) = k_n(z, w)/\sqrt{k_n(w, w)}$. Therefore, we also introduced a normalized form of $s_w(z)$, which was given before. We can write it explicitly as

$$S_w(z) = \frac{\sqrt{1 - |\zeta_0(w)|^2}}{(1 - \zeta_0(z)\overline{\zeta_0(w)})}\frac{|\sigma(w)|}{\sigma(z)\overline{\sigma(w)}} = \eta(w)\frac{\varpi_0(z)\sqrt{\varpi_w(w)/\varpi_0(\alpha_0)}}{\varpi_w(z)\sigma(z)},$$

(9.6)

where the factor $\eta(w)$ of modulus 1 is chosen to satisfy $S_w(w) > 0$. This is obtained by setting

$$\eta(w) = \frac{\overline{\varpi_0(w)}|\sigma(w)||\varpi_0(\alpha_0)|}{|\varpi_0(w)|\overline{\sigma(w)\varpi_0(\alpha_0)}} \in \mathbb{T}.$$

Thus we have the explicit expressions

$$S_w(z) = \begin{cases} \eta(w)\dfrac{\sqrt{1 - |w|^2}}{1 - \overline{w}z}\dfrac{1}{\sigma(z)}, & \eta(w)/\sigma(w) > 0 & \text{for } \mathbb{D}, \\[3mm] \eta(w)\dfrac{(z + \mathbf{i})\sqrt{\operatorname{Im} w}}{z - \overline{w}}\dfrac{1}{\sigma(z)}, & \eta(w)(1 - w\mathbf{i})/\sigma(w) > 0 & \text{for } \mathbb{U}. \end{cases}$$

Note that

$$S_w(w) = \frac{1}{|\sigma(w)|\sqrt{1 - |\zeta_0(w)|^2}} = \frac{|\varpi_0(w)|/|\varpi_0(\alpha_0)|}{|\sigma(w)|\sqrt{\varpi_w(w)/\varpi_0(\alpha_0)}}$$

$$= \begin{cases} \dfrac{1}{|\sigma(w)|\sqrt{1 - |w|^2}} & \text{for } \mathbb{D}, \\[3mm] \dfrac{|w + \mathbf{i}|/2}{|\sigma(w)|\sqrt{\operatorname{Im} w}} & \text{for } \mathbb{U}. \end{cases}$$

In this notation the Szegő kernel can be written as $s_w(z) = S_w(w)S_w(z)$ and conversely $S_w(z) = s_w(z)/\sqrt{s_w(w)} = s_w(z)/S_w(w)$.

Finally we note that

$$|S_w(t)|^2 = \frac{|\varpi_0(t)|^2 P(t,w)}{|\sigma(t)|^2}, \quad t \in \partial\mathbb{O}, \tag{9.7}$$

which brings about the following interesting result.

Lemma 9.1.3. Let $\log \mu' \in L_1(\mathring{\lambda})$. Then the normalized Szegő kernel $S_w(z)$ is an outer spectral factor associated with the density $\mu'(t)[P(t,w)|\varpi_0(t)|^2]^{-1}$ with $P(t,w)$ the Poisson kernel. Thus

$$\frac{1}{S_w(z)} = \eta(w)\exp\left\{\frac{1}{2}\int D(t,z)\log\left[\frac{\mu'(t)}{P(t,w)|\varpi_0(t)|^2}\right]d\mathring{\lambda}(t)\right\},$$

where $\eta(w) \in \mathbb{T}$ is such that $S_w(w) > 0$.

Proof. We know that

$$\sigma(z) = \exp\left\{\frac{1}{2}\int D(t,z)\log\mu'(t)\,d\mathring{\lambda}(t)\right\}.$$

Furthermore,

$$P(t,w)|\varpi_0(t)|^2 = |X_w(t)|^2 = X_w(t)X_{w*}(t),$$

with

$$X_w(z) = \frac{\varpi_0(z)\sqrt{\varpi_w(w)/\varpi_0(\alpha_0)}}{\varpi_w(z)}$$

an outer function in H_2. Hence the result follows. $\qquad\square$

For the normalized kernel we have the following:

Lemma 9.1.4. Let $\log \mu' \in L_1(\mathring{\lambda})$ and let S_w be the normalized Szegő kernel defined in (9.6) and $K_n(z,w)$ the normalized reproducing kernel for $\mathcal{L}_n$. Then

$$\|K_n(z,w) - S_w(z)\|_{\mathring{\mu}}^2 = 2[1 - K_n(w,w)/S_w(w)]. \tag{9.8}$$

Proof. We evaluate the norm in the left-hand side. From the reproducing property of the kernels we readily find

$$\|k_n(z,w)\|_{\mathring{\mu}}^2 = k_n(w,w).$$

Since $K_n(z,w) = k_n(z,w)/\sqrt{k_n(w,w)}$, we find that $\|K_n(z,w)\|_{\mathring{\mu}}^2 = 1$.

By similar arguments we get for the normalized kernel $\|S_w\|_{\mathring\mu}^2 = 1$.
Furthermore,

$$\text{Re}\,\langle K_n(z, w), S_w(z)\rangle_{\mathring\mu} = \text{Re}\,\langle K_n(z, w), s_w(z)/S_w(w)\rangle_{\mathring\mu}$$
$$= K_n(w, w)/S_w(w) > 0.$$

Hence the assertion is proved since

$$\|K_n(z, w) - S_w(z)\|_{\mathring\mu}^2 = \|K_n(\cdot, w)\|_{\mathring\mu}^2 + \|S_w\|_{\mathring\mu}^2 - 2\,\text{Re}\,\langle K_n(z, w), S_w(z)\rangle_{\mathring\mu}.$$

$\square$

Theorem 9.1.5. *Let* $\log \mu' \in L_1(\mathring\lambda)$ *and let* $S_w(z)$ *be the normalized Szegő kernel satisfying* $S_w(w) > 0$ *and* $K_n(z, w)$ *the normalized reproducing kernel for* $\mathcal{L}_n$. *Suppose also that the Blaschke product diverges. Then*

$$\lim_{n\to\infty} \|K_n(z, w) - S_w(z)\|_{\mathring\mu}^2 = 0 \tag{9.9}$$

and also

$$\lim_{n\to\infty} \left\|1 - \frac{K_n(z, w)}{S_w(z)}\right\|_{\lambda_w}^2 = 0 \tag{9.10}$$

and

$$\lim_{n\to\infty} \left\|1 - \frac{k_n(z, w)}{s_w(z)}\right\|_{\lambda_w}^2 = 0, \tag{9.11}$$

where $d\lambda_w(t) = P(t, w)\,d\lambda(t)$ *with* P *the Poisson kernel.*

Proof. From Theorem 9.1.1, we easily derive by taking the square roots that $K_n(w, w) \to S_w(w)$. The first result now follows easily from the previous lemma.

To prove the second one, note that (recall (9.7))

$$d\lambda_w(t) = \frac{P(t, w)}{|\sigma(t)|^2}\,d\mu_a(t) = \frac{|\varpi_0(t)|^2 P(t, w)}{|\sigma(t)|^2}\,d\mathring\mu_a(t) = |S_w(t)|^2\,d\mathring\mu_a(t),$$

with $d\mathring\mu_a = \mu'\,d\mathring\lambda = |\sigma|^2\,d\mathring\lambda$. Thus

$$\left\|1 - \frac{K_n(z, w)}{S_w(z)}\right\|_{\lambda_w}^2 = \|S_w(z) - K_n(z, w)\|_{\mathring\mu_a}^2$$
$$\leq \|S_w(z) - K_n(z, w)\|_{\mathring\mu}^2.$$

Since the latter converges to zero, the assertion (9.10) follows.

The last relation is shown similarly.

$\square$

The result (9.10) is a generalization of Theorem 5.8 of Ref. [94].

Some direct consequences are

Corollary 9.1.6. *Let* $\log \mu' \in L_1(\mathring{\lambda})$ *and suppose the Blaschke product diverges. Let* $S_w(z)$ *be the normalized Szegő kernel and* $K_n(z, w)$ *the normalized reproducing kernels for* $\mathcal{L}_n$. *Then the following convergence results hold:*

$$\lim_{n \to \infty} \int \left| \left| \frac{K_n(t, w)}{S_w(t)} \right|^2 - 1 \right| d\lambda_w(t) = 0, \qquad (9.12)$$

$$\lim_{n \to \infty} \int \left| \frac{1}{|K_n(t, w)|} - \frac{1}{|S_w(t)|} \right| d\lambda_w(t) = 0, \qquad (9.13)$$

where $d\lambda_w(t) = P(t, w) \, d\lambda(t)$ *and* P *is the Poisson kernel.*

Proof. First note that it follows from (9.10) by using $|a - b|^2 \geq (|a| - |b|)^2$ that

$$\lim_{n \to \infty} \int \left(\left| \frac{K_n(t, w)}{S_w(t)} \right| - 1 \right)^2 d\lambda_w(t) = 0. \qquad (9.14)$$

Denoting the integral in (9.12) as I_n, we have by the Schwarz inequality

$$I_n^2 = \left(\int ||K_n/S_w| - 1|\,||K_n/S_w| + 1|\, d\lambda_w \right)^2$$

$$\leq \left(\int ||K_n/S_w| - 1|^2 \, d\lambda_w \right) \cdot \left(\int ||K_n/S_w| + 1|^2 \, d\lambda_w \right).$$

The first integral goes to zero for $n \to \infty$ by (9.14). For the second integral use $|a + b|^2 \leq 2(|a|^2 + |b|^2)$ to find that it is bounded by

$$2 \left(\int \left| \frac{K_n}{S_w} \right|^2 d\lambda_w + \int d\lambda_w \right) = 2 \left(\int \frac{|K_n|^2 |\sigma|^2}{|\varpi_0|^2 P} P d\lambda + \int P d\lambda \right)$$

$$= 2 \left(\int |K_n|^2 |\sigma|^2 \, d\mathring{\lambda} + 1 \right)$$

$$\leq 2 \left(\int |K_n|^2 d\mathring{\mu} + 1 \right) = 4.$$

This proves (9.12).

For the relation (9.13), we note that it can be written as

$$I = \int \left| \frac{|K_n/S_w| - 1}{K_n} \right| d\lambda_w.$$

After squaring this and using the Schwarz inequality, we find that

$$I^2 \leq \left(\int |K_n|^{-2} d\lambda_w \right) \cdot \left(\int \left| |K_n/S_w|^2 - 1 \right|^2 d\lambda_w \right).$$

The second integral goes to zero for $n \to \infty$ by (9.14), whereas for the first integral we can use Theorem 6.4.3 to find that it is

$$\int \frac{P(t, w)}{|K_n(t, w)|^2} d\lambda(t) = \|1\|_{\hat{\mu}}^2 = 1.$$

This proves (9.13). □

In the case of the disk and for $w = 0$, the latter results can be found as Theorem 3.3 in Ref. [168].

9.2. Further convergence results and asymptotic behavior

This section includes a number of convergence results of the approximants we obtained, such as local uniform convergence in $\mathbb{O}$. It is a well-known fact that an infinite Blaschke product $B(z) = B_\infty(z)$ will converge to zero locally uniformly (i.e., uniformly on compact subsets of $\mathbb{O}$) if (7.5) is satisfied. See, for example, Ref. [200, p. 281 ff]. This can be used to obtain some other convergence results of the same type.

Theorem 9.2.1. *Let ϕ_n be the orthonormal functions for $\mathcal{L}_n$ and ψ_n the functions of the second kind. Define $\Omega_n = \psi_n^*/\phi_n^*$ and let the positive real function $\Omega_\mu(z) = \int D(t, z) d\hat{\mu}(t)$ be the Riesz–Herglotz–Nevanlinna transform of the measure μ. Then $\Omega_n = \psi_n^*/\phi_n^*$ converges to Ω_μ uniformly on compact subsets of $\mathbb{O}$ if (7.5) is satisfied. Under the same conditions, $\Omega_n^\times(z) = -\psi_n/\phi_n$ converges to Ω_μ uniformly on compact subsets of $\mathbb{O}^e$.*

Proof. First, we note that $\Omega_n = \psi_n^*/\phi_n^* = \Psi_n^*/\Phi_n^* \in \mathcal{C}$, where Φ_n and Ψ_n are the rotated functions as in Section 6.5.

Let T_n be the recurrence matrix for the rotated functions, that is, $T_n = t_n t_{n-1} \ldots t_1$ with t_n the elementary recurrence matrices as in (4.15). Then, we have by (4.29)

$$T_n = \frac{1}{2} \begin{bmatrix} \Phi_n + \Psi_n & \Phi_n - \Psi_n \\ \Phi_n^* - \Psi_n^* & \Phi_n^* + \Psi_n^* \end{bmatrix}.$$

Hence $\Phi_n^*[1 - \Omega_n \quad 1 + \Omega_n] = 2[0 \quad 1]T_n$ and thus

$$(\Omega_\mu - \Omega_n) = \frac{1}{2}[1 - \Omega_n \quad 1 + \Omega_n]\begin{bmatrix} \Omega_\mu + 1 \\ \Omega_\mu - 1 \end{bmatrix}$$

$$= \frac{1}{\Phi_n^*}[0 \quad 1]T_n\begin{bmatrix} \Omega_\mu + 1 \\ \Omega_\mu - 1 \end{bmatrix}$$

$$= \frac{1}{\Phi_n^*}[0 \quad 1]\begin{bmatrix} \Phi_n\Omega_\mu + \Psi_n \\ \Phi_n^*\Omega_\mu - \Psi_n^* \end{bmatrix}$$

$$= \frac{1}{\Phi_n^*}[0 \quad 1]\begin{bmatrix} \tilde{B}_{n-1}R_{n1} \\ \tilde{B}_n R_{n2} \end{bmatrix} = \frac{\tilde{B}_n}{\Phi_n^*}R_{n2},$$

where $\tilde{B}_k = \zeta_0 B_k$ and R_{n1} and R_{n2} are as defined in (6.23). Thus

$$\Omega_\mu - \Omega_n = \tilde{B}_n R_{n2}/\Phi_n^* \quad \text{in } \mathbb{O}. \tag{9.15}$$

Now define the Schur functions Γ and Γ_n by Cayley transforms of Ω_μ and Ω_n:

$$\Gamma = \frac{1 - \Omega_\mu}{1 + \Omega_\mu} \in \mathcal{B} \quad \text{and} \quad \Gamma_n = \frac{1 - \Omega_n}{1 + \Omega_n} \in \mathcal{B}.$$

Then,

$$\Gamma - \Gamma_n = 2\frac{\Omega_n - \Omega_\mu}{(1 + \Omega_\mu)(1 + \Omega_n)},$$

which in view of (9.15) gives

$$\frac{\Gamma - \Gamma_n}{2\tilde{B}_n} = \frac{-R_{n2}}{\Phi_n^*(1 + \Omega_\mu)(1 + \Omega_n)} \in H(\mathbb{O}).$$

On the boundary $\partial\mathbb{O}$, we have $|\tilde{B}_n| = 1$ a.e. and $|\Gamma| \leq 1$ and $|\Gamma_n| \leq 1$, so that

$$\left| \frac{\Gamma - \Gamma_n}{2\tilde{B}_n} \right| \leq 1 \quad \text{on } \partial\mathbb{O}.$$

The maximum modulus theorem then gives

$$|\Gamma - \Gamma_n| \leq 2|\tilde{B}_n| \quad \text{in } \mathbb{O}.$$

The right-hand side, and hence also the left-hand side, converges to zero uniformly on compact subsets of $\mathbb{O}$ if (7.5) holds. With inverse Cayley transforms we now find that

$$\Omega_\mu - \Omega_n = \frac{1 - \Gamma}{1 + \Gamma} - \frac{1 - \Gamma_n}{1 + \Gamma_n} = 2\frac{\Gamma_n - \Gamma}{(1 + \Gamma)(1 + \Gamma_n)},$$

which will converge exactly as $\Gamma - \Gamma_n$ does. Indeed, the denominator is bounded away from 0 in a compact subset $K \subset \mathbb{O}$ because Ω_μ is continuous in K and thus maps K to a compact subset of the finite plane. Therefore Γ maps K to a compact subset of $\mathbb{D}$ so that $|\Gamma(z)| \le r < 1$ and hence $|1 + \Gamma(z)| \ge 1 - r > 0$ in K. Because Γ_n converges uniformly to Γ in K, then also $|\Gamma_n(z)| \le \rho < 1$ in K for all n. Hence $|1 + \Gamma_n(z)| \ge 1 - \rho > 0$ for all n and $z \in K$. This gives the proof in $\mathbb{O}$.

The proof that $-\psi_n/\phi_n$ converges locally uniformly in $\mathbb{O}^e$ is given along the same lines, knowing that $\lim_{n\to\infty} 1/B_n(z)$ goes locally uniformly to zero in $\mathbb{O}^e$. □

Practically the same proof can be repeated for the following theorem.

Theorem 9.2.2. *Let $K_n(z, w)$ be the normalized kernels for $\mathcal{L}_n$ and $L_n(z, w)$ the associated kernels. Define $\Omega_n(z, w) = L_n(z, w)/K_n(z, w)$. Let $\Omega_\mu(z, w)$ be as defined in Lemma 6.2.2. Consider these as functions in z for fixed $w \in \mathbb{O}$. Then, if (7.5) holds, $\Omega_n(z, w)$ converges uniformly to $\Omega_\mu(z, w)$ on compact subsets of $\mathbb{O}$.*

Proof. We can use the result of Section 6.4 to find that

$$\frac{\Omega_n - \Omega_\mu}{B_n} = \frac{\Delta_{n1}}{K_n} = g_n \in H(\mathbb{O}).$$

From then on the proof runs exactly as in the previous theorem. □

The previous theorem actually corresponds to Lemma 3.4 in Ref. [60].

Again by the same arguments, we also have

Theorem 9.2.3. *Let $Q_n(z) = \phi_n(z) + \tau_n\phi_n^*(z)$, $\tau_n \in \mathbb{T}$ be the para-orthogonal functions defined in (5.2), and $P_n(z) = \psi_n(z) - \tau_n\psi_n^*(z)$ the associated functions of the second kind defined in (5.3), and $R_n(z) = -P_n(z)/Q_n(z)$ the rational approximants of Theorem 6.1.3. Then, if the Blaschke product diverges, $R_n(z)$ converges to $\Omega_\mu(z)$ uniformly on compact subsets of $\mathbb{O} \cup \mathbb{O}^e = \mathbb{C} \setminus \partial\mathbb{O}$.*

Proof. Now we have to use Theorem 6.1.3. □

9.3. Convergence of ϕ_n^*

In this section we discuss the convergence of the ϕ_n^* under the more restrictive condition that the sequence $\{\alpha_n\}$ is bounded away from the boundary $\overline{\partial\mathbb{O}}$.

This means that it is completely included in a compact subset of $\mathbb{O}$. Thus the distance to the boundary $\overline{\partial\mathbb{O}}$ is then some finite positive value. We shall say that $A = \{\alpha_1, \alpha_2, \ldots\}$ is compactly included in $\mathbb{O}$. In this section we shall use the notation $\mathbb{O}_r$ to denote such a subset of $\mathbb{O}$ for which $\varpi_z(z)/\varpi_0(\alpha_0) \geq r$ for all $z \in \mathbb{O}_r$. Note that this means that $1 - |z|^2 \geq r$ for $z \in \mathbb{D}_r$ and that $\operatorname{Im} z \geq r$ for $z \in \mathbb{U}_r$. Thus we suppose that there is some $\delta > 0$ such that

$$\alpha_k \in \mathbb{O}_\delta, \quad k = 0, 1, 2, \ldots.$$

It is clear that this condition is stronger than the divergence of the Blaschke product guaranteed by (7.5), which was our main condition in earlier sections. Since $\mathbb{O}_\delta$ is compact, we have for the case $\mathbb{O} = \mathbb{U}$ that the α_k can not go to ∞. Thus the Blaschke product diverges in that case when $\sum \operatorname{Im} \alpha_k = \infty$ (see Section 1.3). Note that under this more restrictive condition we can make use of the following bounds for all $z \in \mathbb{O}_r$:

$$|\zeta_n(z)| \leq m(r) < 1, \tag{9.16}$$

and for the Poisson kernel we have for some positive m_δ and M_δ

$$0 < \frac{m_\delta}{|\varpi_0(t)|^2} \leq P(t, \alpha_n) \leq \frac{M_\delta}{|\varpi_0(t)|^2} < \infty, \quad \forall t \in \partial\mathbb{O}. \tag{9.17}$$

The main result will be that whenever a subsequence $\alpha_{n(s)} \to \alpha \in \mathbb{O}$, then $\phi^*_{n(s)}$ converges to some function F_α, depending on α. It will turn out that this $F_\alpha(z)$ is the normalized Szegő kernel $S_\alpha(z)$ that appeared in previous sections.

In Section 9.6, we shall give stronger results that rely on convergence properties of orthogonal rational functions with respect to varying measures. We think, however, that the results of this section are interesting in their own right. They also include the case where all the poles of the rational function coincide at some point α. This is precisely what is considered in Section 9.5. The rational functions orthogonal with respect to varying measures will have all their poles in $\hat{\alpha}_0$. This means that they are polynomials in the case of the disk and functions of the form $p_n(z)/(z + i)^n$ with $p_n \in \mathcal{P}_n$ in the case of the half plane.

We start with the following dichotomy:

Lemma 9.3.1. Suppose that $A = \{\alpha_1, \alpha_2, \ldots\}$ is compactly included in $\mathbb{O}$. Then every subsequence $\{\phi^*_{n(s)}\}$ of $\{\phi^*_n\}$ has a subsequence $\{\phi^*_{n(s(q))}\}$ that either converges locally uniformly in $\mathbb{O}$ to an analytic function F, without zeros, or diverges locally uniformly to ∞.

Proof. From the Christoffel–Darboux relation (3.2) with $w = z$, we get

$$|\phi_n^*(z)|^2 \geq 1 - |\zeta_n(z)|^2 \geq 1 - m(r)^2 > 0$$

for $z \in \mathbb{O}_r$. The result of a divergent subsequence then follows from the theory of normal families (Montel's theorem). The limit function F has no zeros by Hurwitz's theorem because ϕ_n^* has no zeros in $\mathbb{O}$. $\qquad\square$

Another dichotomy is

Lemma 9.3.2. Let $A = \{\alpha_1, \alpha_2, \ldots\}$ be compactly included in $\mathbb{O}$. Then the sequence $\{\phi_n^*\}$ is either locally uniformly bounded in $\mathbb{O}$ or diverges locally uniformly to ∞.

In the first case, that is, when $\{\phi_n^*\}$ is locally uniformly bounded in $\mathbb{O}$, then the sequence $\{k_n(z, z)\}$ with k_n the reproducing kernels converges locally uniformly in $\mathbb{O}$ and $\{\phi_n\}$ converges locally uniformly to zero in $\mathbb{O}$.

Proof. Suppose $\{\phi_n^*\}$ does not diverge locally uniformly to ∞. Then, by the previous lemma, there exists a subsequence $\{\phi_{n(s)}^*\}$ that converges locally uniformly to a bounded function in $\mathbb{O}$. Again by the Christoffel–Darboux relation (3.2) with $w = z$, and (9.16), we get

$$k_{n(s)}(z, z) = \sum_{k=0}^{n(s)} |\phi_k(z)|^2 \leq \frac{|\phi_{n(s)}^*(z)|^2}{1 - m(r)^2}$$

for $z \in \mathbb{O}_r$. Since $\{k_{n(s)}(z, z)\}$ is nondecreasing, and $\{\phi_{n(s)}^*(z)\}$ is locally uniformly bounded, it follows that the whole sequence $\{k_n(z, z)\}$ is locally uniformly bounded and in particular $\{\phi_n(z)\}$ converges locally uniformly to zero. Hence, the monotonically increasing sequence $\{k_n(z, z)\}$ converges to a continuous limit. Consequently, $\{k_n(z, z)\}$ converges locally uniformly, $\{\phi_n(z)\}$ converges locally uniformly to zero, and $\{\phi_n^*(z)\}$ is locally uniformly bounded. $\qquad\square$

We now get immediately

Corollary 9.3.3. Let $A = \{\alpha_1, \alpha_2, \ldots\}$ be compactly included in $\mathbb{O}$. Then the sequence of leading coefficients $\{\kappa_n\}$ is either bounded or diverges to ∞.

Proof. If $\{\kappa_n\}$ does not diverge to ∞, then, since $\kappa_n = \phi_n^*(\alpha_n)$, this implies that $\{\phi_n^*\}$ does not diverge locally uniformly to ∞. Therefore, $\{\phi_n^*\}$ is locally uniformly bounded by Lemma 9.3.1, implying that the sequence $\{\kappa_n\}$ is bounded.   □

As we shall see in the next section, if A is compactly included in $\mathbb{O}$, then the condition that the sequence $\{\kappa_n\}$ is bounded and the condition that $\{\phi_n^*\}$ is locally uniformly bounded in $\mathbb{O}$ are equivalent and they are both equivalent with the Szegő condition $\log \mu' \in L_1(\mathring{\lambda})$. Below we frequently use the phrase that for a subsequence $\alpha_{n(s)}$ converging to $\alpha \in \mathbb{O}$, there is some F_α to which $\phi_{n(s)}^*$ will converge. This assumes that $\{\phi_n^*\}$ does not diverge locally uniformly to ∞, which, by our previous observation, actually means that $\log \mu' \in L_1(\mathring{\lambda})$. Thus unless Szegő's condition is satisfied, there can not be such an F_α. However, we shall not mention the Szegő condition and give an interpretation to our results as follows. If Szegő's condition holds, then F_α is actually a function; if Szegő's condition does not hold, then the results still hold if F_α is interpreted as being ∞, $1/F_\alpha$ as zero, etc.

We now can prove

Theorem 9.3.4. *Let $\alpha_{n(s)} \to \alpha \in \mathbb{O}$ and suppose that $\phi_{n(s)}^*(z) \to F_\alpha(z)$, locally uniformly in $\mathbb{O}$. Then $[F_\alpha(z)]^{-1} \in \mathring{H}_2$ and hence F_α has no zeros in $\mathbb{O}$.*

Proof. We prove only the disk case; but similar observations hold for the half plane. We have to prove that there exists a constant B such that

$$\int |F_\alpha(rt)|^{-2} \, d\mathring{\lambda}(t) \le B \quad \text{for } 0 \le r < 1.$$

For $\alpha_{n(s)} \to \alpha$, we know that $|\phi_{n(s)}^*(rt)|^{-2}$ converges uniformly to $|F_\alpha(rt)|^{-2}$. Thus

$$\int \frac{d\mathring{\lambda}(t)}{|\phi_{n(s)}^*(rt)|^2} \quad \text{converges to} \quad \int \frac{d\mathring{\lambda}(t)}{|F_\alpha(rt)|^2}.$$

Thus we should prove that

$$\int \frac{d\mathring{\lambda}(t)}{|\phi_n^*(rt)|^2} \le B.$$

Since $|\phi_n^*(z)|^{-2}$ is harmonic in $\mathbb{D}$, this integral is nondecreasing with r, and it is sufficient to prove that this integral is bounded for $r = 1$. Now by Theorem 6.1.9

$$\int \frac{P(t, \alpha_n)}{|\phi_n^*(t)|^2} \, d\lambda(t) = \|1\|_{\hat{\mu}}^2 = 1.$$

The result now follows by (9.17).   □

We want to prove next that

$$\lim_{s \to \infty} \kappa_{n(s)}^2 = F_\alpha(\alpha)^2.$$

Before we can do this, we need several lemmas.

Lemma 9.3.5. With ϕ_n the orthonormal functions and ψ_n the associated functions of the second kind, we have

$$\mathrm{Re} \left[\frac{\psi_n^*(z)}{\phi_n^*(z)} \right] \geq \frac{|\varpi_0(z)|^2}{|\phi_n^*(z)|^2} \frac{\varpi_n(\alpha_n)/\varpi_0(\alpha_0)}{|\varpi_n(z)|^2} \quad \text{for } z \in \mathbb{O}. \tag{9.18}$$

Proof. First we note that both

$$\mathrm{Re} \frac{\psi_n^*}{\phi_n^*} \quad \text{and} \quad \log \frac{|\varpi_0(z)|^2 \varpi_n(\alpha_n)/\varpi_0(\alpha_0)}{|\varpi_n(z)\phi_n^*(z)|^2}$$

are harmonic in $\mathbb{O} \cup \partial\mathbb{O}$ and hence by Poisson's formula,

$$\log \frac{|\varpi_0(z)|^2 \varpi_n(\alpha_n)/\varpi_0(\alpha_0)}{|\varpi_n(z)\phi_n^*(z)|^2} = \int P(t, z) \log \frac{P(t, \alpha_n)|\varpi_0(t)|^2}{|\phi_n^*(t)|^2} \, d\lambda(t),$$
$$\tag{9.19}$$

whereas by Theorem 4.2.6

$$\mathrm{Re} \left[\frac{\psi_n^*(z)}{\phi_n^*(z)} \right] = \int P(t, z) \frac{P(t, \alpha_n)|\varpi_0(t)|^2}{|\phi_n^*(t)|^2} \, d\lambda(t). \tag{9.20}$$

Jensen's formula for the inequality of arithmetic and geometric mean then yields

$$\int P(t, z) \frac{P(t, \alpha_n)|\varpi_0(t)|^2}{|\phi_n^*(t)|^2} \, d\lambda(t)$$
$$\geq \exp \left\{ \int P(t, z) \log \left[\frac{P(t, \alpha_n)|\varpi_0(t)|^2}{|\phi_n^*(t)|^2} \right] \, d\lambda(t) \right\}.$$

Hence, using (9.20) for the left-hand side and (9.19) for the right-hand side, we get the result. $\qquad\square$

Letting $n(s) \to \infty$ we get the following consequence:

Lemma 9.3.6. With $\alpha_{n(s)} \to \alpha \in \mathbb{O}$, so that $\phi_{n(s)}^*(z) \to F_\alpha(z)$, and with $\Omega_\mu(z) = \int D(t, z) \, d\mathring{\mu}(t)$, we have

$$\frac{|\varpi_0(z)|^2}{|F_\alpha(z)|^2} \frac{\varpi_\alpha(\alpha)/\varpi_0(\alpha_0)}{|\varpi_\alpha(z)|^2} \leq \mathrm{Re}\, \Omega_\mu(z) = \int P(t, z) \, d\mu(t). \tag{9.21}$$

Proof. This follows immediately from the previous lemma since $\psi_n^*(z)/\phi_n^*(z) = \Omega_n(z)$ converges locally uniformly to Ω_μ by Theorem 9.2.1. □

Lemma 9.3.7. With the notation of Lemma 9.3.6 we have

$$F_\alpha(\alpha)^2 \geq \exp\left\{-\int P(t,\alpha)\log\left[\frac{\mu'(t)}{|\varpi_0(t)|^2 P(t,\alpha)}\right]d\lambda(t)\right\}. \qquad (9.22)$$

Proof. Taking nontangential limits to the boundary in (9.21), we get

$$\frac{P(t,\alpha)|\varpi_0(t)|^2}{|F_\alpha(t)|^2} \leq \mu'(t) \quad \text{a.e.} \qquad (9.23)$$

Consequently, we have

$$\int P(t,\alpha)\log\left[\frac{\mu'(t)}{|\varpi_0(t)|^2 P(t,\alpha)}\right]d\lambda(t) \geq -\int P(t,\alpha)\log|F_\alpha(t)|^2\,d\lambda(t).$$

By Poisson's formula, the right-hand side is equal to $-\log F_\alpha(\alpha)^2$. (Note $F_\alpha(\alpha) > 0$.) This is equivalent with the result. □

We have one more lemma:

Lemma 9.3.8. With the notation of Lemma 9.3.6 we have

$$\lim_{s\to\infty} \kappa_{n(s)}^2 \leq \exp\left\{-\int P(t,\alpha)\log\left[\frac{\mu'(t)}{|\varpi_0(t)|^2 P(t,\alpha)}\right]d\lambda(t)\right\}. \qquad (9.24)$$

Proof. Recalling Theorem 6.1.9, we have

$$\begin{aligned}
1 = \|\phi_n^*\|_{\mathring{\mu}}^2 &= \int |\phi_n^*(t)|^2 d\mathring{\mu}(t) \\
&\geq \int |\phi_n^*(t)|^2 \frac{\mu'(t)}{|\varpi_0(t)|^2} d\lambda(t) \\
&= \int P(t,\alpha_n)\frac{|\phi_n^*(t)|^2\mu'(t)}{P(t,\alpha_n)|\varpi_0(t)|^2} d\lambda(t).
\end{aligned}$$

By Jensen's inequality for the arithmetic and geometric mean we then get

$$1 \geq \exp\left\{\int P(t,\alpha_n)\log\left[\frac{|\phi_n^*(t)|^2\mu'(t)}{P(t,\alpha_n)|\varpi_0(t)|^2}\right]d\lambda(t)\right\}.$$

Splitting the logarithm and using

$$\exp\left\{\int P(t,\alpha_n)\log|\phi_n^*(t)|^2\,d\lambda(t)\right\} = |\phi_n^*(\alpha_n)|^2 = \kappa_n^2$$

we get the result. □

Knitting the previous results together gives us the following convergence result for the coefficients of highest degree κ_n

Theorem 9.3.9. *If* $\alpha_{n(s)} \to \alpha \in \mathbb{O}$, *and if* $\log \mu' \in L_1(\mathring{\lambda})$, *then we have for the leading coefficients* κ_n *of the orthonormal functions*

$$\lim_{s \to \infty} \kappa_{n(s)}^2 = \exp\left\{ - \int P(t, \alpha) \log \left[\frac{\mu'(t)}{P(t, \alpha)|\varpi_0(t)|^2} \right] d\lambda(t) \right\}, \quad (9.25)$$

where $P(t, z)$ *represents the Poisson kernel. If* $\int \log \mu' \, d\mathring{\lambda} = -\infty$, *then* (9.25) *still holds if the right-hand side is interpreted as being* ∞.

Proof. Denoting the right-hand side as $S(\alpha)$, we know from Lemma 9.3.7 and the proof of Lemma 9.3.8 that

$$\kappa_n^2 \le S(\alpha) \le F_\alpha(\alpha)^2.$$

Now since $\kappa_{n(s)}^2 = |\phi_{n(s)}^*(\alpha_{n(s)})|^2 \to F_\alpha(\alpha)^2$, the result follows directly. $\quad\square$

A further convergence result is

Theorem 9.3.10. *When* $\log \mu' \in L_1(\mathring{\lambda})$, *and* $\alpha_{n(s)} \to \alpha \in \mathbb{O}$, *then*

$$\frac{1}{|F_\alpha(t)|^2} = \frac{\mu'(t)}{P(t, \alpha)|\varpi_0(t)|^2} \quad \text{a.e. on } \partial\mathbb{O}. \quad (9.26)$$

Proof. By (9.23)

$$\frac{P(t, \alpha)|\varpi_0(t)|^2}{|F_\alpha(t)|^2} \le \mu'(t) \quad \text{a.e.} \quad t \in \partial\mathbb{O}.$$

If strict inequality holds on a set of positive measure, then we would have strict inequality,

$$F_\alpha(\alpha)^2 > S(\alpha),$$

in the previous proof and this is a contradiction with $\kappa_n^2 \le S(\alpha)$. $\quad\square$

Finally, we relate $F_\alpha(z)$ to the normalized Szegő kernel.

Theorem 9.3.11. *Suppose* $\log \mu' \in L_1(\mathring{\lambda})$ *and* $\alpha_{n(s)} \to \alpha \in \mathbb{O}$. *Then* $\phi_{n(s)}^*(z)$ *converges locally uniformly in* $\mathbb{O}$ *to* $S_\alpha(z)$, *the normalized Szegő function as* *defined in (9.6). Thus*

$$S_\alpha(z) = \exp\left\{-\frac{1}{2}\int D(t, z) \log\left[\frac{\mu'(t)}{P(t, \alpha)|\varpi_0(t)|^2}\right] d\mathring{\lambda}(t)\right\}.$$

If $\int \log \mu' \, d\mathring{\lambda} = -\infty$, then this relation still holds if the right-hand side is interpreted as being ∞.

Proof. We know that $\phi^*_{n(s)}(z)$ will converge to some $F_\alpha(z)$ in $\mathbb{O}$. We have to show that it is equal to the normalized Szegő kernel. Since $[F_\alpha(z)]^{-1}$ is in $\mathring{H}_2$ and has no zeros in $\mathbb{O}$, its inner–outer factorization takes the form

$$\frac{1}{F_\alpha(z)} = \eta(\alpha)I_S(z) \exp\left\{\frac{1}{2}\int D(t, z) \log|F_\alpha(t)|^{-2} d\mathring{\lambda}(t)\right\},$$

where $\eta \in \mathbb{T}$ arranges for a normalization $F_\alpha(\alpha) > 0$. The factor I_S is a singular inner function and the last factor is the outer factor. The outer factor is, in view of (9.26), equal to

$$\exp\left\{\frac{1}{2}\int D(t, z) \log\left[\frac{\mu'(t)}{P(t, \alpha)|\varpi_0(t)|^2}\right] d\mathring{\lambda}(t)\right\}.$$

The inner factor is of the form

$$I_S(z) = \exp\left\{\int D(t, z) \, d\mathring{\omega}(t)\right\},$$

where $\mathring{\omega}(t)$ is some singular (positive) measure. Thus we also have

$$\frac{1}{|F_\alpha(z)|} = \exp\left\{\int P(t, z) \, d\omega(t)\right\}$$
$$\cdot \exp\left\{\frac{1}{2}\int P(t, z) \log\left[\frac{\mu'(t)}{P(t, \alpha)|\varpi_0(t)|^2}\right] d\lambda(t)\right\}.$$

But by Theorem 9.3.9

$$\frac{1}{F_\alpha(\alpha)} = \exp\left\{\int P(t, z) \, d\omega(t)\right\} \cdot \frac{1}{F_\alpha(\alpha)}.$$

Thus $\int P(t, \alpha) \, d\omega(t) = 0$, so that the singular measure ω is just zero. This proves the theorem. $\qquad\square$

9.4. Equivalence of conditions

We shall here show the equivalence of the Szegő condition and certain other conditions of boundedness, convergence conditions, and density conditions when the interpolation points α_n are bounded away from the boundary $\partial \mathbb{O}$. We have

Theorem 9.4.1. *Suppose* $A = \{\alpha_1, \alpha_2, \ldots\}$ *is compactly included in* $\mathbb{O}$. *Denote by* ϕ_n *the orthonormal functions, by* $\kappa_n > 0$ *their coefficient of* B_n, *by* ε_n *and* δ_n *the parameters from their recurrence relation (4.8), and by* $k_n(z, w)$ *the reproducing kernels. Then the following statements are equivalent:*

(I) $\log \mu' \in L_1(\mathring{\lambda})$ *(Szegő's condition).*
(II) *The sequence* $\{\kappa_n\}$ *is bounded.*
(III) *The sequence* $\{\phi_n^*(z)\}$ *is locally uniformly bounded in* $\mathbb{O}$.
(IV) *The sequence* $\{\phi_n^*(w)\}$ *is bounded for some* $w \in \mathbb{O}$.
(V) *The series* $\sum_{k=0}^{\infty} |\phi_k(z)|^2$ *converges locally uniformly in* $\mathbb{O}$.
(VI) *The sequence* $\{k_n(z, w)\}$ *converges locally uniformly for* $z, w \in \mathbb{O}$.
(VII) *The sequence* $\{\prod_{k=1}^{n} [|\varepsilon_k|^2 - |\delta_k|^2]\}$ *is bounded away from 0.*
(VIII) *The set* $\{\phi_n\}$ *is not complete in* $L_2(\hat{\mu})$.

Proof. By the relation (4.12), we get

$$\lim_{n \to \infty} \prod_{k=1}^{n} [|\varepsilon_k|^2 - |\delta_k|^2] = \lim_{n \to \infty} \frac{\varpi_n(\alpha_n)}{\kappa_n^2},$$

and hence it follows that (II) $\Leftrightarrow$ (VII).

The equivalence (VIII) $\Leftrightarrow$ (I) was shown in Corollary 7.2.4.

By the dichotomy of Lemma 9.3.2, we get (III) $\Leftrightarrow$ (IV).

The implications (VI) $\Rightarrow$ (V) $\Rightarrow$ (III) $\Rightarrow$ (II) are simple: We just observe that (VI) $\Rightarrow$ (V) by putting $z = w$. This implies that $\{\phi_n\}$ goes locally uniformly to 0 and the Christoffel–Darboux relation then implies (III). By setting $z = \alpha_n$ in (III), we get (II).

We presently show (II) $\Rightarrow$ (VI), which implies with all the previous results that (II)–(VII) are equivalent.

From (II): $\phi_n^*(\alpha_n) = \kappa_n$ does not diverge to ∞ and hence by the dichotomy of Lemma 9.3.2, we get (III). As in the proof of Lemma 9.3.2, we get also (V). Finally, we apply the Schwartz inequality

$$\sum_{k=0}^{n} |\phi_k(z)\overline{\phi_k(w)}| \leq \left(\sum_{k=0}^{n} |\phi_k(z)|^2 \right)^{1/2} \left(\sum_{k=0}^{n} |\phi_k(w)|^2 \right)^{1/2}$$

to see that (VI) holds.

The remaining link is the equivalence (II) $\Leftrightarrow$ (I). Suppose $\alpha_{n(s)} \to \alpha$. Then Theorem 9.3.9 shows that (II) is equivalent with

$$\log \frac{\mu'(t)}{P(t, \alpha)|\varpi_0(t)|^2} \in L_1,$$

where $P(t, z)$ is the Poisson kernel. By (9.17), this is equivalent with (I). $\square$

We also have the following equivalence for a more restrictive situation.

Theorem 9.4.2. *With the same notation as in the previous theorem but now under the condition that $\alpha_n \to \alpha \in \mathbb{O}$, the following statements are equivalent with the conditions (I)–(VIII):*

  (II*) The sequence $\{\kappa_n\}$ converges.
  (III*) The sequence $\{\phi_n^*\}$ converges locally uniformly in $\mathbb{O}$.
  (VII*) The infinite product $\prod_{k=1}^{\infty} [|\varepsilon_k|^2 - |\delta_k|^2]$ converges.

Proof. (II*) $\Rightarrow$ (II), (III*) $\Rightarrow$ (III), and (II*) $\Leftrightarrow$ (VII*) are obvious. (III) $\Rightarrow$ (III*) because (III) $\Leftrightarrow$ (VI) and the Christoffel–Darboux relation (3.2). (II) $\Rightarrow$ (II*) because (II) $\Rightarrow$ (III) $\Rightarrow$ (III*) $\Rightarrow$ (II*). $\square$

9.5. Varying measures

Much stronger results than the ones of Section 9.3 were obtained by Li and Pan [135] in the case of the disk. They used the relation between orthogonal rational functions in $\mathcal{L}_n$ and orthogonal polynomials with respect to varying measures. The theory of orthogonal polynomials with respect to varying measures was mainly developed by G. Lopéz [138, 139, 140, 141]. In this chapter, we shall give an overview of his work. His results and proofs are for the case of the disk. However, the adaptations needed to include the half plane are only minor. We restrict ourselves to the basic results and include here the definitions and summarize without proof the convergence properties that we shall use later on.

We consider rational functions of degree m that have all their poles in $\hat{\alpha}_0$. We construct an orthonormal basis with respect to a measure $d\mathring{\mu}_n$ that depends on n. So we consider the spaces

$$\mathcal{L}_0^m = \mathrm{span}\{\zeta_0(z)^k : k = 0, 1, \ldots, m\} = \left\{ \frac{p_m(z)}{\varpi_0(z)^m} : p_m \in \mathcal{P}_m \right\}$$

and the measure

$$d\mathring{\mu}_n(t) = |w_n(t)|^2 \, d\mathring{\mu}(t), \quad \text{with } w_0 = 1, \quad w_n(z) = \frac{\varpi_0(z)^n}{\pi_n(z)}, \quad n > 0,$$

where as always $\pi_n(z) = \varpi_1(z) \cdots \varpi_n(z)$. In the case of the disk we have obviously $\mathcal{L}_0^n = \mathcal{P}_n$.

We can use our general theory for this special case and construct orthogonal rational functions $\phi_{n,k}$, $k = 0, 1, \ldots$ with $\phi_{n,k} \in \mathcal{L}_0^k$. Since we used a double index here, there should be no confusion with our previous ϕ_n. Thus we have

$$\langle \phi_{n,k}, \phi_{n,l} \rangle_{\hat{\mu}_n} = \delta_{k,l}, \quad k, l = 0, 1, \ldots.$$

The functions are uniquely defined by making their leading coefficient positive:

$$\phi_{n,k}(z) = \upsilon_{n,k} \zeta_0(z)^k + \cdots, \quad \upsilon_{n,k} > 0. \tag{9.27}$$

These functions satisfy the following recurrence relations:

$$\phi_{n,m}(z) = e_{n,m}[\zeta_0(z)\phi_{n,m-1}(z) + \bar{\lambda}_{n,m}\phi_{n,m-1}^{\tilde{*}}(z)],$$
$$\phi_{n,m}^{\tilde{*}}(z) = e_{n,m}[\lambda_{n,m}\zeta_0(z)\phi_{n,m-1}(z) + \phi_{n,m-1}^{\tilde{*}}(z)],$$

with

$$\lambda_{n,m} = \frac{\overline{\phi_{n,m}(\alpha_0)}}{\upsilon_{n,m}}, \quad e_{n,m} = \frac{1}{\sqrt{1 - |\lambda_{n,m}|^2}} = \frac{\upsilon_{n,m}}{\upsilon_{n,m-1}}. \tag{9.28}$$

We have used the special superstar notation $f^{\tilde{*}}$ for $f \in \mathcal{L}_0^n$ to mean

$$f^{\tilde{*}}(z) = \zeta_0(z)^n f_*(z), \quad f \in \mathcal{L}_0^n.$$

Thereby we can distinguish between a superstar in $\mathcal{L}_n$ and a superstar in $\mathcal{L}_0^n$. Of course, this should also be clear from the context.

Furthermore, it is very easy to obtain that

$$\upsilon_{n,m+1}^2 = \upsilon_{n,m}^2 + |\phi_{n,m+1}(\alpha_0)|^2$$

and it is known [141, (6) p. 201] that

$$|\lambda_{n,m}| \leq C \int \left| \frac{|\phi_{n,m-1}(t)|^2}{|\phi_{n,m}(t)|^2} - 1 \right| d\mathring{\lambda}(t),$$

with C a constant that can be chosen independent of n and m.

Obviously, these $\{\phi_{n,m}\}$ have the following relation with the orthonormal basis for $\mathcal{L}_n$. Set

$$F_{n,m}(z) = t_{n,m}\phi_{n,m}(z)w_n(z) \in \mathcal{L}_n, \quad t_{n,m} \in \mathbb{T}.$$

Then

$$\langle \phi_{n,k}, \phi_{n,l} \rangle_{\hat{\mu}_n} = \int \phi_{n,k}(t)\overline{\phi_{n,l}(t)}|w_n(t)|^2 d\hat{\mu}(t)$$

$$= t_{n,k,l} \int F_{n,k}(t)\overline{F_{n,l}(t)}\, d\hat{\mu}(t) = t_{n,k,l} \langle F_{n,k}F_{n,l} \rangle_{\hat{\mu}},$$

with $t_{n,k,l} = t_{n,l}/t_{n,k} \in \mathbb{T}$, so that $t_{n,k,k} = 1$. Thus the $F_{n,k}$ are orthonormal rational basis functions for $\mathcal{L}_n$, which are obtained by orthonormalizing the basis

$$\{\zeta_0(z)^k w_n(z) : k = 0, 1, \ldots, n\}$$

with respect to $d\hat{\mu}$. Thus, we see here yet another basis for $\mathcal{L}_n$, which has the unfortunate property that when n changes, the whole set of basis functions changes. The Blaschke products as basis functions, and their orthonormal derivates ϕ_n, did not give this problem, since we only have to add one basis function and keep all the others when passing from n to $n + 1$. Thus the $F_{n,k}$ are definitely different from the ϕ_k. However, for the reproducing kernel, we have the two expressions

$$k_n(z, w) = \sum_{k=0}^{n} \phi_k(z)\overline{\phi_k(w)} = \sum_{k=0}^{n} F_{n,k}(z)\overline{F_{n,k}(w)}.$$

In the following theorems we shall assume that we select the measure and the point set $A = \{\alpha_1, \alpha_2, \ldots\} \subset \overline{\mathbb{O}}$ such that they satisfy the following conditions:

1. $\mu' > 0$ a.e. (Erdős–Turán condition).
2. $\int d\hat{\mu} = 1, m_n = \int d\hat{\mu}_n < \infty, n > 0$.
3. $\sum_{n=1}^{\infty} m_n^{-1/2n} = \infty$ (Carleman condition).

We shall indicate this briefly by writing $(A, \mu) \in \mathcal{AM}$.

The condition $\mu' > 0$ a.e. was considered by Erdős for the interval $[-1, 1]$. Erdős and Turán used it to prove certain convergence results in L_2. Therefore, $\mu' > 0$ a.e. is usually called the Erdős–Turán condition. Rahmanov used the same condition to show that the recurrence coefficients for the associated orthogonal polynomials behave asymptotically as the recurrence coefficients of Chebyshev polynomials.

In Ref. [141], Lopéz considers a non-Newtonian triangular array of interpolation points $\{\alpha_{i,n} : i = 0, \ldots, n; n = 0, 1, \ldots\}$, which we do not need, in view of the other material treated here. Thus we give the formulation only for the situation of a Newton sequence of interpolation points. Lopéz calls the conditions defining $\mathcal{AM}$ the C-conditions because condition 3 is a Carleman-type condition for the moments m_n. In an earlier paper [138], the properties to follow were proved under the stronger assumption that the set A is compactly included in $\mathbb{O}$. An intermediate condition is considered in Ref. [140]:

3′. The Blaschke product with zeros α_k diverges,

instead of the Carleman condition 3. For results where the spectral factor σ is needed, this factor should be defined, which requires that we replace the condition $\mu' > 0$ a.e. by the stronger one:

1′. $\log \mu' \in L_1(\overset{\circ}{\lambda})$ (Szegő condition).

We indicate that A and μ satisfy conditions 1′, 2, and 3 by $(A, \mu) \in \mathcal{AM}'$. If A and μ satisfy conditions 1′, 2, and 3′, we write $(A, \mu) \in \mathcal{AM}''$. Since 3′ implies 3 and 1′ implies 1, $\mathcal{AM}'' \subset \mathcal{AM}' \subset \mathcal{AM}$. Note that here the points of A are in general points in the *closed* set $\overline{\mathbb{O}} = \mathbb{O} \cup \partial\mathbb{O}$. For the moment, we only need them to be in the open set $\mathbb{O}$. However, these results are more general and when we treat the boundary case in Chapter 11, where the points of A are all in $\partial\mathbb{O}$, practically all the results from this section and from the two sections to follow will be applicable. Note also that for $A \subset \mathbb{O}$, the functions $w_n = \varpi_0^n / \pi_n$ are continuous on $\partial\mathbb{O}$ and thus condition 2 will be satisfied for any normalized measure. Thus if $A \subset \mathbb{O}$, then condition 2 is actually superfluous. Thus $\mathcal{AM}''$ is basically defined by 1′ and 3′ and the normalization $\int d\overset{\circ}{\mu} = 1$. A fortiori this holds when A is compactly included in $\mathbb{O}$. If we allow some $\alpha_i \in \partial\mathbb{O}$, then, in the case of the line, such an α_i is allowed to be ∞. In that case a factor $z - \overline{\alpha}_i$ in the product $\pi_n(z)$ should be replaced by 1. For example, if all $\alpha_i = \infty$, then $w_n(z) = (z + \mathbf{i})^n$ and condition 2 is necessary to ensure that the measure is sufficiently weak at ∞.

The following theorems are readily obtained from Lopéz [138, 139, 140]. We give them without proof.

Theorem 9.5.1. *Let* $(A, \mu) \in \mathcal{AM}$. *Then the measure*

$$d\overset{\circ}{\nu}_n(t) = \frac{d\overset{\circ}{\lambda}(t)}{|w_n(t)\phi_{n,n+k}(t)|^2},$$

with integer k, converges in weak star topology to $d\overset{\circ}{\mu}(t)$, *that is,*

$$\lim_{n \to \infty} \int f(t) \, d\overset{\circ}{\nu}_n(t) = \int f(t) \, d\overset{\circ}{\mu}(t), \qquad \forall f \in C(\overline{\partial\mathbb{O}}).$$

Theorem 9.5.2. Let $(A, \mu) \in \mathcal{AM}$. Then for integers k and m

$$\lim_{n \to \infty} \int \left| \frac{|\phi_{n,n+k}(t)|^2}{|\phi_{n,n+k+m}(t)|^2} - 1 \right| d\mathring{\lambda}(t) = 0.$$

Theorem 9.5.3. Let $(A, \mu) \in \mathcal{AM}$. The $\lambda_{n,m}$ are the recurrence coefficients from (9.28) and the $\upsilon_{n,m}$ are the leading coefficients from (9.27). Then the following convergence results hold for k integer:

1. $\lim_{n \to \infty} \lambda_{n,n+k+1} = 0$.
2. $\lim_{n \to \infty} \frac{\upsilon_{n,n+k+1}}{\upsilon_{n,n+k}} = 1$.
3. $\lim_{n \to \infty} \frac{\phi_{n,n+k+1}(z)}{\phi_{n,n+k}(z)} = \zeta_0(z)$,   locally uniformly $z \in \mathbb{O}^e$.
4. $\lim_{n \to \infty} \frac{\phi^*_{n,n+k+1}(z)}{\phi^*_{n,n+k}(z)} = 1$,   locally uniformly $z \in \mathbb{O}$.
5. $\lim_{n \to \infty} \frac{\phi^*_{n,n+k}(z)}{\phi_{n,n+k}(z)} = 0$,   locally uniformly $z \in \mathbb{O}^e$.

Let $(A, \mu) \in \mathcal{AM}'$ and let σ be the spectral factor of μ, normalized by $\sigma(\alpha_0) > 0$. Then for k integer

6. $\lim_{n \to \infty} \phi^*_{n,n+k}(z) w_n(z) \frac{\overline{w_n(\alpha_0)}}{|w_n(\alpha_0)|} = \frac{1}{\sigma(z)}$,   locally uniformly $z \in \mathbb{O}$.

9.6. Stronger results

We shall now give a relation between the $F_{n,n}$ defined in the previous section and our orthogonal rational functions ϕ_n. This relation will lead us to asymptotic results for the case where $(A, \mu) \in \mathcal{AM}$. More specifically, the α_k need not be contained in a compact subset of $\mathbb{O}$. Let us simplify the notation by setting $F_n = F_{n,n}$. Furthermore, we choose the normalizing constants $t_{n,n} \in \mathbb{T}$ in our definition of F_n such that it gets the same normalization as the ϕ_n. That is, it is chosen such that

$$F_n(z) = \kappa_{n,n} B_n(z) + \cdots \quad \text{with } \kappa_{n,n} > 0.$$

This means that (setting $\eta_n = \prod_{k=1}^n z_k$)

$$t_{n,n} = \eta_n \frac{|\phi^*_{n,n}(\alpha_n) w_n(\alpha_n)|}{\phi^*_{n,n}(\alpha_n) w_n(\alpha_n)}.$$

Define the functions f_n and g_n both in $\mathcal{L}_0^n$ by

$$g_n(z) = \frac{\phi_n(z)}{w_n(z)} \quad \text{and} \quad f_n(z) = \frac{F_n(z)}{w_n(z)}. \tag{9.29}$$

In our previous notation, $f_n = t_{n,n}\phi_{n,n}$. We shall also need the leading coefficients of g_n and f_n in $\mathcal{L}_0^n$:

$$\overline{\tau}_n = g_n^{\bar{*}}(\alpha_0) = \overline{\eta}_n \frac{\phi_n^*(\alpha_0)}{w_n(\alpha_0)} \quad \text{and} \quad \overline{\upsilon}_n = f_n^{\bar{*}}(\alpha_0) = \overline{\eta}_n \frac{F_n^*(\alpha_0)}{w_n(\alpha_0)}. \tag{9.30}$$

This means that

$$g_n(z) = \tau_n \zeta_0(z)^n + \cdots \quad \text{and} \quad f_n(z) = \upsilon_n \zeta_0(z)^n + \cdots.$$

First we need the following lemma (see Ref. [135]).

Lemma 9.6.1. With the previous definitions, it holds that

$$\frac{g_n(z)}{\tau_n}[\zeta_0(z) - \zeta_0(\alpha_n)] = \frac{f_n(z)}{\upsilon_n}\zeta_0(z) - \frac{f_n(\alpha_n)}{f_n^{\bar{*}}(\alpha_n)}\zeta_0(\alpha_n)\frac{f_n^{\bar{*}}(z)}{\upsilon_n}.$$

Proof. A simple calculation shows that

$$h_n(z) = \frac{1}{\upsilon_n} \frac{f_n(z)\zeta_0(z) - \frac{f_n(\alpha_n)}{f_n^{\bar{*}}(\alpha_n)}\zeta_0(\alpha_n)f_n^{\bar{*}}(z)}{\zeta_0(z) - \zeta_0(\alpha_n)}$$

is a monic function in $\mathcal{L}_0^n$. Next consider $H_n(z) = h_n(z)w_n(z) \in \mathcal{L}_n$. We shall show below that it is orthogonal to $\mathcal{L}_{n-1}$ with respect to $\mathring{\mu}$. Since g_n/τ_n and h_n are both monic in $\mathcal{L}_0^n$, the functions H_n and ϕ_n/τ_n must have the same normalization and hence $H_n = \phi_n/\tau_n$.

The orthogonality is shown as follows: Let $G_{n-1}(z) = t_{n-1}(z)w_{n-1}(z)$ be an arbitrary function from $\mathcal{L}_{n-1}$ obtained by choosing $t_{n-1} \in \mathcal{L}_0^{n-1}$ arbitrarily. The c_i appearing below are irrelevant constants. Then

$$\langle H_n, G_{n-1}\rangle_{\mathring{\mu}} = c_1 \int [f_n(t)\zeta_0(t) - c_2 f_n^{\bar{*}}(t)]\overline{t_{n-1}(t)\zeta_0(t)}\, d\mathring{\mu}_n(t)$$

$$= c_1 \langle f_n t_{n-1}\rangle_{\mathring{\mu}_n} + c_3 \langle f_n^{\bar{*}}, \zeta_0 t_{n-1}\rangle_{\mathring{\mu}_n}.$$

The first inner product is zero by the orthogonality of f_n and the second one by Theorem 2.2.1(5), applied in the case where all $\alpha_k = \alpha_0$. This proves the lemma. $\square$

As a last preparation for the main convergence results, we state

Lemma 9.6.2. Let $(A, \mu) \in \mathcal{AM}$ and define with the notation of this section the functions

$$G_n(z) = \frac{g_n(z)}{\tau_n} \in \mathcal{L}_0^n \quad \text{and} \quad F_n(z) = \frac{f_n(z)}{\upsilon_n} \in \mathcal{L}_0^n.$$

Then

$$\lim_{n\to\infty} \frac{G_n(z)[\zeta_0(z) - \zeta_0(\alpha_n)]}{F_n(z)\zeta_0(z)} = 1$$

locally uniformly in $\mathbb{O}^e$.

Proof. Since we have by the previous lemma

$$G_n(z)[\zeta_0(z) - \zeta_0(\alpha_n)] = F_n(z)\zeta_0(z) - \frac{F_n(\alpha_n)F_n^*(z)}{F_n^*(\alpha_n)}\zeta_0(\alpha_n),$$

then also

$$\frac{G_n(z)[\zeta_0(z) - \zeta_0(\alpha_n)]}{F_n(z)\zeta_0(z)} = 1 - \frac{F_n(\alpha_n)\zeta_0(\alpha_n)F_n^*(z)}{F_n^*(\alpha_n)\zeta_0(z)F_n(z)}.$$

To find the limit of the second term in the right-hand side, we observe that

$$\left|\frac{F_n(z)}{F_n^*(z)}\right| = \left|\frac{\phi_{n,n}(z)}{\phi_{n,n}^*(z)}\right| < 1 \quad \text{in } \mathbb{O}.$$

Hence

$$\left|\frac{F_n(\alpha_n)}{F_n^*(\alpha_n)}\right| < 1.$$

And observe that

$$\left|\frac{\zeta_0(\alpha_n)}{\zeta_0(z)}\right| < 1 \quad \text{in } \mathbb{O}^e,$$

whereas by Theorem 9.5.3(5) with $k = 0$

$$\lim_{n\to\infty} \frac{F_n^*(z)}{F_n(z)} = \lim_{n\to\infty} \frac{F_n^*(z)}{F_n(z)} = 0$$

locally uniformly in $\mathbb{O}^e$. The lemma thus follows.   □

Now we can give the main results of this section.

Theorem 9.6.3. *Let* $(A, \mu) \in \mathcal{AM}'$. *Then*

$$\lim_{n\to\infty} \frac{\phi_n^*(z)\varpi_n(z)}{\phi_n^*(\alpha_0)\varpi_n(\alpha_0)} = \frac{\sigma(\alpha_0)\varpi_0(z)}{\sigma(z)\varpi_0(\alpha_0)}$$

locally uniformly in $\mathbb{O}$.

Proof. First note that

$$\phi_n^*(z) = [g_n(z)w_n(z)]^* = \eta_n w_n(z) g_n^{\bar *}(z), \quad \eta_n = z_1 \cdots z_n,$$

so that for $z = \alpha_0$ we obtain

$$\phi_n^*(\alpha_0) = \eta_n w_n(\alpha_0) \bar\tau_n.$$

Thus the expression in the previous lemma can be rewritten as

$$\frac{G_n(z)[\zeta_0(z) - \zeta_0(\alpha_n)]}{F_n(z)\zeta_0(z)} = \frac{\overline{w_n(\alpha_0)} g_n(z) v_n [\zeta_0(z) - \zeta_0(\alpha_n)]}{\eta_n \overline{\phi_n^*(\alpha_0)} f_n(z) \zeta_0(z)}.$$

This goes to 1 locally uniformly in $\mathbb{O}^e$, so that we also have local uniform convergence in $\mathbb{O}$ of

$$\frac{\eta_n w_n(\alpha_0) g_n^{\bar *}(z) \bar v_n [\zeta_{0*}(z) - \overline{\zeta_0(\alpha_n)}]}{\phi_n^*(\alpha_0) f_n^{\bar *}(z) \zeta_{0*}(z)}$$

$$= \frac{[\eta_n w_n(z) g_n^{\bar *}(z)][\eta_n w_n(\alpha_0) \bar v_n][1 - \zeta_0(z)\overline{\zeta_0(\alpha_n)}]}{\phi_n^*(\alpha_0)[\eta_n w_n(z) f_n^{\bar *}(z)]}$$

$$= \frac{\phi_n^*(z) F_n^*(\alpha_0)[1 - \zeta_0(z)\overline{\zeta_0(\alpha_n)}]}{\phi_n^*(\alpha_0) F_n^*(z)}.$$

By Theorem 9.5.3(6), there exist $c_n \in \mathbb{T}$ such that

$$\lim_{n \to \infty} c_n F_n^*(z) = \lim_{n \to \infty} c_n \eta_n f_n^{\bar *}(z) w_n(z) = 1/\sigma(z)$$

locally uniformly in $\mathbb{O}$; hence also $\lim_{n \to \infty} c_n F_n^*(\alpha_0) = 1/\sigma(\alpha_0)$. If $\sigma(\alpha_0) > 0$, then we choose the c_n such that $c_n F_n^*(\alpha_0) > 0$. Using this result and

$$1 - \zeta_0(z)\overline{\zeta_0(\alpha_n)} = \frac{\varpi_0(\alpha_0) \varpi_n(z)}{\varpi_n(\alpha_0) \varpi_0(z)},$$

we get the convergence result of this theorem. □

Note that when $\alpha_n \to \alpha \in \mathbb{O}$, then the previous result can also be derived from Theorem 9.3.11, which says that

$$\lim_{n \to \infty} \frac{\phi_n^*(z) \varpi_\alpha(z)}{\sqrt{\varpi_\alpha(\alpha)/\varpi_0(\alpha_0)}} = \eta(\alpha) \frac{\varpi_0(z)}{\sigma(z)},$$

where $\eta(\alpha) \in \mathbb{T}$ as in (9.6).

The point α_0 in our previous theorem is arbitrary and used for normalization only. It could be any other point $w \in \mathbb{O}$, since indeed by applying the previous theorem in z and w, we get for the ratio

$$\lim_{n\to\infty} \frac{\phi_n^*(z)\varpi_n(z)}{\phi_n^*(w)\varpi_n(w)} = \frac{\sigma(w)\varpi_0(z)}{\sigma(z)\varpi_0(w)}.$$

This holds locally uniformly for $z \in \mathbb{O}$ and for w fixed, but by the symmetry in the relation, it is immediately seen that this must hold locally uniformly for $(z, w) \in \mathbb{O} \times \mathbb{O}$.

We can also prove a local uniform convergence result for the reproducing kernels. For the case of the unit disk, see Ref. [135].

Theorem 9.6.4. *Suppose* $(A, \mu) \in \mathcal{AM}'$ *and let* $k_n(z, w)$ *be the reproducing kernel for* $\mathcal{L}_n$ *and* $s_w(z)$ *the Szegő kernel (9.3). Then we have local uniform convergence of*

$$\lim_{n\to\infty} k_n(z, w) = s_w(z), \quad (z, w) \in \mathbb{O} \times \mathbb{O}.$$

Proof. We know that

$$k_n(z, w) = \sum_{k=0}^{n} F_{n,k}(z)\overline{F_{n,k}(w)} = \left[\sum_{k=0}^{n} \phi_{n,k}(z)\overline{\phi_{n,k}(w)}\right] w_n(z)\overline{w_n(w)}.$$

With the Christoffel–Darboux relation for the $\phi_{n,k}$, we get

$$k_n(z, w) = \frac{\phi_{n,n}^{\bar{*}}(z)\overline{\phi_{n,n}^{\bar{*}}(w)} - \zeta_0(z)\overline{\zeta_0(w)}\phi_{n,n}(z)\overline{\phi_{n,n}(w)}}{1 - \zeta_0(z)\overline{\zeta_0(w)}} w_n(z)\overline{w_n(w)}.$$

Thus

$$\frac{k_n(z, w)}{F_n^*(z)\overline{F_n^*(w)}} = \frac{1}{1 - \zeta_0(z)\overline{\zeta_0(w)}} - \frac{\zeta_0(z)\overline{\zeta_0(w)}}{1 - \zeta_0(z)\overline{\zeta_0(w)}}\left(\frac{\phi_{n,n}(z)}{\phi_{n,n}^{\bar{*}}(z)}\right)\overline{\left(\frac{\phi_{n,n}(w)}{\phi_{n,n}^{\bar{*}}(w)}\right)}.$$

Since we can use Theorem 9.5.3(6), we get local uniform convergence in $\mathbb{O}$ for some $c_n \in \mathbb{T}$

$$\lim_{n\to\infty} c_n F_n^*(z) = \frac{1}{\sigma(z)},$$

so that locally uniformly for $(z, w) \in \mathbb{O} \times \mathbb{O}$:

$$\lim_{n\to\infty} F_n^*(z)\overline{F_n^*(w)} = \frac{1}{\sigma(z)\overline{\sigma(w)}}.$$

By Theorem 9.5.3(5), the same type of convergence holds in

$$\lim_{n \to \infty} \left(\frac{\phi_{n,n}(z)}{\phi_{n,n}^*(z)} \right) \overline{\left(\frac{\phi_{n,n}(w)}{\phi_{n,n}^*(w)} \right)} = 0.$$

Thus the result of the theorem follows because $s_w(z) = 1/[(1-\zeta_0(z)\overline{\zeta_0(w)})\sigma(z)\overline{\sigma(w)}]$. $\square$

As an immediate corollary we have

Corollary 9.6.5. *Let* $(A, \mu) \in \mathcal{AM}'$. *Then*

$$\lim_{n \to \infty} \phi_n(z) = 0 \quad and \quad \lim_{n \to \infty} \frac{\phi_n(z)\phi_n^*(\alpha_0)}{\phi_n^*(z)} = 0$$

locally uniformly in $\mathbb{O}$.

Proof. The first limit holds because by the previous theorem with $w = z$ we have local uniform convergence of $k_n(z, z) = \sum_{k=0}^n |\phi_k(z)|^2$ for $n \to \infty$. Thus $\phi_n(z) \to 0$ locally uniformly.

For the second limit, we observe that by Theorem 9.6.3 we have, after inverting the expression and multiplying it by $\phi_n(z)$,

$$\lim_{n \to \infty} \frac{\phi_n(z)\phi_n^*(\alpha_0)\varpi_n(\alpha_0)}{\phi_n^*(z)\varpi_n(z)} = 0$$

locally uniformly in $\mathbb{O}$ since indeed, by the first part of this corollary, $\phi_n(z)$ converges to zero. When z is in a compact subset of $\mathbb{O}$, then $\varpi_n(\alpha_0)/\varpi_n(z)$ is bounded away from zero. Thus the second limit follows. $\square$

Next we extend the well-known property for orthogonal polynomials $\phi_n/\phi_n^* \to 0$ locally uniformly in $\mathbb{O}$. We show this under the more restrictive condition that $A = \{\alpha_1, \alpha_2, \ldots\}$ is compactly included in $\mathbb{O}$. Recall that this condition in combination with Szegő's condition implies that $(A, \mu) \in \mathcal{AM}'$.

Theorem 9.6.6. *Let* A *be compactly included in* $\mathbb{O}$ *and* $\log \mu' \in L_1(\mathring{\lambda})$. *Then the following limits hold locally uniformly in the indicated regions:*

$$\lim_{n \to \infty} \frac{\phi_n(z)}{\phi_n^*(z)} = 0, \quad z \in \mathbb{O} \quad and \quad \lim_{n \to \infty} \frac{\phi_n^*(z)}{\phi_n(z)} = 0, \quad z \in \mathbb{O}^e.$$

Proof. The second property follows from the first one by taking superstar conjugates. We shall prove that $|\phi_n/\phi_n^*| \to 0$ locally uniformly in $\mathbb{O}$.

By the Christoffel–Darboux relation (3.1), we have for $z = w = \alpha_0$

$$|\phi_n^*(\alpha_0)|^2 - |\phi_n(\alpha_0)|^2 = (1 - |\zeta_n(\alpha_0)|^2) \sum_{k=0}^{n-1} |\mathring{\phi}_k(\alpha_0)|^2. \qquad (9.31)$$

Thus

$$|\phi_n^*(\alpha_0)|^2 \geq (1 - |\zeta_n(\alpha_0)|^2)|\phi_0|^2 = 1 - |\zeta_n(\alpha_0)|^2.$$

Since A is compactly included in $\mathbb{O}$, we have $|\zeta_n(z)| \leq m(r) < 1$ for all $z \in \mathbb{O}_r$. Hence

$$|\phi_n^*(\alpha_0)|^2 \geq 1 - m(r)^2 = C > 0. \qquad (9.32)$$

This inequality, in combination with the second limit of the previous corollary, leads to the result.   □

For the kernels, we can prove the following theorem, which was given by Li and Pan [135] for the unit disk. It gives a result when no conditions on the divergence or convergence of the Blaschke product are given.

Theorem 9.6.7. *Let $\log \mu' \in L_1(\mathring{\lambda})$ and let σ be an outer spectral factor of μ. The sequence of reproducing kernels $\{k_n(z, w)\}$ for $\mathcal{L}_n$ is locally uniformly bounded in $\mathbb{O} \times \mathbb{O}$. Every limit function $k(z, w)$ will be analytic in z and $\overline{w}$ for $(z, w) \in \mathbb{O} \times \mathbb{O}$ and for fixed $w \in \mathbb{O}$, $\sigma(z)k(z, w) \in \mathring{H}_2$.*

Proof. Clearly, $k_n(z, w)\sigma(z) \in \mathring{H}_2$. Thus

$$\frac{k_n(z, w)\sigma(z)}{\varpi_0(z)} \in H_2.$$

By Cauchy's formula, we find

$$\left| \frac{k_n(z, w)\sigma(z)}{\varpi_0(z)} \right| = \left| \int C(t, z) \frac{k_n(t, w)\sigma(t)}{\varpi_0(t)} d\lambda(t) \right|$$

$$\leq \int \frac{|C(t, z)k_n(t, w)\sigma(t)|}{|\varpi_0(t)|} d\lambda(t).$$

Noting that

$$\int |k_n(t, w)|^2 \mu'(t) \, d\mathring{\lambda}(t) \leq \int |k_n(t, w)|^2 \, d\mathring{\mu}(t) = k_n(w, w)$$

and that for $t \in \partial \mathbb{O}$

$$|C(t, z)|^2 = \frac{P(t, z)}{|\varpi_z(z)\varpi_0(\alpha_0)|},$$

so that

$$\int |C(t, z)|^2 \, d\lambda(t) = |\varpi_z(z)\varpi_0(\alpha_0)|^{-1},$$

we get by an application of the Schwarz inequality

$$\left| \frac{k_n(z, w)\sigma(z)}{\varpi_0(z)} \right|^2 \leq \left\{ \int |C(t, z)|^2 \, d\lambda(t) \right\} \cdot \left\{ \int |k_n(t, w)|^2 \mu'(t) \, d\mathring{\lambda}(t) \right\}$$

$$\leq \frac{k_n(w, w)}{|\varpi_z(z)\varpi_0(\alpha_0)|}.$$

Setting $z = w$, we get

$$k_n(w, w) \leq \frac{|\varpi_0(w)|^2}{|\varpi_w(w)\varpi_0(\alpha_0)\sigma(w)^2|}.$$

Plugging this into the previous inequality, we get

$$\left| \frac{k_n(z, w)}{\varpi_0(z)\varpi_0(w)} \right|^2 \leq \frac{1}{|\varpi_z(z)\varpi_w(w)\varpi_0(\alpha_0)^2\sigma(z)^2\sigma(w)^2|}.$$

Now suppose as before that $\mathbb{O}_r$ is a compact subset of $\mathbb{O}$ with $|\varpi_z(z)\varpi_0(\alpha_0)| \geq r$ for all $z \in \mathbb{O}_r$. Define $M(r)$ as the supremum of $|\sigma(z)|^{-2}$ for $z \in \mathbb{O}_r$. This is a finite value since σ is outer. Thus for z and w in $\mathbb{O}_r$ we find that

$$\left| \frac{k_n(z, w)}{\varpi_0(z)\varpi_0(w)} \right|^2 \leq \frac{M(r)^2}{r^2}.$$

Since obviously $|\varpi_0(z)|$ is bounded in $\mathbb{O}_r$, we have found that $\{k_n(z, w)\}$ is locally uniformly bounded as claimed. Thus it is a normal family (in two variables) and this means that there must exist a convergent subsequence with limit function say $k(z, w)$:

$$\lim_{s \to \infty} k_{n(s)}(z, w) = k(z, w),$$

locally uniformly for $(z, w) \in \mathbb{O} \times \mathbb{O}$. We have to prove that $k(z, w)\sigma(z) \in \mathring{H}_2$ and thus that $k(z, w)\sigma(z)/\varpi_0(z) \in H_2$. This follows from the following relations (the formulation rt is for the disk, but an easy adaptation can be made for the real line):

$$\int \left| \frac{k(rt,w)\sigma(rt)}{\varpi_0(rt)} \right|^2 d\lambda(t) = \lim_{s\to\infty} \int \left| \frac{k_{n(s)}(rt,w)\sigma(rt)}{\varpi_0(rt)} \right|^2 d\lambda(t)$$

$$\leq \lim_{s\to\infty} \int \left| \frac{k_{n(s)}(t,w)\sigma(t)}{\varpi_0(t)} \right|^2 d\lambda(t)$$

$$\leq \lim_{s\to\infty} \int |k_{n(s)}(t,w)|^2 \, d\mathring{\mu}(t)$$

$$= \lim_{s\to\infty} k_{n(s)}(w,w) = k(w,w) < \infty.$$

This proves the theorem. □

Corollary 9.6.8. *Under the conditions of the previous theorem, the sequence $\{\phi_n^*\}$ is locally uniformly bounded in $\mathbb{O}$. Every limit function F is an analytic function without zeros in $\mathbb{O}$.*

Proof. By the previous theorem, for $w = z$, the sequence $\{k_n(z,z)\}$ is locally uniformly bounded. Hence, by $k_n(z,z) = \sum_{i=0}^n |\phi_i(z)|^2$, it follows that $\{\phi_n\}$ is locally uniformly bounded. Now by the Christoffel–Darboux formula,

$$|\phi_n^*(z)|^2 = (1 - |\zeta_n(z)|^2)k_n(z,z) + |\zeta_n(z)|^2|\phi_n(z)|^2 \leq k_n(z,z) + |\phi_n(z)|^2,$$

it follows that $\{|\phi_n^*|^2\}$ and hence also $\{\phi_n^*\}$ are locally uniformly bounded. None of the limit functions can have a zero in $\mathbb{O}$ by Hurwitz's theorem because ϕ_n^* has no zeros in $\mathbb{O}$. □

In Theorem 9.3.11, we have shown that if $\alpha_{n(s)} \to \alpha \in \mathbb{O}$, then $\phi_n^*(z) \to S_\alpha(z)$ with $S_w(z)$ the normalized Szegő kernel. Following Pan [173], we now prove a similar result when it is only assumed that $A = \{\alpha_1, \alpha_2, \ldots\}$ is compactly included in $\mathbb{O}$.

Theorem 9.6.9. *Suppose $(A,\mu) \in \mathcal{AM}'$ and let σ be the outer spectral factor normalized by $\sigma(\alpha_0) > 0$. Assume also that A is compactly included in $\mathbb{O}$. Then we have locally uniformly in $\mathbb{O}$*

$$\lim_{n\to\infty} \eta_n \frac{\phi_n^*(z)\varpi_n(z)}{\sqrt{\varpi_n(\alpha_n)/\varpi_0(\alpha_0)}} = \frac{\varpi_0(z)}{\sigma(z)}, \qquad \eta_n = \frac{|\phi_n^*(\alpha_0)\varpi_n(\alpha_0)/\varpi_0(\alpha_0)|}{\phi_n^*(\alpha_0)\varpi_n(\alpha_0)/\varpi_0(\alpha_0)} \in \mathbb{T}.$$

Proof. We first prove that

$$\lim_{n\to\infty} \frac{|\phi_n^*(\alpha_0)|}{\sqrt{1-|\zeta_n(\alpha_0)|^2}} = \frac{1}{\sigma(\alpha_0)\sqrt{1-|\zeta_0(\alpha_0)|^2}}. \tag{9.33}$$

To this end, we first note that by the Christoffel–Darboux relation

$$\frac{|\phi_n^*(\alpha_0)|^2}{1-|\zeta_n(\alpha_0)|^2} = k_{n-1}(\alpha_0,\alpha_0) + \frac{|\phi_n(\alpha_0)|^2}{1-|\zeta_n(\alpha_0)|^2}. \tag{9.34}$$

Theorem 9.6.4 implies that the limit of the second term in (9.34) is given by

$$\lim_{n\to\infty} k_{n-1}(\alpha_0,\alpha_0) = \frac{1}{(1-|\zeta_0(\alpha_0)|^2)|\sigma(\alpha_0)|^2}.$$

The denominator of the last term in (9.34) is bounded away from zero and the numerator goes to zero by Corollary 9.6.5. This proves (9.33). Noting that

$$1-|\zeta_n(\alpha_0)|^2 = \frac{\overline{\varpi_0(\alpha_0)}\varpi_n(\alpha_n)}{|\varpi_n(\alpha_0)|^2},$$

we can rewrite this as

$$\lim_{n\to\infty} \frac{|\phi_n^*(\alpha_0)\varpi_n(\alpha_0)|}{\sqrt{\varpi_n(\alpha_n)/\varpi_0(\alpha_0)}} = \frac{|\varpi_0(\alpha_0)|}{\sigma(\alpha_0)}.$$

The result now follows by combining this with the result of Theorem 9.6.3. □

Note that the factor η_n in this theorem is equal to $|\phi_n^*(\alpha_0)|/\phi_n^*(\alpha_0)$ in the case of the disk. It rotates ϕ_n^* such that it becomes normalized by the condition $\phi_n^*(\alpha_0) > 0$ instead of our usual $\phi_n^*(\alpha_n) = \kappa_n > 0$. In the case of the half plane, an extra rotation with $\varpi_0(\alpha_0)/|\varpi_0(\alpha_0)|$ is needed.

Note also that the extra factor given by the ratio in the left-hand side of this theorem is in the case of the disk simply $1/\sqrt{1-|\alpha_n|^2}$ whereas in the case of the half plane it is $1/\sqrt{\text{Im }\alpha_n}$.

As a consequence we can extend the sequence of equivalent conditions given in Theorem 9.4.1.

Corollary 9.6.10. *Under the same conditions as in Theorem 9.4.1, and with η_n as defined in Theorem 9.6.9, the following statements are equivalent:*

(B) The conditions (I)–(VIII) of Theorem 9.4.1.
(IX) Local uniform convergence of the sequence

$$\eta_n \frac{\phi_n^*(z)\varpi_n(z)}{\sqrt{\varpi_n(\alpha_n)/\varpi_0(\alpha_0)}}.$$

(X) Convergence of the sequence

$$\frac{|\phi_n^*(\alpha_0)\varpi_n(\alpha_0)|}{\sqrt{\varpi_n(\alpha_n)/\varpi_0(\alpha_0)}}.$$

Proof. The condition (I) implies (IX) by the previous theorem. By setting $z = \alpha_0$ in (IX) we get (X) apart from a constant unimodular factor. Hence the sequence of (X) is bounded, but because A is compactly included in $\mathbb{O}$, this means that $\{\phi_n^*(\alpha_0)\}$ is bounded. This is condition (IV) and that closes the circle.   □

The previous theorems assumed the Szegő condition. Following the work of Pan, it is possible to replace this condition by $\mu' > 0$ a.e. This Erdős–Turán condition will then give rise to several ratio asymptotic results. This will be investigated in Section 9.8.

9.7. Weak convergence

We also have a weak star convergence result for the measures considered in Theorem 6.1.9.

Theorem 9.7.1. *Suppose that the Blaschke product diverges. Let $P(t, z)$ be the Poisson kernel. Then the measure defined by $d\mathring{\mu}_n(t) = P(t, \alpha_n)/|\phi_n(t)|^2\, d\lambda(t)$ converges to $d\mathring{\mu}$ in the weak star topology, that is, for any continuous $f \in C(\overline{\partial\mathbb{O}})$,*

$$\lim_{n\to\infty} \int f(t)\, d\mathring{\mu}_n(t) = \int f(t)\, d\mathring{\mu}(t).$$

Proof. The proof is by standard arguments. We know that $\langle f, g\rangle_{\mathring{\mu}_n} = \langle f, g\rangle_{\mathring{\mu}}$ for any $f, g \in \mathcal{L}_n$ by Theorem 6.1.9. Since $f\overline{g} \in \mathcal{R}_n$ ($t \in \partial\mathbb{O}$) and because the divergence of the Blaschke product implies that the space $\mathcal{R}_\infty$ is dense in $C(\overline{\partial\mathbb{O}})$ by Theorems 7.1.2 and 7.1.4, the result follows.   □

The same kind of proof can be repeated for the measures considered in Theorem 6.4.3.

Theorem 9.7.2. *Suppose that the Blaschke product diverges. Let $P(t, z)$ be the Poisson kernel. Then the measure defined by $d\mathring{\mu}_n(t) = P(t, w)/|K_n(t, w)|^2\, d\lambda(t)$ converges to $d\mathring{\mu}$ in the weak star topology, that is, for any continuous*

$f \in C(\overline{\partial \mathbb{O}})$,

$$\lim_{n \to \infty} \int f(t) \, d\mathring{\mu}_n(t) = \int f(t) \, d\mathring{\mu}(t).$$

For $w = \alpha_0$ and setting $K_n(z) = K_n(z, \alpha_0)$, we find as a special case:

$$\lim_{n \to \infty} \int f(t) \frac{d\mathring{\lambda}(t)}{|K_n(t)|^2} = \int f(t) \, d\mathring{\mu}(t) \qquad \forall f \in C(\overline{\partial \mathbb{O}}).$$

Other weak convergence results in the Erdős–Turán class are given in Theorem 9.8.5 and Theorem 9.8.18.

It is now possible to give a characterization theorem for the Szegő condition, now assuming a Carleman condition, rather than assuming that the point set $A = \{\alpha_1, \alpha_2, \ldots\}$ is compactly included in $\mathbb{O}$, as we did in Section 9.4. We recall that the Carleman condition is that

$$\sum_{n=1}^{\infty} m_n^{-1/2n} = \infty \quad \text{with} \quad m_n = \int \frac{\varpi_0(t)^n}{\pi_n(t)} \, d\mathring{\mu}(t).$$

Theorem 9.7.3. *Assume that the Carleman condition holds. Let $k_n(z, w)$ be the reproducing kernels for the spaces $\mathcal{L}_n$. Then the following conditions are equivalent:*

(I) $\log \mu' \in L_1(\mathring{\lambda})$ (Szegő's condition).
(VI) $k_n(z, w)$ converges locally uniformly in $\mathbb{O} \times \mathbb{O}$.
(XI) $k_n(z, \alpha_0)$ converges locally uniformly in $\mathbb{O}$.
(XII) $\lim_{n \to \infty} \sum_{k=0}^{n} |\phi_k(\alpha_0)|^2 < \infty$.

Proof. The implication (I) $\Rightarrow$ (VI) is given in Theorem 9.6.4.

The implications (VI) $\Rightarrow$ (XI) $\Rightarrow$ (XII) are trivial. Thus it remains to show the implication (XII) $\Rightarrow$ (I). Set

$$K_n(z) = K_n(z, \alpha_0) = k_n(z, \alpha_0) / \sqrt{k_n(\alpha_0, \alpha_0)}.$$

By the reproducing property, we have

$$1 = \int |K_n(t)|^2 \, d\mathring{\mu}(t) \geq \int |K_n(t)|^2 \frac{\mu'(t)}{|\varpi_0(t)|^2} \, d\mathring{\lambda}(t).$$

If $P(t, z)$ is the Poisson kernel and $t \in \partial \mathbb{O}$ then $P(t, \alpha_0) |\varpi_0(t)|^2 = 1$, so that we can write

$$1 \geq \int |K_n(t)|^2 P(t, \alpha_0) \mu'(t) \, d\mathring{\lambda}(t),$$

and by the inequality for arithmetic and geometric mean

$$1 \geq \exp\left\{ \int P(t, \alpha_0) \log[|K_n(t)|^2 \mu'(t)] \, d\lambda(t) \right\}.$$

By splitting the logarithm and replacing $P(t, \alpha_0)$ again by $1/|\varpi_0(t)|^2$, we find

$$1 \geq \exp\left\{ \int P(t, \alpha_0) \log |K_n(t)|^2 \, d\lambda(t) \right\} \exp\left\{ \int \log \mu'(t) \, d\mathring{\lambda}(t) \right\}.$$

The first integral is by Poisson's formula equal to $|K_n(\alpha_0)|^2$. Thus we have shown that

$$|K_n(\alpha_0)|^2 \leq \exp\left\{ -\int \log \mu'(t) \, d\mathring{\lambda}(t) \right\}.$$

Since the left-hand side converges to a finite limit, which is obviously larger than 1, it follows that $\log \mu' \in L_1(\mathring{\lambda})$. This proves the theorem. $\square$

9.8. Erdős–Turán class and ratio asymptotics

Using our previous results, it is not difficult to obtain ratio asymptotics when we assume that Szegő's condition $\log \mu' \in L_1(\mathring{\lambda})$ is satisfied. We give a simple example.

Lemma 9.8.1. Let $(A, \mu) \in \mathcal{AM}'$. Then

$$\lim_{n \to \infty} \frac{\phi_{n+1}^*(z)}{\phi_n^*(z)} \frac{\varpi_{n+1}(z)\varpi_n(\alpha_0)}{\varpi_n(z)\varpi_{n+1}(\alpha_0)} \frac{\phi_n^*(\alpha_0)}{\phi_{n+1}^*(\alpha_0)} = 1, \quad z \in \mathbb{O}$$

and

$$\lim_{n \to \infty} \frac{1}{\zeta_{n+1}(z)} \frac{\phi_{n+1}(z)}{\phi_n(z)} \frac{\varpi_{n+1}^*(z)\overline{\varpi_n(\alpha_0)}}{\varpi_n^*(z)\overline{\varpi_{n+1}(\alpha_0)}} \frac{\overline{\phi_n^*(\alpha_0)}}{\overline{\phi_{n+1}^*(\alpha_0)}} = 1, \quad z \in \mathbb{O}^e,$$

where convergence is locally uniform in the indicated regions.

Proof. The first relation, which holds in $\mathbb{O}$, follows immediately from Theorem 9.6.3. The second relation is derived from the first one by taking the substar conjugate. $\square$

Note that in the second limit of this lemma,

$$\frac{\varpi_{n+1}^*(z)}{\zeta_{n+1}(z)\varpi_n^*(z)} = \bar{z}_{n+1} \frac{\varpi_{n+1}(z)}{\varpi_n^*(z)}.$$

When $\mathbb{O} = \mathbb{D}$ and all $\alpha_k = 0$, that is, in the polynomial case, then the first part of Lemma 9.8.1 becomes

$$\lim_{n \to \infty} \frac{\phi_{n+1}^*(z)\phi_n^*(0)}{\phi_n^*(z)\phi_{n+1}^*(0)} = 1.$$

But in the polynomial case, we also have $\lim_{n \to \infty} \phi_{n+1}^*(0)/\phi_n^*(0) = 1$ and thus we have then $\lim_{n \to \infty} \phi_{n+1}^*(z)/\phi_n^*(z) = 1$ locally uniformly in $\mathbb{D}$. Similarly, part (2) of the lemma implies that in the polynomial case $\lim_{n \to \infty} \phi_{n+1}(z)/\phi_n(z) = z$ locally uniformly in $\mathbb{E}$.

The previous lemma illustrated that when Szegő's condition holds, then the ratio asymptotics are easy to obtain. However, ratio asymptotics are typically obtained not assuming the Szegő condition $\log \mu' \in L_1(\mathring{\lambda})$, but using instead the much weaker Erdős–Turán condition $\mu' > 0$ a.e.

For the polynomial case, several such results are obtained in a paper of Maté, Nevai, and Totik [145]. K. Pan has extended their results to the rational case in several papers. The rest of this section is mostly an account of Pan's generalizations. His results rely on a lemma from the paper [145] that we shall formulate first. Originally it is proved for the unit circle, but a Cayley transform allows us to formulate it for the general case. We give it without proof.

Lemma 9.8.2. Let μ be a measure on $\overline{\partial\mathbb{O}}$ with $\mu' > 0$ a.e. and let $p > 0$ and A be real numbers. Assume that

$$\left(\int [f(t)\mu'(t)]^p \, d\mathring{\lambda}(t) \right)^{1/p} \leq A \int f(t) \, d\mathring{\mu}(t)$$

holds for every nonnegative continuous function $f \in C(\overline{\partial\mathbb{O}})$. Then $A \geq 1$.

Before we arrive at a reformulation of Lemma 9.8.1 under the weaker condition $\mu' > 0$ a.e., we shall have to go through several other results. We start with the following theorem, which was proved by K. Pan in [168] in the case of the unit disk.

Theorem 9.8.3. Suppose $\mu' > 0$ a.e. on $\overline{\partial\mathbb{O}}$ and assume that the Blaschke product diverges. Let $K_n(z, w)$ be the normalized reproducing kernels and $\mathrm{K}_n(z) = K_n(z, \alpha_0)$. Then we have

$$\lim_{n \to \infty} \int (|\mathrm{K}_n(t)| \sqrt{\mu'(t)} - 1)^2 \, d\mathring{\lambda}(t) = 0.$$

Proof. Expanding the square in the integrand gives

$$0 \le \int (|\mathrm{K}_n(t)| \sqrt{\mu'(t)} - 1)^2 \, d\overset{\circ}{\lambda}(t) = \int |\mathrm{K}_n(t)|^2 \mu'(t) \, d\overset{\circ}{\lambda}(t)$$
$$- 2 \int |\mathrm{K}_n(t)| \sqrt{\mu'(t)} \, d\overset{\circ}{\lambda}(t) + 1.$$

Furthermore, by Theorem 9.7.2,

$$\int |\mathrm{K}_n(t)|^2 \mu'(t) \, d\overset{\circ}{\lambda}(t) \le \int |\mathrm{K}_n(t)|^2 \, d\overset{\circ}{\mu}(t) = 1. \tag{9.35}$$

Hence it has to be shown that

$$\liminf_{n \to \infty} \int |\mathrm{K}_n(t)| \sqrt{\mu'(t)} \, d\overset{\circ}{\lambda}(t) \ge 1. \tag{9.36}$$

Now, for any nonnegative function $f \in C(\overline{\partial \mathbb{O}})$, we have by the Schwarz inequality

$$\left(\int (f\mu')^{1/4} \, d\overset{\circ}{\lambda} \right)^2 = \left(\int [(\mu')^{1/4} |\mathrm{K}_n|^{1/2}][f^{1/4} |\mathrm{K}_n|^{-1/2}] \, d\overset{\circ}{\lambda} \right)^2$$
$$\le \left(\int (\mu')^{1/2} |\mathrm{K}_n| \, d\overset{\circ}{\lambda} \right) \left(\int f^{1/2} |\mathrm{K}_n|^{-1} \, d\overset{\circ}{\lambda} \right).$$

Squaring both sides and applying the Schwarz's inequality once more on the second integral of the right-hand side gives

$$\left(\int (f\mu')^{1/4} \, d\overset{\circ}{\lambda} \right)^4 \le \left(\int (\mu')^{1/2} |\mathrm{K}_n| \, d\overset{\circ}{\lambda} \right)^2 \left(\int f |\mathrm{K}_n|^{-2} \, d\overset{\circ}{\lambda} \right).$$

Because the last integral converges to $\int f \, d\overset{\circ}{\mu}$ by Theorem 9.7.2, we get

$$\left(\int (f\mu')^{1/4} \, d\overset{\circ}{\lambda} \right)^4 \le \left(\liminf_{n \to \infty} \int \sqrt{\mu'} |\mathrm{K}_n| \, d\overset{\circ}{\lambda} \right)^2 \left(\int f \, d\overset{\circ}{\mu} \right).$$

We now apply Lemma 9.8.2 to find that (9.36) holds, which proves the theorem.

$\square$

Similar arguments can be used to prove the following theorem, which holds under the more restrictive Szegő condition.

Theorem 9.8.4. *Let $K_n(t, w)$ be the normalized reproducing kernel for $\mathcal{L}_n$ and let ϕ_n be the nth orthonormal rational function. Suppose that $(A, \mu) \in \mathcal{AM}''$.*

Then

$$\limsup_{n \to \infty} \int \left| \frac{|K_n(t, w)|^2}{|K_{n+l}(t, w)|^2} - 1 \right| d\lambda_w(t) = 0, \quad w \in \mathbb{O}, \qquad (9.37)$$

where $d\lambda_w(t) = P(t, w)\, d\lambda(t)$, with P the Poisson kernel.

Proof. Note that for $l \geq 0$, it follows from Theorem 6.4.3 that

$$\int \left| \frac{K_n(t, w)}{K_{n+l}(t, w)} \right|^2 d\lambda_w(t) = \int |K_n(t, w)|^2 \, d\mathring{\mu}(t) = 1.$$

However,

$$\int \left| \frac{K_n(t, w)}{K_{n+l}(t, w)} \right| d\lambda_w(t) = \int |K_n(t, w)\overline{K_{n+l}(t, w)}| \frac{d\lambda_w(t)}{|K_{n+l}(t, w)|^2}$$

$$= \int |K_n(t, w)\overline{K_{n+l}(t, w)}| \, d\mathring{\mu}(t)$$

$$\geq |\langle K_n(t, w), K_{n+l}(t, w) \rangle_{\mathring{\mu}}| = \frac{K_n(w, w)}{K_{n+l}(w, w)}.$$

Since $K_{n+l}(w, w)$ is monotonically increasing with l, and $\lim_{l \to \infty} K_{n+l}(w, w) = S_w(w)$, where S_w is the normalized Szegő kernel, we have

$$\inf_{l \geq 0} \int \left| \frac{K_n(t, w)}{K_{n+l}(t, w)} \right| d\lambda_w(t) \geq \frac{K_n(w, w)}{S_w(w)}.$$

Now

$$\int \left(\left| \frac{K_n}{K_{n+l}} \right| - 1 \right)^2 d\lambda_w = \int \left| \frac{K_n}{K_{n+l}} \right|^2 d\lambda_w + \int d\lambda_w - 2 \int \left| \frac{K_n}{K_{n+l}} \right| d\lambda_w.$$

As we have seen, the first integral equals 1. The second integral is also 1, so that we get

$$0 \leq \sup_{l \geq 0} \int \left(\left| \frac{K_n}{K_{n+l}} \right| - 1 \right)^2 d\lambda_w$$

$$\leq 2 \left(1 - \frac{K_n(w, w)}{S_w(w)} \right).$$

Taking the limit for $n \to \infty$, knowing that $K_n(w, w) \to S_w(w)$, we find

$$\limsup_{n \to \infty} \int \left(\left| \frac{K_n}{K_{n+l}} \right| - 1 \right)^2 d\lambda_w = 0.$$

Now, by Schwarz's inequality, we may bound the integral I in (9.37) by

$$I^2 \le \left\{ \int \left(\left| \frac{K_n}{K_{n+l}} \right| + 1 \right)^2 d\lambda_w \right\} \cdot \left\{ \int \left(\left| \frac{K_n}{K_{n+l}} \right| - 1 \right)^2 d\lambda_w \right\}.$$

Using $(|a| + |b|)^2 \le 2(|a|^2 + |b|^2)$, we can bound the first factor by 4. Taking the supremum for $l \ge 0$ and the limit for $n \to \infty$, we know that the second factor goes to zero, and hence the theorem follows. $\square$

Note that for $w = \alpha_0$, we will be able to replace the Szegő condition by the condition $\mu' > 0$ a.e. See Theorem 9.8.6.

     The following theorem is obtained as a direct consequence of Theorem 9.8.3. See Pan [168]. Note that it also includes a weak convergence type of result for the Erdős–Turán class.

Theorem 9.8.5. *Assume that $\mu' > 0$ a.e. on $\overline{\partial \mathbb{O}}$. Define $\kappa_n(z) = K_n(z, \alpha_0)$ with K_n the normalized reproducing kernel. Then*

$$\lim_{n \to \infty} \int \left| |\kappa_n(t)|^2 \mu'(t) - 1 \right| d\mathring{\lambda}(t) = 0 \tag{9.38}$$

and

$$\lim_{n \to \infty} \int \left| |\kappa_n(t)|^{-1} - \sqrt{\mu'(t)} \right| d\mathring{\lambda}(t) = 0. \tag{9.39}$$

Furthermore, we have the following weak convergence theorem. For any bounded measurable function $f \in L_1(\mathring{\lambda})$

$$\lim_{n \to \infty} \int f(t) |\kappa_n(t)|^2 \mu'(t) \, d\mathring{\lambda}(t) = \int f(t) \, d\mathring{\lambda}(t) \tag{9.40}$$

and

$$\lim_{n \to \infty} \int f(t) |\kappa_n(t)|^2 \, d\mathring{\mu}(t) = \int f(t) \, d\mathring{\lambda}(t). \tag{9.41}$$

Proof. To prove the first limit (9.38), we observe that by the Schwarz inequality

$$\left(\int \left| |\kappa_n|^2 \mu' - 1 \right| d\mathring{\lambda} \right)^2 = \left(\int \left| |\kappa_n| \sqrt{\mu'} - 1 \right| \left| |\kappa_n| \sqrt{\mu'} + 1 \right| d\mathring{\lambda} \right)^2$$

$$\le \left(\int \left| |\kappa_n| \sqrt{\mu'} - 1 \right|^2 d\mathring{\lambda} \right) \left(\int \left| |\kappa_n| \sqrt{\mu'} + 1 \right|^2 d\mathring{\lambda} \right).$$

By Theorem 9.8.3, the first integral on the right-hand side converges to zero as $n \to \infty$. For the second integral we use $(|a| + |b|)^2 \leq 2(|a|^2 + |b|^2)$ and (9.35) to find that it is bounded by 4. Thus the first limit (9.38) follows.

For the second limit (9.39), we use again Schwarz's inequality to get

$$\left(\int |\mathrm{K}_n|^{-1} - \sqrt{\mu'} | \, d\overset{\circ}{\lambda} \right)^2 = \left(\int |\mathrm{K}_n|^{-1} |\mathrm{K}_n| \sqrt{\mu'} - 1| \, d\overset{\circ}{\lambda} \right)^2$$

$$\leq \left(\int \frac{d\overset{\circ}{\lambda}}{|\mathrm{K}_n|^2} \right) \left(\int |\mathrm{K}_n| \sqrt{\mu'} - 1|^2 \, d\overset{\circ}{\lambda} \right).$$

The first integral on the right-hand side is 1 by Theorem 6.4.3 whereas the second integral converges to zero as $n \to \infty$ by Theorem 9.8.3.

The weak convergence result (9.40) is a direct consequence of (9.38).

For the formula (9.41) we notice that by (9.40) for $f = 1$, we find

$$\lim_{n \to \infty} \int |\mathrm{K}_n(t)|^2 \mu'(t) \, d\overset{\circ}{\lambda}(t) = 1.$$

Hence $\lim_{n \to \infty} |\mathrm{K}_n(t)|^2 \, d\overset{\circ}{\mu}_s(t) = 0$ if $d\overset{\circ}{\mu}_s = d\overset{\circ}{\mu} - \mu' \, d\overset{\circ}{\lambda}$ is the singular part of $d\overset{\circ}{\mu}$ and thus also for every bounded $f \in L_1(\overset{\circ}{\lambda})$

$$\lim_{n \to \infty} \int f |\mathrm{K}_n(t)|^2 \, d\overset{\circ}{\mu}_s(t) = 0,$$

and this implies (9.41) as a consequence of (9.40). □

The next theorem is also in one of Pan's papers [168] for the case of the circle. It gives a Nevai type characterization for measures satisfying $\mu' > 0$ a.e.

Theorem 9.8.6. *Let $K_n(t, w)$ be the normalized reproducing kernel for $\mathcal{L}_n$ and let $\mathrm{K}_n(z) = K_n(z, \alpha_0)$. Suppose that the Blaschke product diverges to 0. Then $\mu' > 0$ a.e. on $\overline{\partial \mathbb{O}}$ iff*

$$\lim_{n \to \infty} \sup_{l \geq 0} \int \left| \frac{|\mathrm{K}_n(t)|^2}{|\mathrm{K}_{n+l}(t)|^2} - 1 \right| d\overset{\circ}{\lambda}(t) = 0. \tag{9.42}$$

Proof. First suppose that $\mu' > 0$ a.e. Note that for $l \geq 0$, it follows from Theorem 6.4.3 that

$$\int \left| \frac{\mathrm{K}_n(t)}{\mathrm{K}_{n+l}(t)} \right|^2 d\overset{\circ}{\lambda}(t) = \int |\mathrm{K}_n(t)|^2 \, d\overset{\circ}{\mu}(t) = 1.$$

Now, expanding the square gives

$$\int \left(\left| \frac{K_n}{K_{n+l}} \right| - 1 \right)^2 d\overset{\circ}{\lambda} = \int \left| \frac{K_n}{K_{n+l}} \right|^2 d\overset{\circ}{\lambda} + \int d\overset{\circ}{\lambda} - 2 \int \left| \frac{K_n}{K_{n+l}} \right| d\overset{\circ}{\lambda}.$$

As we have seen, the first integral equals 1. Also, the second integral is 1. We show below that

$$\liminf_{n \to \infty} \inf_{l \geq 0} \int \left| \frac{K_n}{K_{n+l}} \right| d\overset{\circ}{\lambda} \geq 1. \tag{9.43}$$

This will imply that

$$\limsup_{n \to \infty} \sup_{l \geq 0} \int \left(\left| \frac{K_n}{K_{n+l}} \right| - 1 \right)^2 d\overset{\circ}{\lambda} = 0.$$

So, by Schwarz's inequality, we may bound the integral I in (9.42) by

$$I^2 \leq \left\{ \int \left(\left| \frac{K_n}{K_{n+l}} \right| + 1 \right)^2 d\overset{\circ}{\lambda} \right\} \cdot \left\{ \int \left(\left| \frac{K_n}{K_{n+l}} \right| - 1 \right)^2 d\overset{\circ}{\lambda} \right\}.$$

Using $(|a| + |b|)^2 \leq 2(|a|^2 + |b|^2)$, we can bound the first factor by 4. Taking the supremum for $l \geq 0$ and the limit for $n \to \infty$, we know that by (9.43) the second factor goes to zero, and hence the theorem follows.

Thus it remains to show that (9.43) holds. Therefore we use again Lemma 9.8.2. Assume $f \in C(\partial \mathbb{O})$ is nonnegative. Then using Hölder's inequality a couple of times, we get:

$$\int (f\mu')^{1/4} d\overset{\circ}{\lambda} = \int \left| \frac{K_n}{K_{n+l}} \right|^{1/2} (|K_{n+l}|^2 \mu')^{1/4} \frac{f^{1/4}}{|K_n|^{1/2}} d\overset{\circ}{\lambda}$$

$$\leq \left(\int \left| \frac{K_n}{K_{n+l}} \right| d\overset{\circ}{\lambda} \right)^{1/2} \left(\int |K_{n+l}|^2 \mu' \, d\overset{\circ}{\lambda} \right)^{1/4} \left(\int \frac{f}{|K_n|^2} d\overset{\circ}{\lambda} \right)^{1/4}.$$

The second integral on the right-hand side is bounded by 1, since

$$\int |K_{n+l}|^2 \mu' \, d\overset{\circ}{\lambda} \leq \int |K_{k+l}|^2 \, d\overset{\circ}{\mu} = 1.$$

The third integral converges to $\int f \, d\overset{\circ}{\mu}$ by Theorem 9.7.2. Thus we have

$$\left(\int (f\mu')^{1/4} d\overset{\circ}{\lambda} \right)^{1/4} \leq \left(\liminf_{n \to \infty} \inf_{l \geq 0} \int \left| \frac{K_n}{K_{n+l}} \right| d\overset{\circ}{\lambda} \right)^2 \left(\int f \, d\overset{\circ}{\mu} \right).$$

Application of Lemma 9.8.2 gives then (9.43). This completes the proof of one direction of this theorem.

For the converse statement, we refer to Pan's paper [168]. He in turn refers to a proof given by Li and Saff [136] for the polynomial case on the circle. The adaptations for the real line and for the rational case are simple. □

The following lemma corresponds to Theorem 2.3 in Ref. [166].

Lemma 9.8.7. Let $K_n(z, w)$ be the normalized reproducing kernels for $\mathcal{L}_n$ and define $\kappa_n(z) = K_n(z, \alpha_0)$ and set $\upsilon_n = \kappa_n(\alpha_0)$. Then

$$\lim_{n \to \infty} \frac{\upsilon_{n-1}}{\upsilon_n} = 1 \quad \text{iff} \quad \lim_{n \to \infty} \frac{\kappa_{n-1}(z)}{\kappa_n(z)} = 1 \quad \text{locally uniformly in } \mathbb{O}.$$

Proof. Note that by Theorem 6.4.3

$$1 = \int |\kappa_{n-1}(t)|^2 \, d\mathring{\mu}(t) = \int |\kappa_{n-1}(t)|^2 \, d\mathring{\mu}_n(t)$$

$$\text{for} \quad d\mathring{\mu}_n(t) = \frac{P(t, \alpha_0) \, d\lambda(t)}{|\kappa_n(t)|^2} = \frac{d\mathring{\lambda}(t)}{|\kappa_n(t)|^2}.$$

Thus

$$\int \left| \frac{\kappa_{n-1}(t)}{\kappa_n(t)} \right|^2 \, d\mathring{\lambda}(t) = 1.$$

Now set $c_n = \upsilon_{n-1}/\upsilon_n$ and $f_n(z) = \kappa_{n-1}(z)/\kappa_n(z)$. Note that $f_n(\alpha_0) = c_n$. Set $g_n(z) = f_n(z) - c_n$. Then $g_n(z)$ is analytic in $\mathbb{O}$ since f_n is analytic in $\mathbb{O}$ and thus by Poisson's formula

$$g_n(z) = \int P(t, z) g_n(t) \, d\lambda(t).$$

Since $\varpi_0(t) P(t, z)$ is locally uniformly bounded for $z \in \mathbb{O}$ (and $t \in \partial\mathbb{O}$), we get

$$|g_n(z)| \leq M \int |g_n(t)| \, d\mathring{\lambda}(t)$$

and thus

$$|g_n(z)|^2 \leq M^2 \int |g_n(t)|^2 \, d\mathring{\lambda}(t) = M^2 \|g_n\|^2.$$

Continuing to use the norm in $\mathring{H}_2$, we also have

$$1 = \|f_n\|^2 = \|c_n + g_n\|^2 = c_n^2 + \|g_n\|^2 + 2c_n \operatorname{Re} \int g_n(t) \, d\mathring{\lambda}(t).$$

Because $g_n(\alpha_0) = 0$, the last integral is zero, so that $\|g_n\|^2 = 1 - c_n^2$. Thus, if $\lim_{n\to\infty} c_n = 1$, then it follows from

$$|g_n(z)|^2 \le M^2 \|g_n\|^2 \le M^2 \left(1 - c_n^2\right)$$

that $g_n(z)$ converges locally uniformly to 0, that is, $f_n(z)$ converges locally uniformly to 1.

The converse statement in the theorem is trivial.   □

Note that $\mathrm{K}_n(z) = \mathrm{F}_n^*(z)$, where $\mathrm{F}_n = \mathrm{F}_{n,n}$ is defined in the beginning of Section 9.6.

Now we finally can formulate the first result on ratio asymptotics for the functions K_n.

Theorem 9.8.8. *Suppose that the Blaschke product diverges and that $\mu' > 0$ a.e. on $\overline{\partial\mathbb{O}}$. Let $K_n(z, w)$ be the normalized reproducing kernels and set $\mathrm{K}_n(z) = K_n(z, \alpha_0)$. Then*

$$\lim_{n\to\infty} \frac{\mathrm{K}_n(z)}{\mathrm{K}_{n+1}(z)} = 1 \quad \text{locally uniformly in } \mathbb{O}.$$

Proof. By the previous lemma, it is sufficient to show that

$$\lim_{n\to\infty} \frac{\upsilon_n}{\upsilon_{n+1}} = 1 \quad \text{for} \quad \upsilon_n = \mathrm{K}_n(\alpha_0).$$

Define the function

$$F(z) = \frac{\mathrm{K}_n(z)\mathrm{K}_n^*(z)\zeta_{n+1}(z) - \mathrm{K}_{n+1}(z)\mathrm{K}_{n+1}^*(z)}{\mathrm{K}_{n+1}^2(z)}.$$

Note that for $t \in \partial\mathbb{O}$

$$|F(t)| = \left| \frac{\mathrm{K}_n(t)\mathrm{K}_{n*}(t)B_{n+1}(t) - \mathrm{K}_{n+1}(t)\mathrm{K}_{(n+1)*}(t)B_{n+1}(t)}{\mathrm{K}_{n+1}^2(t)} \right|$$

$$= \left| \frac{|\mathrm{K}_n(t)|^2}{|\mathrm{K}_{n+1}(t)|^2} - 1 \right|.$$

Since $F(z)$ is analytic in $\overline{\mathbb{O}}$, we can apply Cauchy's theorem to get

$$|F(\alpha_0)| \le \int \left| \frac{|\mathrm{K}_n(t)|^2}{|\mathrm{K}_{n+1}(t)|^2} - 1 \right| d\mathring{\lambda}(t).$$

Because

$$k_{n+1}(z, \alpha_0) = k_n(z, \alpha_0) + \phi_{n+1}(z)\overline{\phi_{n+1}(\alpha_0)},$$

we get after taking the superstar conjugate

$$\phi_{n+1}^*(z)\phi_{n+1}(\alpha_0) = k_{n+1}^*(z, \alpha_0) - \zeta_{n+1}(z)k_n^*(z, \alpha_0)$$
$$= \upsilon_{n+1}K_{n+1}^*(z) - \zeta_{n+1}(z)\upsilon_n K_n^*(z).$$

Call the last of these expressions $S_n(z)$. Then, evaluating $F(z)$ for $z = \alpha_0$, one finds

$$|F(\alpha_0)| = \frac{|S_n(\alpha_0)|}{\upsilon_{n+1}^2} = \frac{|\phi_{n+1}(\alpha_0)\phi_{n+1}^*(\alpha_0)|}{\upsilon_{n+1}^2}.$$

Thus we have

$$\left| \frac{\upsilon_n^2}{\upsilon_{n+1}^2} - 1 \right| = \left| \frac{\upsilon_n^2 - \upsilon_{n+1}^2}{\upsilon_{n+1}^2} \right| = \frac{|\phi_{n+1}(\alpha_0)|^2}{\upsilon_{n+1}^2}$$
$$\leq \frac{|\phi_{n+1}^*(\alpha_0)\phi_{n+1}(\alpha_0)|}{\upsilon_{n+1}^2}$$
$$= \frac{|S_n(\alpha_0)|}{\upsilon_{n+1}^2}$$
$$\leq \int \left| \frac{|K_n(t)|^2}{|K_{n+1}(t)|^2} - 1 \right| d\mathring{\lambda}(t).$$

Because the last integral converges to zero for $n \to \infty$ by Theorem 9.8.6, it follows that $\upsilon_n/\upsilon_{n+1} \to 1$. □

This entails immediately

Corollary 9.8.9. *Suppose that the Blaschke product diverges and that $\mu' > 0$ a.e. Let $k_n(z, w)$ be the (nonnormalized) reproducing kernels. Then*

$$\lim_{n \to \infty} \frac{k_{n-1}(z, \alpha_0)}{k_n(z, \alpha_0)} = 1, \quad \text{locally uniformly in } \mathbb{O}.$$

Proof. Since by definition $K_n(z) = k_n(z, \alpha_0)/\sqrt{k_n(\alpha_0, \alpha_0)}$ and $\upsilon_n = \sqrt{k_n(\alpha_0, \alpha_0)}$, it follows that

$$\frac{K_{n-1}(z)}{K_n(z)} = \frac{k_{n-1}(z, \alpha_0)}{k_n(z, \alpha_0)} \cdot \frac{\upsilon_n}{\upsilon_{n-1}}.$$

Since $\upsilon_{n-1}/\upsilon_n \to 1$, the result follows. □

Now we want to move toward ratio asymptotics for the orthogonal functions. We start with the following lemmas from Ref. [171].

Lemma 9.8.10. Let $\mu' > 0$ a.e. on $\partial \mathbb{O}$ and suppose the Blaschke product with zeros $A = \{\alpha_1, \alpha_n, \ldots\}$ diverges. Let $k_n(z, w)$ be the reproducing kernels and ϕ_n the orthonormal functions. Then

$$\lim_{n \to \infty} \frac{|\phi_n(\alpha_0)|^2}{k_n(\alpha_0, \alpha_0)} = 0.$$

Proof. Note that $k_n(\alpha_0, \alpha_0) = \upsilon_n^2$, so that

$$\frac{|\phi_n(\alpha_0)|^2}{k_n(\alpha_0, \alpha_0)} = 1 - \frac{k_{n-1}(\alpha_0, \alpha_0)}{k_n(\alpha_0, \alpha_0)} = 1 - \frac{\upsilon_{n-1}^2}{\upsilon_n^2}.$$

Because $\upsilon_{n-1}/\upsilon_n \to 1$ for $n \to \infty$, the result follows.   □

Lemma 9.8.11. Let $\mu' > 0$ a.e. on $\partial \mathbb{O}$ and suppose the points $A = \{\alpha_1, \alpha_2, \ldots\}$ are compactly included in $\mathbb{O}$. Then

$$\lim_{n \to \infty} \frac{\phi_n(\alpha_0)}{\phi_n^*(\alpha_0)} = 0.$$

Proof. Because A is bounded away from the boundary, $|\zeta_n(z)| \leq m < 1$ and thus

$$\frac{1 - |\zeta_n(\alpha_0)|^2}{1 - m^2} > 1.$$

Furthermore, we know that $|\phi_n(z)/\phi_n^*(z)| \leq 1$ in $\mathbb{O}$ and thus

$$\frac{1}{1 - |\zeta_n(\alpha_0)|^2 |\phi_n(\alpha_0)/\phi_n^*(\alpha_0)|^2} \geq 1.$$

Thus, using the Christoffel–Darboux relation,

$$k_n(\alpha_0, \alpha_0) = \frac{|\phi_n^*(\alpha_0)|^2 - |\zeta_n(\alpha_0)|^2 |\phi_n(\alpha_0)|^2}{1 - |\zeta_n(\alpha_0)|^2},$$

we get

$$\lim_{n \to \infty} \frac{|\phi_n(\alpha_0)|^2}{|\phi_n^*(\alpha_0)|^2} \leq \lim_{n \to \infty} \left\{ \frac{|\phi_n(\alpha_0)|^2}{|\phi_n^*(\alpha_0)|^2} \frac{1 - |\zeta_n(\alpha_0)|^2}{1 - m^2} \frac{1}{1 - |\zeta_n(\alpha_0)|^2 |\phi_n(\alpha_0)/\phi_n^*(\alpha_0)|^2} \right\}$$

$$\leq \frac{1}{1 - m^2} \lim_{n \to \infty} \frac{|\phi_n(\alpha_0)|^2}{k_n(\alpha_0, \alpha_0)} = 0,$$

where the last equality follows from the previous lemma.   □

Lemma 9.8.12. Let $(A, \mu) \in \mathcal{AM}$ and let $k_n(z, w)$ and $K_n(z, w)$ be the non-normalized and normalized reproducing kernels respectively and let $\mathrm{K}_n(z) = K_n(z, \alpha_0)$. Then

$$\lim_{n \to \infty} \frac{k_n(z, \alpha_0)}{k_n^*(z, \alpha_0)} = \lim_{n \to \infty} \frac{\mathrm{K}_n(z)}{\mathrm{K}_n^*(z)} = 0, \quad \text{locally uniformly in } \mathbb{O}^e.$$

Proof. Using the notation of Section 9.5 on varying measures, it can be checked that

$$\frac{k_n(z, \alpha_0)}{\sqrt{k_n(\alpha_0, \alpha_0)}} = \phi_{n,n}^{\tilde{*}}(z) w_n(z) \eta_n^1, \quad \eta_n^1 \in \mathbb{T},$$

where $\phi_{n,n}$ are the orthonormal functions in $\mathcal{L}_0^m$, $w_n(z) = \varpi_0(z)^n / \pi_n(z)$, $\pi_n = \varpi_1 \cdots \varpi_n$, and $f^{\tilde{*}} = \zeta_0^n f_*$. Consequently,

$$\frac{k_n^*(z, \alpha_0)}{\sqrt{k_n(\alpha_0, \alpha_0)}} = \phi_{n,n}(z) w_n(z) \eta_n^2, \quad \eta_n^2 \in \mathbb{T}.$$

Thus

$$\frac{k_n(z, \alpha_0)}{k_n^*(z, \alpha_0)} = \frac{\phi_{n,n}^{\tilde{*}}(z)}{\phi_{n,n}(z)} \tau_n, \quad \tau_n \in \mathbb{T}.$$

By Theorem 9.5.1(5), the right-hand side goes to 0, locally uniformly in $\mathbb{O}^e$ when $n \to \infty$. This proves the lemma. $\qquad \square$

Lemma 9.8.13. Let $K_n(z, w)$ be the reproducing kernels, $\mathrm{K}_n(z) = K_n(z, \alpha_0)$, and $\upsilon_n = \mathrm{K}_n(\alpha_0)$. Then

$$\phi_n^*(z) = \frac{\phi_n^*(\alpha_0)(1 - \overline{\zeta_n(\alpha_0)}\zeta_n(z))\mathrm{K}_n(z) + \overline{\phi_n(\alpha_0)\zeta_n(\alpha_0)}(\zeta_n(z) - \zeta_n(\alpha_0))\mathrm{K}_n^*(z)}{(1 - |\zeta_n(\alpha_0)|^2)\upsilon_n}$$

and

$$\zeta_n(z)\phi_n(z)$$
$$= \frac{\zeta_n(\alpha_0)\phi_n(\alpha_0)(1 - \zeta_n(z)\overline{\zeta_n(\alpha_0)})\mathrm{K}_n(z) + \overline{\phi_n^*(\alpha_0)}(\zeta_n(z) - \zeta_n(\alpha_0))\mathrm{K}_n^*(z)}{(1 - |\zeta_n(\alpha_0)|^2)\upsilon_n}.$$

Proof. This is a matter of simple algebra. First write down the Christoffel–Darboux formula for $k_n(z, \alpha_0)$ and for its superstar conjugate $k_n^*(z, \alpha_0)$. Then either eliminate $\phi_n(z)$ or $\phi_n^*(z)$ between these formulas and the result follows. $\qquad \square$

We are now ready to prove the ratio asymptotics of Lemma 9.8.1 under the weaker assumption $\mu' > 0$ a.e. but with a stronger condition for the set of points A. See Ref. [171].

Theorem 9.8.14. *Assume that $\mu' > 0$ a.e. on $\partial\mathbb{O}$ and that the Blaschke product with zeros $A = \{\alpha_1, \alpha_2, \ldots\}$ diverges. Then*

$$\lim_{n\to\infty} \frac{\phi_{n+1}^*(z)}{\phi_n^*(z)} \frac{\varpi_{n+1}(z)\varpi_n(\alpha_0)}{\varpi_n(z)\varpi_{n+1}(\alpha_0)} \frac{\phi_n^*(\alpha_0)}{\phi_{n+1}^*(\alpha_0)} = 1, \quad z \in \mathbb{O}$$

and

$$\lim_{n\to\infty} \frac{1}{\zeta_{n+1}(z)} \frac{\phi_{n+1}(z)}{\phi_n(z)} \frac{\varpi_{n+1}^*(z)\overline{\varpi_n(\alpha_0)}}{\varpi_n^*(z)\overline{\varpi_{n+1}(\alpha_0)}} \frac{\overline{\phi_n^*(\alpha_0)}}{\overline{\phi_{n+1}^*(\alpha_0)}} = 1, \quad z \in \mathbb{O}^e,$$

where convergence is locally uniform in the indicated regions.

Proof. We only prove the first formula since the second is an immediate consequence.

Some extensive calculations show that

$$\overline{\zeta_n(\alpha_0)} \frac{\zeta_n(z) - \zeta_n(\alpha_0)}{1 - \zeta_n(z)\overline{\zeta_n(\alpha_0)}} = -\zeta_0(z)\overline{\zeta_0(\alpha_n)}$$

and

$$\frac{1 - \overline{\zeta_n(\alpha_0)}\zeta_n(z)}{1 - |\zeta_n(\alpha_0)|^2} = \frac{\varpi_0(z)\varpi_n(\alpha_0)}{\varpi_0(\alpha_0)\varpi_n(z)}.$$

Using this in the first formula of Lemma 9.8.13 leads to

$$\frac{\phi_n^*(z)\varpi_n(z)}{\phi_n^*(\alpha_0)\varpi_n(\alpha_0)} = \frac{\varpi_0(z)}{\varpi_0(\alpha_0)} \frac{K_n(z)}{v_n} X_n(z),$$

with

$$X_n(z) = \left[1 - \zeta_0(z)\overline{\zeta_0(\alpha_n)} \frac{\overline{\phi_n(\alpha_0)}}{\phi_n^*(\alpha_0)} \frac{K_n^*(z)}{K_n(z)} \right].$$

Since for all $z \in \mathbb{O}$ and for all n we have $|\zeta_0(z)\overline{\zeta_0(\alpha_n)}| < 1$, $|\phi_n(z)/\phi_n^*(z)| < 1$, and because the superstar of Lemma 9.8.12 says that $K_n(z)/K_n^*(z)$ goes to zero locally uniformly in $\mathbb{O}$, it follows that $\lim_{n\to\infty} X_n(z) = 1$, locally uniformly in $\mathbb{O}$.

Taking the ratio of two such expressions and letting $n \to \infty$ gives the result we wanted because

$$\lim_{n\to\infty} \frac{\phi_{n+1}^*(z)}{\phi_n^*(z)} \frac{\varpi_{n+1}(z)\varpi_n(\alpha_0)}{\varpi_n(z)\varpi_{n+1}(\alpha_0)} \frac{\phi_n^*(\alpha_0)}{\phi_{n+1}^*(\alpha_0)}$$

$$= \lim_{n\to\infty} \frac{\mathrm{K}_{n+1}(z)}{\mathrm{K}_n(z)} \lim_{n\to\infty} \frac{\upsilon_n(z)}{\upsilon_{n+1}(z)} \lim_{n\to\infty} \frac{X_{n+1}(z)}{X_n(z)}$$

and

$$\lim_{n\to\infty} \frac{\upsilon_{n+1}}{\upsilon_n}, \quad \lim_{n\to\infty} \frac{\mathrm{K}_{n+1}(z)}{\mathrm{K}_n(z)}, \quad \text{and} \quad \lim_{n\to\infty} X_n(z)$$

all converge to 1 (the last two locally uniformly in $\mathbb{O}$) by Lemma 9.8.7, Theorem 9.8.8, and our previous observations. ☐

Under somewhat more restrictive conditions on the set A we have

Theorem 9.8.15. *Let $\mu' > 0$ a.e. in $\partial\mathbb{O}$ and suppose the set A is compactly included in $\mathbb{O}$. Then*

$$\lim_{n\to\infty} \frac{\phi_n^*(z)}{\phi_n(z)\zeta_n(z)} = \lim_{n\to\infty} \frac{\phi_n^*(z)}{\phi_n(z)} = 0, \quad \text{locally uniformly in } \mathbb{O}^e,$$

and

$$\lim_{n\to\infty} \frac{\phi_n(z)\zeta_n(z)}{\phi_n^*(z)} = \lim_{n\to\infty} \frac{\phi_n(z)}{\phi_n^*(z)} = 0, \quad \text{locally uniformly in } \mathbb{O}.$$

Proof. The convergence in $\mathbb{O}$ follows from the convergence in $\mathbb{O}^e$ by taking the superstar conjugates. We only prove the convergence in $\mathbb{O}^e$.

By Lemma 9.8.13 we get by dividing out:

$$\frac{\phi_n^*(z)}{\phi_n(z)\zeta_n(z)} = \frac{[1 - \overline{\zeta_n(\alpha_0)}\zeta_n(z)]g_n(z) + \tau_n \overline{f}_n \overline{\zeta_n(\alpha_0)}[\zeta_n(z) - \zeta_n(\alpha_0)]}{\zeta_n(\alpha_0)[1 - \overline{\zeta_n(\alpha_0)}\zeta_n(z)]f_n g_n(z) + \tau_n[\zeta_n(z) - \zeta_n(\alpha_0)]},$$

$$(9.44)$$

where

$$g_n(z) = \frac{\mathrm{K}_n(z)}{\mathrm{K}_n^*(z)}, \qquad f_n = \frac{\phi_n(\alpha_0)}{\phi_n^*(\alpha_0)}, \qquad \tau_n = \frac{\overline{\phi_n^*(\alpha_0)}}{\phi_n^*(\alpha_0)} \in \mathbb{T}.$$

By Lemma 9.8.11, $\lim_{n\to\infty} f_n = 0$, and by the superstar of Lemma 9.8.12, $\lim_{n\to\infty} g_n(z) = 0$, locally uniformly in $\mathbb{O}^e$. Moreover, $|\zeta_n(\alpha_0)| \le 1$ and because A is compactly included in $\mathbb{O}$ we have $|\zeta_n(z)| \le M < \infty$, locally uniformly in $\mathbb{O}^e$. This implies that

$$1 - \overline{\zeta_n(\alpha_0)}\zeta_n(z) \quad \text{and} \quad \overline{\zeta_n(\alpha_0)}[\zeta_n(z) - \zeta_n(\alpha_0)]$$

are locally uniformly bounded in $\mathbb{O}^e$. Therefore the numerator of the right-hand side of (9.44) and the first term in its denominator converge to zero, locally uniformly in $\mathbb{O}^e$. However, $|\zeta_n(z) - \zeta_n(\alpha_0)| \geq |\zeta_n(z)| - |\zeta_n(\alpha_0)|$ is locally uniformly bounded away from zero for $z \in \mathbb{O}^e$. Thus the ratio in the right-hand side of (9.44) converges to zero locally uniformly in $\mathbb{O}^e$.

Also, because A is compactly included in $\mathbb{O}$, it follows that $1 < m \leq |\zeta_n(z)| \leq M < \infty$, locally uniformly in $\mathbb{O}^e$, so that the factor $\zeta_n(z)$ can be dropped. This proves the theorem. □

Note that this theorem is the same as Theorem 9.6.6, except that the Szegő condition is replaced by the weaker condition $\mu' > 0$ a.e.

Under the same conditions, we also have the ratio asymptotics. See Ref. [171].

Theorem 9.8.16. *Let $\mu' > 0$ a.e. in $\partial\mathbb{O}$ and suppose the set A is compactly included in $\mathbb{O}$. Let $\epsilon_n \in \mathbb{T}$ be such that $\epsilon_n \phi_n^*(\alpha_0) > 0$, (i.e., $\epsilon_n = |\phi_n^*(\alpha_0)|/\phi_n^*(\alpha_0)$). Then*

$$\lim_{n\to\infty} \frac{\epsilon_n \phi_n^*(z) \varpi_n(z) \varpi_{n+1}(\alpha_0) \sqrt{1 - |\zeta_{n+1}(\alpha_0)|^2}}{\epsilon_{n+1} \phi_{n+1}^*(z) \varpi_{n+1}(z) \varpi_n(\alpha_0) \sqrt{1 - |\zeta_n(\alpha_0)|^2}} = 1,$$

locally uniformly in $\mathbb{O}$.

Proof. In view of Theorem 9.8.14, it is sufficient to show that

$$\lim_{n\to\infty} \frac{|\phi_n^*(\alpha_0)|^2 (1 - |\zeta_{n+1}(\alpha_0)|^2)}{|\phi_{n+1}^*(\alpha_0)|^2 (1 - |\zeta_n(\alpha_0)|^2)} = 1,$$

locally uniformly in $\mathbb{O}$. It follows from the Christoffel–Darboux relation that

$$\frac{|\phi_n^*(\alpha_0)|^2}{1 - |\zeta_n(\alpha_0)|^2} = k_n(\alpha_0, \alpha_0) \left[1 + \frac{|\zeta_n(\alpha_0)|^2}{1 - |\zeta_n(\alpha_0)|^2} \frac{|\phi_n(\alpha_0)|^2}{k_n(\alpha_0, \alpha_0)} \right].$$

Because A is compactly included in $\mathbb{O}$, it holds that for every $z \in \mathbb{O}$ (hence also for $z = \alpha_0$) there is a positive constant $m < 1$ such that for all n we have

$$X_n(z) = \frac{|\zeta_n(z)|^2}{1 - |\zeta_n(z)|^2} \leq \frac{1}{1 - m^2}.$$

Therefore, noting that $k_n(\alpha_0, \alpha_0) = v_n^2$ and setting $X_n = X_n(\alpha_0)$ and $Y_n = |\phi_n(\alpha_0)|^2/v_n^2$, we have

$$\frac{|\phi_n^*(\alpha_0)|^2(1 - |\zeta_{n+1}(\alpha_0)|^2)}{|\phi_{n+1}^*(\alpha_0)|^2(1 - |\zeta_n(\alpha_0)|^2)} = \frac{v_n^2}{v_{n+1}^2} \frac{[1 + X_n Y_n]}{[1 + X_{n+1} Y_{n+1}]}.$$

Because $\lim_{n\to\infty} v_n^2/v_{n+1}^2 = 1$ by Lemma 9.8.12, $\lim_{n\to\infty} Y_n = 0$ by Lemma 9.8.10, and X_n is uniformly bounded, it follows that the right-hand side converges to 1 locally uniformly in $\mathbb{O}$. $\square$

In the beginning of this section, we gave several norm convergence theorems that were valid for $\mu' > 0$ a.e. and involved reproducing kernels. We shall conclude this section with similar theorems of Pan [171] that involve the orthogonal functions.

Theorem 9.8.17. *Let A be compactly included in $\mathbb{O}$ and $\mu' > 0$ a.e. on $\partial\mathbb{O}$. Then*

$$\lim_{n\to\infty} \int (|\phi_n(t)x_n(t)|\sqrt{\mu'(t)} - 1)^2 \, d\mathring{\lambda}(t) = 0,$$

where

$$x_n(z) = \frac{\sqrt{1 - |\zeta_n(\alpha_0)|^2}}{1 - \zeta_n(z)\overline{\zeta_n(\alpha_0)}}.$$

Proof. Using the Christoffel–Darboux relation, we can show that the normalized reproducing kernel

$$K_n(z) = K_n(z, \alpha_0) = \frac{k_n(z, \alpha_0)}{\sqrt{k_n(\alpha_0, \alpha_0)}}$$

equals

$$K_n(z) = x_n(z)X_n(z)\frac{\phi_n^*(z)\overline{\phi_n^*(\alpha_0)}}{|\phi_n^*(\alpha_0)|}, \tag{9.45}$$

where

$$X_n(z) = \frac{1 - \overline{\zeta_n(\alpha_0)}\zeta_n(z)r_n(z)\overline{r_n(\alpha_0)}}{[1 - |\zeta_n(\alpha_0)|^2|r_n(\alpha_0)|^2]^{1/2}} \quad \text{and} \quad r_n(z) = \frac{\phi_n(z)}{\phi_n^*(z)}.$$

Note that $\lim_{n\to\infty} r_n(\alpha_0) = 0$ by Lemma 9.8.11. Furthermore, $|r_n(z)| \le 1$ for $z \in \overline{\mathbb{O}}$, $|\zeta_n(t)| = 1$ for $t \in \partial\mathbb{O}$, and $|\zeta_n(\alpha_0)| \le m < 1$. Therefore,

$$\lim_{n\to\infty} X_n(t) = 1, \quad \forall t \in \partial\mathbb{O}. \tag{9.46}$$

Next observe that

$$\int (|\phi_n x_n| \sqrt{\mu'} - 1)^2 \, d\mathring{\lambda} = \int (|\phi_n x_n| \sqrt{\mu'} - |\mathrm{K}_n| \sqrt{\mu'} + |\mathrm{K}_n| \sqrt{\mu'} - 1)^2 \, d\mathring{\lambda}$$

$$\leq 2 \int (|\phi_n x_n| \sqrt{\mu'} - |\mathrm{K}_n| \sqrt{\mu'})^2 \, d\mathring{\lambda}$$

$$+ 2 \int (|\mathrm{K}_n| \sqrt{\mu'} - 1)^2 \, d\mathring{\lambda}.$$

The second integral on the right converges to 0 as $n \to \infty$ by Theorem 9.8.3. Using (9.45), we find that the first integral is

$$\int (1 - |X_n|)^2 \, |x_n \phi_n|^2 \mu' \, d\mathring{\lambda}.$$

By using the definition of ζ_n and of the Poisson kernel $P(t, z)$, it follows after some algebra that for $t \in \overline{\partial \mathbb{O}}$

$$|x_n(t)|^2 = \frac{1}{|\varpi_0(t)|^2 P(t, \alpha_n)}.$$

Because A is compactly included in $\mathbb{O}$, there exist positive constants m and M (see (9.17)) such that

$$0 < m \leq |x_n(t)|^2 \leq M < \infty, \quad \forall t \in \overline{\partial \mathbb{O}}.$$

Furthermore, if we let $M_n = \max\{|1 - |X_n(t)|| : t \in \overline{\partial \mathbb{O}}\}$, then we can bound our last integral by

$$\int (1 - |X_n|)^2 \, |x_n \phi_n|^2 \mu' \, d\mathring{\lambda} \leq M_n^2 M \int |\phi_n|^2 \mu' \, d\mathring{\lambda}$$

$$\leq M_n^2 M \int |\phi_n|^2 \, d\mathring{\mu} = M_n^2 M.$$

Taking the limit for $n \to \infty$, we find by (9.46) that this goes to zero. Thus also

$$\lim_{n \to \infty} \int (|\phi_n x_n| \sqrt{\mu'} - 1)^2 \, d\mathring{\lambda} = 0.$$

This concludes the proof.   □

Note that because $|x_n(t)|^{-2} = |\varpi_0(t)|^2 P(t, \alpha_n)$, we can write the limit of the previous theorem as

$$\lim_{n \to \infty} \int \left(|\phi_n(t)| \sqrt{\mu'_n(t)} - 1 \right)^2 \, d\mathring{\lambda}(t) = 0, \quad \mu'_n(t) = \frac{\mu'(t)}{|\varpi_0(t)|^2 P(t, \alpha_n)}.$$

Finally, we give the general formulation of Theorem 2.6 in Ref. [171].

Theorem 9.8.18. *Let $P(t, z)$ denote the Poisson kernel. Suppose that $\mu' > 0$ a.e. on $\overline{\partial \mathbb{O}}$ and that $A = \{\alpha_1, \alpha_2, \ldots\}$ is compactly included in $\mathbb{O}$. Then*

$$\lim_{n \to \infty} \int \left| |\phi_n(t)|^2 \mu'_n(t) - 1 \right| d\overset{\circ}{\lambda}(t) = 0, \qquad \mu'_n(t) = \frac{\mu'(t)}{|\varpi_0(t)|^2 P(t, \alpha_n)};$$
(9.47)

$$\lim_{n \to \infty} \int \left| \frac{\sqrt{P(t, \alpha_n)}}{|\phi_n(t) \varpi_0(t)|} - \sqrt{\mathring{\mu}'(t)} \right| d\lambda(t) = 0, \qquad \mathring{\mu}'(t) = \frac{\mu'(t)}{|\varpi_0(t)|^2}.$$
(9.48)

For any bounded $f \in L_1(\overset{\circ}{\lambda})$

$$\lim_{n \to \infty} \int f(t) |\phi_n(t)|^2 \mu'_n(t) \, d\overset{\circ}{\lambda}(t) = \int f(t) \, d\overset{\circ}{\lambda}(t), \qquad \mu'_n(t) = \frac{\mu'(t)}{|\varpi_0(t)|^2 P(t, \alpha_n)}.$$
(9.49)

Proof. Let x_n be as in Theorem 9.8.17. Then by the Schwarz inequality, we find for the left-hand side of (9.47)

$$\left(\int \left| |\phi_n x_n|^2 \mu' - 1 \right| d\overset{\circ}{\lambda} \right)^2 = \left(\int \left| |\phi_n x_n| \sqrt{\mu'} - 1 \right| \left| |\phi_n x_n| \sqrt{\mu'} + 1 \right| d\overset{\circ}{\lambda} \right)^2$$

$$\leq \left(\int \left| |\phi_n x_n| \sqrt{\mu'} - 1 \right|^2 d\overset{\circ}{\lambda} \right)$$

$$\times \left(\int \left| |\phi_n x_n| \sqrt{\mu'} + 1 \right|^2 d\overset{\circ}{\lambda} \right).$$

The first of these integrals converges to zero by Theorem 9.8.17 and the second one is bounded, so that (9.47) follows.

For (9.48), we have again by the Schwarz inequality

$$\left(\int \left| |\phi_n x_n|^{-1} - \sqrt{\mu'} \right| d\overset{\circ}{\lambda} \right)^2 = \left(\int |\phi_n x_n|^{-1} \left| |\phi_n x_n| \sqrt{\mu'} - 1 \right| d\overset{\circ}{\lambda} \right)^2$$

$$\leq \left(\int |\phi_n x_n|^{-2} d\overset{\circ}{\lambda} \right) \left(\int \left| |\phi_n x_n| \sqrt{\mu'} - 1 \right|^2 d\overset{\circ}{\lambda} \right).$$

The first of these integrals equals

$$\int \frac{P(t, \alpha_n) |\varpi_0(t)|^2 \, d\overset{\circ}{\lambda}(t)}{|\phi_n(t)|^2} = \int \frac{P(t, \alpha_n) \, d\lambda(t)}{|\phi_n(t)|^2} = \int d\mathring{\mu}(t) = 1$$

by Theorem 9.7.1. The second of these integrals converges to 0 by Theorem 9.8.17, so that (9.48) follows.

The relation (9.49) is an immediate consequence of (9.47). $\qquad \square$

9.9. Root asymptotics

We finally give some examples of root asymptotics. Such asymptotics typically involve potential theory, which we briefly recall. See, for example, Ref. [193].

Let us first consider the normalized counting measure ν_n^A defined by

$$\nu_n^A = \frac{1}{n} \sum_{j=1}^n \delta_{\alpha_j},$$

which assigns a point mass at α_j, taking into account the multiplicity of α_j. It is called the zero distribution of the polynomial $\pi_n^*(z) = \prod_{k=1}^n (z - \alpha_k)$. For any measure ν with compact support in $\mathbb{C}$, the logarithmic potential is defined as

$$V_\nu(z) = -\int \log |z - x| \, d\nu(x).$$

For example,

$$V_{\nu_n^A}(z) = -\int \log |z - x| \, d\nu_n^A(x) = -\frac{1}{n} \sum_{j=1}^n \log |z - \alpha_j|.$$

Thus obviously

$$|\pi_n^*(z)|^{1/n} = \exp\left\{-V_{\nu_n^A}(z)\right\}.$$

Now assume that ν_n^A converges to some measure ν^A with compact support in the weak star topology, which we denote as

$$\nu_n^A \xrightarrow[n]{*} \nu^A,$$

that is,

$$\lim_{n \to \infty} \int f(x) \, d\nu_n^A(x) = \int f(x) \, d\nu^A(x), \quad \forall f \in C(\overline{\mathbb{C}}).$$

This convergence implies

$$\lim_{n \to \infty} |\pi_n^*(z)|^{1/n} = \exp\{-V_{\nu^A}(z)\}, \quad z \in \mathbb{C} \setminus \operatorname{supp}(\nu^A) \qquad (9.50)$$

and

$$\limsup_{n \to \infty} |\pi_n^*(z)|^{1/n} \leq \exp\{-V_{\nu^A}(z)\}, \quad z \in \mathbb{C},$$

where convergence is uniform on each compact set of the indicated region.

Set

$$\pi_n^*(z) = \prod_{j=1}^{n}(z - \overline{\alpha}_j).$$

Let $\nu^{\overline{A}}$ be the measure associated with the point set $\overline{A} = \{\overline{\alpha}_n\}_0^{\infty}$, just as ν^A was associated with the set $A = \{\alpha_k\}_0^{\infty}$. Note that from the definition of $V_{\nu_n^A}$ and $V_{\nu_n^{\overline{A}}}$, we have

$$V_{\nu_n^{\overline{A}}}(z) = -\int \log|z - x|\, d\nu_n^{\overline{A}}(x) = -\frac{1}{n}\sum_{j=1}^{n}\log|z - \overline{\alpha}_j|$$

$$= -\frac{1}{n}\sum_{j=1}^{n}\log|\overline{z} - \alpha_j| = V_{\nu_n^A}(\overline{z}).$$

Therefore $V_{\nu^{\overline{A}}}(z) = V_{\nu^A}(\overline{z})$.

The reason for introducing $\overline{\pi}_n^*(z)$ is that

$$\pi_n(z) = z^n \overline{\pi}_n^*(1/z) \quad \text{for } \mathbb{D}, z \neq 0,$$

$$\pi_n(z) = \overline{\pi}_n^*(z) \qquad \text{for } \mathbb{U}.$$

Thus the introduction of $\nu^{\overline{A}}$ allows us to state that in the case of $\mathbb{D}$ (for $z \neq 0$)

$$\lim_{n\to\infty} |\pi_n(z)|^{1/n} = \lim_{n\to\infty} |z||\overline{\pi}_n^*(1/z)|^{1/n} = |z|\exp\{-V_{\nu^{\overline{A}}}(1/z)\}$$

$$= |z|\exp\{-V_{\nu^A}(\hat{z})\}, \tag{9.51}$$

with $\hat{z} = 1/\overline{z}$, and in the case of $\mathbb{U}$

$$\lim_{n\to\infty} |\pi_n(z)|^{1/n} = \lim_{n\to\infty} |\overline{\pi}_n^*(z)|^{1/n} = \exp\{-V_{\nu^{\overline{A}}}(z)\} = \exp\{-V_{\nu^A}(\hat{z})\}, \tag{9.52}$$

with $\hat{z} = \overline{z}$. This holds for any $z \in \mathbb{C}_0 \setminus \mathrm{supp}(\nu^{\hat{A}})$, where $\mathbb{C}_0 = \mathbb{C} \setminus \{0\}$ for $\mathbb{D}$ and $\mathbb{C}_0 = \mathbb{C}$ for $\mathbb{U}$ and where $\hat{A}$ refers to the set $\hat{A} = \{\hat{\alpha}_1, \hat{\alpha}_2, \ldots\}$. Furthermore, since $\varpi_0^*(z)/\varpi_{0*}(z)$ equals z for $\mathbb{D}$ and 1 for $\mathbb{U}$,

$$\limsup_{n\to\infty} |\pi_n(z)|^{1/n} \leq \left|\frac{\varpi_0^*(z)}{\varpi_{0*}(z)}\right|\exp\{-V_{\nu^A}(\hat{z})\}, \quad z \in \mathbb{C}_0, \tag{9.53}$$

where $\mathbb{C}_0 = \mathbb{C} \setminus \{0\}$ for $\mathbb{D}$ and $\mathbb{C}_0 = \mathbb{C}$ for $\mathbb{U}$.

A combination of (9.50) and (9.51, 9.52) leads to

Lemma 9.9.1. Let B_n denote the Blaschke products with zeros from the set $A_n = \{\alpha_k\}_{k=1}^n$. Suppose that for the zero distributions ν_n^A we have the weak star convergence $\nu_n^A \xrightarrow[n]{*} \nu^A$ with supp(ν^A) compact. Then

$$\lim_{n\to\infty} |B_n(z)|^{1/n} = \exp\{\lambda(z)\} \quad \text{and} \quad \lim_{n\to\infty} |B_n(z)|^{-1/n} = \exp\{\lambda(\hat{z})\}$$

for $z \in \mathbb{C}_0 \setminus \{\text{supp}(\nu^A) \cup \text{supp}(\nu^{\hat{A}})\}$, with

$$\lambda(z) = \int \log|\zeta_z(x)| \, d\nu^A(x), \tag{9.54}$$

where

$$\zeta_z(x) = \frac{x - z}{1 - \bar{z}x} \quad \text{for } \mathbb{D} \quad \text{and} \quad \zeta_z(x) = \frac{x - z}{x - \bar{z}} \quad \text{for } \mathbb{U}.$$

For $z \in \mathbb{C}_0$ we have

$$\limsup_{n\to\infty} |B_n(z)|^{1/n} \le \exp\{\lambda(z)\} \quad \text{and} \quad \limsup_{n\to\infty} |B_n(z)|^{-1/n} \le \exp\{\lambda(\hat{z})\}.$$

Proof. We give the proof for $\mathbb{O} = \mathbb{D}$, since for $\mathbb{O} = \mathbb{U}$ it is even simpler. Because we have for $z \in \mathbb{C}_0 \setminus \{\text{supp}(\nu^A) \cup \text{supp}(\nu^{\hat{A}})\}$

$$\lim_{n\to\infty} |B_n(z)|^{1/n} = |z|^{-1} \exp\{-V_{\nu^A}(z) + V_{\nu^A}(1/\bar{z})\}$$

while

$$V_{\nu^A}(z) = -\int \log|x - z| \, d\nu^A(x)$$

and

$$V_{\nu^A}(1/\bar{z}) = -\int \log|1 - \bar{z}x| \, d\nu^A(x) + \log|z|$$

we get the first result.

For the second formula we note that $|B_n(z)|^{-1} = |B_{n*}(z)| = |B_n(\hat{z})|$ and the result is then immediate.

The proofs for the lim sup results are similar. □

Note that $\lambda(\hat{z}) = -\lambda(z)$.

Since $\lim f_{n+1}/f_n = L$ implies $\lim f_n^{1/n} = L$, we can deduce the root asymptotics for ϕ_n^* from the known ratio asymptotics of Theorem 9.8.14.

Theorem 9.9.2. *If $\mu' > 0$ a.e. in $\overline{\partial \mathbb{O}}$ and if A is compactly included in $\mathbb{O}$, then*

$$\lim_{n \to \infty} |\phi_n^*(z)|^{1/n} = 1, \quad \text{locally uniformly in } \mathbb{O}.$$

Proof. It follows from Theorem 9.8.14 that

$$\lim_{n \to \infty} \left| \frac{\phi_n^*(z) \varpi_n(z)}{\phi_n^*(\alpha_0) \varpi_n(\alpha_0)} \right|^{1/n} = 1, \quad \text{locally uniformly in } \mathbb{O}. \tag{9.55}$$

Because ratio asymptotics imply root asymptotics, it follows from Theorem 9.8.16 that

$$\lim_{n \to \infty} \frac{|\phi_n^*(\alpha_0)|^2}{1 - |\zeta_n(\alpha_0)|^2} = 1.$$

However, because A is compactly included in $\mathbb{O}$, there is some m such that $1 - m^2 \leq 1 - |\zeta_n(\alpha_0)|^2 \leq 1$, and thus

$$\lim_{n \to \infty} [1 - |\zeta_n(\alpha_0)|^2]^{1/n} = 1.$$

Consequently,

$$\lim_{n \to \infty} |\phi_n^*(\alpha_0)|^{1/n} = 1. \tag{9.56}$$

Furthermore, for any $z \in \mathbb{O}_r$, there exist $0 < m(r) < M(r) < \infty$ such that

$$m(r) \leq \left| \frac{\varpi_n(z)}{\varpi_n(\alpha_0)} \right| \leq M(r),$$

which means that

$$\lim_{n \to \infty} \left| \frac{\varpi_n(z)}{\varpi_n(\alpha_0)} \right|^{1/n} = 1, \quad \text{locally uniformly in } \mathbb{O}. \tag{9.57}$$

Finally, from (9.55), (9.56), and (9.57), the proof is achieved. $\square$

For the sequence $\{\phi_n(z)\}$ we have the following result.

Theorem 9.9.3. *Assume $\mu' > 0$ a.e. in $\overline{\partial \mathbb{O}}$ and let A be compactly included in $\mathbb{O}$ and $v_n^A \xrightarrow[n]{*} v^A$ (hence $\operatorname{supp}(v^A) \subset \overline{A}$ compact). Then for the case of the disk*

$$\lim_{n \to \infty} |\phi_n(z)|^{1/n} = |z|^{-1} \exp\{-V_{v^A}(z) + V_{v^A}(1/\bar{z})\} = \exp\{\lambda(z)\},$$

and for the case of the half plane

$$\lim_{n\to\infty} |\phi_n(z)|^{1/n} = \exp\{-V_{\nu^A}(z) + V_{\nu^A}(\bar{z})\} = \exp\{\lambda(z)\},$$

locally uniformly in $\mathbb{O}^e \setminus \mathrm{supp}(\nu^A)$. *The function* $\lambda(z)$ *is as in (9.54).*

Proof. Taking the substar in the previous theorem gives

$$\lim_{n\to\infty} \left| \frac{\phi_n(z)}{B_n(z)} \right|^{1/n} = 1, \quad \text{locally uniformly in } \mathbb{O}^e.$$

Thus by Lemma 9.9.1

$$\lim_{n\to\infty} |B_n(z)|^{1/n} = \frac{|\varpi_{0*}(z)| \exp\{-V_{\nu^A}(z)\}}{|\varpi_0^*(z)| \exp\{-V_{\nu^A}(\hat{z})\}} = \exp\{\lambda(z)\},$$

locally uniformly in $\mathbb{O}^e \setminus \mathrm{supp}(\nu^A)$. This completes the proof because $|\varpi_{0*}(z) / \varpi_0^*(z)|$ is $|z|^{-1}$ for $\mathbb{O} = \mathbb{D}$ and it is 1 for $\mathbb{O} = \mathbb{U}$. □

As an example, we consider the disk situation where $\lim_{n\to\infty} \alpha_n = \alpha \in \mathbb{D}$ so that $\nu^A = \delta_\alpha$. In this case $\mathrm{supp}(\nu^A) = \{\alpha\}$ and $\mathrm{supp}(\nu^{\hat{A}}) = \{\hat{\alpha}\} = \{1/\bar{\alpha}\}$. Furthermore,

$$V_{\delta_\alpha}(z) = -\int \log|z - x| \, d\delta_\alpha(x) = -\log|z - \alpha|$$

and

$$V_{\delta_\alpha}(\hat{z}) = -\log|1/\bar{z} - \alpha| = \log|z| - \log|1 - z\bar{\alpha}|.$$

Thus for $z \in \mathbb{E} \setminus \{\hat{\alpha}\}$,

$$\lim_{n\to\infty} |\phi_n(z)|^{1/n} = |z|^{-1} \exp\{V_{\nu^A}(\hat{z}) - V_{\nu^A}(z)\}$$

$$= |z|^{-1} \exp\{\log|z| - \log|1 - \bar{\alpha}z| + \log|z - \alpha|\}$$

$$= \left| \frac{z - \alpha}{1 - \bar{\alpha}z} \right| = |\zeta_\alpha(z)|.$$

So when $\alpha = 0$, we have $\lim_{n\to\infty} |\phi_n(z)|^{1/n} = |z|$. In particular, if $\alpha_n = 0$ for all n, then $\phi_n(z)$ becomes the Szegő polynomial and the well-known result that

$$\lim_{n\to\infty} |\phi_n(z)|^{1/n} = |z|, \quad \text{locally uniformly in } \mathbb{E},$$

is recovered (compare with Ref. [40]).

A similar computation for the case $\mathbb{O} = \mathbb{U}$ leads to the same result: If $\lim_{n \to \infty} \alpha_n = \alpha \in \mathbb{O}$, then $\lim_{n \to \infty} |\phi_n(z)|^{1/n} = |\zeta_\alpha(z)|$, locally uniformly in $\mathbb{O}^e \setminus \{\mathring{\alpha}\}$.

Next we prove nth root asymptotics for the para-orthogonal rational functions

$$Q_n(z, \tau_n) = \phi_n(z) + \tau_n \phi_n^*(z), \quad \tau_n \in \mathbb{T}.$$

Theorem 9.9.4. *Let* $\log \mu' \in L_1(d\mathring{\lambda})$ *and let* A *be compactly included in* $\mathbb{O}$. *Then*

1. $\lim_{n \to \infty} |Q_n(z, \tau_n)|^{1/n} = 1$, *locally uniformly in* $\mathbb{O}$.
2. *If, moreover,* $\nu_n^A \xrightarrow{*}{}_n \nu^A$ *with* $\operatorname{supp}(\nu^A)$ *compact, then* $\lim_{n \to \infty} |Q_n(z, \tau_n)|^{1/n} = \exp\{\lambda(z)\}$ *locally uniformly in* $\mathbb{O}^e \setminus \operatorname{supp}(\nu^A)$, *where* $\lambda(z)$ *is as in (9.54)*.

Proof. We shall give the proof only for $\mathbb{O} = \mathbb{D}$. For $\mathbb{O} = \mathbb{U}$, the proof is completely similar.

1. Set

$$\chi_n(z) = \frac{(1 - \overline{\alpha}_n z) Q_n(z, \tau_n)}{\phi_n^*(0)}.$$

Then for any $z \in \mathbb{D}$

$$\frac{\chi_{n+1}(z)}{\chi_n(z)} = \left[\frac{(1 - \overline{\alpha}_{n+1} z)\phi_{n+1}^*(z)\phi_n^*(0)}{(1 - \overline{\alpha}_n z)\phi_n^*(z)\phi_{n+1}^*(0)}\right]\left[\frac{\tau_{n+1} + \phi_{n+1}(z)/\phi_{n+1}^*(z)}{\tau_n + \phi_n(z)/\phi_n^*(z)}\right] = \gamma_n \Delta_n.$$

Clearly by Theorem 9.8.14 $\lim_{n \to \infty} \gamma_n = 1$ whereas

$$\frac{1 - |\phi_{n+1}(z)/\phi_{n+1}^*(z)|}{1 + |\phi_n(z)/\phi_n^*(z)|} \leq |\Delta_n| \leq \frac{1 + |\phi_{n+1}(z)/\phi_{n+1}^*(z)|}{1 - |\phi_n(z)/\phi_n^*(z)|}.$$

Since by Theorem 9.6.6 $\lim_{n \to \infty} \phi_n(z)/\phi_n^*(z) = 0$, locally uniformly in $\mathbb{D}$, we find that $\lim_{n \to \infty} |\Delta_n| = 1$. Thus we have proved that $\lim_{n \to \infty} |\chi_{n+1}(z)/\chi_n(z)| = 1$, which implies that $\lim_{n \to \infty} |\chi_n(z)|^{1/n} = 1$. Now, since $Q_n(z, \tau_n) = \phi_n^*(z)\chi_n(z)/(1 - \overline{\alpha}_n z)$, $\lim_{n \to \infty} |1 - \overline{\alpha}_n z|^{1/n} = 1$ locally uniformly in $\mathbb{D}$, and $\lim_{n \to \infty} |\phi_n^*(0)|^{1/n} = 1$, the proof follows.

2. Since $\phi_n(z) \neq 0$ in $\mathbb{E}$, we can write

$$Q_n(z, \tau_n) = \phi_n(z)[1 + \tau_n \phi_n^*(z)/\phi_n(z)].$$

This gives

$$|\phi_n(z)|[1 - |\phi_n^*(z)/\phi_n(z)|] \leq |Q_n(z, \tau_n)| \leq |\phi_n(z)|[1 + |\phi_n^*(z)/\phi_n(z)|]$$

and consequently

$$|\phi_n(z)|^{1/n}[1 - |\phi_n^*(z)/\phi_n(z)|]^{1/n} \leq |Q_n(z, \tau_n)|^{1/n}$$
$$\leq |\phi_n(z)|^{1/n}[1 + |\phi_n^*(z)/\phi_n(z)|]^{1/n}.$$

By Theorem 9.6.6, $\lim_{n \to \infty} \phi_n^*(z)/\phi_n(z) = 0$, locally uniformly in $\mathbb{E}$, which yields immediately $\lim_{n \to \infty} |\phi_n(z)|^{1/n} = \lim_{n \to \infty} |Q_n(z, \tau_n)|^{1/n}$. By Theorem 9.9.3, the proof is completed. $\square$

Let us now see how the sequence $\{Q_n(z, \tau_n)\}$ behaves on the boundary $\partial \mathbb{O}$. We define

$$\|Q_n\|_\infty = \max_{z \in \partial \mathbb{O}} |Q_n(z, \tau_n)| = \max_{z \in \mathbb{O} \cup \partial \mathbb{O}} |Q_n(z, \tau_n)|.$$

Theorem 9.9.5. *Let* $\log \mu' \in L_1(\mathring{\lambda})$ *and assume that A is compactly included in* $\mathbb{O}$*. Then*

$$\lim_{n \to \infty} \|Q_n\|_\infty^{1/n} = 1.$$

Proof. Because the Cayley transform maps $H_\infty(\mathbb{D})$ onto $H_\infty(\mathbb{U})$, it is sufficient to give the proof for $\mathbb{D}$. If we take $z \in \mathbb{D}$, then $|Q_n(z, \tau_n)| \leq \|Q_n\|_\infty$ and therefore

$$|Q_n(z, \tau_n)|^{1/n} \leq \|Q_n\|_\infty^{1/n}$$

and

$$\liminf_{n \to \infty} |Q_n(z, \tau_n)|^{1/n} = \lim_{n \to \infty} |Q_n(z, \tau_n)|^{1/n} = 1 \leq \liminf_{n \to \infty} \|Q_n\|_\infty^{1/n}. \quad (9.58)$$

Since A is compactly included in $\mathbb{D}$ and hence $\hat{A}$ bounded away from $\mathbb{T}$, it follows that $Q_n(z, \tau_n) \in \mathcal{L}_n$, having poles in the set $\hat{A} = \{\hat{\alpha}_1, \hat{\alpha}_2, \ldots\}$, is analytic in a disk of radius $\rho > 1$ for any $n = 1, 2, \ldots$. Hence we have

$$\|Q_n\|_\infty = \max_{z \in \mathbb{T} \cup \mathbb{D}} |Q_n(z, \tau_n)| \leq \max_{z \in T_\rho} |Q_n(z, \tau_n)|, \qquad T_\rho = \{z \in \mathbb{C} : |z| = \rho\}.$$

By Theorem 9.9.4(2), we have for $z \in T_\rho \subset \mathbb{E}$

$$\lim_{n \to \infty} |Q_n(z, \tau_n)|^{1/n} = |z|^{-1} \exp\{V_{\nu^A}(z) - V_{\nu^A}(1/\bar{z})\} = \exp\{\lambda(z)\}.$$

Define $\gamma(z) = \exp\{\lambda(z)\}$. Then for any $\epsilon > 0$, there exists some n_0 such that for all $n > n_0$ and $z \in T_\rho$

$$\|Q_n(z, \tau_n)|^{1/n} - \gamma(z)| < \epsilon$$

or, equivalently,

$$\gamma(z) - \epsilon < |Q_n(z, \tau_n)|^{1/n} < \epsilon + \gamma(z).$$

Hence, for sufficiently large n

$$\left[\max_{z \in T_\rho} |Q_n(z, \tau_n)| \right]^{1/n} \leq \max_{z \in T_\rho} |Q_n(z, \tau_n)|^{1/n} \leq \max_{z \in T_\rho} \{\epsilon + \gamma(z)\}.$$

Let $\rho e^{i\theta}$ be a point in T_ρ where $\gamma(z)$ reaches its maximum. Then

$$\|Q_n\|_\infty^{1/n} \leq \left[\max_{z \in T_\rho} |Q_n(z, \tau_n)| \right]^{1/n} \leq \epsilon + \gamma(\rho e^{i\theta}), \tag{9.59}$$

where

$$\gamma(\rho e^{i\theta}) = \rho^{-1} \exp\{V_{\nu^A}(\rho e^{i\theta}) - V_{\nu^A}(e^{i\theta}/\rho)\}.$$

If we now let ρ tend to 1^+, then we find $\lim_{\rho \to 1^+} \gamma(\rho e^{i\theta}) = 1$. In combination with (9.59) this yields

$$\limsup_{n \to \infty} \|Q_n\|_\infty^{1/n} \leq 1. \tag{9.60}$$

Finally, by (9.58) and (9.60) the proof follows. $\qquad\qquad\qquad\qquad\qquad$ □

The previous results were used in Ref. [44] to obtain estimates of the rate of convergence of $R_n(z, \tau_n)$ to Ω_μ. We give examples of such estimates in the next section.

9.10. Rates of convergence

The root asymptotics of the last section can be used to prove geometric convergence for rational interpolants to Ω_μ and for R-Szegő quadrature formulas.

Let us start with the rational interpolants of Theorem 9.2.1. It was shown that both sequences of rational functions

$$\Omega_n^\times(z) = -\frac{\psi_n(z)}{\phi_n(z)} \in H(\mathbb{O}^e) \quad \text{and} \quad \Omega_n(z) = \frac{\psi_{n*}(z)}{\phi_{n*}(z)} = \frac{\psi_n^*(z)}{\phi_n^*(z)} \in H(\mathbb{O})$$

converge locally uniformly to $\Omega_\mu(z)$. The first one in $\mathbb{O}^e$ and the second one in $\mathbb{O}$. We now add to this that the convergence is geometric. For similar results see also Ref. [172].

Theorem 9.10.1. *With the notation*

$$E_n^\times(z) = \Omega_\mu(z) - \Omega_n^\times(z) = \Omega_\mu(z) + \frac{\psi_n(z)}{\phi_n(z)}$$

and

$$E_n(z) = \Omega_\mu(z) - \Omega_n(z) = \Omega_\mu(z) - \frac{\psi_n^*(z)}{\phi_n^*(z)}$$

and assuming that $v_n^A \xrightarrow[n]{*} v^A$ *with* $\operatorname{supp}(v^A)$ *compact and that the Blaschke product diverges, we have*

1. *For all* $z \in \mathbb{O}$: $\limsup_{n\to\infty} |E_n(z)|^{1/n} \le \exp\{\lambda(z)\} < 1$.
2. *For all* $z \in \mathbb{O}^e \cup \{\infty\}$: $\limsup_{n\to\infty} |E_n^\times(z)|^{1/n} \le \exp\{\lambda(\hat{z})\} < 1$.

Here $\lambda(z)$ *is as in (9.54).*

Proof. By using a substar conjugate, part (2) is immediately obtained from part (1). We thus only have to prove the first part.

From the proof of Theorem 9.2.1, we obtain

$$|E_n(z)| \le |B_n(z)||h(z)|, \qquad h = \frac{4}{(1+\Gamma)(1+\Gamma_n)}.$$

Hence $0 < m(z) \le |h(z)| \le M(z) < \infty$ in compact subsets of $\mathbb{O}$. Therefore, we also have

$$\limsup_{n\to\infty} |E_n(z)|^{1/n} \le \limsup_{n\to\infty} |B_n(z)|^{1/n},$$

and the result follows for $z \in \mathbb{C}_0 \cap \mathbb{O}$ from Lemma 9.9.1. In the case of the disk, the result is also true for $z = \alpha_0 = 0$ because $E_n(0) = 0$ for all n. $\square$

A similar result can be obtained for the interpolants $L_n(z, w)/K_n(z, w)$ of Theorem 9.2.2 and the interpolants $R_n(z) = -P_n(z)/Q_n(z)$ of Theorem 9.2.3. We give the explicit formulation for the latter case. The proof is as the previous one.

Theorem 9.10.2. *Suppose* $R_n(z) = -P_n(z)/Q_n(z)$ *is the rational interpolant to* Ω_μ *of Theorem 9.2.3 with error* $E_n(z) = \Omega_\mu(z) - R_n(z)$. *Suppose that* $v_n^A \xrightarrow[n]{*} v^A$, *with* $\operatorname{supp}(v^A)$ *compact, and that the Blaschke product diverges. Then we have the following estimates for the convergence rates:*

1. *For all* $z \in \mathbb{O}$: $\limsup_{n\to\infty} |E_n(z)|^{1/n} \le \exp\{\lambda(z)\} < 1$.
2. *For all* $z \in \mathbb{O}^e \cup \{\infty\}$: $\limsup_{n\to\infty} |E_n(z)|^{1/n} \le \exp\{\lambda(\hat{z})\} < 1$.

Here again $\lambda(z)$ *is as defined in (9.54).*

To end this section, we give the convergence for the R-Szegő quadrature formulas introduced in Chapter 5.

For each $n = 1, 2, \ldots$, let $\{\xi_{nk}\}_{k=1}^{n}$ be the zeros of the para-orthogonal rational functions $Q_n(z, \tau_n)$ and consider the corresponding R-Szegő formulas

$$I_n\{f\} = \sum_{k=1}^{n} \lambda_{nk} f(\xi_{nk}),$$

which approximate the integral $I_\mu\{f\} = \int f(t)\, d\mathring{\mu}(t)$. We first prove the convergence of the quadrature formula when $f \in C(\overline{\partial\mathbb{O}})$, that is, when f is continuous on $\overline{\partial\mathbb{O}}$.

Theorem 9.10.3. *If the Blaschke product diverges, then, with the previous notation,*

$$\lim_{n \to \infty} I_n\{f\} = I_\mu\{f\}, \quad \forall f \in C(\overline{\partial\mathbb{O}}).$$

Proof. If $\mathcal{R}_n = \mathcal{L}_n \cdot \mathcal{L}_{n*}$ and $\mathcal{R}_\infty = \cup_{n=0}^{\infty} \mathcal{R}_n$, then we know from Chapter 7 that $\mathcal{R}_\infty$ is dense in the space of continuous functions $C(\overline{\partial\mathbb{O}})$ iff the Blaschke product diverges. Let $f \in C(\overline{\partial\mathbb{O}})$ and let $\epsilon > 0$. Take $g \in \mathcal{R}_\infty$ such that

$$\|f - g\|_\infty < \frac{\epsilon}{2}.$$

Then there is a k such that $g \in \mathcal{R}_k$. Since $I_n\{\cdot\}$ is exact in $\mathcal{R}_k$ for $n > k$, we have

$$|I_\mu\{f\} - I_n\{f\}| \leq |I_\mu\{f\} - I_\mu\{g\}| + |I_n\{g\} - I_\mu\{f\}|$$

$$\leq \int |f - g|\, d\mathring{\mu} + \sum_{j=1}^{n} |g(\xi_{nj}) - f(\xi_{nj})| \leq \frac{\epsilon}{2} + \frac{\epsilon}{2} = \epsilon$$

whenever $n > k$ since $\lambda_{nj} > 0$ and $\int d\mathring{\mu} = \sum_{j=1}^{n} \lambda_{nj} = 1$. This proves the convergence. $\qquad\square$

In the case of the disk, one can prove along the lines of Ref. [54, pp. 127–129] that the quadrature converges not only for the continuous functions, but for all integrable functions $f \in L_1(\mathring{\mu})$. Such a theorem in the case where the α_k are cyclically repeated can be found in Ref. [32]. When all $\alpha_k = 0$, the proof was given in Ref. [122].

We add to the previous results the rate of convergence. The rate of convergence of the quadrature formula $I_n\{f\}$, where f is a function analytic in a region containing $\partial \mathbb{O}$, will depend on how large this region is. Thus, to get better estimates, we make the following construction.

Let $\mathcal{G}$ denote the set of all regions (closed and connected) G in $\overline{\mathbb{C}}$ such that $\overline{\partial \mathbb{O}} \subset G$ and $G \cap \{A_0 \cup \hat{A}_0\} = \emptyset$ and that have a boundary $\Gamma = \partial G$ which is a finite union of rectifiable Jordan curves. Let $f \in H(G)$ be an analytic function in G.

In order to simplify the notation, we give a separate treatment for $\mathbb{O} = \mathbb{D}$ and $\mathbb{O} = \mathbb{U}$. Suppose first that $\mathbb{O} = \mathbb{D}$. Then we have by Cauchy's theorem

$$ f(z) = \frac{1}{2\pi \mathrm{i}} \int_\Gamma \frac{z+x}{z-x} \left(-\frac{f(x)}{2x} \right) dx, \quad z \in G \in \mathcal{G}. $$

Since $\mathbb{T} \subset G$, we get by Fubini's theorem

$$ I_\mu\{f\} = \int f(t)\,d\mu(t) = \frac{1}{2\pi \mathrm{i}} \int_\Gamma \Omega_\mu(x) \left(-\frac{f(x)}{2x} \right) dx. $$

Let $R_n(z) = -P(z)/Q_n(z)$ be the rational interpolant for $\Omega_\mu(z)$ constructed from the para-orthogonal functions $Q_n = \phi_n + \tau_n \phi_n^*$ and the associated functions $P_n = \psi_n - \tau_n \psi_n^*$, with $\tau_n \in \mathbb{T}$. As before, $\xi_{nj} \in \overline{\partial \mathbb{O}}$ denote the zeros of Q_n. It holds then that for $f \in H(G)$, $G \in \mathcal{G}$

$$ I_n\{f\} = \sum_{j=1}^n \lambda_{nj} f(\xi_{nj}) = \sum_{j=1}^n \lambda_{nj} \left[\frac{1}{2\pi \mathrm{i}} \int_\Gamma D(\xi_{nj}, x) \left(-\frac{f(x)}{2x} \right) dx \right] $$

$$ = \frac{1}{2\pi \mathrm{i}} \int_\Gamma \left[\sum_{j=1}^n \lambda_{nj} D(\xi_{nj}, x) \right] \left(-\frac{f(x)}{2x} \right) dx. $$

By Theorem 5.3.3 we know that

$$ R_n(x) = \int D(t, x)\,d\mu_n(t) = \sum_{j=1}^n \lambda_{nj} D(\xi_{nj}, x). $$

Hence

$$ I_n\{f\} = \frac{1}{2\pi \mathrm{i}} \int_\Gamma R_n(x) \left(-\frac{f(x)}{2x} \right) dx $$

and an alternative expression for the error is

$$ E_n = I_\mu\{f\} - I_n\{f\} = \frac{1}{2\pi \mathrm{i}} \int_\Gamma (\Omega_\mu(x) - R_n(x)) \left(-\frac{f(x)}{2x} \right) dx. $$

In the case $\mathbb{O} = \mathbb{U}$, we have similarly

$$f(z) = \frac{1}{2\pi i} \int_\Gamma D(z, x) \left(-\frac{2i f(x)}{1 + x^2} \right) dx, \quad z \in G \in \mathcal{G}$$

and

$$I_\mu\{f\} = \int f(t) \, d\mathring{\mu}(t) = \frac{1}{2\pi i} \int_\Gamma \Omega_\mu(x) \left(-\frac{2i f(x)}{1 + x^2} \right) dx$$

and

$$I_n\{f\} = \int f(t) \, d\mathring{\mu}_n(t) = \frac{1}{2\pi i} \int_\Gamma R_n(x) \left(-\frac{2i f(x)}{1 + x^2} \right) dx.$$

Theorem 9.10.4. *Suppose the Blaschke product diverges and $v_n^A \xrightarrow{*}_n v^A$, with* supp$(v^A)$ *compact, $f \in H(G)$, and with $G \in \mathcal{G}$ ($\mathcal{G}$ is defined above) and let the error of the R-Szegő formula $I_n\{f\}$ be given by $e_n = I_\mu\{f\} - I_n\{f\}$. Then*

$$\limsup_{n\to\infty} |e_n|^{1/n} \le \rho < 1, \quad \rho = \max\{\rho_1, \rho_2\},$$

where $\rho_1 = \max_{x \in \Gamma \cap \mathbb{O}} \exp\{\lambda(x)\}$, $\rho_2 = \max_{x \in \Gamma \cap \mathbb{O}^e} \exp\{\lambda(\hat{x})\}$ with $\lambda(z)$ as in (9.54), and $\Gamma = \partial G$.

Proof. By our previous derivations, we have

$$|e_n| = |I_\mu\{f\} - I_n\{f\}| \le \max_{x \in \Gamma} |\Omega_\mu(x) - R_n(x)| \frac{1}{2\pi} \int_\Gamma |F(x)| \, dx$$

with

$$F(x) = -\frac{f(x)}{2x} \quad \text{for } \mathbb{T} \quad \text{and} \quad F(x) = -\frac{2i f(x)}{1 + x^2} \quad \text{for } \mathbb{R}.$$

Because F is bounded on $\Gamma \subset G$ and Γ is rectifiable, there exists a finite positive constant M such that

$$|e_n| \le M \max_{x \in \Gamma} |\Omega_\mu(x) - R_n(x)|.$$

We can use Theorem 9.10.2 (1) for $x \in \Gamma_i = \Gamma \cap \mathbb{O}$ to establish the bound

$$\limsup_{n\to\infty} |e_n|^{1/n} \le \limsup_{n\to\infty} \left[\max_{x \in \Gamma_i} |\Omega_\mu(x) - R_n(x)|^{1/n} \right]$$

$$\le \max_{x \in \Gamma_i} \limsup_{n\to\infty} |\Omega_\mu(x) - R_n(x)|^{1/n}$$

$$= \max_{x \in \Gamma_i} \exp\{\lambda(x)\} = \rho_1,$$

where $\lambda(x)$ is given by (9.54). If $x \in \Gamma_e = \Gamma \cap \mathbb{O}^e$, one can use the second part of Theorem 9.10.2 to deduce in a similar way that

$$\limsup_{n\to\infty} |e_n|^{1/n} \leq \max_{z\in\Gamma_e} \exp\{\lambda(\hat{z})\} = \rho_2,$$

and this proves the theorem.   □

Note that we can always take $\Gamma_e = \Gamma \cap \mathbb{O}^e$ to be the reflection of $\Gamma_i = \Gamma \cap \mathbb{O}$ in the boundary $\partial\mathbb{O}$. That is, $\Gamma_e = \{z \in \overline{\mathbb{C}} : \hat{z} \in \Gamma_i\}$. In that case of course $\rho = \rho_1 = \rho_2$.

Take for example the case of the disk with $\lim_{n\to\infty} \alpha_n = \alpha \in \mathbb{D}$. We know then that $\exp\{\lambda(z)\} = |\zeta_\alpha(z)|$. To find ρ, we have to find its maximum on a curve $\Gamma_i \subset \mathbb{D}$, which contours all the α_n. Thus the more these α_n are concentrated in the neighborhood of α, the better we can make our estimate. If in the extreme case all $\alpha_n = \alpha$, then we could take a small circle around α, so that we can make $|\zeta_\alpha(z)|$ – and hence also $\lambda(z)$ and thus also ρ – as small as we want.

For more results about the rates of convergence for multipoint Padé and multipoint Padé-type approximants and corresponding quadrature formulas see Refs. [44, 40].

10

Moment problems

In this chapter we will study the moment problem. This is equivalent to the Nevanlinna–Pick problem for the disk or the half plane. For a finite measure on $\mathbb{T}$, we may define the moments

$$c_{-k} = \int t^k d\mu, \quad k \in \mathbb{Z}.$$

The trigonometric moment problem is the following: Given the moments c_k, $k \in \mathbb{Z}$ find the corresponding measure on $\mathbb{T}$. Necessary and sufficient conditions for the existence of a solution and for the uniqueness are to be given. If possible, find a way to construct the solution. All this is related to orthogonal polynomials with respect to a linear functional defined on the set of polynomials with the given moments. A quadrature formula, based on these polynomials, then gives a way to construct a solution for the problem.

In our case, we shall consider more general moments, which are related to orthogonal rational functions and we will treat again the unit circle and the real line in parallel.

10.1. Motivation and formulation of the problem

We suppose that we are given a linear functional M defined on $\mathcal{R}_\infty = \mathcal{L}_\infty \cdot \mathcal{L}_{\infty*}$ with $\mathcal{L}_\infty = \cup_{n=0}^\infty \mathcal{L}_n$ and $\mathcal{L}_{\infty*} = \{f : f_* \in \mathcal{L}_\infty\}$. We suppose it satisfies

$$M\{f_*\} = \overline{M\{f\}} \quad \text{and} \quad M\{ff_*\} > 0, \quad \forall f \neq 0, \ f \in \mathcal{L}_\infty.$$

This functional induces an inner product on $\mathcal{L}_\infty$ (or equivalently in $\mathcal{L}_{\infty*}$) that is given by

$$\langle f, g \rangle_M = M\{fg_*\} = \langle g_*, f_* \rangle_M, \quad f, g \in \mathcal{L}_\infty \quad \text{or} \quad f, g \in \mathcal{L}_{\infty*}.$$

By our assumption, this inner product is Hermitian and positive definite. When μ is a positive finite measure on $\overline{\partial \mathbb{O}}$, then

$$M\{f\} = \int f(t)\, d\mathring{\mu}(t), \quad f \in \mathcal{R}_\infty$$

is an example of such a functional. We also assume that M is bounded and normalized by the condition $M\{1\} = 1$.

In this chapter we shall address the problem of finding conditions under which such a measure exists that will represent the functional M defined on the space $\mathcal{R}_\infty$ by an infinite set of generalized moments. Such a measure will be called a solution of the moment problem.

We can motivate this as follows. We recall from Lemma 6.1.5 that the Riesz–Herglotz–Nevanlinna kernel $D(t, z)$ has the following formal Newton series expansion (all $\alpha_k \in \mathbb{O}$):

$$D(t, z) = 1 + 2\sum_{k=1}^{\infty} a_k(t)(z - \alpha_0)\pi^*_{k-1}(z), \quad \pi^*_{k-1}(z) = (z - \alpha_1)\cdots(z - \alpha_{k-1}),$$

with

$$a_k(t) = \frac{\varpi_0(t)}{\varpi_0(\alpha_0)\pi^*_k(t)}, \quad k = 1, 2, \ldots.$$

Thus we have by formal integration

$$\Omega_\mu(z) = \int D(t, z)\, d\mathring{\mu}(t) = 1 + 2\sum_{k=1}^{\infty} \mu_k(z - \alpha_0)\pi^*_{k-1}(z),$$

with

$$\mu_0 = 1, \quad \mu_k = \int \frac{\varpi_0(t)\, d\mathring{\mu}(t)}{\varpi_0(\alpha_0)\pi^*_k(t)}, \quad k = 1, 2, \ldots$$

as the general moments.

These moments $\mu_k, k = 0, 1, 2, \ldots$ define the functional M on $\mathcal{L}_\infty$. However, because

$$\overline{\mu}_k = \int \frac{\varpi_{0*}(t)\, d\mathring{\mu}(t)}{\varpi_0(\alpha_0)[\pi^*_k(t)]_*},$$

we also know (e.g., by partial fraction decomposition) all the values

$$\mu_{kl} = \int \frac{|\varpi_0(t)|^2\, d\mathring{\mu}(t)}{|\varpi_0(\alpha_0)|^2 \pi^*_k(t)[\pi^*_l(t)]_*} = \int \frac{d\mu(t)}{|\varpi_0(\alpha_0)|^2 \pi^*_k(t)[\pi^*_l(t)]_*},$$

$$k, l = 0, 1, 2, \ldots$$

in terms of μ_k, $k = 0, 1, 2, \ldots$, and these of course define M on $\mathcal{R}_\infty = \mathcal{L}_\infty \cdot \mathcal{L}_{\infty*} = \mathcal{L}_\infty + \mathcal{L}_{\infty*}$.

Note that in the case of $\partial \mathbb{O} = \mathbb{R}$, we have formulated the moment problem to include a possible mass point at ∞ (recall the integrals are over the extended real line $\overline{\mathbb{R}} = \mathbb{R} \cup \{\infty\}$). This is necessary because the moments are generated by rational basis functions. In the classical situation of polynomials, there can not be a solution of the moment problem with a mass point at infinity because all the basis functions $\{x, x^2, x^3, \ldots\}$ tend to infinity at ∞. Thus in the classical situation, the moment problem where the integrals are over $\mathbb{R}$ and the moment problem where the integrals are over $\overline{\mathbb{R}}$ are the same. In our situation the basis functions are rational and almost all of them tend to zero at ∞, so that a point mass at ∞ should not be excluded.

Almost all the previous results still hold when $\langle \cdot, \cdot \rangle_\mu$ is replaced by $\langle \cdot, \cdot \rangle_M$ and $\int \cdot \, d\hat{\mu}(t)$ is replaced by $M\{\cdot\}$. Therefore, we shall keep the notation with the indication of μ instead of M, in anticipation of the measure μ we want to find. In fact, the positivity of the inner product we defined in this section guarantees the existence of at least one solution. This follows, for example, from Theorem 9.2.1. See also our Theorem 10.3.1 to be proved later (more precisely Corollaries 10.3.2 and 10.3.3). We also keep the familiar notation of ϕ_n for the orthonormal rational functions, ψ_n for the functions of the second kind, $k_n(z, w)$ for the reproducing kernels, etc.

Thus, if we know that a solution exists, the only problem that remains is whether this solution is unique or not. When the Blaschke products diverge, then the uniqueness of the solution is guaranteed by Theorem 9.2.1, so that in this chapter we shall be mainly interested in the other situation, that is,

$$\sum (1 - |\alpha_k|) < \infty \quad \text{for } \mathbb{D} \quad \text{and} \quad \sum \operatorname{Im} \alpha_k / (1 - |\alpha_k|^2) < \infty \quad \text{for } \mathbb{U}. \tag{10.1}$$

However, we shall formulate the results so that they cover both situations, that is, the case of the divergent as well as the convergent Blaschke product. We shall follow rather closely the analysis given in Akhiezer's book on the classical moment problem [2].

10.2. Nested disks

Let us introduce the notation $\mathbb{O}_0$ for the set $\mathbb{O}$, excluding the points α_k, $k = 1, 2, \ldots$, and similarly $\mathbb{O}_0^e$ excludes the reflected points $\hat{\alpha}_k$, $k = 1, 2, \ldots$ from $\mathbb{O}^e$:

$$\mathbb{O}_0 = \mathbb{O} \setminus A = \{z \in \mathbb{O} : z \neq \alpha_k, \ k = 1, 2, \ldots\},$$
$$\mathbb{O}_0^e = \mathbb{O}^e \setminus \hat{A} = \{z \in \mathbb{O}^e : z \neq \hat{\alpha}_k, \ k = 1, 2, \ldots\}.$$

For fixed $z \in \mathbb{O}_0 \cup \mathbb{O}_0^e$, we consider the values of

$$s = R_n(z, \tau) = -\frac{\psi_n(z) - \tau \psi_n^*(z)}{\phi_n(z) + \tau \phi_n^*(z)} = -\frac{P_n(z, \tau)}{Q_n(z, \tau)}, \quad \tau \in \mathbb{T}, \ n \geq 1. \quad (10.2)$$

Note that $Q_n(z, \tau)$ are the para-orthogonal rational functions and $P_n(z, \tau)$ are the associated functions of the second kind. It turns out that when τ runs over $\mathbb{T}$ and z is fixed, then s will describe a circle $K_n(z)$ in the complex plane. Another equation that describes this is

$$|\psi_n^*(z) - s\phi_n^*(z)| = |\psi_n(z) + s\phi_n(z)|.$$

Since (ϕ_n, ϕ_n^*) and $(\psi_n, -\psi_n^*)$ are both solutions of the recurrence relation (4.1) or, equivalently, of (4.8), then $(\psi_n + s\phi_n, -\psi_n^* + s\phi_n^*)$ with $s \in \mathbb{C}$ arbitrary will also be a solution and we get by Green's formula

$$\frac{|\psi_n^*(z) - s\phi_n^*(z)|^2 - |\psi_n(z) + s\phi_n(z)|^2}{1 - |\zeta_n(z)|^2} - \frac{|1 - s|^2 - |1 + s|^2}{1 - |\zeta_0(z)|^2}$$

$$= \sum_{k=0}^{n-1} |\psi_n(z) + s\phi_n(z)|^2.$$

If $s \in K_n(z)$, then the first term vanishes and the equation of the circle becomes

$$\sum_{k=0}^{n-1} |\psi_k(z) + s\phi_k(z)|^2 = \frac{2(s + \bar{s})}{1 - |\zeta_0(z)|^2}.$$

Recall that (see Section 2.1)

$$1 - |\zeta_0(z)|^2 = \frac{\varpi_0(\alpha_0)\overline{\varpi_z(z)}}{|\varpi_0(z)|^2} = \begin{cases} 1 - |z|^2 & \text{for } \mathbb{D}, \\ \dfrac{4 \operatorname{Im} z}{|z + \mathrm{i}|^2} & \text{for } \mathbb{U}. \end{cases}$$

We shall denote the closed disk bounded by $K_n(z)$ as $\Delta_n(z)$. Thus $s \in \Delta_n(z)$ when in the previous relation the equality sign is replaced by a $\leq$ sign.

Since $z \in \mathbb{O}$ iff $1 - |\zeta_0(z)|^2 > 0$, and $z \in \mathbb{O}^e$ iff $1 - |\zeta_0(z)|^2 < 0$, we find that $K_n(z)$ is completely in the right (left) half plane iff z is in $\mathbb{O}$ (is in $\mathbb{O}^e$).

It follows immediately from the equations for the disks $\Delta_n(z)$ that they are nested: $\Delta_{n+1}(z) \subset \Delta_n(z)$ for $n = 1, 2, \ldots$. The intersection of all the disks is denoted as $\Delta_\infty = \Delta_\infty(z)$. This Δ_∞ is thus either a circular disk (with positive radius) or it reduces to a point.

The center c_n and the radius r_n of K_n are given by

$$c_n = \frac{\psi_n^* \overline{\phi_n^*} + \psi_n \overline{\phi_n}}{|\phi_n^*|^2 - |\phi_n|^2} \quad \text{and} \quad r_n = \left| \frac{\psi_n^* \phi_n + \psi_n \phi_n^*}{|\phi_n^*|^2 - |\phi_n|^2} \right|.$$

By the determinant formula and Christoffel–Darboux relation, we may rewrite the latter as

$$r_n = 2 \left| \frac{\varpi_0(z)\varpi_0^*(z)}{\varpi_0(\alpha_0)\overline{\varpi_z(z)}} \right| \frac{|B_{n-1}(z)|}{k_{n-1}(z,z)}, \tag{10.3}$$

where $k_{n-1}(z,w)$ is the reproducing kernel. More explicitly we have

$$r_n = \begin{cases} \dfrac{2|z|}{|1-|z|^2|} \dfrac{|B_{n-1}(z)|}{k_{n-1}(z,z)} & \text{for } \mathbb{D}, \\[3mm] \dfrac{|1+z^2|}{2|\operatorname{Im} z|} \dfrac{|B_{n-1}(z)|}{k_{n-1}(z,z)} & \text{for } \mathbb{U}. \end{cases}$$

Since the disks are nested, the sequence of $|B_n(z)|/k_n(z,z)$ is nonincreasing. This is obvious for $z \in \mathbb{O}_0$, but it also holds for $z \in \mathbb{O}_0^e$. Moreover, K_∞ is a point iff this sequence tends to 0. Obviously, $k_n(z,z) \geq 1$, so that the sequence tends to zero in $\mathbb{O}$ when $B_n(z)$ does, that is, when the Blaschke product diverges and thus when (10.1) is not satisfied. In this case we have a limiting point. Thus the divergence of the Blaschke product is sufficient to have a limiting point if $z \in \mathbb{O}_0$. We now prove that this is also true for $z \in \mathbb{O}_0^e$.

Lemma 10.2.1. If $z \in \mathbb{O}_0 \cup \mathbb{O}_0^e$ and if the Blaschke product diverges, then $\Delta_\infty(z)$ is a point.

Proof. As we explained above, this is clear for $z \in \mathbb{O}_0$. For $z \in \mathbb{O}_0^e$, we note that by the Christoffel–Darboux relation

$$\sum_{k=0}^{n} |\phi_k(z)|^2 = \frac{|\phi_{n+1}^*(z)|^2 - |\phi_{n+1}(z)|^2}{1 - |\zeta_{n+1}(z)|^2}.$$

Taking the superstar of this gives

$$\sum_{k=0}^{n} |B_{n\backslash k}(z)|^2 |\phi_k^*(z)|^2 = \frac{|\phi_{n+1}^*(z)|^2 - |\phi_{n+1}(z)|^2}{1 - |\zeta_{n+1}(z)|^2},$$

so that

$$\sum_{k=0}^{n} |\phi_k(z)|^2 = \sum_{k=0}^{n} |B_{n\backslash k}(z)|^2 |\phi_k^*(z)|^2.$$

Hence

$$k_n(z,z) = \sum_{k=0}^{n} |B_{n\backslash k}(z)|^2 |\phi_k^*(z)|^2 = |B_n(z)|^2 \sum_{k=0}^{n} |\phi_{k*}(z)|^2.$$

Thus, the expression for the radius gives

$$r_n(z) = 2 \left| \frac{\varpi_0(z)\varpi_0^*(z)}{\varpi_0(\alpha_0)\varpi_z(z)} \right| \frac{1}{|B_{n-1}(z)| \sum_{k=0}^{n-1} |\phi_{k*}(z)|^2},$$

and this goes to zero for $z \in \mathbb{O}_0^e$ because the Blaschke product diverges to infinity in $\mathbb{O}^e$. □

Thus we have proved that if (10.1) is not satisfied, that is, if the Blaschke product diverges (and if it diverges, it will diverge for all $z \in \mathbb{O}_0 \cup \mathbb{O}_0^e$), then we shall have a limiting point. If, however, the conditions (10.1) are satisfied, then the Blaschke product converges (to a nonzero value) for every $z \in \mathbb{O}_0 \cup \mathbb{O}_0^e$. Thus in this case we shall have a limiting point if and only if

$$\sum_{k=0}^{\infty} |\phi_k(z)|^2 = \infty, \quad z \in \mathbb{O}_0 \cup \mathbb{O}_0^e.$$

We shall refer to the case where Δ_∞ is a point as the limit point situation, whereas the case where Δ_∞ is a disk (with positive radius) is the limit disk situation.

We shall now generalize the theorems on invariance and analyticity given in Akhiezer's book [2]. By invariance is meant that if $\Delta_\infty(z)$ is a point (disk) for some $z = w$, then it will be a point (disk) for all z. By analyticity, it is meant that if $\Delta_\infty(z)$ is a point, then it is an analytic function of z everywhere in $\mathbb{O}_0 \cup \mathbb{O}_0^e$.

We shall need some lemmas first.

Lemma 10.2.2. Choose $z \in \mathbb{O}_0 \cup \mathbb{O}_0^e$. Then

1. The recurrence (4.1) has at least one solution (x_n, x_n^+) for which $(x_n) \in \ell_2$, that is, for which $\sum_{k=0}^{\infty} |x_k|^2 < \infty$.
2. If $\Delta_\infty(z)$ is a disk with positive radius, then $(x_n) \in \ell_2$ for every solution (x_n, x_n^+) of (4.1).
3. If the Blaschke product converges, then $(x_n) \in \ell_2$ for every solution (x_n, x_n^+) of (4.1) iff $\Delta_\infty(z)$ is a disk with positive radius.

Proof.

1. Statement 1 is obvious, taking for example

$$x_n(z) = \psi_n(z) + s\phi_n(z), \qquad x_n^+(z) = -\psi_n^*(z) + s\phi_n^*(z),$$

with $s \in \Delta_\infty(z)$.

2. If $\Delta_\infty(z)$ has a positive radius, then $(\phi_k(z)) \in \ell_2$ and for $s \in \Delta_\infty(z)$, also $(\psi_k(z) + s\phi_k(z)) \in \ell_2$. This implies that also $(\psi_k(z)) \in \ell_2$. Since (ϕ_n, ϕ_n^*) and $(\psi_n, -\psi_n^*)$ form a basis for the solution space of (4.1), it follows that all solutions are in ℓ_2.

3. The previous part proves this in one direction (which does not need the convergence of the Blaschke product). Conversely, if $(x_n) \in \ell_2$ for every solution (x_n, x_n^+) of (4.1), then $(\phi_n(z)) \in \ell_2$ and because the Blaschke product converges, the radius of $\Delta_\infty(z)$ will be positive. □

Lemma 10.2.3. Define for $k = 0, 1, \ldots, n-1$; $n = 0, 1, \ldots$

$$a_{kn}(w) = \overline{\phi_k(w)}\psi_n^*(w) + \overline{\psi_k(w)}\phi_n^*(w). \tag{10.4}$$

Then

$$\phi_n(z) = \frac{\overline{\varpi_{n*}(w)}}{2\overline{\varpi_{0*}(w)}B_n(w)\varpi_n(z)}$$

$$\times \left[2\varpi_0(z)\overline{\phi_n^*(w)} - \frac{\overline{\varpi_0(\alpha_0)}\varpi_w(z)}{\overline{\varpi_0(w)}} \sum_{k=0}^{n-1} a_{kn}(w)\phi_k(z) \right] \tag{10.5}$$

and

$$\psi_n(z) = \frac{\overline{\varpi_{n*}(w)}}{2\overline{\varpi_{0*}(w)}B_n(w)\varpi_n(z)}$$

$$\times \left[2\varpi_0(z)\overline{\psi_n^*(w)} - \frac{\overline{\varpi_0(\alpha_0)}\varpi_w(z)}{\overline{\varpi_0(w)}} \sum_{k=0}^{n-1} a_{kn}(w)\psi_k(z) \right]. \tag{10.6}$$

Proof. Consider the Christoffel–Darboux relation

$$\frac{\phi_n^*(z)\overline{\phi_n^*(w)} - \phi_n(z)\overline{\phi_n(w)}}{1 - \zeta_n(z)\overline{\zeta_n(w)}} = \sum_{k=0}^{n-1} \phi_k(z)\overline{\phi_k(w)}. \tag{10.7}$$

Formula 2 of Corollary 4.3.4 is

$$\frac{\psi_n^*(z)\overline{\psi_n^*(w)} - \psi_n(z)\overline{\psi_n(w)}}{1 - \zeta_n(z)\overline{\zeta_n(w)}} = \sum_{k=0}^{n-1} \psi_k(z)\overline{\psi_k(w)}, \tag{10.8}$$

whereas formula 1 of the same corollary yields the following two equalities:

$$\frac{\phi_n^*(z)\overline{\psi_n^*(w)} + \phi_n(z)\overline{\psi_n(w)}}{1 - \zeta_n(z)\overline{\zeta_n(w)}} = \left[\frac{2}{1 - \zeta_0(z)\overline{\zeta_0(w)}} - \sum_{k=0}^{n-1} \phi_k(z)\overline{\psi_k(w)} \right] \tag{10.9}$$

and

$$\frac{\psi_n^*(z)\overline{\phi_n^*(w)} + \psi_n(z)\overline{\phi_n(w)}}{1 - \zeta_n(z)\overline{\zeta_n(w)}} = \left[\frac{2}{1 - \zeta_0(z)\overline{\zeta_0(w)}} - \sum_{k=0}^{n-1} \psi_k(z)\overline{\phi_k(w)} \right].$$

(10.10)

Elimination of $\phi_n^*(z)$ from (10.7) and (10.9) gives

$$a_{nn}(w)\phi_n(z) = [1 - \zeta_n(z)\overline{\zeta_n(w)}] \left[\frac{2\phi_n^*(w)}{1 - \zeta_0(z)\overline{\zeta_0(w)}} - \sum_{k=0}^{n-1} a_{kn}(w)\phi_k(z) \right]$$

(10.11)

while elimination of $\psi_n^*(z)$ from (10.8) and (10.10) gives

$$a_{nn}(w)\psi_n(z) = [1 - \zeta_n(z)\overline{\zeta_n(w)}] \left[\frac{2\psi_n^*(w)}{1 - \zeta_0(z)\overline{\zeta_0(w)}} - \sum_{k=0}^{n-1} a_{kn}(w)\psi_k(z) \right].$$

(10.12)

Now we use

$$1 - \zeta_n(z)\overline{\zeta_n(w)} = \frac{\varpi_n(\alpha_n)\overline{\varpi_z(w)}}{\varpi_n(z)\overline{\varpi_n(w)}}$$

and the determinant formula of Theorem 4.2.6 to get

$$\frac{1 - \zeta_n(z)\overline{\zeta_n(w)}}{a_{nn}(w)} = \frac{\overline{\varpi_0(\alpha_0)}}{\overline{\varpi_0(w)}\varpi_{0*}(w)} \frac{\overline{\varpi_{n*}(w)}\varpi_w(z)}{2\varpi_n(z)\overline{B_n(w)}},$$

so that (10.11) and (10.12) can be transformed into (10.5) and (10.6).   □

From the Christoffel–Darboux relation follows

Lemma 10.2.4. Assume the Blaschke product converges. Choose $z \in \mathbb{O}_0 \cup \mathbb{O}_0^e$. Then $(\phi_n(z)) \in \ell_2 \Leftrightarrow (\phi_n^*(z)) \in \ell_2$ and $(\psi_n(z)) \in \ell_2 \Leftrightarrow (\psi_n^*(z)) \in \ell_2$.

Proof. As in the proof of Lemma 10.2.1, we have

$$\sum_{k=0}^{n} |\phi_k(z)|^2 = \sum_{k=0}^{n} |B_{n \setminus k}(z)|^2 |\phi_k^*(z)|^2.$$

It follows similarly from the formulas in Corollary 4.3.4 that

$$\sum_{k=0}^{n} |\psi_k(z)|^2 = \sum_{k=0}^{n} |B_{n \setminus k}(z)|^2 |\psi_k^*(z)|^2$$

for each n. Now, let $B(z)$ denote the infinite Blaschke product. Then for $z \in \mathbb{O}_0 \cup \mathbb{O}_0^e$ and for $k = 0, 1, \ldots, n$ we have

$$0 < |B(z)| \leq |B_n(z)| \leq |B_{n \setminus k}(z)| \leq 1, \quad z \in \mathbb{O}_0$$

and

$$1 \leq |B_{n \setminus k}(z)| \leq |B_n(z)| \leq |B(z)| < \infty, \quad z \in \mathbb{O}_0^e,$$

so that the lemma follows. □

Finally we are ready to state and prove the invariance theorem.

Theorem 10.2.5 (Invariance). *If $\Delta_\infty(w)$ is a disk with positive radius for some $w \in \mathbb{O}_0 \cup \mathbb{O}_0^e$, then $\Delta_\infty(z)$ is a disk with positive radius for every $z \in \mathbb{O}_0 \cup \mathbb{O}_0^e$ and*

$$\sum_{k=0}^{n} |\phi_k(z)|^2 \quad and \quad \sum_{k=0}^{n} |\psi_k(z)|^2 \tag{10.13}$$

converge locally uniformly in $\mathbb{O} \cup \mathbb{O}^e$ as $n \to \infty$. This situation can only occur if (10.1) is satisfied, that is, if the Blaschke product converges.

If $\Delta_\infty(w)$ is a point for some $w \in \mathbb{O}_0 \cup \mathbb{O}_0^e$, then $\Delta_\infty(z)$ is a point for every $z \in \mathbb{O}_0 \cup \mathbb{O}_0^e$. This situation will certainly occur if (10.1) is not satisfied, that is, if the Blaschke product diverges.

Proof. First notice that if (10.1) is not satisfied (i.e., if the Blaschke product diverges for some $w \in \mathbb{O}_0 \cup \mathbb{O}_0^e$), then it will diverge for every $z \in \mathbb{O}_0 \cup \mathbb{O}_0^e$ and, by Lemma 10.2.1, this implies that $\Delta_\infty(z)$ is a point for every $z \in \mathbb{O}_0 \cup \mathbb{O}_0^e$.

Now, assume that (10.1) is satisfied. Define

$$A_n(z) = \frac{\overline{\varpi_{n*}(w)} \varpi_0(z)}{\overline{\varpi_{0*}(w)} B_n(w) \varpi_n(z)}$$

and

$$C_n(z) = \frac{\overline{\varpi_0(\alpha_0)} \overline{\varpi_{n*}(w)} \varpi_w(z)}{2 \overline{\varpi_0(w)} \overline{\varpi_{0*}(w)} B_n(w) \varpi_n(z)},$$

and let a_{nk} be as in (10.4). Then, because Δ_∞ is a disk (see Lemma 10.2.2(3)), $(\phi_k(z)) \in \ell_2$ and $(\psi_k(z)) \in \ell_2$ so that

$$\sum_{n=0}^{\infty} \sum_{k=0}^{n-1} |a_{kn}|^2 \leq 2 \sum_{n=0}^{\infty} \sum_{k=0}^{n-1} (|\phi_k|^2 |\psi_n^*|^2 + |\psi_k|^2 |\phi_k^*|^2)$$

$$\leq 2 \left(\sum_{n=0}^{\infty} |\psi_n^*|^2 \right) \left(\sum_{k=0}^{\infty} |\phi_k|^2 \right) + 2 \left(\sum_{n=0}^{\infty} |\phi_n^*|^2 \right) \left(\sum_{k=0}^{\infty} |\psi_k|^2 \right)$$

$$< \infty$$

by Lemma 10.2.4.

Let K be a compact subset of $\mathbb{O}_0 \cup \mathbb{O}_0^e$. Then $A_n(z)$ and $C_n(z)$ are uniformly bounded for $z \in K$. Say

$$|A_n(z)| \leq R_1, \qquad |C_n(z)| \leq R_2, \quad z \in K, \ n = 0, 1, \dots.$$

Then Lemma 10.2.3 expresses ϕ_n or ψ_n by a formula of the form

$$x_n(z) = A_n(z)c_n - C_n(z) \sum_{k=0}^{n-1} a_{kn} x_k(z),$$

where $(c_n) \in \ell_2$ and $\sum_{n=0}^{\infty} \sum_{k=0}^{n-1} a_{kn} < \infty$.

We shall now prove that $\sum_{n=0}^{\infty} |x_n(z)|^2$ converges uniformly in K. Clearly

$$\left(\sum_{n=m}^{N} |x_n|^2 \right)^{1/2} \leq R_1 \left(\sum_{n=m}^{N} |c_n|^2 \right)^{1/2} + R_2 \left(\sum_{n=m}^{N} \left| \sum_{k=0}^{n-1} a_{kn} x_k \right|^2 \right)^{1/2}.$$

Let $0 < \epsilon < 1$ and choose $m = m(\epsilon, R_1, R_2)$ such that

$$\left(\sum_{n=m}^{\infty} |c_n|^2 \right)^{1/2} < \frac{\epsilon}{R_1} \quad \text{and} \quad \left(\sum_{n=m}^{\infty} \sum_{k=0}^{n-1} |a_{kn}|^2 \right)^{1/2} < \frac{\epsilon}{R_2}.$$

Then, for $N \geq m$, we have

$$\left(\sum_{n=m}^{N} |x_n|^2 \right)^{1/2} \leq \epsilon + R_2 \left[\sum_{n=m}^{N} \left(\sum_{k=0}^{n-1} |a_{kn}|^2 \sum_{k=0}^{n-1} |x_k|^2 \right) \right]^{1/2}$$

$$\leq \epsilon + R_2 \left(\sum_{k=0}^{N} |x_k|^2 \right)^{1/2} \left(\sum_{n=m}^{\infty} \sum_{k=0}^{n-1} |a_{kn}|^2 \right)^{1/2}$$

$$\leq \epsilon + \epsilon \left(\sum_{k=0}^{N} |x_k|^2 \right)^{1/2}$$

$$\leq \epsilon + \epsilon \left(\sum_{k=m}^{N} |x_k|^2 \right)^{1/2} + \epsilon \left(\sum_{k=0}^{m-1} |x_k|^2 \right)^{1/2},$$

so

$$(1 - \epsilon) \left(\sum_{k=m}^{N} |x_k|^2 \right)^{1/2} \leq \epsilon + \epsilon \left(\sum_{k=0}^{m-1} |x_k|^2 \right)^{1/2}.$$

As $\sum_{k=0}^{m-1} |x_k|^2$ is continuous on K, there is an $M > 0$ such that

$$\left(\sum_{k=0}^{m-1} |x_k|^2 \right)^{1/2} \leq M \quad \text{on } K.$$

Hence

$$\left(\sum_{k=m}^{N} |x_k|^2 \right)^{1/2} \leq \frac{\epsilon(M + 1)}{1 - \epsilon}, \quad N \geq m,$$

implying that $\sum_{n=0}^{\infty} |x_n|^2$ converges uniformly on K.

All this implies with $c_n = \overline{\phi_n^*(w)}$, $x_n = \phi_n(z)$, or $c_n = \overline{\psi_n^*(w)}$ and $x_n = \psi_n(z)$, respectively, that also $(\phi_n(z)) \in \ell_2$ and $(\psi_n(z)) \in \ell_2$ locally uniformly for $z \in K$. Thus the series (10.13) converge locally uniformly on K. In particular, $\Delta_\infty(z)$ is a disk for each $z \in \mathbb{O}_0 \cup \mathbb{O}_0^e$. $\qquad \square$

As a consequence, we have a dichotomy:

- either $\Delta_\infty(z)$ is a disk (with positive radius) for every $z \in \mathbb{O}_0 \cup \mathbb{O}_0^e$
  (and this can only happen if the Blaschke product converges)
- or $\Delta_\infty(z)$ is a point for every $z \in \mathbb{O}_0 \cup \mathbb{O}_0^e$
  (and this will certainly happen when the Blaschke product diverges).

Thus in what follows we can say that Δ_∞ is a disk or a point, without specifying the $z \in \mathbb{O}_0 \cup \mathbb{O}_0^e$.

Corollary 10.2.6. *In the case of a limiting disk, the radius of $K_\infty(z)$ is a continuous function of $z \in \mathbb{O}_0 \cup \mathbb{O}_0^e$.*

Proof. This follows from (10.3) and the previous theorem.   □

Next we prove the theorem on analyticity.

Theorem 10.2.7 (Analyticity). *Let Δ_∞ be a point (limit point situation). Then (see (10.2))*

$$s(z) = \lim_{n \to \infty} R_n(z, \tau)$$

exists for $z \in \mathbb{O}_0 \cup \mathbb{O}_0^e$ and $\tau \in \mathbb{T}$ and is independent of τ. The function s is analytic in $\mathbb{O}_0 \cup \mathbb{O}_0^e$ and

$$\frac{\operatorname{Re} s(z)}{1 - |\zeta_0(z)|^2} > 0, \quad z \in \mathbb{O}_0 \cup \mathbb{O}_0^e.$$

Proof. For $z \in \mathbb{O}_0 \cup \mathbb{O}_0^e$, let $s(z)$ be the point $\Delta_\infty(z)$. Then clearly $R_n(z, \tau) \to s(z)$ as $n \to \infty$ for each $\tau \in \mathbb{T}$. Now take a fixed $\tau \in \mathbb{T}$ and put $s_n(z) = R_n(z, \tau)$. Then s_n is analytic in $\mathbb{O}_0 \cup \mathbb{O}_0^e$. From the equation of $K_n(z)$, we obtain

$$|1 + s_n(z)|^2 = |\psi_0 + s_n(z)\phi_0|^2 \le 2\frac{s_n(z) + \overline{s_n(z)}}{1 - |\zeta_0(z)|^2}.$$

Thus

$$1 + s_n(z) + \overline{s_n(z)} + |s_n(z)|^2 \le 2\frac{s_n(z) + \overline{s_n(z)}}{1 - |\zeta_0(z)|^2}$$

or, using $|s + \bar{s}| \le 2|s|$,

$$|s_n(z)|^2 \le 1 + |s_n(z)|^2 \le 2\frac{1 + |\zeta_0(z)|^2}{|1 - |\zeta_0(z)|^2|}|s_n(z)|.$$

Thus $|s_n(z)| \le 2|A(z)|$ with

$$A(z) = \frac{1 + |\zeta_0(z)|^2}{1 - |\zeta_0(z)|^2}$$

is uniformly bounded on compact subsets of $\mathbb{O}_0 \cup \mathbb{O}_0^e$. Therefore, $s(z)$ is analytic in $\mathbb{O}_0 \cup \mathbb{O}_0^e$.

Moreover, from the inequality defining $\Delta_\infty(z)$, the last inequality of the theorem follows.   □

Note that the denominator in the last inequality of the theorem is positive (negative) iff $z \in \mathbb{O}$ ($z \in \mathbb{O}^e$). Thus the last inequality of the theorem says that $s(z)$ is in the right (left) half plane iff $z \in \mathbb{O}$ ($z \in \mathbb{O}^e$).

10.3. The moment problem

Let $\mathcal{M}$ denote the set of all solutions of the moment problem. As we already mentioned before, the theorem below will imply that our moment problem will always have a solution: $\mathcal{M} \neq \emptyset$. We identify two solutions as being the same in a measure theoretic sense, that is, the solutions are the same whenever $d\mu_1 = d\mu_2$, thus whenever they define the same linear functional on the set $C(\overline{\partial\mathbb{O}})$ of all continuous functions on the boundary $\overline{\partial\mathbb{O}}$.

A $\mu \in \mathcal{M}$ will always have an infinite support. If it were only supported on a finite number of points $t_i \in \partial\mathbb{O}$, $i = 1, \ldots, n$, then we may define the polynomial $N_n(z) = \prod_{k=1}^{n}(z - t_k)$ and set $R(z) = N_n(z)/\pi_n(z) \in \mathcal{L}_n$, where $\pi_n(z) = \prod_{k=1}^{n} \varpi_k(z)$. Obviously

$$0 < M\{RR_*\} = \|R\|_{\hat{\mu}}^2 = \int |R(t)|^2 \, d\hat{\mu}(t) = 0,$$

which is a contradiction.

Theorem 10.3.1. *Fix* $z \in \mathbb{O}_0 \cup \mathbb{O}_0^e$ *and define for* $\mu \in \mathcal{M}$

$$\Omega_\mu(z) = \int D(t, z) \, d\hat{\mu}(t).$$

Then

$$\{\Omega_\mu(z) : \mu \in \mathcal{M}\} = \Delta_\infty(z).$$

Proof. Set $s = \Omega_\mu(z)$ for some $\mu \in \mathcal{M}$. Let $f(t) = \overline{D(t, z)}$, $t \in \partial\mathbb{O}$, and let

$$\sum_{k=0}^{\infty} \gamma_k \phi_k(t)$$

be the generalized Fourier series of $f(t)$. Then

$$\gamma_k = \int f(t)\phi_{k*}(t) \, d\hat{\mu}(t), \quad k = 0, 1, \ldots,$$

especially

$$\gamma_0 = \int f(t) \, d\hat{\mu}(t) = \bar{s}.$$

Moreover,

$$\overline{\gamma}_k = \int D(t, z)[\phi_k(t) - \phi_k(z)] \, d\hat{\mu}(t) + \int D(t, z)\phi_k(z) \, d\hat{\mu}(t)$$
$$= \psi_k(z) + s\phi_k(z), \quad k = 1, 2, \ldots.$$

Bessel's inequality gives

$$\sum_{k=0}^{\infty} |\gamma_k|^2 \le \int |D(t, z)|^2 d\mathring{\mu}(t).$$

But, some computations lead to the identity

$$1 + |D(t, z)|^2 = \frac{1 + |\zeta_0(z)|^2}{1 - |\zeta_0(z)|^2} [D(t, z) + \overline{D(t, z)}], \quad t \in \partial\mathbb{O},$$

so that

$$\int |D(t, z)|^2 d\mathring{\mu}(t) = -1 + \frac{1 + |\zeta_0(z)|^2}{1 - |\zeta_0(z)|^2} (s + \bar{s}).$$

Also,

$$|\psi_0 + s\phi_0|^2 = |1 + s|^2 = 1 + s + \bar{s} + |s|^2 = 1 + s + \bar{s} + |\gamma_0|^2.$$

Hence Bessel's inequality becomes

$$\sum_{k=0}^{\infty} |\psi_k + s\phi_k|^2 \le \left(\frac{1 + |\zeta_0(z)|^2}{1 - |\zeta_0(z)|^2} + 1 \right) (s + \bar{s}).$$

Thus

$$\sum_{k=0}^{\infty} |\psi_k + s\phi_k|^2 \le \frac{2(s + \bar{s})}{1 - |\zeta_0(z)|^2},$$

which means that $s \in \Delta_\infty(z)$.

It remains to show that any $s \in \Delta_\infty(z)$ corresponds to the Riesz–Herglotz–Nevanlinna transform of a $\mu \in \mathcal{M}$. Let us assume first that s is a boundary point of $\Delta_\infty(z)$. Then, for each n, there is a point $s_n \in K_n(z)$ such that $s_n \to s$ as $n \to \infty$. For each n, there is a point $\tau_n \in \mathbb{T}$ such that $R_n(z, \tau_n) = s_n$. Let μ_n be a solution of the truncated moment problem (in $\mathcal{L}_{(n-1)*} \cdot \mathcal{L}_{(n-1)}$) with parameter τ_n. Then

$$s_n = R_n(z, \tau_n) = \int D(t, z) \, d\mathring{\mu}_n(t).$$

By Helly's selection theorem [115, p. 575; 87, p. 56; 94, p. 222], there is a subsequence $(\mathring{\mu}_{n(j)})$ of $(\mathring{\mu}_n)$ and a positive measure $\mathring{\mu}$ on $\overline{\partial\mathbb{O}}$ such that $\mathring{\mu}_{n(j)} \to \mathring{\mu}$. By Helly's convergence theorem [115, p. 573; 87, p. 56; 94, p. 222],

$$\int g(t) \, d\mathring{\mu}_{n(j)}(t) \to \int g(t) \, d\mathring{\mu}(t), \quad j \to \infty$$

for all $g \in C(\overline{\partial \mathbb{O}})$. For the case $\overline{\partial \mathbb{O}} = \mathbb{T}$, Helly's theorems are directly applicable. For the case $\overline{\partial \mathbb{O}} = \overline{\mathbb{R}}$, one can use a Cayley transform of the circle case. By this mapping we obtain a so-called one-point or Alexandroff compactification of the real line [96] and any neighborhood of ∞ is isomorphic to a neighborhood of a finite point, so that Helly's theorems are also applicable in this case.

Clearly $\mu \in \mathcal{M}$. Moreover,

$$s_{n(j)} = \int D(t, z) \, d\mathring{\mu}_{n(j)}(t) \to \int D(t, z) \, d\mathring{\mu}(t), \quad j \to \infty.$$

As $s_n \to s$ for $n \to \infty$, this implies

$$s = \int D(t, z) \, d\mathring{\mu}(t),$$

so that $s = \Omega_\mu(z)$ for some $\mu \in \mathcal{M}$.

Now assume that $\Delta_\infty(z)$ is a disk and let s belong to its interior. Then s is a convex combination $\lambda s_1 + (1 - \lambda)s_2$ ($0 < \lambda < 1$) of points s_1 and s_2 on the boundary $K_\infty(z)$. By the above, there are $\mu_1, \mu_2 \in \mathcal{M}$ such that

$$s_j = \int D(t, z) \, d\mathring{\mu}_j(t), \quad j = 1, 2.$$

Clearly $\mu = \lambda \mu_1 + (1 - \lambda)\mu_2 \in \mathcal{M}$ and $s = \Omega_\mu(z)$. $\square$

Corollary 10.3.2. *In the case of a limiting disk, for each $s \in \Delta_\infty(z)$, $z \in \mathbb{O}_0 \cup \mathbb{O}_0^e$, there are infinitely many $\mu \in \mathcal{M}$ such that $s = \Omega_\mu(z)$. In this case, the moment problem has infinitely many solutions.*

Corollary 10.3.3. *In the case of a limiting point, the moment problem has a unique solution.*

Proof. If $\mu_1, \mu_2 \in \mathcal{M}$, then the functions Ω_{μ_1} and Ω_{μ_2} coincide on $\mathbb{C} \setminus \partial \mathbb{O}$. For $\mu = \mu_1 - \mu_2$, we have that $\mathring{\mu}$ is of bounded variation whereas the Poisson–Stieltjes integral

$$0 = \Omega_{\mu_1}(z) - \Omega_{\mu_2}(z) = \int P(t, z) \, d\mu(t), \quad z \in \mathbb{C} \setminus \partial \mathbb{O}$$

is analytic in $\mathbb{O}$. This implies that μ is absolutely continuous (Theorem of F. and M. Riesz [76, p. 41; 92, p. 61]). Thus

$$0 = \mu'(z) = \int P(t, z)\mu'(t) \, d\lambda(t),$$

so that the difference between the boundary functions of μ_1 and μ_2 is a constant. $\square$

As in Akhiezer's book [2, p. 43], we adopt the following definition:
We say that a solution $\mu \in \mathcal{M}$ is *N-extremal* at a point $z \in \mathbb{O}_0 \cup \mathbb{O}_0^e$ if $s = \Omega_\mu(z)$
belongs to the boundary $K_\infty(z)$ of $\Delta_\infty(z)$.

This implies that $\mu \in \mathcal{M}$ is N-extremal at z iff

$$\sum_{k=0}^{\infty} |\psi_k(z) + s\phi_k(z)|^2 = \frac{2(s + \bar{s})}{1 - |\zeta_0(z)|^2}. \tag{10.14}$$

Let us denote by $\mathcal{L}$ the closure in $L_2(\mathring{\mu})$ of $\mathcal{L}_\infty$. We can now state a density
result.

Theorem 10.3.4. *Let $\mu \in \mathcal{M}$ be a solution of the moment problem. Then*

1. *If $\mathcal{L}_\infty$ is dense in $L_2(\mathring{\mu})$, then μ is an N-extremal at every $z \in \mathbb{O}_0 \cup \mathbb{O}_0^e$.*
2. *If μ is N-extremal at some $z \in \mathbb{O}_0 \cup \mathbb{O}_0^e$, then $\mathcal{L}_\infty$ is dense in $L_2(\mathring{\mu})$, that is,*
   $\mathcal{L} = L_2(\mathring{\mu})$.

Proof. The proof is along the same lines as the proof of the previous theorem.
Let $z \in \mathbb{O}_0 \cup \mathbb{O}_0^e$. Put

$$s = \Omega_\mu(z) = \int D(t, z) \, d\mathring{\mu}(t)$$

and let

$$f_z(t) = D(t, z)_* = -\overline{f_{\bar{z}}(t)}, \qquad t \in \overline{\partial\mathbb{O}} \tag{10.15}$$

(substar w.r.t. t). Let

$$\sum_{k=0}^{\infty} \gamma_k \phi_k(z)$$

be the generalized Fourier series of f_z. Then, as in the proof of Theorem 10.3.1,
the Bessel inequality

$$\sum_{k=0}^{\infty} |\gamma_k|^2 \leq \int |f_z(t)|^2 d\mathring{\mu}(t)$$

becomes

$$\sum_{k=0}^{\infty} |\psi_k(z) + s\phi_k(z)|^2 \leq \frac{2(s + \bar{s})}{1 - |\zeta_0(z)|^2}, \tag{10.16}$$

which is the equation defining $\Delta_\infty(z)$.

1. Now suppose that $\mathcal{L}_\infty$ is dense in $L_2(\mathring{\mu})$. Because obviously $f_z \in L_2(\mathring{\mu})$, we should have equality in (10.16) by Parseval. This means precisely that $s \in K_\infty(z)$ and thus that μ is N-extremal.

2. Conversely, suppose that the solution $\mu \in \mathcal{M}$ of the moment problem is N-extremal at $z \in \mathbb{O}_0 \cup \mathbb{O}_0^e$. From (10.15)

$$\Omega_\mu(\hat{z}) = -\overline{\Omega_\mu(z)},$$

and it follows that $\Delta_\infty(\hat{z}) = -\overline{\Delta_\infty(z)}$ and in particular that $K_\infty(\hat{z}) = -\overline{K_\infty(z)}$. Hence μ is also N-extremal at $\hat{z}$ and $f_{\hat{z}}$ is in the closure $\mathcal{L}$ of $\mathcal{L}_\infty$ in $L_2(\mathring{\mu})$. Because also $1 \in \mathcal{L}_\infty$, we find that both

$$\frac{1}{\overline{\varpi_z(t)}} \quad \text{and} \quad \frac{1}{\varpi_z^*(t)} = \frac{1}{t - z}$$

belong to $\mathcal{L}$. We show that also

$$\frac{1}{[\overline{\varpi_z(t)}]^k} \quad \text{and} \quad \frac{1}{[\varpi_z^*(t)]^k} = \frac{1}{(t - z)^k}$$

are in $\mathcal{L}$ for $k = 1, 2, \ldots$.

For each k and each $p \in \mathcal{P}_n$, there exists a constant A and a polynomial $q \in \mathcal{P}_{n-1}$ such that

$$\frac{1}{t - z}\left[\frac{1}{(t - z)^k} - \frac{p(t)}{\pi_n(t)}\right] = \frac{1}{(t - z)^{k+1}} - \frac{A}{t - z} - \frac{q(t)}{\pi_n(t)}$$

($A = p(z)/\pi_n(z)$ and $q(t) = (p(t) - A\pi_n(t))/(t - z)$), so

$$\left\|\frac{1}{(t - z)^{k+1}} - \frac{A}{t - z} - \frac{q(t)}{\pi_n(t)}\right\|_{\mathring{\mu}} \leq M \left\|\frac{1}{(t - z)^k} - \frac{p(t)}{\pi_n(t)}\right\|_{\mathring{\mu}},$$

where M is a constant (depending on z), given by $M = \sup\{|t - z|^{-1} : t \in \partial\mathbb{O}\}$. For fixed $z \notin \partial\mathbb{O}$, it is a finite constant. As q/π_n and p/π_n are in $\mathcal{L}$, it follows by induction that $(t - z)^{-k} \in \mathcal{L}$ for $k = 1, 2, \ldots$. In a similar way, we also obtain that $[\varpi_z(t)]^{-k} \in \mathcal{L}$ for $k = 1, 2, \ldots$. Now, let $g \in L_2(\mathring{\mu})$ be such that

$$\int f(t)\overline{g(t)} \, d\mathring{\mu}(t) = 0 \tag{10.17}$$

for all $f \in \mathcal{L}$. In order to show that $\mathcal{L}_\infty$ is dense in $L_2(\mathring{\mu})$, it is sufficient to show that $g(t) = 0$ $\mathring{\mu}$-a.e. for $t \in \overline{\partial\mathbb{O}}$. To do so, set

$$d\mathring{\nu}(t) = \overline{g(t)} \, d\mathring{\mu}(t), \quad t \in \overline{\partial\mathbb{O}}.$$

Then $\mathring{v}$ is a complex measure of bounded variation since $g \in L_2(\mathring{\mu}) \subset L_1(\mathring{\mu})$. Therefore, when $C(t, z)$ is the Cauchy kernel,

$$H(z) = \int C(t, z) \, d\mathring{v}(t)$$

is analytic in $\mathbb{C} \setminus \partial \mathbb{O}$. As (10.17) holds for all $f \in \mathcal{L}$, it follows that

$$\int \frac{d\mathring{v}(t)}{(t - z)^k} = 0 \quad \text{and} \quad \int \frac{d\mathring{v}(t)}{[\varpi_z(t)]^k} = 0, \qquad k = 1, 2, \ldots.$$

But then $H(z) \equiv 0$ on $\mathbb{C} \setminus \partial \mathbb{O}$ because H and all its derivatives vanish at z and at $\hat{z}$. As in the proof of Corollary 10.3.3, it now follows that $d\mathring{v}(t) = g(t) \, d\mathring{\mu}(t) = 0$ or that $g(t) = 0$ $\mathring{\mu}$-a.e. on $\partial \mathbb{O}$. $\qquad \square$

The previous theorem says that $\mu \in \mathcal{M}$ is N-extremal at all $z \in \mathbb{O}_0 \cup \mathbb{O}_0^c$ iff μ is N-extremal in at least one $z \in \mathbb{O}_0 \cup \mathbb{O}_0^c$. Thus we can say that μ is N-extremal without specifying the z, so that we have

Theorem 10.3.5. *If $\mu \in \mathcal{M}$ is a solution of the moment problem, then μ is N-extremal iff $\mathcal{L}$ is dense in $L^2(\mathring{\mu})$.*

11

The boundary case

Whereas in the previous chapters, we have considered the situation where all the interpolation points α_k were in $\mathbb{O}$, we shall in this chapter consider the situation where the interpolation points α_k are all *on the boundary* $\overline{\partial\mathbb{O}}$.

We shall also start out from the beginning with an inner product that is defined by a linear functional M:

$$\langle f, g \rangle = M\{fg_*\}, \quad f, g \in \mathcal{L}_\infty.$$

When $\langle f, f \rangle \neq 0$ for all $f \neq 0$ that are in $\mathcal{L}$, then the functional is called quasi-definite and when, moreover, $\langle f, f \rangle > 0$ for $f \neq 0$, it is called positive definite. When $M\{f\} \in \mathbb{R}$ for all $f \in \mathcal{L}_\infty \cdot \mathcal{L}_{\infty*}$, then M is called real. We shall assume that M is real and positive definite.

We suppose that, with the new location of the α_k, the spaces $\mathcal{L}_n$ are as defined before and that the ϕ_n are the corresponding orthogonal rational functions, whenever they exist.

11.1. Recurrence for points on the boundary

The situation where the points are on the boundary is considerably different from the situation where they are not. One simple observation, for example, is that the Blaschke products can not be used as a basis anymore. Indeed, these are all equal to a constant of modulus one.

Recall that $\hat{z} = 1/\overline{z}$ for the circle and $\hat{z} = \overline{z}$ for the line. Let $\alpha_k, k = 1, 2, \ldots$ be a sequence of points that are all in $\overline{\partial\mathbb{O}}$. Note that then $\alpha_k = \hat{\alpha}_k$, whence we can write our definition of the spaces $\mathcal{L}_n$ as

$$\mathcal{L}_n = \left\{ \frac{p_n(z)}{\pi_n^*(z)} : p_n \in \mathcal{P}_n; \pi_n^*(z) = \prod_{k=1}^{n}(z - \alpha_k) \right\}.$$

Since we consider only a countable number of $\alpha_k \in \overline{\partial \mathbb{O}}$, we can always find a point $\alpha_\emptyset \in \partial \mathbb{O}$ such that $\alpha_k \neq \alpha_\emptyset$ for all $k = 0, 1, \ldots$. A simple transformation can bring this $\alpha_\emptyset$ to any position on $\partial \mathbb{O}$ that we would prefer. Thus it is not a real restriction to assume that, in the circular case, $\alpha_k \neq 1$ for all finite k. Thus we shall set by definition $\alpha_\emptyset = 1$ in the case of the circle. For the real line, this corresponds to all α_k being different from zero, and there we set $\alpha_\emptyset = 0$. The slash in the index symbolizes that it is a "forbidden" value for the α_k. In contrast, we had in the previous situation that, for the disk, the polynomials were a special case when all $\alpha_k = \alpha_0 = 0$. For interpolation on the boundary, we have in the situation of the real line that the polynomial case is recovered when all poles are at infinity. Therefore, it seems natural to set there $\alpha_0 = \infty$ and the corresponding point on the circle is $\alpha_0 = -1$. However, we still need the old meaning of α_0 and we shall denote it from now on as β. Thus $\beta = 0$ for the circle and $\beta = i$ for the line. These definitions make the point α_0 "less exceptional" because it now belongs to $\overline{\partial \mathbb{O}}$ just like all the other α_k, whereas β does not. This unfortunate, but necessary, change of notation will be used consistently, meaning that, whenever appropriate, an index 0 (like in the definition of Z_k below) will refer to the use of α_0 (new meaning) in the general definition. If we want to refer to β we shall use the index β (instead of the index 0 as we did before with the old meaning of α_0). For example, from now on $1/\varpi_0^*(z) = 1/(z+1)$ for $\mathbb{T}$ and $1/\varpi_0^*(z) = 0$ for $\mathbb{R}$, whereas $\varpi_\beta^*(z) = z$ for $\mathbb{T}$ and $\varpi_\beta^*(z) = z - i$ for $\mathbb{R}$. The table below summarizes our notational changes.

	Notation alert		
	α_0	β	$\alpha_\emptyset$
	old new	new	new
$\mathbb{T}$	0 -1	0	1
$\mathbb{R}$	i ∞	i	0

A remarkable observation is that in the present situation we have (with the same definition of the substar we had before)

$$\mathcal{L}_{n*} = \{f : f_* \in \mathcal{L}_n\} = \mathcal{L}_n.$$

Therefore, it is natural to consider a basis $\{b_k\}_{k=0}^n$ for $\mathcal{L}_n$ that satisfies $b_{k*} = b_k$. Such a basis is described by

$$b_0 = 1, \qquad b_n = \prod_{k=1}^n Z_k(z), \quad n = 1, 2, \ldots,$$

where (note that we use the definition below also for $k = 0$)

$$Z_k(z) = \frac{\mathbf{i}(1 - z)}{(\alpha_k - z)/(\alpha_k - 1)}, \quad k \geq 0 \quad \text{for } \mathbb{T},$$

$$Z_k(z) = \frac{z}{1 - z/\alpha_k}, \quad k \geq 0 \quad \text{for } \mathbb{R},$$

$$Z_k(z) = \mathbf{i}\frac{(z - \alpha_\emptyset)(\alpha_\emptyset - \alpha_k)}{(z - \alpha_k)(\beta - \alpha_\emptyset)}, \quad k \geq 0 \quad \text{for } \partial\mathbb{O}$$

or in an invariant notation

$$Z_k(z) = \frac{\mathbf{i}\varpi_\emptyset(z)/\varpi_\emptyset(\beta)}{\varpi_k(z)/\varpi_k(\alpha_\emptyset)} = \frac{\mathbf{i}\varpi_\emptyset^*(z)/\varpi_\emptyset^*(\beta)}{\varpi_k^*(z)/\varpi_k^*(\alpha_\emptyset)} = Z_{k*}(z), \quad k \geq 0. \tag{11.1}$$

Thus these basis functions indeed satisfy the relation $b_{n*} = b_n$. We shall use the notation $b(z)$ for the numerator in Z_k, that is, we set

$$b(z) = \mathbf{i}\frac{\varpi_\emptyset(z)}{\varpi_\emptyset(\beta)} = \begin{cases} \mathbf{i}(1 - z) & \text{for } \mathbb{T}, \\ z & \text{for } \mathbb{R} \end{cases}$$

and thus

$$Z_k = \frac{b(z)}{\varpi_k(z)/\varpi_k(\alpha_\emptyset)}.$$

Some very useful observations can be made here. First, note that with this basis, the polynomial case will appear naturally for the situation of the real line, by setting all $\alpha_k = \alpha_0 = \infty$. Indeed,

$$b_n(z) = z^n \prod_{k=1}^{n} \frac{1}{1 - z/\alpha_k} = z^n \quad \text{if } \alpha_k = \infty, \ k = 1, 2, \ldots.$$

We also note that for $n \geq 1$

$$b_n(z) = \frac{[b(z)]^n}{\pi_n^*(z)/\pi_n^*(\alpha_\emptyset)}.$$

Thus writing $f \in \mathcal{L}_n$ as $p_n(z)/\pi_n^*(z)$ or as $q_n(z)/[\pi_n^*(z)/\pi_n^*(\alpha_\emptyset)]$ is just a matter of a constant factor relating the polynomial numerators $p_n(z) = q_n(z)\pi_n^*(\alpha_\emptyset)$. We will use both possibilities, depending on what is the most convenient.

It is also useful to see that $1/Z_k(z)$ vanishes for $z = \alpha_k$.

Now let us consider as before a linear functional M, Hermitian and positive definite, which is defined on the space $\mathcal{R}_\infty = \mathcal{L}_\infty \cdot \mathcal{L}_\infty$, $\mathcal{L}_\infty = \cup_{n=0}^{\infty}\mathcal{L}_n$. This means that

$$M\{f_*\} = \overline{M\{f\}}, \quad f \in \mathcal{L}_\infty \cdot \mathcal{L}_\infty \quad \text{and} \quad M\{ff_*\} > 0, \quad 0 \neq f \in \mathcal{L}_\infty.$$

260   *11. The boundary case*

Note that in the previous situation, where the points α_k are not on the boundary, then $\mathcal{L}_n \cdot \mathcal{L}_{n*}$ is the same as $\mathcal{L}_n + \mathcal{L}_{n*}$, but this is no longer true in general for the boundary situation. We now have $\mathcal{R}_n = \mathcal{L}_n \cdot \mathcal{L}_n$. With the linear functional so defined, we can again introduce an inner product

$$\langle f, g \rangle_M = M\{f g_*\}, \quad f, g \in \mathcal{L}_\infty.$$

We can construct orthonormal functions $\phi_n \in \mathcal{L}_n$ with respect to this inner product and they can be expressed in terms of the b_k we have just introduced:

$$\phi_n(z) = \sum_{k=0}^n \beta_k^{(n)} b_k(z).$$

We assume that the orthonormal functions ϕ_n have a *leading coefficient* in the basis $\{b_k\}$ that is positive. We continue calling it $\kappa_n = \beta_n^{(n)}$. Also, the coefficient $\beta_{n-1}^{(n)}$ will play a special role and we shall also reserve a special notation for it: $\beta_{n-1}^{(n)} = \kappa_n'$. Thus

$$\phi_n(z) = \phi_n(\alpha_\emptyset) + \cdots + \kappa_n' b_{n-1}(z) + \kappa_n b_n(z).$$

It is easily seen that we can get κ_n and κ_n' from

$$\kappa_n = \left[\frac{\phi_n(z)}{b_n(z)}\right]_{z=\alpha_n} \quad \text{and} \quad \kappa_n + \frac{\kappa_n'}{Z_n(\alpha_{n-1})} = \left[\frac{\phi_n(z)}{b_n(z)}\right]_{z=\alpha_{n-1}}, \quad n \geq 1. \quad (11.2)$$

With the normalization $\kappa_n > 0$, we have

Lemma 11.1.1. The orthonormal functions ϕ_n have real coefficients with respect to the basis $\{b_k\}$ and $\phi_{n*} = \phi_n$.

Proof. Because $b_{k*} = b_k$, it is obvious that if the coefficients are real, also $\phi_{n*} = \phi_n$. The proof of real coefficients follows easily by induction. The result is true for $n = 0$. Suppose it is true for $i \leq n - 1$. Then by the Gram–Schmidt procedure

$$\phi_n = \chi_n / \|\chi_n\|, \quad \text{with} \quad \chi_n = b_n - \sum_{i=0}^{n-1} \gamma_i \phi_i, \quad \gamma_i = \langle b_n, \phi_i \rangle.$$

Using $M\{f_*\} = \overline{M\{f\}}$, $\langle f, g \rangle = M\{f g_*\}$, $b_{n*} = b_n$, and $\phi_{i*} = \phi_i$ for $i < n$, it follows that the coefficients $\gamma_i = \langle b_n, \phi_i \rangle = M\{b_n \phi_{i*}\} = \overline{M\{b_{n*}\phi_i\}} = \overline{M\{b_n \phi_{i*}\}} = \overline{\gamma}_i$ are real. Since ϕ_i has real coefficients with respect to the basis $\{b_k\}$, then χ_n and thus also ϕ_n will have real coefficients with respect to the basis $\{b_k\}$. $\qquad \square$

The notions "degenerate" and "exceptional" as defined in Section 4.5 coincide and are replaced by the notion "singular." We shall now call ϕ_n (and also its index n) *singular* when $\phi_n = p_n/\pi_n^*$ and $p_n(\alpha_{n-1}) = 0$ and *regular* otherwise. In the case of the real line, α_k can be ∞. A zero of p_k at ∞ then means that the degree of p_k is less than k.

We are now ready to formulate the recurrence relation.

Theorem 11.1.2. *For* $n = 2, 3, \ldots,$ *let* $\phi_k \in \mathcal{L}_k$, $k = n - 2, n - 1, n$ *be three successive orthonormal rational functions. Then* ϕ_{n-1} *and* ϕ_n *are regular if and only if there exists a recurrence relation of form*

$$\phi_n(z) = \left(A_n Z_n(z) + B_n \frac{Z_n(z)}{Z_{n-2}(z)} \right) \phi_{n-1}(z) + C_n \frac{Z_n(z)}{Z_{n-2}(z)} \phi_{n-2}(z), \quad (11.3)$$

with constants A_n, B_n, C_n *satisfying the conditions*

$$E_n = A_n + B_n/Z_{n-2}(\alpha_{n-1}) \neq 0, \quad (11.4)$$

$$C_n \neq 0. \quad (11.5)$$

Proof. First suppose that ϕ_n and ϕ_{n-1} are regular. Choose A_n arbitrary and define

$$W_n(z) = \frac{Z_{n-2}(z)}{Z_n(z)} \phi_n(z) - A_n Z_{n-2}(z) \phi_{n-1}(z).$$

Let $\phi_n(z) = q_n(z)/[\pi_n^*(z)/\pi_n^*(\alpha_\emptyset)]$. Then

$$W_n(z) = \frac{\alpha_\emptyset - \alpha_{n-2}}{z - \alpha_{n-2}} \frac{q_n(z) - A_n b(z) q_{n-1}(z)}{\pi_{n-1}^*(z)/\pi_{n-1}^*(\alpha_\emptyset)},$$

with $b(z)$ as defined above. Thus, if we choose

$$A_n = q_n(\alpha_{n-2})/[q_{n-1}(\alpha_{n-2}) b(\alpha_{n-2})] \quad (11.6)$$

we obtain that $W_n \in \mathcal{L}_{n-1}$. Recall that ϕ_{n-1} is regular and $\alpha_{n-2} \neq \alpha_\emptyset$, so that A_n is well defined; this is also true if $\alpha_{n-2} = \infty$. This implies that W_n can be written as

$$W_n(z) = B_n \phi_{n-1}(z) + C_n \phi_{n-2}(z) + \sum_{k=0}^{n-3} D_k \phi_k(z).$$

For $n = 2$, the sum is empty and the result is obvious. For $n \geq 3$, it is easily checked that $W_n \perp \mathcal{L}_{n-3}$, hence that all $D_k = 0$, $k = 0, \ldots, n - 3$. What then remains is equivalent with the formula (11.3).

Taking the numerator of this formula and putting $z = \alpha_{n-1}$ gives

$$q_n(\alpha_{n-1}) = b(\alpha_{n-1})[A_n + B_n/Z_{n-2}(\alpha_{n-1})]q_{n-1}(\alpha_{n-1}). \quad (11.7)$$

Because $q_n(\alpha_{n-1}) \neq 0$, this gives (11.4).

Observe then that $f_n(z) = b_{n-1}(z)/Z_n(z)$ is an element from $\mathcal{L}_{n-1}$. Thus, it is orthogonal to ϕ_n and we get

$$0 = \left\langle Z_n \left(A_n + \frac{B_n}{Z_{n-2}} \right) \phi_{n-1}, \frac{b_{n-1}}{Z_n} \right\rangle_M + C_n \left\langle \frac{Z_n}{Z_{n-2}} \phi_{n-2}, \frac{b_{n-1}}{Z_n} \right\rangle_M,$$

which can be written in the form (use $Z_{k*} = Z_k$)

$$0 = \left\langle \left(A_n + \frac{B_n}{Z_{n-2}} \right) b_{n-1}, \phi_{n-1} \right\rangle_M + C_n \left\langle \frac{b_{n-1}}{Z_{n-2}}, \phi_{n-2} \right\rangle_M.$$

The left factor in the first inner product is in $\mathcal{L}_{n-1}$ and its leading coefficient is $A_n + B_n/Z_{n-2}(\alpha_{n-1})$, which is nonzero. Thus the first inner product is nonzero and therefore also the second one will be nonzero. This implies that $C_n \neq 0$.

Conversely, suppose that (11.3)–(11.5) holds for some $n \geq 2$. Since $\phi_{n-1} \in \mathcal{L}_{n-1} \setminus \mathcal{L}_{n-2}$, it follows that $q_{n-1}(\alpha_{n-1}) \neq 0$. Therefore it follows by (11.4) and (11.7) that $q_n(\alpha_{n-1}) \neq 0$. Because A_n is well defined, it follows from (11.6) that $q_{n-1}(\alpha_{n-2}) \neq 0$. Hence ϕ_{n-1} and ϕ_n are regular.   □

Some particular cases in the situation of the real line are worth mentioning.

Corollary 11.1.3. *Suppose that we are in the case of the real line and $\alpha_n = \infty$ for all n. Then the ϕ_n are polynomials and they satisfy a recurrence relation of the form*

$$\phi_n(z) = (A_n z + B_n)\phi_{n-1}(z) + C_n\phi_{n-2}(z), \quad n = 2, 3, \ldots,$$

with

$$A_n \neq 0, \qquad C_n \neq 0, \quad n = 2, 3, \ldots.$$

Proof. Noting that in this case the sequence ϕ_n is always regular, the result follows immediately from the previous theorem.   □

We can also relate the recurrence of Theorem 11.1.2 to orthogonal Laurent polynomials, which are obtained by orthogonalizing the basis $\{1, z^{-1}, z, z^{-2}, z^2, \ldots\}$ with respect to a measure on $\mathbb{R}$. This result is not immediate though. The powers of z can be obtained by setting $\alpha_k = \infty$ in the previous definition of the basis

$\{b_n\}$. However, for the powers of z^{-1}, we need the "forbidden" value $\alpha_k = 0$. We should therefore give an alternative definition. So, setting

$$Z_k(z) = \frac{1 - \alpha_k z}{z - \alpha_k}$$

and

$$b_0 = 1; \qquad b_{2n} = Z_2 Z_4 \cdots Z_{2n}, n \geq 1; \qquad b_{2n+1} = Z_1 Z_3 \cdots Z_{2n+1}, n \geq 0,$$

we get the basis required here by using $\alpha_{2k} = \infty$ and $\alpha_{2k+1} = 0$. The proof of the previous theorem can now be adapted, according to this choice of the basis functions, and one obtains

Theorem 11.1.4. *Suppose we are in the case of the real line. Let ϕ_n be the orthonormal Laurent polynomials obtained by orthogonalization of $1, z^{-1}, z, z^{-2}, z^2, \ldots$. Assume that the sequence ϕ_n is regular. Then these Laurent polynomials satisfy recurrences*

$$\phi_{2k}(z) = (A_{2k}z + B_{2k})\phi_{2k-1}(z) + C_{2k}\phi_{2k-2}(z), \quad k = 1, 2, \ldots,$$

$$\phi_{2k+1}(z) = (A_{2k+1}/z + B_{2k+1})\phi_{2k}(z) + C_{2k+1}\phi_{2k-1}(z), \quad k = 1, 2, \ldots,$$

with $A_n \neq 0$ and $C_n \neq 0$.

Many more details and applications of these Laurent polynomials can be found for example in Refs. [118], [162], [119], [114], and [51].

Note that the latter recurrence relations for regular orthogonal Laurent polynomials are formally obtained from our general recurrences by setting $Z_0 = 1$, $Z_{2k} = z$, and $Z_{2k+1} = z^{-1}$ and $\alpha_{2k} = \infty$ and $\alpha_{2k+1} = 0$. The analog for the unit circle would be obtained with

$$Z_k(z) = \frac{z + \alpha_k}{z - \alpha_k}$$

and $\alpha_{2k} = -1$ and $\alpha_{2k+1} = 1$. This results in orthogonal Laurent polynomials in the variable $(z + 1)/(z - 1)$.

We shall now derive an explicit relation for the coefficient C_n from the recurrence relation. We introduce the expression

$$D_n = \frac{1}{Z_{n-2}(z)} - \frac{1}{Z_{n-1}(z)}.$$

We need the following equality, valid for $n \geq 2$, which can be obtained by some calculations:

$$D_n = i \frac{(\beta - \alpha_\emptyset)(\alpha_{n-1} - \alpha_{n-2})}{(\alpha_\emptyset - \alpha_{n-1})(\alpha_\emptyset - \alpha_{n-2})}. \tag{11.8}$$

Note that D_n is a constant not depending on z. Thus we can use

$$D_n = \frac{1}{Z_{n-2}(\alpha_{n-1})} = -\frac{1}{Z_{n-1}(\alpha_{n-2})} \in \mathbb{R}$$

for any $n \geq 2$ when it will be convenient. One application is to replace Z_{n-2}^{-1} by $D_n + Z_{n-1}^{-1}$ so that we can rewrite the recurrence relation (11.3) as

$$\phi_n(z) = \left(E_n Z_n(z) + B_n \frac{Z_n(z)}{Z_{n-1}(z)} \right) \phi_{n-1}(z) + C_n \frac{Z_n(z)}{Z_{n-2}(z)} \phi_{n-2}(z),$$
$$n = 2, 3, \ldots, \tag{11.9}$$

with E_n as in (11.4). The definition of E_n as given above will define E_n only when the recurrence relation holds, that is, when the system ϕ_n is regular. It is possible to define E_n independent of the regularity of the system as we do in the following lemma.

Lemma 11.1.5. Suppose the expansion of the orthonormal rational functions in the basis $\{b_k(z)\}$ is given by

$$\phi_n(z) = \kappa_n b_n(z) + \kappa_n' b_{n-1}(z) + \cdots.$$

For $n \geq 2$, define

$$E_n = \frac{1}{\kappa_{n-1}} \left[\kappa_n + \frac{\kappa_n'}{Z_n(\alpha_{n-1})} \right]. \tag{11.10}$$

Then $E_n \in \mathbb{R}$.

If the recurrence relation (11.3) holds, so that A_n and B_n are defined, then the E_n of (11.10) coincide with the $E_n = A_n + B_n D_n$ of (11.4). The latter also holds for $n = 1$, if we set by definition $A_1 = \kappa_1$ and $B_1 = \kappa_1'$.

Proof. The E_n are real since κ_n, κ_n', and κ_{n-1} are real and also $Z_n(\alpha_{n-1})$ is real for $n \geq 1$. To show that $E_n = A_n + B_n D_n$, we take the recurrence relation (11.3), divide it by $b_n(z)$, and set $z = \alpha_{n-1}$. With the identities (11.2) and using $1/Z_k(\alpha_k) = 0$, the given relation follows immediately. $\square$

Note that it follows from definition (11.10) and the formula (11.2) that n is a singular index iff $E_n = 0$. This statement does not depend on the existence of the recurrence relation.

We can now prove

Lemma 11.1.6. Suppose that the recurrence relation (11.3) holds with coefficients A_n, B_n, and C_n. Let D_n be as defined above and $E_n = A_n + B_n D_n$. Then

$$E_n = -C_n E_{n-1}, \quad n \geq 2. \tag{11.11}$$

Proof. Clearly $b_{n-1}/Z_n \in \mathcal{L}_{n-1}$ so that $\langle \phi_n, b_{n-1}/Z_n \rangle_M = 0$. Using the recurrence relation for ϕ_n, we get

$$0 = \langle \phi_n, b_{n-1}/Z_n \rangle_M$$
$$= A_n \langle b_{n-1}, \phi_{n-1} \rangle_M + B_n \left\langle \frac{b_{n-1}}{Z_{n-2}}, \phi_{n-1} \right\rangle_M + C_n \left\langle \frac{b_{n-1}}{Z_{n-2}}, \phi_{n-2} \right\rangle_M.$$

Using $\langle b_{n-2}, \phi_{n-1} \rangle_M = 0$ and $\langle b_i, \phi_i \rangle_M = 1/\kappa_i$, we have

$$0 = \frac{A_n}{\kappa_{n-1}} + \frac{B_n D_n}{\kappa_{n-1}} + \frac{C_n}{\kappa_{n-2}} + C_n D_n \langle b_{n-1}, \phi_{n-2} \rangle_M. \tag{11.12}$$

The remaining inner product can be evaluated when we use

$$b_{n-1}(z) = \frac{\phi_{n-1}(z)}{\kappa_{n-1}} - \frac{\kappa'_{n-1}}{\kappa_{n-1}} b_{n-2}(z) + \cdots,$$

so that

$$\langle b_{n-1}, \phi_{n-2} \rangle_M = \left\langle \frac{\phi_{n-1}}{\kappa_{n-1}}, \phi_{n-2} \right\rangle_M - \frac{\kappa'_{n-1}}{\kappa_{n-1}} \langle b_{n-2}, \phi_{n-2} \rangle_M = -\frac{\kappa'_{n-1}}{\kappa_{n-1}\kappa_{n-2}}. \tag{11.13}$$

When we combine (11.12) and (11.13), we get

$$0 = \frac{A_n}{\kappa_{n-1}} + \frac{B_n D_n}{\kappa_{n-1}} + \frac{C_n}{\kappa_{n-2}} - \frac{C_n D_n \kappa'_{n-1}}{\kappa_{n-1}\kappa_{n-2}}$$
$$= \frac{A_n + B_n D_n}{\kappa_{n-1}} + \frac{C_n}{\kappa_{n-2}} \left[1 - D_n \frac{\kappa'_{n-1}}{\kappa_{n-1}} \right]$$
$$= \frac{E_n}{\kappa_{n-1}} + \frac{C_n}{\kappa_{n-1}} \cdot \frac{1}{\kappa_{n-2}} \left[\kappa_{n-1} + \frac{\kappa'_{n-1}}{Z_{n-1}(\alpha_{n-2})} \right]$$
$$= \frac{1}{\kappa_{n-1}} [E_n + C_n E_{n-1}],$$

which gives us the expression we wanted. □

For solutions of the recurrence relation, we can prove a general summation theorem. To formulate this theorem, we define for $n \geq 1$

$$H(z, w) = \frac{1}{Z_{n-1}(z)} - \frac{1}{Z_{n-1}(w)} \qquad (11.14)$$

and find after some computations that

$$H(z, w) = \begin{cases} \mathrm{i} \dfrac{(z - w)(\beta - \alpha_\emptyset)}{(w - \alpha_\emptyset)(z - \alpha_\emptyset)} & \text{for } \partial \mathbb{O}, \\[3ex] -\mathrm{i} \dfrac{z - w}{(w - 1)(z - 1)} & \text{for } \mathbb{T}, \\[3ex] -\dfrac{z - w}{zw} & \text{for } \mathbb{R}. \end{cases}$$

Note that this expression does not depend on n. Furthermore, we set

$$H_n(z, w) = \frac{1}{Z_{n-2}(w)Z_{n-1}(z)} - \frac{1}{Z_{n-2}(z)Z_{n-1}(w)},$$

and hence

$$\begin{aligned} H_n(z, w) &= \left[\frac{1}{Z_{n-2}(w)Z_{n-1}(z)} - \frac{1}{Z_{n-2}(w)Z_{n-2}(z)} \right] \\ &\quad + \left[\frac{1}{Z_{n-2}(w)Z_{n-2}(z)} - \frac{1}{Z_{n-2}(z)Z_{n-1}(w)} \right] \\ &= \frac{-D_n}{Z_{n-2}(w)} + \frac{D_n}{Z_{n-2}(z)} = D_n H(z, w). \qquad (11.15) \end{aligned}$$

We now have the following analog of Theorem 4.3.1.

Theorem 11.1.7. *Let $x_n(z)$ and $y_n(z)$ be two solutions of the recurrence relation (11.3) and define*

$$F_n(z, w) = \frac{x_n(w)y_{n-1}(z)}{Z_n(w)Z_{n-1}(z)} - \frac{y_n(z)x_{n-1}(w)}{Z_n(z)Z_{n-1}(w)}.$$

Then, with $H(z, w)$ as in (11.14) and E_n as in (11.10),

$$\begin{aligned} F_n(z, w) &= y_{n-1}(z)x_{n-1}(w)H(z, w)E_n - C_n F_{n-1}(z, w) \\ &= \left[\sum_{k=1}^{n-1} y_k(z)x_k(w) \right] H(z, w)E_n + (-1)^n C_n C_{n-1} \cdots C_2 F_1(z, w). \end{aligned}$$

Proof. We use the recurrence relation for x_n and y_n in the definition of $F_n(z, w)$, which gives

$$F_n(z, w) = A_n x_{n-1}(w) y_{n-1}(z) \left[\frac{1}{Z_{n-1}(z)} - \frac{1}{Z_{n-1}(w)} \right]$$

$$+ B_n x_{n-1}(w) y_{n-1}(z) \left[\frac{1}{Z_{n-1}(z) Z_{n-2}(w)} - \frac{1}{Z_{n-1}(w) Z_{n-2}(z)} \right]$$

$$- C_n F_{n-1}(z, w).$$

Using the expressions (11.14) and (11.15), we find

$$F_n(z, w) = x_{n-1}(w) y_{n-1}(z) H(z, w) [A_n + B_n D_n] - C_n F_{n-1}(z, w)$$

$$= x_{n-1}(w) y_{n-1}(z) H(z, w) E_n - C_n F_{n-1}(z, w).$$

An induction argument leads to the result. □

It is possible to derive from this formula the Christoffel–Darboux type formulas given below. However, this would require that the system ϕ_n be regular, since it is based on the existence of the recurrence relation. It is possible, however, to prove the Christoffel–Darboux formulas without using the recurrence relation and only relying on the orthogonality properties of the ϕ_n. This is what we shall do in Section 11.3. We first introduce the functions of the second kind in the next section.

11.2. Functions of the second kind

As we did before, we shall associate with our orthonormal functions ϕ_n some functions $\psi_n \in \mathcal{L}_n$, which are called *functions of the second kind*. They are defined by

$$\psi_n(z) = M_t \{ D(t, z) [\phi_n(t) - \phi_n(z)] \} - \delta_{n0} \frac{Z_0(z)}{Z_0(\beta)},$$

where we used the notation M_t to indicate that M works on the argument as a function of t, considering z as a parameter. We first want to show that these functions ψ_n also satisfy the recurrence relation (11.3) whenever the ϕ_n do.

Theorem 11.2.1. *Suppose that the system of orthogonal rational functions ϕ_n is regular and let ψ_n be the functions of the second kind associated with them. Then these ψ_n satisfy the same recurrence relation (11.3) as the ϕ_n.*

Proof. We use the recurrence relation for $\phi_n(t)$ and $\phi_n(z)$ in the definition of ψ_n. This gives for $n \geq 2$

$$\psi_n(z) = A_n M_t \{ D(t, z) [Z_n(t)\phi_{n-1}(t) - Z_n(z)\phi_{n-1}(z)] \}$$

$$+ B_n M_t \left\{ D(t, z) \left[\frac{Z_n(t)}{Z_{n-2}(t)} \phi_{n-1}(t) - \frac{Z_n(z)}{Z_{n-2}(z)} \phi_{n-1}(z) \right] \right\}$$

$$+ C_n M_t \left\{ D(t, z) \left[\frac{Z_n(t)}{Z_{n-2}(t)} \phi_{n-2}(t) - \frac{Z_n(z)}{Z_{n-2}(z)} \phi_{n-2}(z) \right] \right\}$$

$$= A_n Z_n(z)\psi_{n-1}(z) + B_n \frac{Z_n(z)}{Z_{n-2}(z)} \psi_{n-1}(z) + C_n \frac{Z_n(z)}{Z_{n-2}(z)} \psi_{n-2}(z)$$

$$+ M_t \{ D(t, z) f_n(t, z) \} + \delta_{n2} \frac{Z_2(z)}{Z_0(\beta)} C_2, \tag{11.16}$$

with

$$f_n(t, z) = A_n [Z_n(t) - Z_n(z)]\phi_{n-1}(t)$$

$$+ \left[\frac{Z_n(t)}{Z_{n-2}(t)} - \frac{Z_n(z)}{Z_{n-2}(z)} \right] [B_n \phi_{n-1}(t) + C_n \phi_{n-2}(t)].$$

We note that

$$Z_n(t) - Z_n(z) = \mathbf{i} \frac{(t - z)(\alpha_\emptyset - \alpha_n)^2}{(t - \alpha_n)(z - \alpha_n)(\beta - \alpha_\emptyset)}$$

and

$$\frac{Z_n(t)}{Z_{n-2}(t)} - \frac{Z_n(z)}{Z_{n-2}(z)} = \frac{(t - z)(\alpha_\emptyset - \alpha_n)(\alpha_{n-2} - \alpha_n)}{(t - \alpha_n)(z - \alpha_n)(\alpha_\emptyset - \alpha_{n-2})}.$$

Therefore,

$$f_n(t, z) = \frac{(t - z)(\alpha_\emptyset - \alpha_n)}{(t - \alpha_n)(z - \alpha_n)} \left[A_n \frac{\mathbf{i}(\alpha_\emptyset - \alpha_n)}{\beta - \alpha_\emptyset} \phi_{n-1}(t) \right.$$

$$\left. + \frac{\alpha_{n-2} - \alpha_n}{\alpha_\emptyset - \alpha_{n-2}} [B_n \phi_{n-1}(t) + C_n \phi_{n-2}(t)] \right].$$

Note that in the argument of the linear functional in (11.16), the factor $t - z$ in the numerator of $f_n(t, z)$ cancels against the same factor in the denominator of $D(t, z)$.

Next we split $D(t, z)$ as $D_1(t, z) + D_2(t, z)$:

$$D(t, z) = \begin{cases} \dfrac{t - \alpha_n}{t - z} + \dfrac{z + \alpha_n}{t - z} & \text{for } \mathbb{T}, \\[3mm] \dfrac{-\mathrm{i}z(t - \alpha_n)}{t - z} + \dfrac{-\mathrm{i}(1 + z\alpha_n)}{t - z} & \text{for } \mathbb{R}, \\[3mm] \dfrac{c_n(t - \alpha_n)}{t - z} + \dfrac{c_n'}{t - z} & \text{for } \partial\mathbb{O}, \end{cases}$$

with $c_n = 1$ and $c_n' = z + \alpha_n$ for $\mathbb{T}$, whereas $c_n = -\mathrm{i}z$ and $c_n' = -\mathrm{i}(1 + z\alpha_n)$ for $\mathbb{R}$.
Using the orthogonality of the ϕ_k, we find

$$M_t\{D_1(t, z)f_n(t, z)\} = \begin{cases} 0 & \text{for } n \geq 3, \\[3mm] c_2 C_2 \dfrac{(\alpha_\emptyset - \alpha_2)(\alpha_0 - \alpha_2)}{(z - \alpha_2)(\alpha_\emptyset - \alpha_0)} & \text{for } n = 2. \end{cases}$$

For the second term with $D_2(t, z)$, we use again the recurrence relation to write $f_n(t, z)$ as

$$f_n(t, z) = \frac{t - z}{z - \alpha_n} \left[-\phi_n(t) + \mathrm{i}\frac{\alpha_\emptyset - \alpha_n}{\beta - \alpha_\emptyset} A_n \phi_{n-1}(t) \right.$$
$$\left. + \frac{\alpha_\emptyset - \alpha_n}{\alpha_\emptyset - \alpha_{n-2}} (B_n \phi_{n-1}(t) + C_n \phi_{n-2}(t)) \right].$$

Again, by the orthogonality of the ϕ_k, we get

$$M_t\{D_2(t, z)f_n(t, z)\} = \begin{cases} 0 & \text{for } n \geq 3, \\[3mm] c_2' C_2 \dfrac{\alpha_\emptyset - \alpha_2}{(z - \alpha_2)(\alpha_\emptyset - \alpha_0)} & \text{for } n = 2. \end{cases}$$

This proves the recurrence relation for $n \geq 3$ directly. For $n = 2$, we can put together all the terms involved to find that the recurrence relation is again satisfied because

$$\frac{Z_2(z)}{Z_0(\beta)} C_2 + c_2 C_2 \frac{(\alpha_\emptyset - \alpha_2)(\alpha_0 - \alpha_2)}{(z - \alpha_2)(\alpha_\emptyset - \alpha_0)} + c_2' C_2 \frac{\alpha_\emptyset - \alpha_2}{(z - \alpha_2)(\alpha_\emptyset - \alpha_0)} = 0. \quad \square$$

In analogy with Lemma 4.2.2, we can prove

Lemma 11.2.2. Let ϕ_n be the orthonormal system and ψ_n the associated functions of the second kind. For $n > 0$ and for any f such that (as a function of t) $D(t, z)[f(t) - f(z)] \in \mathcal{L}_{n-1}$ we have

$$\psi_n(z)f(z) = M_t\{D(t, z)[\phi_n(t)f(t) - \phi_n(z)f(z)]\}.$$

Proof. The proof is as in Lemma 4.2.2. $\hfill\square$

In particular, we can take for f any function in $\mathcal{L}_{n-1}$ or we can take it of the form $f(t) = g(t)(t - \alpha_n)/(t + z)$ for the case of $\mathbb{T}$ or of the form $f(t) = g(t)(t - \alpha_n)/(1 + tz)$ for the case of $\mathbb{R}$, where $g \in \mathcal{L}_n$.

We shall now give a Liouville–Ostrogradskii type determinant formula. Note that here we do not assume the system $\{\phi_n\}$ to be regular.

Theorem 11.2.3 (Determinant formula). *Let ϕ_n be the orthonormal functions and ψ_n the functions of the second kind. Define*

$$F_n(z, w) = \frac{\psi_n(w)}{Z_n(w)} \frac{\phi_{n-1}(z)}{Z_{n-1}(z)} - \frac{\phi_n(z)}{Z_n(z)} \frac{\psi_{n-1}(w)}{Z_{n-1}(w)}.$$

Then for $n \geq 1$ we have, with E_n as in (11.10),

$$F_n(z, z) = -E_n Z_0(\beta) \frac{\varpi_\beta(z)\varpi_\beta^*(z)}{\varpi_\emptyset(z)\varpi_\emptyset^*(z)},$$

or more explicitly,

$$F_n(z, z) = \frac{2iz}{(1 - z)^2} E_n \quad for\ \mathbb{T} \quad and \quad F_n(z, z) = -\frac{\mathbf{i}(1 + z^2)}{z^2} E_n \quad for\ \mathbb{R}.$$

Proof. We note that

$$\phi_{n-1}(z)[\phi_n(t) - \phi_n(z)] - \phi_n(z)[\phi_{n-1}(t) - \phi_{n-1}(z)]$$
$$= \phi_{n-1}(t)[\phi_n(t) - \phi_n(z)] - \phi_n(t)[\phi_{n-1}(t) - \phi_{n-1}(z)].$$

Multiply with $D(t, z)$ and apply M_t to get for the left-hand side:

$$\phi_{n-1}(z)\psi_n(z) - \phi_n(z)\psi_{n-1}(z).$$

for the right-hand side, we find

$$M_t\{\phi_{n-1}(t)D(t, z)[\phi_n(t) - \phi_n(z)]\} - M_t\{\phi_n(t)D(t, z)[\phi_{n-1}(t) - \phi_{n-1}(z)]\}.$$

Note that in the second term $D(t, z)[\phi_{n-1}(t) - \phi_{n-1}(z)] \in \mathcal{L}_{n-1}$, so that this term is zero by the orthogonality of ϕ_n. To compute the first term, we write

$$h(t) = D(t, z)[\phi_n(t) - \phi_n(z)] = \gamma_n b_n(t) + \gamma_n' b_{n-1}(t) + \cdots,$$

where

$$\gamma_n = \left[\frac{h(t)}{b_n(t)}\right]_{t=\alpha_n} = D(\alpha_n, z)\kappa_n$$

and

$$\gamma_n + \frac{\gamma_n'}{Z_n(\alpha_{n-1})} = \left[\frac{h(t)}{b_n(t)}\right]_{t=\alpha_{n-1}} = D(\alpha_{n-1}, z)\left[\kappa_n + \frac{\kappa_n'}{Z_n(\alpha_{n-1})}\right]$$

$$= D(\alpha_{n-1}, z)\kappa_{n-1}E_n.$$

Then, by the orthogonality of $\phi_{n-1} = \phi_{(n-1)*}$,

$$M\{\phi_{n-1}h\} = \gamma_n M\{\phi_{n-1}b_n\} + \gamma_n' M\{\phi_{n-1}b_{n-1}\}.$$

Because $\phi_k = \kappa_k b_k + \kappa_k' b_{k-1} + \cdots$, it follows by orthogonality that

$$M\{\phi_{n-1}b_n\} = -\frac{\kappa_n'}{\kappa_n\kappa_{n-1}} \quad \text{and} \quad M\{\phi_{n-1}b_{n-1}\} = \frac{1}{\kappa_{n-1}}.$$

Thus we obtain

$$M\{\phi_{n-1}h\} = -\gamma_n \frac{\kappa_n'}{\kappa_n\kappa_{n-1}} + \frac{\gamma_n'}{\kappa_{n-1}}$$

$$= -\frac{\gamma_n\kappa_n'}{\kappa_n\kappa_{n-1}} + \left[D(\alpha_{n-1}, z)E_n - \frac{\gamma_n}{\kappa_{n-1}}\right]Z_n(\alpha_{n-1})$$

$$= -D(\alpha_n, z)\frac{\kappa_n'}{\kappa_{n-1}} + D(\alpha_{n-1}, z)E_n Z_n(\alpha_{n-1})$$

$$\quad - D(\alpha_n, z)\kappa_n Z_n(\alpha_{n-1})$$

$$= -D(\alpha_n, z)E_n Z_n(\alpha_{n-1}) + D(\alpha_{n-1}, z)E_n Z_n(\alpha_{n-1})$$

$$= E_n Z_n(\alpha_{n-1})[D(\alpha_{n-1}, z) - D(\alpha_n, z)].$$

Working this out gives the result. Indeed, for the case $\mathbb{T}$ we have

$$D(t, z) - D(w, z) = \frac{t+z}{t-z} - \frac{w+z}{w-z} = \frac{2z(w-t)}{(t-z)(w-z)}.$$

With $t = \alpha_{n-1}$ and $w = \alpha_n$, this shows that

$$M\{\phi_{n-1}h\} = E_n Z_n(\alpha_{n-1})\frac{2z(\alpha_n - \alpha_{n-1})}{(\alpha_n - z)(\alpha_{n-1} - z)} = \frac{2iz}{(1-z)^2}E_n Z_n(z)Z_{n-1}(z).$$

For the case $\mathbb{R}$, we have

$$D(t, z) - D(w, z) = -i\left[\frac{1+tz}{t-z} - \frac{1+wz}{w-z}\right] = -i\frac{(w-t)(1+z^2)}{(t-z)(w-z)}.$$

With $t = \alpha_{n-1}$ and $w = \alpha_n$, this gives

$$M\{\phi_{n-1}h\} = -\mathrm{i}E_n Z_n(\alpha_{n-1})\frac{(\alpha_n - \alpha_{n-1})(1 + z^2)}{(\alpha_n - z)(\alpha_{n-1} - z)}$$

$$= -\frac{\mathrm{i}(1 + z^2)}{z^2} E_n Z_n(z) Z_{n-1}(z).$$

This completes the proof.   □

11.3. Christoffel–Darboux relation

We shall now derive the Christoffel–Darboux relation for the boundary situation.

Theorem 11.3.1 (Christoffel–Darboux relation). *Let ϕ_n be the orthonormal functions in $\mathcal{L}_n$ and let $H(z, w)$ and E_n be defined by (11.14) and (11.10). Then*

$$\frac{\phi_n(w)\phi_{n-1}(z)}{Z_n(w)Z_{n-1}(z)} - \frac{\phi_n(z)\phi_{n-1}(w)}{Z_n(z)Z_{n-1}(w)} = H(z, w)E_n \sum_{k=0}^{n-1} \phi_k(z)\phi_k(w).$$

Consequently, if we let $z \to w$, then we obtain

$$\frac{f_n'(z)f_{n-1}(z) - f_n(z)f_{n-1}'(z)}{\mathrm{i}\varpi_\emptyset^*(\beta)} = E_n \sum_{k=0}^{n-1} [\phi_k(z)]^2 = E_n k_{n-1}(z, z),$$

where

$$f_k(z) = \frac{\varpi_k^*(z)}{\varpi_k^*(\alpha_\emptyset)}\phi_k(z)$$

and the prime means derivative.

Proof. Define

$$g_n(z, w) = \frac{\phi_n(w)}{Z_n(w)}\frac{\phi_{n-1}(z)}{Z_{n-1}(z)} \quad \text{and} \quad F_n(z, w) = g_n(z, w) - g_n(w, z).$$

Set

$$F(w) = \frac{F_n(z, w)}{H(z, w)} = c\tilde{F}(w) \in \mathcal{L}_{n-1},$$

$$c = \mathrm{i}\frac{(\beta - \alpha_\emptyset)^2}{(\beta - \alpha_\emptyset)(\alpha_\emptyset - \alpha_n)(\alpha_\emptyset - \alpha_{n-1})},$$

$$\tilde{F}(w) = \frac{\tilde{g}_n(z, w) - \tilde{g}_n(w, z)}{z - w},$$

$$\tilde{g}_n(z, w) = \phi_n(w)\phi_{n-1}(z)(z - \alpha_{n-1})(w - \alpha_n).$$

We have to prove that $F(w) = \sum_{k=0}^{n-1} \gamma_k(z)\phi_k(w)$ with $\gamma_k(z) = E_n\phi_k(z)$. Since $F \in \mathcal{L}_{n-1}$ we can set

$$F(w) = \sum_{k=0}^{n-1} \gamma_k(z)\phi_k(w), \qquad \gamma_k(z) = \tilde{\gamma}_k(z)c, \qquad \tilde{\gamma}_k(z) = M\{\tilde{F}\,\phi_k\}.$$

But

$$\tilde{\gamma}_k(z) = M\{\tilde{F}\,\phi_k\} = M_w\{\tilde{F}(w)[\phi_k(w) - \phi_k(z)]\} + \phi_k(z)M\{\tilde{F}\}.$$

For the first term we find

$$M_w\{\tilde{F}(w)[\phi_k(w) - \phi_k(z)]\} = \phi_{n-1}(z)(z - \alpha_{n-1})q_n - \phi_n(z)(z - \alpha_n)q_{n-1},$$

with for $i = n, n - 1$

$$q_i = M\{\phi_i\,p_i\}, \qquad p_i(w) = \frac{w - \alpha_i}{z - w}[\phi_k(w) - \phi_k(z)] \in \mathcal{L}_{i-1}.$$

Hence, by orthogonality $q_i = 0$ and thus $\tilde{\gamma}_k(z) = \phi_k(z)M\{\tilde{F}\}$. It remains to show that $M\{F\} = cM\{\tilde{F}\} = E_n$. Note that

$$M\{\tilde{F}\} = (z - \alpha_{n-1})\phi_{n-1}(z)f_n(z) - (z - \alpha_n)\phi_n(z)f_{n-1}(z),$$

$$f_i(z) = M_w\left\{\frac{w - \alpha_i}{z - w}\phi_i(w)\right\}.$$

For $\partial\mathbb{O} = \mathbb{T}$ we add and subtract

$$(z - \alpha_n)(z - \alpha_{n-1})\phi_{n-1}(z)\phi_n(z)D(z, w)\frac{1}{2z}.$$

This leads to

$$M\{\tilde{F}\} = (z - \alpha_{n-1})\phi_{n-1}(z)h_n(z) - (z - \alpha_n)\phi_n(z)h_{n-1}(z), \quad (11.17)$$

where, using Lemma 11.2.2

$$h_i(z) = M_w\left\{D(z, w)\left[\frac{w - \alpha_i}{z + w}\phi_i(w) - \frac{z - \alpha_i}{2z}\phi_i(z)\right]\right\} = -\frac{z - \alpha_i}{2z}\psi_i(z).$$

For $\partial\mathbb{O} = \mathbb{R}$ we add and subtract

$$(z - \alpha_n)(z - \alpha_{n-1})\phi_{n-1}(z)\phi_n(z)D(z, w)\frac{\mathbf{i}}{1 + z^2},$$

which leads to the same formula (11.17), but now with

$$h_i(z) = \mathbf{i}M_w\left\{D(z, w)\left[\frac{w - \alpha_i}{1 + zw}\phi_i(w) - \frac{z - \alpha_i}{1 + z^2}\phi_i(z)\right]\right\} = -\frac{z - \alpha_i}{1 + z^2}\psi_i(z).$$

Or, in an invariant notation,

$$h_i(z) = -\frac{\mathbf{i}\varpi_i^*(z)\varpi_\emptyset(\beta)}{\varpi_\beta(z)\varpi_\beta^*(z)Z_0(\beta)}\psi_i(z).$$

Therefore,

$$M\{\tilde{F}\} = -\mathbf{i}\frac{\varpi_{n-1}^*(z)\varpi_n^*(z)\varpi_\emptyset(\beta)}{\varpi_\beta(z)\varpi_\beta^*(z)Z_0(\beta)}[\phi_{n-1}(z)\psi_n(z) - \psi_{n-1}(z)\phi_n(z)].$$

Finally, using the determinant formula of Theorem 11.2.3 to substitute inside the square brackets, we obtain after some computations

$$M\{F\} = cM\{\tilde{F}\} = E_n,$$

which is what remained to be shown.

The proof for the confluent formula is obtained by dividing out $H(z, w)$ and letting $w \to z$. We leave the details to the reader. □

Note that for the case of the real line and when all α_i are at infinity, we should find the Christoffel–Darboux relation for polynomials on the line as a special case. Noting that then $Z_k(z) = z$ and $1/Z_k(\alpha_i) = 0$, we get the classical relation for polynomials

$$\phi_n(w)\phi_{n-1}(z) - \phi_n(z)\phi_{n-1}(w) = (w - z)\frac{\kappa_n}{\kappa_{n-1}}\overline{k_{n-1}(z, w)},$$

where $k_{n-1}(z, w) = \sum_{k=0}^{n-1}\phi_k(z)\overline{\phi_k(w)}$ is the reproducing kernel for $\mathcal{L}_{n-1} = \mathcal{P}_{n-1}$. For the confluent formula, the factor $\varpi_k^*(z)/\varpi_k^*(\beta)$ is in the polynomial case just 1, so that f_k is equal to the orthogonal polynomial ϕ_k, and $\mathbf{i}\varpi_\emptyset^*(\beta) = -1$. Thus we find again a classical formula.

A bivariate generalization of the determinant formula is formulated as follows.

Theorem 11.3.2. *Let ϕ_n be the orthogonal functions for $\mathcal{L}_n$ and ψ_n the associated functions of the second kind. Let $H(z, w)$ be as in (11.14), E_n as in (11.10),* and $D(z, w)$ *the usual Riesz–Herglotz–Nevanlinna kernel. Then*

$$\frac{\psi_n(w)\phi_{n-1}(z)}{Z_n(w)Z_{n-1}(z)} - \frac{\phi_n(z)\psi_{n-1}(w)}{Z_n(z)Z_{n-1}(w)} = \left[\sum_{k=1}^{n-1} \phi_k(z)\psi_k(w) - D(z, w)\right] H(z, w)E_n.$$

Note that for $w \to z$, this reduces to the determinant formula.

Proof. Define as in the previous proof

$$\tilde{g}_n(z, w) = (w - \alpha_n)(z - \alpha_{n-1})\phi_n(w)\phi_{n-1}(z)$$

and set $G(z, w) = \tilde{g}_n(z, w) - \tilde{g}_n(w, z)$. Now apply M_t to the expression

$$F(t, z, w) = \begin{cases} \dfrac{D(t, w)}{t + w}G(t, z) - \dfrac{D(t, w)}{2w}G(w, z) & \text{for } \mathbb{T}, \\[3mm] \dfrac{D(t, w)}{-i(1 + tw)}G(t, z) - \dfrac{D(t, w)}{-i(1 + w^2)}G(w, z) & \text{for } \mathbb{R}. \end{cases}$$

Then we obtain by Lemma 11.2.2

$$M_t\{F(t, z, w)\} = (z - \alpha_n)(w - \alpha_{n-1})\phi_n(z)\psi_{n-1}(w)$$
$$- (w - \alpha_n)(z - \alpha_{n-1})\psi_n(w)\phi_{n-1}(z).$$

But, noting that with the notation of the previous theorem

$$G(z, w) = (z - w)\tilde{F}(w) = \frac{z - w}{cH(z, w)}\left[\frac{\phi_n(w)\phi_{n-1}(z)}{Z_n(w)Z_{n-1}(z)} - \frac{\phi_n(z)\phi_{n-1}(w)}{Z_n(z)Z_{n-1}(w)}\right],$$

we find by the Christoffel–Darboux formula that

$$G(z, w) = dE_n \sum_{k=0}^{n-1} \phi_k(z)\phi_k(w),$$

with

$$d = \frac{z - w}{c} = -i\frac{(z - w)(\alpha_\emptyset - \alpha_n)(\alpha_\emptyset - \alpha_{n-1})}{\beta - \alpha_\emptyset}.$$

Using this expression for $G(z, w)$ in the definition of $F(t, z, w)$ leads as before by Lemma 11.2.2 to the following result in the case that $\partial\mathbb{O} = \mathbb{T}$:

$$M_t\{F(t, z, w)\} = \frac{1}{c}E_n\left[\sum_{k=1}^{n-1}\phi_k(z)(w - z)\psi_k(w) + (z + w)\right]$$
$$= -dE_n\left[\sum_{k=1}^{n-1}\phi_k(z)\psi_k(w) - D(z, w)\right].$$

For $\partial \mathbb{O} = \mathbb{R}$ the final result is the same, but in the derivation, the last term $(z + w)$ at the end of the first line is replaced by $-\mathrm{i}(1 + zw)$. Equating the two expressions for $M_t\{F(t, z, w)\}$ gives a formula that is equivalent to the formula that had to be proved.

For $z = w$, the determinant formula follows because $H(z, z) = 0$ whereas $H(z, w)D(z, w)$ gives indeed the required expression since the $z - w$ in the numerator of $H(z, w)$ cancels against the $z - w$ in the denominator of $D(z, w)$.
$\qquad\qquad\qquad\qquad\qquad\qquad\qquad\qquad\qquad\qquad\qquad\qquad\qquad\qquad$ $\square$

Also for the functions of the second kind, we can obtain Christoffel–Darboux relations by applying the trick of the previous proof once more.

Theorem 11.3.3. *Let ψ_n be the functions of the second kind. Then, in analogy with the Christoffel–Darboux relation, we have*

$$\frac{\psi_n(w)\psi_{n-1}(z)}{Z_n(w)Z_{n-1}(z)} - \frac{\psi_n(z)\psi_{n-1}(w)}{Z_n(z)Z_{n-1}(w)} = H(z, w)E_n\left[\sum_{k=1}^{n-1}\psi_k(z)\psi_k(w) - 1\right],$$

with $H(z, w)$ as in (11.14) and E_n as in (11.10).

Proof. We set now

$$G(z, w) = (w - \alpha_n)(z - \alpha_{n-1})\phi_n(w)\psi_{n-1}(z)$$
$$\qquad\qquad - (z - \alpha_n)(w - \alpha_{n-1})\phi_{n-1}(w)\psi_n(z).$$

With this $G(z, w)$, we define $F(t, z, w)$ as in the previous proof. By applying M_t on $F(t, z, w)$ we find with the help of Lemma 11.2.2

$$M_t\{F(t, z, w)\} = (w - \alpha_n)(z - \alpha_{n-1})\psi_n(w)\psi_{n-1}(z)$$
$$\qquad\qquad - (z - \alpha_n)(w - \alpha_{n-1})\psi_{n-1}(w)\psi_n(z).$$

However, we have by the previous theorem (we use the notation introduced there)

$$G(z, w) = dE_n\left[\sum_{k=1}^{n-1}\phi_k(w)\psi_k(z) - D(z, w)\right].$$

Substituting this in the definition of $F(t, z, w)$ and applying M_t to it gives, as in the previous proof,

$$M_t\{F(t, z, w)\} = dE_n\left[\sum_{k=1}^{n-1}\psi_k(z)\psi_k(w) - 1\right].$$

Setting the two expressions for $M_t\{F(t, z, w)\}$ equal to each other gives a formula that is equivalent with the required identity. □

Note that again here we can let $z \to w$, to obtain a formula containing derivatives just as in the Christoffel–Darboux case.

Finally, we give an identity that is obtained by combination of the three previous theorems.

Theorem 11.3.4. *Let ϕ_n be the orthonormal functions and let ψ_n be the functions of the second kind. Define for some complex number s*

$$\chi_n(z; s) = \psi_n(z) + s\phi_n(z).$$

Then for arbitrary complex s and t,

$$\frac{\chi_n(w; t)\chi_{n-1}(z; s)}{Z_n(w)Z_{n-1}(z)} - \frac{\chi_n(z; s)\chi_{n-1}(w; t)}{Z_n(z)Z_{n-1}(w)}$$

$$= H(z, w)E_n \left[\sum_{k=1}^{n-1} \chi_k(z; s)\chi_k(w; t) + [st - 1 + D(z, w)(t - s)] \right],$$

with $H(z, w)$ as in (11.14) and E_n as in (11.10).

Proof. This is directly obtained by working out the left-hand side and using the three previous theorems. □

11.4. Green's formula

We can give a complex version of the formulas given in the previous section. The same method is used in the proofs with complex conjugates at the appropriate places. Therefore, we shall not include the details but only give the main steps. First we need the analogs of the expressions (11.14) and (11.15). They are

$$\tilde{H}(z, w) = \frac{1}{Z_n(z)} - \frac{1}{\overline{Z_n(w)}} = \begin{cases} -\mathrm{i}\dfrac{\varpi_w(z)\overline{\varpi_\emptyset(\beta)}}{\varpi_\emptyset(z)\overline{\varpi_\emptyset(w)}} & \text{for } \partial\mathbb{O}, \\[2ex] \dfrac{-\mathrm{i}(1 - z\overline{w})}{(1 - z)(1 - \overline{w})} & \text{for } \mathbb{T}, \\[2ex] \dfrac{-(z - \overline{w})}{z\overline{w}} & \text{for } \mathbb{R}. \end{cases} \quad (11.18)$$

This relation holds for any $n \geq 0$. Note that $\tilde{H}(z, w) = H_*(z, w)$, where the substar is with respect to z. For $w = z$, we have $\tilde{H}(z, z) = 2P(\alpha_\emptyset, z)/Z_0(\beta)$, where $P(t, z)$ is the Poisson kernel.

Also, for $n \geq 2$,

$$\tilde{H}_n(z, w) = \frac{1}{Z_{n-2}(w)Z_{n-1}(z)} - \frac{1}{Z_{n-1}(w)Z_{n-2}(z)} = D_n\tilde{H}(z, w).$$

(11.19)

We now give without proof the following complex analog of Theorem 11.1.7.

Theorem 11.4.1 (Green's formula). *Let $x_n(z)$ and $y_n(z)$ both be solutions of the recurrence relation (11.3) and define*

$$G_n(z, w) = \overline{\left(\frac{x_n(w)}{Z_n(w)}\right)}\left(\frac{y_{n-1}(z)}{Z_{n-1}(z)}\right) - \left(\frac{y_n(z)}{Z_n(z)}\right)\overline{\left(\frac{x_{n-1}(w)}{Z_{n-1}(w)}\right)}.$$

Then, with $\tilde{H}(z, w)$ as in (11.18) and E_n as in (11.10),

$$G_n(z, w) = y_{n-1}(z)\overline{x_{n-1}(w)}\tilde{H}(z, w)E_n - C_nG_{n-1}(z, w)$$

$$= \left[\sum_{k=1}^{n-1} y_k(z)\overline{x_k(w)}\right]\tilde{H}(z, w)E_n + (-1)^nC_nC_{n-1}\cdots C_2G_1(z, w).$$

This result holds because the numbers A_n, B_n, and C_n are real. As a corollary of this formula, one can find the summation formulas in the next theorem. However, when derived from the Green formula, they would only be proved for the case of a regular system, that is, when the recurrence relation holds. With the techniques of the previous section, the same result can be derived using only the orthogonality of the ϕ_n. However, a much simpler technique is to take the substar of the corresponding relations in Theorems 11.3.1–11.3.3 of the previous section, taking into account that

$$\phi_{n*} = \phi_n \quad \text{and} \quad \psi_{n*} = -\psi_n.$$

We give the result without further proof.

Theorem 11.4.2. *With the notation for $\tilde{H}(z, w)$ and E_n of the previous theorem and with ϕ_n the orthonormal functions and ψ_n the functions of the second kind, we get*

$$\overline{\left(\frac{\phi_n(w)}{Z_n(w)}\right)}\left(\frac{\phi_{n-1}(z)}{Z_{n-1}(z)}\right) - \left(\frac{\phi_n(z)}{Z_n(z)}\right)\overline{\left(\frac{\phi_{n-1}(w)}{Z_{n-1}(w)}\right)}$$

$$= \left[\sum_{k=0}^{n-1} \phi_k(z)\overline{\phi_k(w)}\right]\tilde{H}(z, w)E_n,$$

$$\overline{\left(\frac{\psi_n(w)}{Z_n(w)}\right)} \left(\frac{\psi_{n-1}(z)}{Z_{n-1}(z)}\right) - \left(\frac{\psi_n(z)}{Z_n(z)}\right) \overline{\left(\frac{\psi_{n-1}(w)}{Z_{n-1}(w)}\right)}$$

$$= \left[\sum_{k=1}^{n-1} \psi_k(z)\overline{\psi_k(w)} + 1\right] \tilde{H}(z, w) E_n,$$

$$\overline{\left(\frac{\psi_n(w)}{Z_n(w)}\right)} \left(\frac{\phi_{n-1}(z)}{Z_{n-1}(z)}\right) - \left(\frac{\phi_n(z)}{Z_n(z)}\right) \overline{\left(\frac{\psi_{n-1}(w)}{Z_{n-1}(w)}\right)}$$

$$= \left[\sum_{k=1}^{n-1} \phi_k(z)\overline{\psi_k(w)} - D_*(z, w)\right] \tilde{H}(z, w) E_n,$$

where in the last equation, the substar is with respect to z.

Note that this also holds for $w = z$. In that case

$$D_*(z, z) = \frac{1 + |z|^2}{1 - |z|^2} \quad \text{for } \mathbb{T} \quad \text{and} \quad D_*(z, z) = \mathbf{i}\frac{1 + |z|^2}{z - \bar{z}} \quad \text{for } \mathbb{R},$$

which corresponds to (recall that $\zeta_\beta(z) = z$ for $\mathbb{T}$ and $\zeta_\beta(z) = (z-\mathbf{i})/(z+\mathbf{i})$ for $\mathbb{R}$)

$$D_*(z, z) = \frac{1 + |\zeta_\beta(z)|^2}{1 - |\zeta_\beta(z)|^2}.$$

The previous relations can be combined as in the real case to give the following:

Theorem 11.4.3. *Let ϕ_n be the orthonormal functions and ψ_n the functions of the second kind. For any complex s, we set*

$$\chi_n(z; s) = \psi_n(z) + s\phi_n(z).$$

Then, for arbitrary complex s and t,

$$\overline{\left(\frac{\chi_n(w; t)}{Z_n(w)}\right)} \left(\frac{\chi_{n-1}(z; s)}{Z_{n-1}(z)}\right) - \left(\frac{\chi_n(z; s)}{Z_n(z)}\right) \overline{\left(\frac{\chi_{n-1}(w; t)}{Z_{n-1}(w)}\right)}$$

$$= \left[\sum_{k=1}^{n-1} \overline{\chi_k(w; t)}\chi_k(z; s) + 1 + s\bar{t} - (s + \bar{t})D_*(z, w)\right] \tilde{H}(z, w) E_n.$$

In particular, for $z = w$ and $s = t$

$$\overline{\left(\frac{\chi_n(z; s)}{Z_n(z)}\right)} \left(\frac{\chi_{n-1}(z; s)}{Z_{n-1}(z)}\right) - \left(\frac{\chi_n(z; s)}{Z_n(z)}\right) \overline{\left(\frac{\chi_{n-1}(z; s)}{Z_{n-1}(z)}\right)}$$

$$= \left[\sum_{k=1}^{n-1} |\chi_k(z; s)^2 + |1 - s|^2\right] \tilde{H}(z, z) E_n + (s + \bar{s}) E_n X(z),$$

$$(11.20)$$

where

$$X(z) = Z_0(\beta)\frac{|z - \beta|^2}{|z - \alpha_\emptyset|^2}. \tag{11.21}$$

Proof. By the previous theorem, we can evaluate the left-hand side and obtain the first formula. For $z = w$ and $s = t$, this equals

$$\left[\sum_{k=1}^{n-1} |\chi_k(z; s)|^2\right] \tilde{H}(z, z)E_n + E_n Y(z; s),$$

with

$$Y(z; s) = \tilde{H}(z, z)[1 - (s + \bar{s})D_*(z, z) + |s|^2]$$

$$= \tilde{H}(z, z)|1 - s|^2 + (s + \bar{s})\tilde{H}(z, z)[1 - D_*(z, z)].$$

In the case of $\mathbb{T}$ we get

$$\tilde{H}(z, z)[1 - D_*(z, z)] = \mathbf{i}\frac{1 - |z|^2}{|1 - z|^2}\left[\frac{1 + |z|^2}{1 - |z|^2} - 1\right] = \mathbf{i}\frac{2|z|^2}{|1 - z|^2},$$

which gives the result. Similarly, we find for $\mathbb{R}$

$$\tilde{H}(z, z)[1 - D_*(z, z)] = \frac{z - \bar{z}}{|z|^2}\left[-\mathbf{i}\frac{1 + |z|^2}{\bar{z} - z} - 1\right]$$

$$= \frac{\bar{z} - z + \mathbf{i}(1 + |z|^2)}{|z|^2} = \mathbf{i}\frac{|z - \mathbf{i}|^2}{|z|^2}.$$

In both cases we get the last term in (11.20).   □

11.5. Quasi-orthogonal functions

We shall introduce in this section the quasi-orthogonal functions, which are the analogs of the para-orthogonal functions in the situation where the α_k are not on the boundary.

Let ϕ_n be the orthonormal functions with poles on the boundary $\partial\mathbb{O}$. We define the quasi-orthogonal functions as

$$Q_n(z, \tau) = \phi_n(z) + \tau\frac{Z_n(z)}{Z_{n-1}(z)}\phi_{n-1}(z), \quad \tau \in \mathbb{R}, \ n \geq 1. \tag{11.22}$$

We set by definition

$$Q_n(z, \infty) = \frac{Z_n(z)}{Z_{n-1}(z)}\phi_{n-1}(z).$$

Our immediate aim is to prove that $Q_n(z, \tau)$ has n simple zeros that are all on $\overline{\partial \mathbb{O}}$. We shall need several lemmas before we will be able to prove this. For the formulations to follow, it will be convenient to say that a polynomial $p_n \in \mathcal{P}_n$ has a zero at $z = \infty$ when p_n^* has a zero at $z = 0$, that is, when its degree is less than n. Thus a polynomial is understood to have a zero at ∞ when its degree is defective. So, in the proofs below, where it is allowed that a polynomial has a zero at ∞, the reader should modify the proofs given for finite zeros whenever it is necessary. Usually this is dealt with by setting in a factorization of a polynomial $p_n(z) = a(z - t_1) \cdots (z - t_n)$, the factor $z - t_i$ for $t_i = \infty$ equal to 1.

The following lemma is a simple observation.

Lemma 11.5.1. Let $Q_n(z, \tau) = p_n(z, \tau)/\pi_n^*(z)$ be a quasi-orthogonal function. Then we have

1. $Q_n(z, \tau) = Q_{n*}(z, \tau)$.
2. If $\xi \in \overline{\mathbb{C}} \setminus \overline{\partial \mathbb{O}}$ is a zero of order m for $p_n(z, \tau)$, then $\hat{\xi}$ is also a zero of order m.

Proof. We check each of the claims.

1. This equation is obvious from the definition.
2. Since $\xi \notin \overline{\partial \mathbb{O}}$, it is a zero of $p_n(z, \tau)$ iff it is a zero of $Q_n(z, \tau)$. For a zero ξ that is not 0 or ∞, with multiplicity $m = 1$, the result is a direct consequence of the previous relation. Assuming $m > 1$ it is clear that $h(z) = Q_n(z, \tau)/[(z - \xi)(z - \hat{\xi})]$ will again have the property that $h(\zeta) = 0$ implies $h(\hat{\zeta}) = 0$, etc.

   A simple adaptation of this argument will show that the result also holds for $\xi = 0$ or ∞. $\square$

We are now in a position to prove the following:

Lemma 11.5.2. Let $Q_n(z, \tau) = p_n(z, \tau)/\pi_n^*(z)$ be a quasi-orthogonal function. Then the polynomial $p_n(z, \tau)$ has n simple zeros on $\overline{\partial \mathbb{O}}$.

Proof. For notational reasons, it is awkward to give an invariant proof. Therefore we give separate derivations for the case $\mathbb{R}$ and the case $\mathbb{T}$.

1. The case $\mathbb{R}$. This is the simplest case. Let $t_1, \ldots, t_l$ be all the zeros lying on $\mathbb{R}$, which have odd multiplicity $2m_i + 1$, with $m_i \geq 0$. All the other zeros (on $\mathbb{R}$ or not) will come in pairs $(\xi_i, \overline{\xi}_i)$, $i = 1, \ldots, k$. The couple of zeros $(\xi_i, \overline{\xi}_i)$ is repeated if its order is larger than 1. (We recall that t_i can be ∞, in which case we replace in what follows $z - t_i$ by 1, and a similar observation holds for $\xi_i = \infty$.) Thus there is some nonzero constant c such that

$$p_n(z, \tau) = c \prod_{i=1}^{l} (z - t_i)^{2m_i+1} \prod_{i=1}^{k} [(z - \xi_i)(z - \overline{\xi}_i)]$$

$$= c \prod_{i=1}^{l} (z - t_i)^{2m_i+1} \prod_{i=1}^{k} [(z - \xi_i)(z - \xi_i)_*]. \qquad (11.23)$$

Set $M = m_1 + \cdots + m_l$. Then $n = l + 2M + 2k$. We have to prove that all $m_i = 0$ and $k = 0$, or, equivalently, because all these numbers are nonnegative integers, we have to prove that $M + k \geq 1$ leads to a contradiction. Assume therefore that $M + k \geq 1$. Define the function

$$T(z) = \frac{(z - t_1) \cdots (z - t_l)(z - \alpha_n)}{(z - \alpha_1) \cdots (z - \alpha_{n-1})}.$$

We shall prove first that T is orthogonal to Q_n. Because $l + 1 = n - 2(M + k) + 1 \leq n - 1$, it follows that $T \in \mathcal{L}_{n-1}$. Thus $\langle T, \phi_n \rangle_M = 0$. Clearly

$$S(z) = \frac{z - \alpha_{n-1}}{z - \alpha_n} T(z) \in \mathcal{L}_{n-2},$$

so that

$$\langle S(z), \phi_{n-1}(z) \rangle_M = \left\langle T(z), \frac{z - \alpha_{n-1}}{z - \alpha_n} \phi_{n-1}(z) \right\rangle_M = 0. \qquad (11.24)$$

Therefore (use the definition of $Q_n(z, \tau)$), we obtain $\langle T, Q_n \rangle_M = 0$. This is however impossible because (make use of (11.23))

$$\langle Q_n, T \rangle_M = c \langle W, W \rangle_M \neq 0,$$

where

$$W(z) = \frac{(z - t_1)^{m_1+1} \cdots (z - t_l)^{m_l+1} (z - \xi_1) \cdots (z - \xi_k)}{(z - \alpha_1) \cdots (z - \alpha_{n-1})} \not\equiv 0.$$

Thus we come to a contradiction and this means that $M + k = 0$, that is, all $m_i = 0$, $i = 1, \ldots, l$ and $k = 0$. Thus all the zeros are simple and are in $\overline{\mathbb{R}}$.

2. *The case* $\mathbb{T}$. The difficulty is here that the substar transformation introduces powers of z, which should be taken care of by considering the multiplicity of $z = 0$ as a zero of $p_n(z, \tau)$.

As in part 1, we let t_i, $i = 1, \ldots, l$ denote the zeros of $p_n(z, \tau)$ of odd multiplicity $2m_i + 1$, which are on $\mathbb{T}$. Let $z = 0$ be a zero of multiplicity m.

Hence, by Lemma 11.5.1, also $z = \infty$ will be a zero of order m for $Q_n(z, \tau)$, that is, $p_n(z, \tau)$ will have degree at most $n - m$. All the other zeros will come in (possibly repeated) pairs $(\xi_i, 1/\overline{\xi}_i)$, $i = 1, \ldots, k$. Thus, again setting $M = m_1 + \cdots + m_l$, we have $2k + 2M + 2m + l = n$. We have to prove that $M + m + k \geq 1$ leads to a contradiction. Define the function

$$T(z) = \frac{(z - t_1) \cdots (z - t_l)(z - \alpha_n)z^{M+m+k-1}}{(z - \alpha_1) \cdots (z - \alpha_{n-1})}.$$

If $M + k + m \geq 1$, then the numerator has degree $l + M + m + k = n - (M + m + k) \leq n - 1$, and thus $T \in \mathcal{L}_{n-1}$, so that $\langle T, \phi_n \rangle_M = 0$. Again setting $S(z) = (z - \alpha_{n-1})T(z)/(z - \alpha_n) \in \mathcal{L}_{n-2}$, we arrive at (11.24), so that T is orthogonal to Q_n. However, because for a nonzero constant c

$$p_n(z, \tau) = cz^m \prod_{i=1}^{l}(z - t_i)^{2m_i+1} \prod_{i=1}^{k}[(z - \xi_i)(z - 1/\overline{\xi}_i)]$$

$$= c\frac{(-1)^{k+M}t_1^{m_1} \cdots t_l^{m_l}}{\overline{\xi}_1 \cdots \overline{\xi}_k}z^{m+k+M} \prod_{i=1}^{l}[(z - t_i)^{m_i+1}(z - t_i)_*^{m_i}$$

$$\times \prod_{i=1}^{k}[(z - \xi_i)(z - \xi_i)_*]$$

and also

$$z - \alpha_n = -(z - \alpha_n)_*/(z\overline{\alpha}_n),$$

so that we may write again $\langle T, Q_n \rangle_M = c'\langle W, W \rangle_M \neq 0$, where c' is some (nonzero) constant and

$$W(z) = \frac{(z - t_1)^{m_1+1} \cdots (z - t_l)^{m_l+1}(z - \xi_1) \cdots (z - \xi_k)}{(z - \alpha_1) \cdots (z - \alpha_{n-1})} \not\equiv 0.$$

This contradicts $\langle T, Q_n \rangle_M = 0$, so that we may conclude that $M + m + k = 0$ and hence that $m = k = 0$ and all $m_i = 0$, $i = 1, \ldots, l$. Thus all the zeros are on $\mathbb{T}$ and they are simple. □

Note that the previous theorem holds for any real value of τ and thus also for $\tau = 0$, so that we have also proved that the numerators $p_n(z) = p_n(z, 0)$ of the orthogonal rational functions ϕ_n have n simple zeros, which are all on $\partial\mathbb{O}$. The problem is that some of these zeros may coincide with one of the $\{\alpha_k\}_{k=1}^{n-2}$, and this is something that can not be excluded. Since the zeros of p_n are simple, such a zero will cancel against the corresponding factor in the denominator of ϕ_n and so ϕ_n will not have a zero at that point.

We have, however, the following lemma.

Lemma 11.5.3. If ϕ_n is regular, then the numerators of ϕ_n and ϕ_{n-1} have no common zero.

Proof. Set $\phi_n(z) = p_n(z)/\pi_n^*(z)$. The adaptations for the real line when some of the α_k are ∞ can be made as mentioned above. We leave the details of that to the reader. We have to prove that a common zero w, that is, $p_n(w) = p_{n-1}(w) = 0$, leads to a contradiction. Obviously $w \neq \alpha_n$, for otherwise ϕ_n would be in $\mathcal{L}_{n-1}$. For similar reasons, w can not be α_{n-1}. It turns out that w can only be one of $\alpha_1, \ldots, \alpha_{n-2}$. Indeed, if w were in $\partial\mathbb{O} \setminus \{\alpha_1, \ldots, \alpha_n\}$, then, recalling the definition $f_k(z) = \phi_k(z)\varpi_k^*(z)/\varpi_k^*(\alpha_0)$ from the confluent form of the Christoffel–Darboux Theorem 11.3.1, this would mean that f_n and f_{n-1} had a common zero. Thus the left-hand side in the confluent Christoffel–Darboux formula would vanish for $z = w$ and this is impossible since the right-hand side is not zero because $E_n \neq 0$ if ϕ_n is regular.

Thus if $p_n(w) = p_{n-1}(w) = 0$, then the only possibility is that $w \in \{\alpha_1, \ldots, \alpha_{n-2}\}$. Assume $w = \alpha_k$ with $1 \le k \le n - 2$. Then we see immediately that

$$\phi_n(z)\frac{z - \alpha_n}{z - \alpha_k} \in \mathcal{L}_{n-1} \quad \text{and} \quad \phi_{n-1}(z)\frac{z - \alpha_n}{z - \alpha_k} \in \mathcal{L}_{n-1}.$$

However, if we denote $h(z) = (z - \alpha_n)/(z - \alpha_k)$, then

$$\langle \phi_{n-1}, h\phi_n \rangle = \langle h\phi_{n-1}, \phi_n \rangle = 0$$

because $\phi_{n-1}h \in \mathcal{L}_{n-1} \perp \phi_n$. Therefore, $h\phi_n \in \mathcal{L}_{n-2}$. If we denote the $n - 1$ zeros of $p_n(z)$ that are not equal to α_k by $t_1, \ldots, t_{n-1}$, then

$$h(z)\phi_n(z) = c\frac{(z - t_1) \cdots (z - t_{n-1})(z - \alpha_n)}{(z - \alpha_1) \cdots (z - \alpha_n)} \in \mathcal{L}_{n-2},$$

with c a nonzero constant, and this can only be when $\alpha_{n-1} \in \{t_1, \ldots, t_{n-1}\}$. Thus $p_n(\alpha_{n-1}) = 0$, and this contradicts the regularity of ϕ_n.   □

Now we turn to the zeros of the quasi-orthogonal functions.

Lemma 11.5.4. Suppose that ϕ_n is regular and let $Q_n(z, \tau)$ be a quasi-orthogonal function. Then for a fixed $w \in \overline{\partial\mathbb{O}}$, the numerator of $Q_n(z, \tau)$ will vanish at $z = w$ for exactly one $\tau \in \overline{\mathbb{R}}$.

Proof. Set $Q_n(z, \tau) = p_n(z, \tau)/\pi_n^*(z)$ and $\phi_n(z) = p_n(z)/\pi_n^*(z)$. Then

$$p_n(z, \tau) = p_n(z) + \tau \frac{\varpi_{n-1}(z)}{\varpi_{n-1}(\alpha_\emptyset)} p_{n-1}(z).$$

Suppose $p_n(w, \tau) = 0$, If $w = \alpha_{n-1}$, then $\tau = \infty$ is the only possibility because otherwise $p_n(\alpha_{n-1})$ would be zero, contradicting the regularity of ϕ_n.

For $w \neq \alpha_{n-1}$, there are two possibilities: either $p_{n-1}(w) \neq 0$, and then there is a unique solution for τ making $p_n(w, \tau) = 0$, namely

$$\tau = - \frac{p_n(w)\varpi_{n-1}(\alpha_\emptyset)}{p_{n-1}(w)\varpi_{n-1}(w)},$$

or $p_{n-1}(w) = 0$, in which case $\tau = \infty$ is the only possibility because for any finite τ, we would have $p_n(w, \tau) = p_n(w)$ and this can not be zero by Lemma 11.5.3. □

The following result is an immediate consequence, which is obvious without proof.

Lemma 11.5.5. Assume that ϕ_n is regular and let $Q_n(z, \tau) = p_n(z, \tau)/\pi_n^*(z)$ be a corresponding quasi-orthogonal function. Then, except for at most $n + 1$ values of $\tau \in \overline{\mathbb{R}}$, none of the points in $\{\alpha_0, \alpha_1, \ldots, \alpha_n\}$ are zeros of $p_n(z, \tau)$.

We shall call the values of τ for which the conclusion in the previous lemma is true, *regular values of* τ. Note that $\tau = \infty$ can never be a regular value, because a regular value implies that $p_n(\alpha_{n-1}, \tau) \neq 0$, whereas by definition this is always zero for $\tau = \infty$. Thus excluding the zeros $\{\alpha_0, \alpha_1, \ldots, \alpha_{n-2}, \alpha_n\}$ gives another set of n conditions that exclude at most n finite values as being regular. Thus all (finite) real values of, except at most n ones, are regular. We shall call $Q_n(z, \tau)$ regular when ϕ_n is regular and τ is a regular value for $Q_n(z, \tau)$.

From Lemmas 11.5.2 and 11.5.5 and recalling that all, except n values of $\tau \in \mathbb{R}$, are regular values for $Q_n(z, \tau)$, we can formulate the following:

Corollary 11.5.6. Assume that $Q_n(z, \tau)$ is regular and thus that the orthogonal functions ϕ_n is regular and that τ is a regular value. Then $Q_n(z, \tau)$ has n simple zeros, all lying on $\overline{\partial\mathbb{O}} \setminus \{\alpha_0, \alpha_1, \ldots, \alpha_n\}$.

Proof. This follows immediately from the previous lemmas and the definition of a regular value τ. □

11.6. Quadrature formulas

We shall construct in this section quadrature formulas for the linear functional $M\{f\} = \int f \, d\hat{\mu}$. Assume that the quasi-orthogonal function $Q_n(z, \tau)$ is regular and thus that the sequence $\{\phi_n\}$ is regular and that τ is a regular value for $Q_n(z, \tau)$. Let us denote by $\xi_i = \xi_{ni}(\tau)$, $i = 1, \ldots, n$ the n simple zeros of the regular quasi-orthogonal function $Q_n(z) = Q_n(z, \tau)$. We know that $\xi_i \in \overline{\partial \mathbb{O}} \setminus \{\alpha_0, \alpha_1, \ldots, \alpha_n\}$ for all i. Our aim is to construct quadrature formulas of the form

$$M\{f\} \approx \sum_{i=1}^{n} \lambda_{ni}(\tau) f(\xi_{ni}(\tau)),$$

where the $\lambda_{ni}(\tau)$ are appropriate positive weights.

The fundamental interpolating functions are exactly as in Chapter 5:

$$L_{ni}(z) = \frac{\pi_{n-1}^*(\xi_i)}{\pi_{n-1}^*(z)} \prod_{k \neq i}^{n} \frac{z - \xi_k}{\xi_i - \xi_k} = \frac{\varpi_n^*(z)}{\varpi_n^*(\xi_i)} \frac{Q_n(z)}{(z - \xi_i) Q_n'(\xi_i)} \in \mathcal{L}_{n-1}$$

(prime is derivative). For the situation where one of the ξ_i is at ∞, recall the comments at the beginning of Section 11.5.

The weights $\lambda_{ni} = \lambda_{ni}(\tau)$ are defined by

$$\lambda_{ni} = M\{L_{ni}\}.$$

Since we have in the present situation $\mathcal{L}_n = \mathcal{L}_{n*}$, the spaces $\mathcal{R}_n = \mathcal{L}_n \cdot \mathcal{L}_n$ are

$$\mathcal{R}_n = \left\{ \frac{p(z)}{[\pi_n^*(z)]^2} : p \in \mathcal{P}_{2n} \right\} = \left\{ \frac{p_n(z) q_n(z)}{\pi_n^*(z) \pi_n^*(z)}, \ p_n, q_n \in \mathcal{P}_n \right\}.$$

We have the following theorem.

Theorem 11.6.1. *The quadrature formula*

$$I_n\{f\} = \sum_{i=1}^{n} \lambda_{ni} f(\xi_i),$$

with nodes and weights as defined above, has domain of validity $\mathcal{R}_{n-1}$. *The weights are positive.*

Proof. The proof is practically the same as the proof of Theorem 5.3.1. Define the auxiliary function

$$h(z) = f(z) - \sum_{i=1}^{n} f(\xi_i) L_{ni}(z).$$

Clearly, it can be written as

$$h(z) = \frac{p_{2n-2}(z)}{\pi^*_{n-1}(z)\pi^*_{n-1}(z)} = \frac{(z-\alpha_n)p_{2n-2}(z)}{\pi^*_n(z)\pi^*_{n-1}(z)}, \qquad p_{2n-2} \in \mathcal{P}_{2n-2}.$$

Since $h(\xi_i) = 0$ for $i = 1, \ldots, n$ (interpolation property), and since the numerator of $Q_n(z)$ is a nonzero constant times the nodal polynomial $(z - \xi_1) \cdots (z - \xi_n)$, we can write

$$h(z) = Q_n(z)g(z), \qquad g(z) = \frac{(z-\alpha_n)q_{n-2}(z)}{\pi^*_{n-1}(z)}, \qquad q_{n-2} \in \mathcal{P}_{n-2}.$$

By its form, $g \in \mathcal{L}_{n-1}$. Hence $g_* \in \mathcal{L}_{n-1}$ while

$$\frac{z - \alpha_{n-1}}{z - \alpha_n} g(z) \in \mathcal{L}_{n-2}$$

and thus

$$\left[\frac{z - \alpha_{n-1}}{z - \alpha_n} g(z)\right]_* \in \mathcal{L}_{n-2}.$$

This implies that

$$M\{h\} = c_1 \langle \phi_n, g_* \rangle_M + c_2 \left\langle \phi_{n-1}(z), \frac{z - \alpha_{n-1}}{z - \alpha_n} g_*(z) \right\rangle_M = 0$$

(where c_1 and c_2 are constants). Thus the quadrature is exact for all $f \in \mathcal{R}_{n-1}$.

To prove that the weights are positive, note that $L_{ni*}(z) = L_{ni}(z)$. Set

$$k(z) = L_{ni}(z)[L_{ni}(z) - 1] \in \mathcal{R}_{n-1}.$$

Because $k(\xi_i) = 0$, for all $i = 1, \ldots, n$, we have $M\{k\} = 0$ because the quadrature formula is exact. Therefore,

$$\lambda_{ni} = M\{L_{ni}\} = M\{L_{ni}L_{ni*}\} = \langle L_{ni}, L_{ni} \rangle_M > 0.$$

This concludes the proof. $\qquad\qquad\qquad\qquad\qquad\qquad\qquad\qquad\qquad$ □

Theorem 11.6.2. *Assume that the system $\{\phi_n\}$ is regular and that $\tau = 0$ is a regular value. Then ϕ_n has n simple zeros on $\overline{\partial\mathbb{O}}$ and the corresponding quadrature formula*

$$\sum_{k=1}^{n} \lambda_{nk}(0) f(\xi_{nk}(0))$$

is exact in $\mathcal{L}_n \cdot \mathcal{L}_{n-1}$.

Proof. Indeed, as in the previous proof, it holds for any $f \in \mathcal{L}_n \cdot \mathcal{L}_{n-1}$, that is, for f of the form

$$f(z) = \frac{p_n(z) p_{n-1}(z)}{\pi_n^*(z) \pi_{n-1}^*(z)}, \quad p_k \in \mathcal{P}_k,$$

that

$$h(z) = f(z) - \sum_{i=1}^{n} f(\xi_i) L_{ni}(z) = \frac{q_n(z) q_{n-1}(z)}{\pi_n^*(z) \pi_{n-1}^*(z)} \in \mathcal{L}_n \cdot \mathcal{L}_{n-1}, \quad q_k \in \mathcal{P}_k.$$

Because $\tau = 0$, the $\xi_i = \xi_{ni}(0)$ are the zeros of ϕ_n. Hence

$$h(z) = \phi_n(z) \frac{r_{n-1}(z)}{\pi_{n-1}^*(z)} = \phi_n(z) g(z), \quad r_{n-1} \in \mathcal{P}_{n-1}, \ g \in \mathcal{L}_{n-1}.$$

Therefore, $M\{h\} = \langle \phi_n, g_* \rangle_M = 0$ since $g \in \mathcal{L}_{n-1} = \mathcal{L}_{(n-1)*}$. This means that

$$M\{f\} = \sum_{k=1}^{n} f(\xi_i) M\{L_{ni}\}, \quad \forall f \in \mathcal{L}_n \cdot \mathcal{L}_{n-1}.$$

This concludes the proof. □

The situation of the previous theorem is comparable with the classical situation where all $\alpha_k = \infty$ in the case of the real line. It is known that constructing quadrature formulas with $\tau = 0$ gives formulas exact in the set of polynomials $\mathcal{P}_{2n-1}$ rather than in the set $\mathcal{P}_{2n-2}$, obtained when $\tau \neq 0$. See Akhiezer [2, p. 21]. The quadrature with $\tau \neq 0$ integrates exactly in a subspace of one dimension lower than in the case $\tau = 0$. This can be explained by the fact that, for $\tau \neq 0$, the orthogonal functions are replaced by quasi-orthogonal functions, which have an orthogonality defect of one.

As in Section 5.4, we can derive alternative expressions for the weights. Therefore, we introduce functions $P_n(z)$ of the second kind, associated with the quasi-orthogonal functions $Q_n(z)$ as

$$P_n(z) = P_n(z, \tau) = \psi_n(z) + \tau \frac{Z_n(z)}{Z_{n-1}(z)} \psi_{n-1}(z), \quad \tau \in \overline{\mathbb{R}}, \ n \geq 1,$$

with the same convention as for the quasi-orthogonal functions to define

$$P_n(z, \infty) = \psi_n(z) \frac{Z_n(z)}{Z_{n-1}(z)}.$$

Note that although we call them functions of the second kind, they are *not* given by $\int D(t, z)[Q_n(t) - Q_n(z)] d\hat{\mu}(t)$.

The formulation and the proof of Theorem 5.4.1 can be given without change. It still holds that the following is true.

Theorem 11.6.3. *The weights of the quadrature formula are given by*

$$\lambda_{nk} = \frac{1}{2} \frac{\varpi_\beta(\beta)}{\varpi_\beta(\xi_k)\varpi_\beta^*(\xi_k)} \frac{P_n(\xi_k)}{Q_n'(\xi_k)},$$

where the prime means differentiation with respect to z, Q_n is a regular quasi-orthogonal function with zeros ξ_k, $k = 1, \ldots, n$, and the P_n is the associated function of the second kind.

We need an adaptation of the proof of Theorem 5.4.2, to cope with the boundary case.

Theorem 11.6.4. *With the notation of the previous theorem*

$$\lambda_{nk} = \left[\sum_{j=0}^{n-1} |\phi_j(\xi_k)|^2 \right]^{-1}.$$

Proof. Since $Q_n(\xi_k) = 0$, we get from the first formula in Theorem 11.4.2 with $w = \xi_k$ and $z = t$

$$\frac{\overline{\phi_{n-1}(\xi_k)}}{\overline{Z_{n-1}(\xi_k)}Z_n(t)}[Q_n(\xi_k) - Q_n(t)] = \left[\sum_{j=0}^{n-1} \phi_j(t)\overline{\phi_j(\xi_k)} \right] \tilde{H}(t, \xi_k) E_n.$$

Multiplying by $\overline{Z_{n-1}(\xi_k)}Z_n(t)/(\xi - t)$ and letting t tend to ξ_k, we get

$$\overline{\phi_{n-1}(\xi_k)}Q_n'(\xi_k) = k_{n-1}(\xi_k, \xi_k) E_n \lim_{t \to \xi_k} \frac{\tilde{H}(t, \xi_k)\overline{Z_{n-1}(\xi_k)}Z_n(t)}{\xi_k - t},$$

where $k_{n-1}(z, z) = \sum_{j=0}^{n-1} |\phi_j(z)|^2$. Let L be the limit in the right-hand side. Then

$$L = \lim_{t \to \xi_k} \frac{\tilde{H}(t, \xi_k)\overline{Z_{n-1}(\xi_k)}Z_n(t)}{\xi_k - t} = \begin{cases} \dfrac{-\mathbf{i}}{\varpi_\beta^*(\alpha_\emptyset)} \dfrac{\varpi_n^*(\alpha_\emptyset)\varpi_{n-1}^*(\alpha_\emptyset)}{\varpi_n^*(\xi_k)\varpi_{n-1}^*(\xi_k)} & \text{for } \partial\mathbb{O}, \\[4mm] -\mathbf{i}\dfrac{(\alpha_n - 1)(\alpha_{n-1} - 1)}{(\alpha_n - \xi_k)(\alpha_{n-1} - \xi_k)} & \text{for } \mathbb{T}, \\[4mm] \dfrac{\alpha_n\alpha_{n-1}}{(\alpha_n - \xi_k)(\alpha_{n-1} - \xi_k)} & \text{for } \mathbb{R}. \end{cases}$$

$$(11.25)$$

Similarly, using the complex conjugate of the third formula in Theorem 11.4.2 we obtain

$$\frac{\overline{\phi_{n-1}(\xi_k)}}{\overline{Z_{n-1}(\xi_k)}Z_n(t)}P_n(t) = \left[\sum_{j=0}^{n-1}\psi_j(t)\overline{\phi_j(\xi_k)} - \overline{D_*(\xi_k,t)}\right]\overline{\tilde{H}}\,(\xi_k,t)E_n.$$

We multiply by $\overline{Z_{n-1}(\xi_k)}Z_n(t)$ and let $t \to \xi_k$. Because

$$\lim_{t\to\xi_k}\overline{\tilde{H}}\,(\xi_k,t)\overline{Z_{n-1}(\xi_k)}Z_n(t) = 0,$$

the sum immediately drops out in the right-hand side. So there remains

$$\overline{\phi_{n-1}(\xi_k)}P_n(\xi_k) = -E_n\lim_{t\to\xi_k}[\overline{D_*(\xi_k,t)}\overline{\tilde{H}}\,(\xi_k,t)\overline{Z_{n-1}(\xi_k)}Z_n(t)],$$

and this limit equals (L is the limit in (11.25))

$$\lim_{t\to\xi_k}[\overline{D_*(\xi_k,t)}\overline{\tilde{H}}\,(\xi_k,t)\overline{Z_{n-1}(\xi_k)}Z_n(t)] = \begin{cases} -2L\dfrac{\varpi_\beta(\xi_k)\varpi_\beta^*(\xi_k)}{\varpi_\beta(\beta)} & \text{for } \partial\mathbb{O}, \\[3mm] \dfrac{2i\xi_k(\alpha_n-1)(\alpha_{n-1}-1)}{(\alpha_n-\xi_k)(\alpha_{n-1}-\xi_k)} & \text{for } \mathbb{T}, \\[3mm] \dfrac{i(1+\xi_k^2)\alpha_n\alpha_{n-1}}{(\alpha_n-\xi_k)(\alpha_{n-1}-\xi_k)} & \text{for } \mathbb{R}. \end{cases}$$

$$(11.26)$$

Plugging the limits (11.25) and (11.26) into the expressions for $Q_n'(\xi_k)$ and $P_n(\xi_k)$ and then substituting the results in the expression for λ_{nk} of the previous theorem gives the result. $\qquad\square$

For further reference, we shall denote by $\mu_n(\cdot,\tau)$ the discrete measure of these quadrature formulas, that is, the measure that takes masses $\lambda_{ni}(\tau)$ at the nodes $\xi_{ni}(\tau)$. We also recall that such a measure exists for all, except at most n, values $\tau \in \mathbb{R}$.

11.7. Nested disks

If $Q_n(z,\tau)$ are the quasi-orthogonal functions, then we recall the functions of the second kind associated with them are given by

$$P_n(z,\tau) = \psi_n(z) + \tau\frac{Z_n(z)}{Z_{n-1}(z)}\psi_{n-1}(z).$$

Furthermore, we define the rational functions

$$R_n(z, \tau) = -\frac{P_n(z, \tau)}{Q_n(z, \tau)}. \tag{11.27}$$

When n is a regular index, then the mapping from τ to $s = s_n(z) = R_n(z, \tau)$ will transform (for a fixed $z \in \overline{\mathbb{C}} \setminus \partial\mathbb{O}$) the extended real line onto a circle $K_n(z)$. The closed disk defined by $K_n(z)$ and its interior will be denoted as $\Delta_n(z)$. Note that this is well defined as a circle with finite radius because $Q_n(z, \tau)$ has all its zeros on $\partial\mathbb{O}$, so that for $z \notin \overline{\partial\mathbb{O}}$, $Q_n(z, \tau) \neq 0$ for all $\tau \in \overline{\mathbb{R}}$ and thus $s_n(z)$ will never be infinite.

When n is a singular index, then the transformation is degenerate. In that case, the whole plane is mapped to a point. Indeed, since for a singular index n we have $E_n = 0$, it follows from the Christoffel–Darboux relation that

$$\frac{\phi_n(w)}{Z_n(w)}\frac{\phi_{n-1}(z)}{Z_{n-1}(z)} = \frac{\phi_n(z)}{Z_n(z)}\frac{\phi_{n-1}(w)}{Z_{n-1}(w)}$$

for any z and w. Choose w such that $\phi_{n-1}(w)/Z_{n-1}(w) \neq 0$. Then we see that there should be a constant c_n such that

$$\phi_n(z) = c_n \frac{Z_n(z)}{Z_{n-1}(z)}\phi_{n-1}(z).$$

Since it follows similarly from Theorem 11.3.3 that the same holds for the functions of the second kind ψ_n, we get

$$R_n(z, \tau) = -\frac{P_n(z, \tau)}{Q_n(z, \tau)} = -\frac{(c_n Z_n/Z_{n-1} + \tau)\psi_{n-1}(z)}{(c_n Z_n/Z_{n-1} + \tau)\phi_{n-1}(z)} = -\frac{\psi_{n-1}(z)}{\phi_{n-1}(z)}.$$

This is independent of τ and thus we have a point instead of a disk.

We have the following theorem:

Theorem 11.7.1. *Suppose that the index n is regular. Then for $z \in \overline{\mathbb{C}} \setminus \partial\mathbb{O}$*

1. The equation of the circle $K_n(z)$ is given by

$$\sum_{k=1}^{n-1} |\psi_k(z) + s\phi_k(z)|^2 + |1 - s|^2 = (s + \bar{s})\frac{2|\zeta_\beta(z)|^2}{1 - |\zeta_\beta(z)|^2}.$$

2. The (closed) disk $\Delta_n(z)$ is obtained by replacing the equality sign by $\leq$.
3. The center c_n and the radius r_n of the disk are

$$c_n = -\frac{\left(\frac{\psi_n}{Z_n}\right)\overline{\left(\frac{\phi_{n-1}}{Z_{n-1}}\right)} - \left(\frac{\psi_{n-1}}{Z_{n-1}}\right)\overline{\left(\frac{\phi_n}{Z_n}\right)}}{\left(\frac{\phi_n}{Z_n}\right)\overline{\left(\frac{\phi_{n-1}}{Z_{n-1}}\right)} - \left(\frac{\phi_{n-1}}{Z_{n-1}}\right)\overline{\left(\frac{\phi_n}{Z_n}\right)}}$$

$$= \frac{1}{k_{n-1}(z, z)}\left[\frac{1 + |\zeta_\beta(z)|^2}{1 - |\zeta_\beta(z)|^2} - \sum_{k=1}^{n-1}\overline{\phi_k(z)}\psi_k(z)\right]$$

and

$$r_n = \frac{\left| \left(\frac{\psi_n}{Z_n}\right)\left(\frac{\phi_{n-1}}{Z_{n-1}}\right) - \left(\frac{\psi_{n-1}}{Z_{n-1}}\right)\left(\frac{\phi_n}{Z_n}\right) \right|}{\left| \left(\frac{\phi_n}{Z_n}\right)\left(\frac{\phi_{n-1}}{Z_{n-1}}\right) - \left(\frac{\phi_{n-1}}{Z_{n-1}}\right)\left(\frac{\phi_n}{Z_n}\right) \right|}$$

$$= \frac{1}{k_{n-1}(z,z)} \left| \frac{2\zeta_\beta(z)}{1 - |\zeta_\beta(z)|^2} \right|,$$

where $k_{n-1}(z,z) = \sum_{k=0}^{n-1} |\phi_k(z)|^2$.

4. The circle $K_n(z)$ is in the right (left) half plane iff $z \in \mathbb{O}$ ($z \in \mathbb{O}^e$).

5. If m is also a regular index and $m > n$, then $\Delta_m(z) \subset \Delta_n(z)$.

6. The circles $K_n(z)$ and $K_{n-1}(z)$ touch.

Proof. We can solve the relation $s = R_n(z, \tau)$ for τ to get

$$\tau = -\frac{Z_{n-1}(z)}{Z_n(z)} \frac{\psi_n(z) + s\phi_n(z)}{\psi_{n-1}(z) + s\phi_{n-1}(z)}$$

$$= -\frac{|Z_{n-1}|^2}{Z_n \overline{Z}_{n-1}} \frac{[\psi_n + s\phi_n][\overline{\psi_{n-1} + s\phi_{n-1}}]}{|\psi_{n-1} + s\phi_{n-1}|^2}.$$

Taking the imaginary part gives

$$\operatorname{Im} \tau = \frac{1}{2i} \frac{|Z_{n-1}|^2}{|\psi_{n-1} + s\phi_{n-1}|^2} g_n(z, s),$$

with

$$g_n(z, s) = \frac{(\psi_n + s\phi_n)(\overline{\psi_{n-1} + s\phi_{n-1}})}{\overline{Z}_n Z_{n-1}} - \frac{(\overline{\psi_n + s\phi_n})(\psi_{n-1} + s\phi_{n-1})}{Z_n \overline{Z}_{n-1}}.$$

Now suppose that the index n is regular. Then with the help of Green's formula (11.20) (since n is a regular index and hence $E_n \neq 0$) we can write the equation of the circle $K_n(z)$ as

$$\sum_{k=1}^{n-1} |\psi_k(z) + s\phi_k(z)|^2 + |1 - s|^2 = (s + \overline{s}) \frac{-X(z)}{\tilde{H}(z, z)},$$

with $X(z)$ as given in Theorem 11.4.3 by (11.21). Thus

$$\frac{-X(z)}{\tilde{H}(z, z)} = \begin{cases} \dfrac{2|z|^2}{1 - |z|^2} & \text{for } \mathbb{T}, \\[2ex] \dfrac{|z - i|^2}{2 \operatorname{Im} z} & \text{for } \mathbb{R}, \\[2ex] \dfrac{2\varpi_\emptyset^*(\beta)|\varpi_\beta^*(z)|^2}{\varpi_z(z)} = \dfrac{2|\zeta_\beta(z)|^2}{1 - |\zeta_\beta(z)|^2} & \text{for } \partial\mathbb{O}. \end{cases}$$

This is (1).

Recall that the denominator $1 - |\zeta_\beta(z)|^2$ is positive, negative, or zero, iff z belongs to $\mathbb{O}$, $\mathbb{O}^e$, or $\partial\mathbb{O}$ respectively. This means that the circle will be in the right or left half plane depending on z being in $\mathbb{O}$ or $\mathbb{O}^e$ respectively. This is (4).

The closed disk $\Delta_n(z)$ with boundary $K_n(z)$ is given by using an inequality sign instead of an equality. To find out in which sense the inequality has to be set to get the interior of the circle, we make the following observations. Since

$$(s+\bar s)\frac{2|\zeta_\beta(z)|^2}{1-|\zeta_\beta(z)|^2} - \sum_{k=1}^{n-1}|\psi_k(z)+s\phi_k(z)|^2 - |1-s|^2 = A|s|^2 + Bs + \overline{B}\bar s + D,$$

with $A = -\sum_{k=0}^{n-1}|\phi_k|^2 < 0$, it follows that this expression will become negative for $|s|$ sufficiently large. Thus this expression is negative outside the disk. Therefore, the closed disk is described by

$$\sum_{k=1}^{n-1}|\psi_k(z)+s\phi_k(z)|^2 + |1-s|^2 \leq (s+\bar s)\frac{2|\zeta_\beta(z)|^2}{1-|\zeta_\beta(z)|^2}. \quad (11.28)$$

This is (2).

Since the sum in the left-hand side of (11.28) is nondecreasing with n, it follows that we have nested disks, that is, for any regular index $m > n$, we have $\Delta_m(z) \subset \Delta_n(z)$. This is (5).

Since $R_n(z,\infty) = R_{n-1}(z,0)$, the circles will touch, even if the index n is singular. In the latter case, $K_n(z)$ is a point on $K_{n-1}(z)$.

The expressions for center and radius for a general linear fractional transform ($\tau \in \mathbb{R}$)

$$s = -\frac{a-\tau b}{c-\tau d} = -\frac{a\bar d - b\bar c}{c\bar d - d\bar c} + \frac{ad-bc}{c\bar d - d\bar c}\cdot\frac{\bar c - \tau\bar d}{c-\tau d}$$

are obviously

$$\text{center} = -\frac{a\bar d - b\bar c}{c\bar d - d\bar c} \quad \text{and} \quad \text{radius} = \left|\frac{ad-bc}{c\bar d - d\bar c}\right|.$$

Using the Green and Christoffel–Darboux formulas, the expressions for c_n and r_n as in (3) will follow. □

Corollary 11.7.2. *For $z \in \{\beta, \hat\beta\}$, we find the following special cases. All the circles $K_n(\beta)$, $n = 1, 2, \ldots$ reduce to the same point $s = 1$ and all the circles $K_n(\hat\beta)$ reduce to the same point $s = -1$.*

Proof. Recall that $\zeta_\beta(\beta) = 0$ and $\zeta_\beta(\hat\beta) = \infty$. Let n be regular. It then follows from the expression for the radius r_n that for $z \in \{\beta, \hat\beta\}$ this radius is zero.

Writing out the equation for $K_n(z)$ for $z = \beta$ and $z = \hat{\beta}$, we find that only $s = 1$ satisfies the equation when $z = \beta$ and only $z = -1$ satisfies the equation when $z = \hat{\beta}$.

Since successive circles touch, also for singular indices, we get the same point for all n.   □

Assume from now on that there are infinitely many regular indices. Suppose these are $n(\nu)$, $\nu = 1, 2, \ldots$. Because the disks $\Delta_{n(\nu)}$ are nested, it follows that $\Delta_\infty = \lim_{\nu \to \infty} \Delta_{n(\nu)}$ is a disk or reduces to a point. Its radius is

$$r(z) = \lim_{\nu \to \infty} r_{n(\nu)}(z).$$

We have the following lemma.

Lemma 11.7.3. Suppose $z \in \mathbb{C}_\beta = \overline{\mathbb{C}} \setminus (\partial \mathbb{O} \cup \{\beta, \hat{\beta}\})$.
If $\Delta_\infty(z)$ is a disk (with positive radius), then

$$\sum_{k=0}^{\infty} |\phi_k(z)|^2 < \infty \quad \text{and} \quad \sum_{k=0}^{\infty} |\psi_k(z)|^2 < \infty.$$

If $\Delta_\infty(z)$ is a point, then

$$\sum_{k=0}^{\infty} |\phi_k(z)|^2 = \infty \quad \text{and} \quad \sum_{k=0}^{\infty} |\psi_k(z)|^2 = \infty.$$

Proof. It follows from the expression for the radii that $r(z)$ is positive (zero) iff $k_\infty(z, z)$ is finite (infinite), that is, iff $\sum_{k=0}^{\infty} |\phi_k(z)|^2$ is finite (infinite).

Let s be a point from the disk $\Delta_\infty(z)$. Then it follows from (11.28) that in any case (disk or point)

$$\sum_{k=1}^{\infty} |\psi_k(z) + s(z)\phi_k(z)|^2 < \infty.$$

Thus $\sum_{k=1}^{\infty} |\phi_k(z)|^2 < \infty$ iff $\sum_{k=1}^{\infty} |\psi_k(z)|^2 < \infty$.   □

Before we can prove an invariance theorem, we also need the following lemma.

Lemma 11.7.4. If n is a regular index, then for some parameter $w \in \mathbb{C}_\beta = \overline{\mathbb{C}} \setminus (\partial \mathbb{O} \cup \{\beta, \hat{\beta}\})$, the functions ϕ_n and ψ_n can be written as

$$\phi_n(z) = \phi_n(w) + \frac{z - w}{\alpha_n - z} \left[A_{0n}(w) + \sum_{k=1}^{n-1} \phi_k(z) A_{kn}(w) \right],$$

$$\psi_n(z) = \psi_n(w) + \frac{z-w}{\alpha_n - z} \left[B_{0n}(w) + \sum_{k=1}^{n-1} \psi_k(z) A_{kn}(w) \right]$$

where

$$A_{0n}(w) = V_n(w)\phi_n(w) - Y_n(w)\psi_n(w),$$

$$B_{0n}(w) = V_n(w)\psi_n(w) - Y_n(w)\phi_n(w),$$

$$A_{kn}(w) = Y_n(w)a_{kn}(w),$$

$$a_{kn}(w) = \psi_k(w)\phi_n(w) - \phi_k(w)\psi_n(w), \quad k = 1, \dots, n-1,$$

$$Y_n(w) = \frac{w - \alpha_n}{2w} \quad \text{for } \mathbb{T} \quad \text{and} \quad Y_n(w) = \mathbf{i}\frac{w - \alpha_n}{1 + w^2} \quad \text{for } \mathbb{R},$$

$$V_n(w) = \frac{w + \alpha_n}{2w} \quad \text{for } \mathbb{T} \quad \text{and} \quad V_n(w) = \frac{1 + \alpha_n w}{1 + w^2} \quad \text{for } \mathbb{R}.$$

Proof. From the Christoffel–Darboux formula in Theorem 11.3.1 and the mixed formula in Theorem 11.3.2, we get

$$\frac{\phi_n(z)\phi_{n-1}(w)}{Z_n(z)Z_{n-1}(w)} - \frac{\phi_n(w)\phi_{n-1}(z)}{Z_n(w)Z_{n-1}(z)} = -H(z,w)E_n \sum_{k=0}^{n-1} \phi_k(z)\phi_k(w),$$

$$\frac{\phi_n(z)\psi_{n-1}(w)}{Z_n(z)Z_{n-1}(w)} - \frac{\psi_n(w)\phi_{n-1}(z)}{Z_n(w)Z_{n-1}(z)} = -H(z,w)E_n$$

$$\times \left[\sum_{k=1}^{n-1} \phi_k(z)\psi_k(w) - D(z,w) \right].$$

Elimination of $\phi_{n-1}(z)$ gives

$$\frac{\phi_n(z)}{Z_n(z)Z_{n-1}(w)} a_{n-1,n}(w)$$

$$= -H(z,w)E_n \left[\sum_{k=1}^{n-1} \phi_k(z)a_{k,n}(w) - [\psi_n(w) + D(z,w)\phi_n(w)] \right].$$

From the determinant formula of Theorem 11.3.2, we find

$$a_{n-1,n}(w) = -E_n Z_{n-1}(w)Z_n(w)X(w),$$

with

$$X(w) = \frac{2\mathbf{i}w}{(1-w)^2} \quad \text{for } \mathbb{T} \quad \text{and} \quad X(w) = -\mathbf{i}\frac{1+w^2}{w^2} \quad \text{for } \mathbb{R}.$$

Hence

$$-\frac{H(z,w)E_n Z_n(z)Z_{n-1}(w)}{a_{n-1,n}(w)} = \begin{cases} \dfrac{z-w}{\alpha_n-z}\left[-\dfrac{\alpha_n-w}{2w}\right] & \text{for } \mathbb{T}, \\[3mm] \dfrac{z-w}{\alpha_n-z}\left[-\mathbf{i}\dfrac{\alpha_n-w}{1+w^2}\right] & \text{for } \mathbb{R}, \\[3mm] \dfrac{z-w}{\alpha_n-z}Y_n(w) & \text{for } \partial\mathbb{O}. \end{cases}$$

Thus

$$\phi_n(z) = \frac{z-w}{\alpha_n-z}Y_n(w)\left[\sum_{k=1}^{n-1}\phi_k(z)a_{kn}(w) - [\psi_n(w)+D(z,w)\phi_n(w)]\right].$$

Next we compute

$$-\frac{z-w}{\alpha_n-z}Y_n(w)D(z,w) = \begin{cases} \dfrac{(z+w)(\alpha_n-w)}{2w(\alpha_n-z)} = 1 + \dfrac{(z-w)(w+\alpha_n)}{(\alpha_n-z)2w} & \text{for } \mathbb{T}, \\[3mm] \dfrac{(1+zw)(\alpha_n-w)}{(1+w^2)(\alpha_n-z)} = 1 + \dfrac{(z-w)(1+\alpha_n w)}{(\alpha_n-z)(1+w^2)} & \text{for } \mathbb{R}, \\[3mm] 1 + \dfrac{z-w}{\alpha_n-z}V_n(w) & \text{for } \partial\mathbb{O}. \end{cases}$$

Thus

$$\phi_n(z) = \phi_n(w) + \frac{z-w}{\alpha_n-z}[V_n(w)\phi_n(w) - Y_n(w)\psi_n(w)]$$

$$+ \frac{z-w}{\alpha_n-z}\left[\sum_{k=1}^{n-1}\phi_k(z)[Y_n(w)a_{kn}(w)]\right].$$

This is the first formula required.

The second formula is proved similarly.   □

We can now prove the following invariance theorem.

Theorem 11.7.5 (Invariance). *Suppose* $w \in \mathbb{C}_\beta = \overline{\mathbb{C}} \setminus (\overline{\partial\mathbb{O}} \cup \{\beta, \hat{\beta}\})$ *and suppose that* $\Delta_\infty(w)$ *is a disk with positive radius. Then* $\Delta_\infty(z)$ *is a disk with positive radius for every* $z \in \mathbb{C}_\beta$ *and*

$$\sum_{k=0}^{n}|\phi_k(z)|^2 \quad \text{and} \quad \sum_{k=0}^{n}|\psi_k(z)|^2$$

converge locally uniformly in $\mathbb{C}_\beta$ *as* $n \to \infty$.

Proof. From Lemma 11.7.4, we know that with $x_n = \phi_n$ and $y_n = \psi_n$ or vice versa, we have, with the notation as in Lemma 11.7.4,

$$x_n(z) = x_n(w) + \frac{z-w}{\alpha_n - z} Y_n(w) \sum_{k=1}^{n-1} x_k(z) a_{kn}(w)$$

$$+ \frac{z-w}{\alpha_n - z} [V_n(w)x_n(w) - Y_n(w)y_n(w)].$$

From the definition of $a_{kn}(w)$, it follows that

$$\sum_{n=1}^{\infty} \sum_{k=1}^{n-1} |a_{kn}(w)|^2 \leq 2 \sum_{n=1}^{\infty} \sum_{k=1}^{n-1} (|\psi_k(w)|^2 |\phi_n(w)|^2 + |\phi_k(w)|^2 |\psi_n(w)|^2)$$

$$\leq 4 \left(\sum_{n=1}^{\infty} |\psi_n(w)|^2 \right) \left(\sum_{n=1}^{\infty} |\phi_n(w)|^2 \right) < \infty$$

by Lemma 11.7.3. Now suppose that C is a compact subset of $\mathbb{C}_\beta$. Then for arbitrary $z \in C$ and $w \in \mathbb{C}_\beta$ fixed

$$\frac{z-w}{\alpha_n - z} Y_n(w) \quad \text{and} \quad \frac{z-w}{\alpha_n - z} V_n(w)$$

are uniformly bounded, say

$$\left| \frac{z-w}{\alpha_n - z} Y_n(w) \right| \leq R_1 \quad \text{and} \quad \left| \frac{z-w}{\alpha_n - z} V_n(w) \right| \leq R_2.$$

So

$$\left(\sum_{n=m}^{N} |x_n(z)|^2 \right)^{1/2} \leq \left(\sum_{n=m}^{N} |x_n(w)|^2 \right)^{1/2} + R_1 \left(\sum_{n=m}^{N} |y_n(w)|^2 \right)^{1/2}$$

$$+ R_2 \left(\sum_{n=m}^{N} |x_n(w)|^2 \right)^{1/2}$$

$$+ R_1 \left(\sum_{n=m}^{N} \left| \sum_{k=1}^{n-1} a_{kn}(w)x_k(z) \right|^2 \right)^{1/2}.$$

For any $\epsilon \in (0, 1)$, choose $m = m(\epsilon, R_1, R_2) > 1$ such that

$$\left(\sum_{n=m}^{\infty} |y_n(w)|^2 \right)^{1/2} < \frac{\epsilon}{R_1}, \qquad \left(\sum_{n=m}^{\infty} |x_n(w)|^2 \right)^{1/2} < \frac{\epsilon}{1 + R_2},$$

and

$$\left(\sum_{n=m}^{\infty}\sum_{k=1}^{n-1}|a_{kn}(w)|^2\right)^{1/2} < \frac{\epsilon}{R_1}.$$

Then

$$\left(\sum_{n=m}^{N}|x_n(z)|^2\right)^{1/2} \le \epsilon + \epsilon + R_1 \left[\sum_{n=m}^{N}\left(\sum_{k=1}^{n-1}|a_{kn}(w)|^2\right)\left(\sum_{k=1}^{n-1}|x_k(z)|^2\right)\right]^{1/2}$$

$$\le 2\epsilon + R_1 \left(\sum_{k=1}^{N}|x_k(z)|^2\right)^{1/2}\left(\sum_{n=m}^{\infty}\sum_{k=1}^{n-1}|a_{kn}(w)|^2\right)^{1/2}$$

$$\le 2\epsilon + \epsilon \left(\sum_{k=1}^{N}|x_k(z)|^2\right)^{1/2}$$

$$\le 2\epsilon + \epsilon \left(\sum_{k=m}^{N}|x_k(z)|^2\right)^{1/2} + \epsilon \left(\sum_{k=1}^{m-1}|x_k(z)|^2\right)^{1/2}.$$

Hence

$$(1-\epsilon)\left(\sum_{n=m}^{N}|x_n(z)|^2\right)^{1/2} \le 2\epsilon + \epsilon\left(\sum_{k=1}^{m-1}|x_k(z)|^2\right)^{1/2}.$$

Because $\sum_{k=1}^{m-1}|x_k(z)|^2$ is a continuous function in C, there is an M_m, depending on m, such that

$$\left(\sum_{k=1}^{m-1}|x_k(z)|^2\right)^{1/2} \le M_m$$

holds uniformly in C. Thus

$$\left(\sum_{k=m}^{N}|x_k(z)|^2\right)^{1/2} \le \frac{(2+M_m)\epsilon}{1-\epsilon}, \quad N \ge m.$$

Now we first fix an ϵ, for example, $\epsilon = 1/2$, and let m_0 be the corresponding index m. It follows that $\sum_{k=m_0}^{\infty}|x_k(z)|^2$ has a finite upper bound $(2+M_{m_0})^2$ in C, and consequently, $\sum_{k=1}^{\infty}|x_k(z)|^2$ has a finite upper bound M^2 in C.

We now return to the argument with an arbitrary ϵ and corresponding m. It follows that

$$(1-\epsilon)\left(\sum_{k=m}^{N}|x_k(z)|^2\right)^{1/2} \le \left[2 + \left(\sum_{k=1}^{\infty}|x_k(z)|^2\right)^{1/2}\right]\epsilon$$

and hence from the foregoing that

$$\left(\sum_{k=m}^{N} |x_k(z)|^2 \right)^{1/2} \le \frac{(2+M)\epsilon}{1-\epsilon}, \quad N \ge m.$$

This shows that $\{(\sum_{k=1}^{N} |x_k(z)|^2)^{1/2}\}$ is a uniform Cauchy sequence in C, which proves the uniform convergence of $\sum_{k=0}^{\infty} |x_k(z)|^2$ in C. $\square$

We can now also prove the analyticity theorem.

Theorem 11.7.6 (Analyticity). *Let $\Delta_{\infty}(z)$ be a point and let R_n be defined by (11.27). Then the limit*

$$s(z) = \lim_{n \to \infty} R_n(z, \tau), \quad z \in \mathbb{C} \setminus \partial\mathbb{O}$$

is an analytic function of z not depending on τ. Moreover,

$$\frac{\text{Re } s(z)}{1 - |\zeta_\beta(z)|^2} > 0, \quad z \in \mathbb{C} \setminus \partial\mathbb{O}.$$

Proof. By Theorem 11.7.1 (6), it follows that if $\Delta_{\infty}(z)$ is a point then

$$\lim_{n \to \infty} R_n(z, \tau) = \lim_{n \to \infty} s_n(z) = s(z)$$

exists and is independent of τ.

Obviously, $s_n(z) = R_n(z, \tau)$ is analytic for $z \in \mathbb{C} \setminus \partial\mathbb{O}$. Thus the analyticity of $s(z)$ will follow if the functions $s_n(z)$ are uniformly bounded in compact subsets of $\mathbb{C} \setminus \partial\mathbb{O}$. This is shown as follows. We know that

$$|1 - s_n(z)|^2 + \sum_{k=1}^{n-1} |\psi_k(z) + s_n(z)\phi_k(z)|^2 = (s_n(z) + \bar{s}_n(z)) \frac{2|\zeta_\beta(z)|^2}{1 - |\zeta_\beta(z)|^2}.$$

Thus

$$1 + |s_n|^2 + \sum_{k=1}^{n-1} |\psi_k + s_n\phi_k|^2 = (s_n + \bar{s}_n) \frac{1 + |\zeta_\beta|^2}{1 - |\zeta_\beta|^2}$$

or, using $|s_n + \bar{s}_n| \le 2|s_n|$,

$$|s_n|^2 < 1 + |s_n|^2 + \sum_{k=1}^{n-1} |\psi_k + s_n\phi_k|^2 \le 2|s_n| \left| \frac{1 + |\zeta_\beta|^2}{1 - |\zeta_\beta|^2} \right|.$$

Therefore,

$$|s_n(z)| \le 2 \left| \frac{1 + |\zeta_\beta(z)|^2}{1 - |\zeta_\beta(z)|^2} \right|.$$

Since the right-hand side is uniformly bounded in compact subsets of $\mathbb{C} \setminus \partial \mathbb{O}$, the analyticity of $s(z)$ follows.

The last inequality is a direct consequence of Theorem 11.7.1 (4).   □

11.8. Moment problem

Recall that in the previous chapters we had the relation $d\mathring{\mu}(t) = |\varpi_0(t)|^2 d\mu(t)$. For $\mathbb{R}$ we had $\varpi_0(t) = t + i$ whereas for $\mathbb{T}$ this factor was just 1. In this chapter, we have changed the meaning of α_0 and renamed it as β. Therefore, for t on the boundary $\partial \mathbb{O}$, we shall now need $|\varpi_\beta(t)|^2 = |t - \overline{\beta}|^2$ to replace $|\varpi_0(t)|^2$. It is just 1 for $t \in \mathbb{T}$ and it is $1 + t^2$ for $t \in \mathbb{R}$. After this introductory note, we can state our moment problem.

It has been outlined in the introduction that the moment problem is different from the moment problem in the previous chapters. In fact we should consider two kinds of moment problems. We are given a functional M defined on $\mathcal{R}_\infty = \mathcal{L}_\infty \cdot \mathcal{L}_\infty$, which is real and positive definite, that is,

$$M\{f_*\} = \overline{M\{f\}}, \quad f \in \mathcal{L}_\infty \cdot \mathcal{L}_\infty \quad \text{and} \quad M\{ff_*\} > 0, \quad 0 \ne f \in \mathcal{L}_\infty.$$

To represent the functional in $\mathcal{L}_\infty$ as an integral, we suppose it is defined there by the moments

$$\mu_{n0} = M\{b_n(z)\}, \quad n = 0, 1, \dots. \tag{11.29}$$

The problem is to find a representation of this functional by a measure μ on $\overline{\partial \mathbb{O}}$ such that

$$\int b_n(t)\, d\mathring{\mu}(t) = \mu_{n0}, \quad n = 0, 1, \dots, \tag{11.30}$$

and thus such that

$$M\{f\} = \int f(t)\, d\mathring{\mu}(t), \quad \forall f \in \mathcal{L}_\infty. \tag{11.31}$$

We call this the moment problem in $\mathcal{L}_\infty$. Note that when $\partial \mathbb{O} = \mathbb{R}$ and all $\alpha_i = \infty$, then this is the classical Hamburger moment problem [104, 105, 106].

If we want to represent the functional in $\mathcal{R}_\infty$ as an integral, we assume that it is defined by the moments

$$\mu_{nm} = M\{b_n b_m\}, \quad n, m = 0, 1, \dots.$$

The problem is to find a measure μ on $\overline{\partial\mathbb{O}}$ such that

$$\mu_{nm} = \int b_n(t)b_m(t)\,d\mathring{\mu}(t), \quad n, m = 0, 1, \ldots.$$

This problem is equivalent to finding a measure such that

$$M\{f\} = \int f(t)\,d\mathring{\mu}(t), \quad \forall f \in \mathcal{R}_\infty$$

or also such that

$$\langle f, g \rangle_M = \int f(t)\overline{g(t)}\,d\mathring{\mu}(t) = \langle f, g \rangle_{\mathring{\mu}}, \quad \forall f, g \in \mathcal{L}_\infty. \quad (11.32)$$

We call this the moment problem in $\mathcal{R}_\infty$. Note, as already explained in the introduction, both the moment problem in $\mathcal{L}_\infty$ and the moment problem in $\mathcal{R}_\infty$ are rational generalizations of the classical Hamburger moment problem in the sense that they both reduce to the Hamburger moment problem when $\partial\mathbb{O} = \mathbb{R}$ and all $\alpha_i = \infty$.

We note that when there is only a finite number of different $\alpha_i \neq \infty$ that are repeated cyclically: $\alpha_1, \ldots, \alpha_p, \alpha_1, \ldots, \alpha_p, \alpha_1, \ldots$, then the study of our orthogonal rational functions would simplify considerably. Indeed, in that case $\mathcal{R}_\infty = \mathcal{L}_\infty \cdot \mathcal{L}_\infty = \mathcal{L}_\infty$ since a product of rational functions whose poles are among the points $\alpha_1, \ldots, \alpha_p$ is again a rational function of the same type. Thus, in that case, the moment problems in $\mathcal{R}_\infty$ and in $\mathcal{L}_\infty$ are the same.

Another remark to make is that we could as well have used the moments

$$\mu'_{nm} = M\left\{ \frac{1}{\pi_n^*[\pi_m^*]_*} \right\}$$

to define the functional in $\mathcal{R}_\infty$, as was done in [157, 158, 160, 45]. However, this representation does not allow an elegant treatment of the points $\alpha_i = \infty$ in the case of the real line. However, when there are only a finite number of different finite α_i, then the choice of μ'_{nm} would be more natural to solve a multipoint Hamburger moment problem.

As in the introduction of Section 10.3, it can be shown that a solution should have infinitely many points on which it is supported.

We have the following existence theorem.

Theorem 11.8.1. *Let M be a real positive functional defined on $\mathcal{R}_\infty$. Then $\langle f, g \rangle = M\{fg_*\}$ defines an inner product for $f, g \in \mathcal{L}_\infty$. Assume that the corresponding sequence of orthogonal rational functions ϕ_n has infinitely many regular indices. Then there exists at least one measure μ on $\overline{\partial\mathbb{O}}$ that solves* *the moment problem in $\mathcal{L}_\infty$, that is, that satisfies (11.30), or equivalently, the problem (11.31).*

Proof. The proof is based on the fact that by the regularity assumption, we can find for a regular index k a regular τ_k, such that there is a quadrature formula. Thus there is an infinite sequence of such quadrature formulas. Denote by $\mathring{\mu}_k = \mathring{\mu}_k(\cdot, \tau_k)$ the discrete measure representing the quadrature formula

$$\sum_{i=1}^{k} \lambda_{ki}(\tau_k) f(\xi_{ki}(\tau_k)) = \int f(t) \, d\mathring{\mu}_k(t),$$

which is exact in $\mathcal{R}_{k-1}$ and hence also in $\mathcal{L}_{k-1}$. In particular, $\int d\mathring{\mu}_k(t) = M\{1\}$ for all k. Thus there is an infinite sequence of these measures and it is uniformly bounded by $\int d\mathring{\mu}_k(t) = M\{1\} = 1$. Hence by Helly's selection principle, the sequence (μ_k) will have a convergent subsequence $\mathring{\mu}_{k(j)} \to \mathring{\mu}$. Next we prove that such a $\mathring{\mu}$ solves the moment problem in $\mathcal{L}_\infty$, that is, that

$$M\{b_n\} = \int b_n(t) \, d\mathring{\mu}(t), \qquad n = 0, 1, \ldots.$$

For $n = 0$, we can apply Helly's convergence theorem since $b_0 = 1$ is continuous. Recall that for the extended real line we can use the one-point compactification [96] to make Helly's convergence theorem work since ∞ may be considered as any other (finite) point of the line. See Section 10.3. This observation also implies that the argumentation below will hold for $\overline{\partial\mathbb{O}} = \overline{\mathbb{R}}$.

We now prove for an arbitrary but fixed $n > 0$ that the moment relation holds. Let us denote the elements of the subsequence $k(j)$ by k for simplicity. Note that in this case Helly's convergence theorem is not applicable because the b_n are not continuous on $\overline{\partial\mathbb{O}}$. However, consider any compact subset $J \subset \overline{\partial\mathbb{O}}$ that does not contain the points $\alpha_1, \ldots, \alpha_n$. Then b_n is indeed continuous on J and this means that by Helly's convergence theorem we can find a K large enough such that for a given $\epsilon > 0$ we have for all $k = k(j) > K$

$$\left| \int_J b_n(t) [d\mathring{\mu}(t) - d\mathring{\mu}_k(t)] \right| < \frac{\epsilon}{3}. \tag{11.33}$$

In contrast, setting $I = \overline{\partial\mathbb{O}} \setminus J$, we have for $k = k(j) > n$,

$$\left| \int_I b_n(t) \, d\mathring{\mu}_k(t) \right| = \left| \int_I \frac{|b_n(t)|^2}{b_{n*}(t)} \, d\mathring{\mu}_k(t) \right|$$

$$\leq \max_I \frac{1}{|b_n(t)|} \int_I |b_n(t)|^2 d\mathring{\mu}_k(t) \leq \max_I |1/b_n(t)| \cdot \mu_{nn}.$$

We can always choose J large enough such that the maximum of $|1/b_n|$ in I is arbitrarily small. Indeed, the set I will then contain small neighborhoods of the $\alpha_1, \ldots, \alpha_n$, none of which is $\alpha_\emptyset$. The zeros of $1/b_n$ are precisely in the points $\alpha_1, \ldots, \alpha_n$ and its pole is at $z = \alpha_\emptyset$. Because μ_{nn} is finite, it follows that for any $\epsilon > 0$, we can make J large enough to satisfy

$$\left| \int_I b_n(t) \, d\mathring{\mu}_k(t) \right| < \frac{\epsilon}{3}. \tag{11.34}$$

Note that this holds for any $k > n$, that is, J is independent of k. Next we want to show that J can also be made large enough to satisfy

$$\left| \int_I b_n(t) \, d\mathring{\mu}(t) \right| < \frac{\epsilon}{3}. \tag{11.35}$$

To obtain this, we consider sets J_p, none of which contain $\alpha_1, \ldots, \alpha_n$, and such that $J_q \subset J_p$ for $p > q$ and $\cup_p J_p = \overline{\partial\mathbb{O}} \setminus \{\alpha_1, \ldots, \alpha_n\}$. Note that for $k = k(j) > n$ and for any p

$$\int_{J_p} b_n(t) \, d\mathring{\mu}(t) = \int_{J_p} b_n(t)[d\mathring{\mu}(t) - d\mathring{\mu}_k(t)] + \int_{J_p} b_n(t) \, d\mathring{\mu}_k(t).$$

Thus if $p > q$ and $k > n$

$$\left| \int_{J_p - J_q} b_n(t) \, d\mathring{\mu}(t) \right| \leq \left| \int_{J_p - J_q} b_n(t)[d\mathring{\mu}(t) - d\mathring{\mu}_k(t)] \right| + \left| \int_{J_p - J_q} b_n(t) \, d\mathring{\mu}_k(t) \right|. \tag{11.36}$$

By (11.34), there is always a p and q large enough such that for any $\eta > 0$

$$\left| \int_{J_p - J_q} b_n(t) \, d\mathring{\mu}_k(t) \right| < \frac{\eta}{2} \tag{11.37}$$

for all large k. By (11.33), we can make k so large that for any $\eta > 0$

$$\left| \int_{J_p - J_q} b_n(t)[d\mathring{\mu}(t) - d\mathring{\mu}_k(t)] \right| < \frac{\eta}{2}. \tag{11.38}$$

Combining (11.36), (11.37), and (1138) shows that it is possible to make

$$\left| \int_{J_p} b_n(t) \, d\mathring{\mu}(t) - \int_{J_q} b_n(t) \, d\mathring{\mu}(t) \right| < \eta$$

for any $\eta > 0$. This means that

$$\left(\int_{J_p} b_n(t)\, d\mathring{\mu}(t) \right)_p$$

is a Cauchy sequence, so that its limit for $p \to \infty$ (which is the integral over $\overline{\partial\mathbb{O}} \setminus \{\alpha_1, \ldots, \alpha_n\}$) exists. We find that

$$\int_I d\mathring{\mu}_k(t) = \int_I \frac{b_n(t) b_{n*}(t)}{|b_n(t)|^2} d\mathring{\mu}_k(t)$$

$$= \int_I \frac{b_n(t) b_{n*}(t)}{|b_n(t)|^2} d\mathring{\mu}(t)$$

$$\leq \max_{t \in I} \frac{1}{|b_n(t)|^2} \mu_{nn}$$

for all $k > n$. Thus $\int_I d\mathring{\mu}_k(t)$ can be made arbitrarily small for all $k > n$ by choosing I small enough; hence $\int_I d\mathring{\mu}(t)$ can also be made arbitrarily small by choosing I small enough. Hence $\mathring{\mu}\{\alpha_1, \ldots, \alpha_n\} = 0$. It thus follows that for any $\epsilon > 0$ there exists a p large enough such that

$$\left| \int_I b_n(t)\, d\mathring{\mu}(t) \right| = \left| \int b_n(t)\, d\mathring{\mu}(t) - \int_{J_p} b_n(t)\, d\mathring{\mu}(t) \right| < \frac{\epsilon}{3},$$

which proves (11.35). Finally, by (11.34), (11.35), and (11.33),

$$\left| \int b_n(t)[d\mathring{\mu}_k(t) - d\mathring{\mu}(t)] \right| \leq \left| \int_I b_n(t)\, d\mathring{\mu}_k(t) \right| + \left| \int_I b_n(t) d\mathring{\mu}(t) \right|$$

$$+ \left| \int_J b_n(t)[d\mathring{\mu}_k(t) - d\mathring{\mu}(t)] \right| < \epsilon$$

because each of the terms in the right-hand side can be bounded by $\epsilon/3$. Thus

$$\lim_{k(j) \to \infty} \int b_n(t)\, d\mathring{\mu}_{k(j)}(t) = \int b_n(t)\, d\mathring{\mu}(t).$$

Since

$$\int b_n(t)\, d\mathring{\mu}_k(t) = \mu_{n0} = M\{b_n\}, \quad \text{for } k > n,$$

this proves the theorem. $\qquad\qquad\square$

We now use our framework of nested disks to obtain information about when the solution is unique. Let us denote by $\mathcal{M}^{\mathcal{L}}$ the set of solutions of the moment problem in $\mathcal{L}_\infty$ and $\mathcal{M}^{\mathcal{R}}$ the set of solutions of the moment problem in $\mathcal{R}_\infty$. Then we have

Theorem 11.8.2. *Assume that the sequence of orthonormal functions* ϕ_n *has infinitely many regular indices, and hence* $\mathcal{M}^{\mathcal{L}} \neq \emptyset$. *Fix* $z \in \overline{\mathbb{C}} \setminus \partial\mathbb{O}$ *and define for* $\mu \in \mathcal{M}^{\mathcal{L}}$ *the Riesz–Herglotz–Nevanlinna transform*

$$\Omega_\mu(z) = \int D(t, z)\, d\mathring{\mu}(t).$$

Then

$$\{\Omega_\mu(z) : \mu \in \mathcal{M}^{\mathcal{R}}\} \subset \Delta_\infty(z) \subset \{\Omega_\mu(z) : \mu \in \mathcal{M}^{\mathcal{L}}\}.$$

Proof. Let $s = \Omega_\mu(z)$ for some $\mu \in \mathcal{M}^{\mathcal{R}}$. Note that the system $\{\phi_n\}$ is then orthonormal with respect to the inner product defined by the measure μ. Let $f(z) = \overline{D(t, z)}$, $t \in \overline{\partial\mathbb{O}}$. Writing the generalized Fourier series of $f(z)$ as

$$f(z) \sim \sum_{k=0}^{\infty} \gamma_k \phi_k(z), \quad \gamma_k = \langle f, \phi_k \rangle$$

and then using

$$\gamma_k = \int \overline{D(t, z)\phi_k(t)}\, d\mathring{\mu}(t),$$

we then get that $\overline{\gamma}_0 = s$ and $\overline{\gamma}_k = \psi_k(z) + s\phi_k(z)$ for $k \geq 1$ because

$$\overline{\gamma}_k = \int D(t, z)\phi_k(t)\, d\mathring{\mu}(t)$$

$$= \int D(t, z)[\phi_k(t) - \phi_k(z)]\, d\mathring{\mu}(t) + \phi_n(z)\Omega_\mu(z), \quad t \in \overline{\partial\mathbb{O}}.$$

However, it can be shown that for $t \in \overline{\partial\mathbb{O}}$

$$1 + |D(t, z)|^2 = \frac{1 + |\zeta_\beta(z)|^2}{1 - |\zeta_\beta(z)|^2}[D(t, z) + \overline{D(t, z)}].$$

Using Bessel's inequality, it then follows that

$$|s|^2 + \sum_{k=1}^{\infty} |\psi_k + s\phi_k|^2 = \sum_{k=0}^{\infty} |\gamma_k|^2 \leq \int |D(t, z)|^2 d\mathring{\mu}(t)$$

$$= -1 + \frac{1 + |\zeta_\beta(z)|^2}{1 - |\zeta_\beta(z)|^2}(s + \overline{s}),$$

which can be rearranged as

$$|1 - s|^2 + \sum_{k=1}^{\infty} |\psi_k + s\phi_k|^2 \leq \frac{2|\zeta_\beta(z)|^2}{1 - |\zeta_\beta(z)|^2}(s + \bar{s}).$$

This means that $s \in \Delta_\infty(z)$. Note that for $z = \beta$, the inequality reduces to $|1 - s| \leq 0$, whereas for $z = \hat{\beta}$, it reduces to $|1 + s|^2 \leq 0$. This is consistent with the fact that for all normalized measures μ, $\Omega_\mu(\beta) = 1$ and $\Omega_\mu(\hat{\beta}) = -1$.

Next we show that if $s \in \Delta_\infty(z)$, then it is the Riesz–Herglotz–Nevanlinna transform of some $\mu \in \mathcal{M}^\mathcal{L}$. This is readily shown by using the quadrature formulas we have discussed. We consider the limiting point and the limiting disk cases separately.

If $\Delta_\infty(z)$ is a point, then since $s \in \Delta_\infty(z)$, there must exist $s_n \in K_n(z)$ such that $s_n \to s$. Since there is for each regular n some τ_n such that

$$s_n = R_n(z, \tau_n) = \int D(t, z) \, d\mathring{\mu}_n(t),$$

Helly's selection criterion then yields that there is a subsequence $\mathring{\mu}_{n(j)} \to \mathring{\mu}$. By the proof of the previous theorem, $\mu \in \mathcal{M}^\mathcal{L}$, and by Helly's convergence theorem,

$$\lim_{j \to \infty} \int D(t, z) \, d\mathring{\mu}_{n(j)}(t) = \int D(t, z) \, d\mathring{\mu}(t) = s.$$

Thus s is the Riesz–Herglotz–Nevanlinna transform of a $\mu \in \mathcal{M}^\mathcal{L}$.

If $\Delta_\infty(z)$ is a disk, let s be a point on the boundary $K_\infty(z)$. Recall that for a fixed n, we can, except for finitely many values of τ, associate a quadrature formula with $R_n(z, \tau)$. Let us denote the discrete measure that is associated with this quadrature by $\mathring{\mu}_n(\cdot, \tau)$. It depends on n but also on the choice of τ. We can then, for every regular index n, choose an $s_n \in K_n(z)$ such that these s_n tend to s and such that $s_n = \Omega_{\mu_n}(z)$, where $\mu_n = \mu_n(\cdot, \tau_n)$ and where τ_n is chosen such that $s_n = R_n(z, \tau_n)$. By Helly's theorems and the proof of the previous theorem, there exists a $\mu \in \mathcal{M}^\mathcal{L}$ such that $\Omega_\mu(z) = s$.

Thus every s on the boundary $K_\infty(z)$ is of the form $\Omega_\mu(z)$ with $\mu \in \mathcal{M}^\mathcal{L}$. Now let s be an interior point of $\Delta_\infty(z)$. Then it can be found as a convex combination $s = \lambda s_1 + (1 - \lambda)s_2$ $(0 < \lambda < 1)$ of points s_1, s_2 on the boundary $K_\infty(z)$. Thus there exist $\mu_1, \mu_2 \in \mathcal{M}^\mathcal{L}$ such that

$$s_j = \int D(t, z) \, d\mathring{\mu}_j(t).$$

Thus $\mu = \lambda\mu_1 + (1 - \lambda)\mu_2 \in \mathcal{M}^\mathcal{L}$ and $s = \Omega_\mu(z)$.   □

Now the following corollary is obvious.

Corollary 11.8.3. *In the case of a limiting disk, for each* $s \in \Delta_\infty(z)$, $z \in \overline{\mathbb{C}} \setminus \partial\mathbb{O}$, *there is a* $\mu \in \mathcal{M}^{\mathcal{L}}$ *such that* $s = \Omega_\mu(z)$. *The moment problem in* $\mathcal{L}_\infty$ *has infinitely many solutions. In the case of a limiting point, a solution of the moment problem in* $\mathcal{R}_\infty$ *is unique.*

11.9. Favard type theorem

Here we want to show that if we generate rational functions by the recurrence relation of Theorem 11.1.2, then they are orthogonal with respect to some functional. Thus if

$$\phi_0 \in \mathbb{R} \setminus \{0\},$$

$$\phi_1(z) = (E_1 Z_1(z) + B_1)\,\phi_0(z),$$

and

$$\phi_n(z) = \left(A_n Z_n(z) + B_n \frac{Z_n(z)}{Z_{n-2}(z)}\right)\phi_{n-1}(z) + C_n \frac{Z_n(z)}{Z_{n-2}(z)}\phi_{n-2}(z),$$
$$n = 2, 3, \ldots \quad (11.39)$$

or, by (11.9),

$$\phi_n(z) = \left(E_n Z_n(z) + B_n \frac{Z_n(z)}{Z_{n-1}(z)}\right)\phi_{n-1}(z) + C_n \frac{Z_n(z)}{Z_{n-2}(z)}\phi_{n-2}(z),$$
$$n = 2, 3, \ldots, \quad (11.40)$$

with E_n and C_n nonzero, then they are orthogonal with respect to some functional M, meaning that

$$\langle \phi_k, \phi_{l*} \rangle_M = M\{\phi_k \phi_{l*}\} = M\{\phi_k \phi_l\} = 0, \quad \text{for } k \neq l.$$

If we want this to be a real inner product, that is, if we want $M\{f_*\} = \overline{M\{f\}}$, then the coefficients A_n, B_n, C_n, and E_n should be real:

$$A_n, B_n, E_n \in \mathbb{R}, \quad E_n = A_n + B_n D_n \neq 0, \quad n = 1, 2, \ldots,$$

$$C_n \in \mathbb{R}, \quad C_n \neq 0, \quad n = 2, 3, \ldots, \quad (11.41)$$

which is shown in Section 11.1 (see Lemma 11.1.1), and if we want it to be positive ($M\{f f_*\} > 0$ for $f \neq 0$), then by a small adaptation of Lemma 11.1.6, we obtain

$$M\{\phi_{n-1}^2\} = -\frac{C_n E_{n-1}}{E_n} M\{\phi_{n-2}^2\}, \quad n = 2, 3, \ldots, \quad (11.42)$$

so that the positivity is guaranteed if

$$M\{\phi_0^2\} > 0 \quad \text{and} \quad C_n E_{n-1}/E_n < 0, \quad n = 2, 3, \ldots. \quad (11.43)$$

Here we assume that the leading coefficients κ_k are positive. If we want the rational functions to be normalized, then we should choose

$$E_n = -C_n E_{n-1}, \quad n = 2, 3, \ldots \quad (11.44)$$

with $M\{\phi_0^2\} = 1$. Thus we assume in this section that the coefficients of the recurrence satisfy the conditions (11.41) and (11.44).

We start with an extension of the above recurrence relation. To this end we use the following lemma.

Lemma 11.9.1.

1. For all constants A and B and for all j, k, n integer such that $\alpha_n \neq \alpha_k$, there exist unique constants a and b such that

$$A + \frac{B}{Z_j} = \frac{a}{Z_n} + \frac{b}{Z_k}.$$

2. For all constants A and B and for all j, k integer, there exist unique constants a and b such that

$$A + \frac{B}{Z_j} = a + \frac{b}{Z_k}.$$

Proof. To have the first relation, we note first that, by the definitions of Z_k and Z_n, there exist nonzero constants c_1, c_2 and nonzero constants d_1, d_2 such that it is rewritten as

$$\frac{c_1 A(z - 1) + c_2 B(z - \alpha_j)}{z - 1} = \frac{d_1 a(z - \alpha_n) + d_2 b(z - \alpha_k)}{z - 1}.$$

Thus the constants a and b exist if the following system can be solved for a and b:

$$c_1 A + c_2 B = d_1 a + d_2 b,$$
$$c_1 A + c_2 B \alpha_j = d_1 a \alpha_n + d_2 b \alpha_k.$$

Since the determinant of the system is $d_1 d_2 (\alpha_k - \alpha_n) \neq 0$, there is a unique solution.

For the second relation, we find a similar system:

$$c_1 A + c_2 B = d_1 a + d_2 b,$$
$$c_1 A + c_2 B \alpha_j = d_1 a + d_2 b \alpha_k.$$

Now the determinant is $d_1 d_2 (\alpha_k - 1) \neq 0$ since none of the α_k is assumed to be 1. □

This lemma is used to derive the following extension of our recurrence relation.

Theorem 11.9.2. *Let the ϕ_n be generated by the previous recurrence relation. Consider an integer $n \geq j + 2 \geq 3$ and assume that $\alpha_n \notin \{\alpha_{n-1}, \alpha_{n-2}, \ldots, \alpha_{j+1}\}$. Then there exist constants a_k, $k = 1, \ldots, n - j - 1$ and constants b_k, $k = n - j - 1, n - j - 2$, all depending on n and j, such that*

$$\phi_n(z) = a_1 \phi_{n-1}(z) + \cdots + a_{n-j-1} \phi_{j+1}(z)$$
$$+ b_{n-j-1} \frac{Z_n(z)}{Z_{j+1}(z)} \phi_{j+1}(z) + b_{n-j-2} C_{j+2} \frac{Z_n(z)}{Z_j(z)} \phi_j(z).$$

Proof. For $n = j + 2 \geq 3$, the formula reduces to

$$\phi_n(z) = a_1 \phi_{n-1}(z) + b_1 \frac{Z_n(z)}{Z_{n-1}(z)} \phi_{n-1}(z) + b_0 C_n \frac{Z_n(z)}{Z_{n-2}(z)} \phi_{n-2}(z),$$

which, by the previous lemma, is seen to be the recurrence itself in the form (11.40). To prove the induction step, assume that for some k with $n - 2 \geq k \geq j + 2$ we have

$$\phi_n(z) = a_1 \phi_{n-1}(z) + \cdots + a_{n-k} \phi_k(z)$$
$$+ b_{n-k} \frac{Z_n(z)}{Z_k(z)} \phi_k(z) + b_{n-k-1} C_{k+1} \frac{Z_n(z)}{Z_{k-1}(z)} \phi_{k-1}(z).$$

We apply (11.40) to ϕ_k in the right-hand side, which gives

$$\phi_n - a_1 \phi_{n-1} - \cdots - a_{n-k} \phi_k$$
$$= b_{n-k} \frac{Z_n}{Z_k} \left[\left(E_k + \frac{B_k}{Z_{k-1}} \right) Z_k \phi_{k-1} + C_k \frac{Z_k}{Z_{k-2}} \phi_{k-2} \right]$$
$$+ b_{n-k-1} C_{k+1} \frac{Z_n}{Z_{k-1}} \phi_{k-1}$$
$$= \left[b_{n-k} E_k + \frac{b_{n-k} B_k + b_{n-k-1} C_{k+1}}{Z_{k-1}} \right] Z_n \phi_{k-1} + b_{n-k} \frac{Z_n}{Z_{k-2}} C_k \phi_{k-2}.$$

Since $\alpha_n \neq \alpha_k$, we can apply part 1 of the previous lemma for the bracketed expression to write it as

$$\frac{a_{n-k-1}}{Z_n} + \frac{b_{n-k+1}}{Z_{k-1}}.$$

Hence

$$\phi_n - a_1 \phi_{n-1} - \cdots - a_{n-k}\phi_k = a_{n-k-1}\phi_{k-1} + b_{n-k+1}\frac{Z_n}{Z_{k-1}}\phi_{k-1}$$

$$+ b_{n-k}\frac{Z_n}{Z_{k-2}}\phi_{k-2}.$$

This proves the induction step.   □

Applying the recurrence once more on the extended recurrence gives another form of the extended recurrence as in the following theorem.

Theorem 11.9.3. *Let the ϕ_n be generated by the previous recurrence relation. Consider an integer $n \geq j + 2 \geq 3$ and assume that $\alpha_n \notin \{\alpha_{n-1}, \alpha_{n-2}, \ldots, \alpha_{j+1}\}$. Then there exist constants a_k, $k = 1, \ldots, n - j - 1$, b_{n-j-1}, b'_{n-j}, and a'_{n-j}, all depending on n and j, such that*

$$\phi_n(z) = a_1 \phi_{n-1}(z) + \cdots + a_{n-j-1}\phi_{j+1}(z) + a'_{n-j}\phi_j(z)$$

$$+ b'_{n-j}Z_n(z)\phi_j(z) + b_{n-j-1}C_{j+1}\frac{Z_n(z)}{Z_{j-1}(z)}\phi_{j-1}(z).$$

Proof. Substitute in the next to last term of the extended recurrence of the previous theorem

$$\frac{\phi_{j+1}}{Z_{j+1}} = \left(E_{j+1} + \frac{B_{j+1}}{Z_j} \right) \phi_j + \frac{C_{j+1}}{Z_{j-1}}\phi_{j-1},$$

so that the last two terms become

$$Z_n \left(b_{n-j-1}E_{j+1} + \frac{b_{n-j-1}B_{j+1} + b_{n-j-2}C_{j+2}}{Z_j} \right) \phi_j + b_{n-j-1}C_{j+1}\frac{Z_n}{Z_{j-1}}\phi_{j-1}.$$

Now apply part 2 of Lemma 11.9.1 to the bracketed expression to write it as

$$b'_{n-j} + \frac{a'_{n-j}}{Z_n}$$

and the desired formula follows as in the previous theorem.   □

Now we will prove the Favard theorem. We recall all the basic assumptions made:

(A1) $\alpha_k \neq \alpha_\emptyset, k = 1, 2, \ldots$.
(A2) ϕ_n is generated by the recurrence (11.39) or, equivalently, by (11.40).
(A3) $\phi_n \in \mathcal{L}_n \setminus \mathcal{L}_{n-1}, n = 1, 2, \ldots, 0 \neq \phi_0 \in \mathbb{R}$.
(A4) $E_n, B_n \in \mathbb{R}, n = 1, 2, \ldots$.
(A5) $E_n \neq 0, n = 1, 2, \ldots$.
(A6) $E_n = -C_n E_{n-1}, n = 2, 3, \ldots$.

If we set $\phi_n = p_n/\pi_n^*$, with $p_n \in \mathcal{P}_n$, then (A3) implies that $p_n(\alpha_n) \neq 0$ whereas (A5) implies regularity, that is, $p_n(\alpha_{n-1}) \neq 0$. Assume that we have constructed a functional M such that $\{\phi_n\}$ is orthogonal with respect to m. The conditions (A4) and (A6) imply that all the coefficients in the recurrence relations are real and this guarantees that the functional is real. Assumption (A6) together with $M\{\phi_0^2\} = 1$ imply positivity of M as well as the normalization $M\{\phi_n^2\} = 1, n = 0, 1, 2, \ldots$. Assumptions (A5) and (A6) imply that $C_n \neq 0$ for $n = 2, 3, \ldots$.

We first note that we should only define the functional M such that we have orthogonality, because if we define $M\{\phi_0^2\} = 1$, then the assumption (A6) guarantees that, if we have orthogonality, then we have also defined all the norms to be 1 because of (11.42).

Defining M on $\mathcal{L}_\infty$ is simple. We set by definition

$$M\{\phi_0\} = 1/\phi_0 \quad \text{and} \quad M\{\phi_k\} = 0, \quad k = 1, 2, \ldots.$$

Note that the definition of $M\{\phi_0\}$ takes care of the normalization. The problem is to extend M to $\mathcal{R}_\infty = \mathcal{L}_\infty \cdot \mathcal{L}_\infty$ such that $M\{\phi_k\phi_l\} = 0$ for $k \neq l$. Therefore, we consider the following table of products:

$$
\begin{array}{cccccc}
\phi_0\phi_1 & \phi_0\phi_2 & \phi_0\phi_3 & \phi_0\phi_4 & \phi_0\phi_5 & \cdots \\
& \phi_1\phi_2 & \phi_1\phi_3 & \phi_1\phi_4 & \phi_1\phi_5 & \cdots \\
& & \phi_2\phi_3 & \phi_2\phi_4 & \phi_2\phi_5 & \cdots \\
& & & \phi_3\phi_4 & \phi_3\phi_5 & \cdots \\
& & & & \ddots &
\end{array}
$$

When M is applied to these entries, we should always get zero. The first row (row 0) is dealt with by our previous definition of M on $\mathcal{L}_\infty$. For the subsequent rows, we consider $B_{-1} = \{\phi_0\}$ and

$$B_n = B_{n-1} \cup \{\phi_n\phi_l : l = n+1, n+2, \ldots\}, \quad n \geq 0.$$

Thus B_n contains the first $n+1$ rows in the above scheme (and ϕ_0). Furthermore, define

$$B_{n,n} = B_{n-1} \quad \text{and} \quad B_{n,k} = B_{n,k-1} \cup \{\phi_n\phi_k\}, \quad k > n,$$

that is, $B_{n,k}$, $k = n, n+1, \ldots$ adds to B_{n-1} (the first n rows) the elements of the $(n+1)$st row, one by one. These sets generate the following spaces:

$$\mathcal{R}_{n,k} = \text{span } B_{n,k}, \quad n = 0, 1, \ldots; \; k = n, n+1, \ldots.$$

(Note that $\mathcal{L}_\infty = \mathcal{R}_{1,1} = \cup_{k=0}^\infty \mathcal{R}_{0,k}$.) The strategy is to obtain a definition of M on the successive spaces

$$\mathcal{R}_{1,2}, \mathcal{R}_{1,3}, \ldots; \mathcal{R}_{2,2}, \mathcal{R}_{2,3}, \mathcal{R}_{2,4}, \ldots; \mathcal{R}_{3,3}, \mathcal{R}_{3,4}, \mathcal{R}_{3,5}, \ldots; \mathcal{R}_{4,4}, \ldots$$

such that we have the required orthogonality relations. Thus we walk through the table above row by row and in each from left to right. Each time we consider the next product $\phi_n\phi_k$, we have to check whether that product is in a subspace on which M has already been defined or not. If the new product is not in a previous subspace, we can just *define* $M\{\phi_n\phi_k\}$ to be zero; otherwise we have to *prove* that it gives zero.

Eventually we have M defined on the subspace

$$\mathcal{R}'_\infty = \text{span } \{\cup B_{n,k} : n = 0, 1, \ldots; k = n+1, n+2, \ldots\} \subset \mathcal{R}_\infty$$

such that

$$M\{\phi_0^2\} = 1, \qquad M\{\phi_n\phi_k\} = 0, \quad n = 0, 1, \ldots; \; k = n+1, n+2, \ldots$$

It will turn out that in fact $\mathcal{R}'_\infty = \mathcal{R}_\infty$ and by assumption (A6) we then also have $M\{\phi_n^2\} = 1$, $n = 0, 1, \ldots$. We now elaborate these successive steps.

Row 0

We set by definition

$$M\{\phi_0\} = 1/\phi_0, \qquad M\{\phi_k\} = 0, \quad k = 1, 2, \ldots,$$

which defines M on $\mathcal{L}_\infty$.

Row 1

Initialization: $M\{\phi_1\phi_2\}$

If $\phi_1\phi_2 \notin \mathcal{L}_\infty$, we set by definition $M\{\phi_1\phi_2\} = 0$.

If $\phi_1\phi_2 \in \mathcal{L}_\infty$, then there is a k such that

$$\phi_1\phi_2 = \frac{p_1 p_2}{\pi_1^* \pi_2^*} = \frac{q_k}{\pi_k^*}, \quad q_k \in \mathcal{P}_k.$$

This k should be at least 3 because otherwise we would find that $p_1(\alpha_1)p_2(\alpha_1) = 0$ while by our assumptions both $p_1(\alpha_1)$ and $p_2(\alpha_1)$ are nonzero. Thus

$$\frac{\pi_k^*}{\pi_2^*} p_1 p_2 = \pi_1^* q_k.$$

Therefore π_k^*/π_2^* has a zero in α_1. This means that there is some $m \geq 3$ such that $\alpha_m = \alpha_1$. Let m be the smallest such m. We distinguish between $m = 3$ and $m \geq 4$.

(a) $m = 3$

Thus $\alpha_1 = \alpha_3$ and we get from the recurrence relation (11.39)

$$\phi_3 = \left(A_3 + \frac{B_3}{Z_1}\right) Z_3 \phi_2 + C_3 \frac{Z_3}{Z_1} \phi_1 = (A_3 Z_1 + B_3)\phi_2 + C_3 \phi_1.$$

Note that $A_3 \neq 0$ because otherwise ϕ_3 would be a linear combination of ϕ_1 and ϕ_2, contradicting assumption (A3). Apply M to the previous relation and this gives

$$M\{\phi_3\} = A_3 M\{Z_1 \phi_2\} + B_3 M\{\phi_2\} + C_3 M\{\phi_1\}.$$

Because $M\{\phi_3\} = M\{\phi_2\} = M\{\phi_1\} = 0$, it follows that $M\{Z_1\phi_2\} = 0$ and consequently that also $M\{\phi_1\phi_2\} = 0$.

(b) $m \geq 4$

In this case we know that $\alpha_1 = \alpha_m \notin \{\alpha_{m-1}, \alpha_{m-2}, \ldots, \alpha_2\}$. We now use the extended recurrence

$$\phi_m = a_1\phi_{m-1} + \cdots + a_{m-3}\phi_3 + a'_{m-2}\phi_2 + b'_{m-2}Z_1\phi_2 + b_{m-3}C_3\phi_1.$$

Note that $b'_{m-2} \neq 0$; otherwise $\phi_m \in \mathcal{L}_{m-1}$, contradicting assumption (A3). Applying M to this relation and using $M\{\phi_k\} = 0$ for $k = 1, 2, \ldots, m$, we find that $M\{Z_1\phi_2\} = 0$, which implies $M\{\phi_1\phi_2\} = 0$.

Induction step for row 1

We have to prove that $M\{\phi_1\phi_k\} = 0$, given that $M\{\phi_i\phi_k\} = 0$ for $\phi_i\phi_k \in B_{1,k-1}$. This is treated as in the general induction step for row n. We refer to that step below.

Row n, $n \geq 2$

Initialization: $M\{\phi_n\phi_{n+1}\} = 0$

If $\phi_n\phi_{n+1} \notin \mathcal{R}_{n,n}$ then we define $M\{\phi_n\phi_{n+1}\} = 0$.

If $\phi_n\phi_{n+1} \in \mathcal{R}_{n,n}$ then there is some k such that

$$\phi_n\phi_{n+1} + \phi_1\frac{q_{k1}}{\pi_k^*} + \cdots + \phi_{n-1}\frac{q_{k,n-1}}{\pi_k^*} = \frac{q_{k0}}{\pi_k^*}, \quad q_{ki} \in \mathcal{P}_k,$$

that is,

$$\frac{p_n p_{n+1}}{\pi_n^*\pi_{n+1}^*} + \frac{p_n q_{k1}}{\pi_1^*\pi_k^*} + \cdots + \frac{p_{n-1}q_{k,n-1}}{\pi_{n-1}^*\pi_k^*} = \frac{q_{k0}}{\pi_k^*}$$

and thus

$$p_n p_{n+1} + \frac{\pi_n^*\pi_{n+1}^*}{\pi_1^*\pi_k^*}p_1 q_{k1} + \cdots + \frac{\pi_n^*\pi_{n+1}^*}{\pi_{n-1}^*\pi_k^*}p_{n-1}q_{k,n-1} = \pi_n^*\frac{\pi_{n+1}^*}{\pi_k^*}q_{k0}.$$

If $k \leq n+1$, then for $z = \alpha_n$, we get $p_n(\alpha_n)p_{n+1}(\alpha_n) = 0$, which is contradicting our assumptions, and hence $k > n+1$. But then

$$\frac{\pi_k^*}{\pi_{n+1}^*}p_n p_{n+1} + \frac{\pi_n^*}{\pi_1^*}p_1 q_{k1} + \cdots + \frac{\pi_n^*}{\pi_{n-1}^*}p_{n-1}q_{k,n-1} = \pi_n^*q_{k0}$$

shows that π_k^*/π_{n+1}^* should have a zero at α_n. This means that there is some $m \geq n+2$ such that $\alpha_m = \alpha_n$. Let m be the smallest such index. We distinguish again between (a) $m = n+2$ and (b) $m > n+2$.

(a) $m = n+2$

Thus $\alpha_m = \alpha_n = \alpha_{n+2}$. We use the recurrence relation (11.39):

$$\phi_{n+2} = \left(A_{n+2} + \frac{B_{n+2}}{Z_n}\right)Z_{n+2}\phi_{n+1} + C_{n+2}\frac{Z_{n+2}}{Z_n}\phi_n$$

$$= A_{n+2}Z_n\phi_{n+1} + B_{n+2}\phi_{n+1} + C_{n+2}\phi_n.$$

The coefficient A_{n+2} is nonzero because otherwise $\phi_{n+2} \in \mathcal{L}_{n+1}$. We multiply this relation by b_{n-1} and apply M to the result to get

$$M\{b_{n-1}\phi_{n+2}\} = A_{n+2}M\{b_n\phi_{n+1}\} + B_{n+2}M\{b_{n-1}\phi_{n+1}\} + C_{n+2}M\{b_{n-1}\phi_n\}.$$

Using the orthogonalities given by the induction hypothesis, we find that the left-hand side and the last two terms on the right-hand side vanish so that $M\{b_n\phi_{n+1}\} = 0$ and thus also $M\{\phi_n\phi_{n+1}\} = 0$.

(b) $m \geq n + 3$

In this case $\alpha_m \notin \{\alpha_{m-1}, \alpha_{m-2}, \ldots, \alpha_{n+2}\}$. We use the extended recurrence relation to write (note $Z_m = Z_n$)

$$\phi_m = a_1\phi_{m-1} + \cdots + a_{m-n-2}\phi_{n+2} + a'_{m-n-1}\phi_{n+1}$$
$$+ b'_{m-n-1}Z_n\phi_{n+1} + b_{m-n-2}C_{n+2}\phi_n.$$

The coefficient $b'_{m-n-1} \neq 0$; otherwise $\phi_m \in \mathcal{L}_{m-1}$. We multiply this again by b_{n-1} and apply M to the result:

$$M\{b_{n-1}\phi_m\} = a_1 M\{b_{n-1}\phi_{m-1}\} + \cdots + a_{m-n-2}M\{b_{n-1}\phi_{n+2}\}$$
$$+ a'_{m-n-1}M\{b_{n-1}\phi_{n+1}\} + b'_{m-n-1}M\{b_n\phi_{n+1}\}$$
$$+ b_{m-n-2}C_{n+2}M\{b_{n-1}\phi_n\}.$$

By the induction hypothesis all the terms vanish except the first one on the second row. Thus $M\{b_n\phi_{n+1}\} = 0$ so that also $M\{\phi_n\phi_{n+1}\} = 0$.

Induction step for row n

We now consider $n \geq 2$ and $j \geq n + 2$ for which we know that

$$M\{\phi_m\phi_k\} = 0, \quad m = 0, 1, \ldots, n-1, \; k \geq m+1,$$
$$M\{\phi_n\phi_k\} = 0, \quad k = n+1, n+2, \ldots, j-1.$$

We have to prove that $M\{\phi_n\phi_j\} = 0$.

From the recurrence relation (11.40), we get

$$\phi_j = Z_j\left(E_j + \frac{B_j}{Z_{j-1}}\right)\phi_{j-1} + C_j\frac{Z_j}{Z_{j-2}}\phi_{j-2}$$

and thus

$$\frac{b_n}{Z_j}\phi_j = \left(E_j + \frac{B_j}{Z_{j-1}}\right)b_n\phi_{j-1} + C_j\frac{b_n}{Z_{j-2}}\phi_{j-2}.$$

Applying M to this expression gives

$$M\left\{\frac{b_n}{Z_j}\phi_j\right\} = E_j M\{b_n\phi_{j-1}\} + B_j M\left\{\frac{b_n}{Z_{j-1}}\phi_{j-1}\right\} + C_j M\left\{\frac{b_n}{Z_{j-2}}\phi_{j-2}\right\}.$$

Because b_n and b_n/Z_{j-1} are both in $\mathcal{L}_n$ and $n \leq j-2$, it follows by the induction hypothesis that the first two terms in the right-hand side are zero. For the third term, we note that for $j = n+2, b_n/Z_{j-2} = b_{n-1}$, so that again by the induction hypothesis the third term is zero. However, if $j > n+2$, then $b_n/Z_{j-2} \in \mathcal{L}_n$ with $n < j-2$, so that again the third term is zero by the induction hypothesis. Thus in any case, the right-hand side vanishes completely. Therefore, we conclude that

$$M\{b_n\phi_{j-1}\} = 0.$$

Now we distinguish two cases: (A) $\alpha_n \neq \alpha_j$ and (B) $\alpha_n = \alpha_j$.

(A) $\alpha_n \neq \alpha_j$

We may then set by Lemma 11.9.1 (2) that $1/Z_j = 1/Z_n + c$, with c a nonzero constant. Hence

$$M\{b_n\phi_{j-1}\} = M\{b_{n-1}\phi_j\} + cM\{b_n\phi_j\} = 0.$$

The first term in the middle part is zero by the induction hypothesis and thus we find that $M\{b_n\phi_j\} = 0$ and thus also $M\{\phi_n\phi_j\} = 0$.

(B) $\alpha_n = \alpha_j$

Here again, we distinguish two cases (1) $\phi_n\phi_j \notin \mathcal{R}_{n,j-1}$ and (2) $\phi_n\phi_j \in \mathcal{R}_{n,j-1}$.

(1) If $\phi_n\phi_j \notin \mathcal{R}_{n,j-1}$, then we may define $M\{\phi_n\phi_j\} = 0$ and we are done.

(2) If $\phi_n\phi_j \in \mathcal{R}_{n,j-1}$ then there exist an integer k and constants c_i and polynomials $q_{k,i} \in \mathcal{P}_k$ such that

$$\phi_n\phi_j + c_{j-1}\phi_n\phi_{j-1} + \cdots + c_{n+1}\phi_n\phi_{n+1}$$
$$+ \phi_{n-1}\frac{q_{k,n-1}}{\pi_k^*} + \cdots + \phi_1\frac{q_{k,1}}{\pi_k^*} = \frac{q_{k,0}}{\pi_k^*}$$

or, setting $\phi_i = p_i/\pi_i^*$,

$$p_n p_j + c_{j-1}\frac{\pi_j^*}{\pi_{j-1}^*}p_n p_{j-1} + \cdots + c_{n+1}\frac{\pi_j^*}{\pi_{n+1}^*}p_n p_{n+1}$$
$$+ \frac{\pi_n^*\pi_j^*}{\pi_{n-1}^*\pi_k^*}p_{n-1}q_{k,n-1} + \cdots + \frac{\pi_n^*\pi_j^*}{\pi_1^*\pi_k^*}p_1 q_{k,1} = \pi_n^*\frac{\pi_j^*}{\pi_k^*}q_{k,0}.$$

The index k can not be less than $j+1$, for otherwise, putting $z = \alpha_j$ in the previous relation would yield $p_n(\alpha_j)p_j(\alpha_j) = p_n(\alpha_n)p_j(\alpha_j) = 0$, which contradicts our assumptions. By rewriting the previous identity as

$$\frac{\pi_k^*}{\pi_j^*}p_n p_j + c_{j-1}\frac{\pi_k^*}{\pi_{j-1}^*}p_n p_{j-1} + \cdots + c_{n+1}\frac{\pi_k^*}{\pi_{n+1}^*}p_n p_{n+1}$$
$$+ \frac{\pi_n^*}{\pi_{n-1}^*}p_{n-1}q_{k,n-1} + \cdots + \frac{\pi_n^*}{\pi_1^*}p_1 q_{k,1} = \pi_n^* q_{k,0}$$

and putting $z = \alpha_n$ we find that π_k^*/π_j^* is zero for $z = \alpha_n$. Thus there must be an index $m \geq j + 1$ such that $\alpha_m = \alpha_n$. Let m be the smallest such index. We once more distinguish two cases: (a) $m = j + 1$ and (b) $m \geq j + 2$.

(a) $m = j + 1$: Thus we have now that $\alpha_n = \alpha_j = \alpha_{j+1}$.
By the recurrence (11.40), we have

$$\phi_{j+1} = \left(E_{j+1} + \frac{B_{j+1}}{Z_j} \right) Z_{j+1}\phi_j + C_{j+1}\frac{Z_{j+1}}{Z_{j-1}}\phi_{j-1}.$$

Because $Z_{j+1} = Z_n = Z_j$, we get after multiplication with b_{n-1}

$$b_{n-1}\phi_{j+1} = E_{j+1}b_n\phi_j + B_{j+1}b_{n-1}\phi_j + C_{j+1}\frac{b_n}{Z_{j-1}}\phi_{j-1}.$$

After applying M, this gives

$$M\{b_{n-1}\phi_{j+1}\} = E_{j+1}M\{b_n\phi_j\} + B_{j+1}M\{b_{n-1}\phi_j\}$$
$$+ C_{j+1}M\left\{ \frac{b_n}{Z_{j-1}}\phi_{j-1} \right\}.$$

Since $j \geq n + 2$, the left-hand side and the second term in the right-hand side are zero by the induction hypothesis, but also the last term is zero by the same argument because $b_n/Z_{j-1} \in \mathcal{L}_n$. Thus it follows that $M\{b_n\phi_j\} = 0$ so that $M\{\phi_n\phi_j\} = 0$.

(b) $m \geq j + 2$: Thus we have $\alpha_m = \alpha_n = \alpha_j \notin \{\alpha_{m-1}, \alpha_{m-2}, \ldots, \alpha_{j+1}\}$.
By the extended recurrence relation of Theorem 11.9.2, we have

$$\phi_m = a_1\phi_{m-1} + \cdots + a_{m-j-1}\phi_{j+1} + b_{m-j-1}\frac{Z_m}{Z_{j+1}}\phi_{j+1} + b_{m-j-2}C_{j+2}\phi_j.$$

Note that $b_{m-j-1} \neq 0$, for otherwise $\phi_m \in \mathcal{L}_{m-1}$. By the extended recurrence relation of Theorem 11.9.3, we have

$$\phi_m = a_1\phi_{m-1} + \cdots + a_{m-j-1}\phi_{j+1} + a'_{m-j}\phi_j$$
$$+ b'_{m-j}Z_n\phi_j + b_{m-j-1}C_{j+1}\frac{\phi_{j-1}}{Z_{j-1}}$$

with (see the proof of Theorem 11.9.3) $b'_{m-j} = cE_{j+1}b_{m-j-1}$ and c being a nonzero constant. Because $b_{m-j-1} \neq 0$, also $b'_{m-j} \neq 0$. Now we multiply this relation with b_{n-1} and apply M to obtain

$$M\{b_{n-1}\phi_m\} = a_1M\{b_{n-1}\phi_{m-1}\} + \cdots + a_{m-j-1}M\{b_{n-1}\phi_{j+1}\}$$
$$+ a'_{m-j}M\{b_{n-1}\phi_j\} + b'_{m-j}M\{b_n\phi_j\}$$
$$+ b_{m-j-1}C_{j+1}M\left\{ \frac{b_n}{Z_{j-1}}\phi_{j-1} \right\}.$$

11. The boundary case

By the induction hypothesis (recall $m \geq j + 1$), all the terms vanish, except for the first term on the second line. Thus $M\{b_n\phi_j\} = 0$ and hence $M\{\phi_n\phi_j\} = 0$.

This concludes the induction step for row n.

The diagonal $M\{\phi_n^2\}$

So far we have defined M on $\mathcal{R}'_\infty$ such that

$$M\{\phi_0^2\} = 1 \quad \text{and} \quad M\{\phi_i\phi_j\} = 0, \quad i \neq j.$$

We also have to satisfy $M\{\phi_n^2\} = 1, n \geq 1$.

By the recurrence relation, we have

$$\frac{Z_1}{Z_2}\phi_2 = E_2 Z_1 \phi_1 + B_2 \phi_1 + C_2 \phi_0.$$

Noting that, by Lemma 11.9.1, $Z_1/Z_2 = c_1 + c_2 Z_1$, it follows that the left-hand side is in the span of $\{\phi_1\phi_1, \phi_2\}$. Hence it follows that

$$\phi_1 Z_1 \in \text{span}\{\phi_0, \phi_1\phi_2, \phi_1\phi_2\} \subset \mathcal{R}'_\infty.$$

Thus also $\phi_1^2 \in \mathcal{R}'_\infty$. Similarly, it follows in general, for $n \geq 3$, that from the recurrence relation we get

$$\frac{b_n}{Z_{n+1}}\phi_{n+1} = E_{n+1}b_n\phi_n + B_{n+1}b_{n-1}\phi_n + C_{n+1}\frac{b_n}{Z_{n-1}}\phi_{n-1}.$$

Again by Lemma 11.9.1 we can replace $1/Z_{n+1}$ and $1/Z_{n-1}$ by $1/Z_n$ plus some constant. Thus the previous relation implies that

$$b_n\phi_n \in \text{span}\{\phi_{n-1}\phi_n, \phi_{n-1}\phi_{n-1}, \phi_{n-1}\phi_{n+1}, \phi_n\phi_{n+1}\}.$$

Therefore, $\phi_n^2 \in \mathcal{R}'_\infty$ for all $n \geq 0$, which means that $\mathcal{R}'_\infty = \mathcal{R}_\infty = \mathcal{L}_\infty \cdot \mathcal{L}_\infty$. In other words, we have defined M on the whole space $\mathcal{R}_\infty$ and by our assumption (A6) we know that we have automatically $M\{\phi_n^2\} = 1$. Thus we have now proved the Favard theorem.

Theorem 11.9.4 (Favard). *Let $\{\phi_n\}$ be a sequence of rational functions generated by the recurrence (11.39) or, equivalently, by (11.40) and assume that (A1)–(A6) are satisfied. Then there exists a functional M on $\mathcal{R}_\infty = \mathcal{L}_\infty \cdot \mathcal{L}_\infty$ such that*

$$\langle f, g \rangle = M\{fg_*\}$$

defines a real positive inner product on $\mathcal{L}_\infty$ for which the ϕ_n form an orthonormal system.

Proof. By the previous analysis, the definition is given such that the orthonormality is satisfied. It is easily proved that $M\{h_*\} = \overline{M\{h\}}$ for any $h \in \mathcal{R}_\infty$ since $h = f g_*$, with $f = \sum a_i \phi_i \in \mathcal{L}_\infty$ and $g = \sum b_i \phi_i \in \mathcal{L}_\infty$, so that

$$M\{h_*\} = M\left\{\sum \overline{a}_i \phi_{i*} \cdot \sum b_j \phi_j\right\}$$

$$= \sum \overline{a}_i b_i = \overline{\sum \overline{b}_i a_i} = \overline{M\{h\}}.$$

Also, positivity is guaranteed by

$$M\{f f_*\} = \sum |b_i|^2 > 0. \qquad \square$$

11.10. Interpolation

Suppose now that we have a solution of the moment problem in $\mathcal{R}_\infty$ so that we have the inner product defined as before by

$$\langle f, g \rangle = \int f(t)\overline{g(t)}\, d\mathring{\mu}(t), \quad f, g \in \mathcal{L}_\infty, \tag{11.45}$$

where we assume the normalization $\int \mathring{\mu}(t) = 1$.

We make the following observation (see also the appendix of Ref. [62]):

Lemma 11.10.1. If μ is a positive measure on $\mathbb{T}$, such that

$$\int \frac{d\mu(t)}{|\pi_n^*(t)|^2} < \infty,$$

then $\mathcal{L}_n$, as a subspace of $L_2(\mu)$, is in the closure $\mathcal{P}$ of

$$\text{span}\{1, e^{i\theta}, e^{2i\theta}, \ldots\}.$$

Proof. Define

$$f(t) = \frac{p_n(t)}{\pi_n^*(t)}, \quad p_n \in \mathcal{P}_n, \ t \in \mathbb{T}$$

and

$$f_\epsilon(t) = \frac{p_n(t)}{\pi_n^*(rt)}, \quad \text{with} \quad 0 \le r = 1 - \epsilon < 1.$$

For a fixed $\alpha \in \mathbb{T}$ and $t \in \mathbb{T}$ arbitrary, we have

$$\frac{1}{|rt - \alpha|} - \frac{1}{|t - \alpha|} \leq \left| \frac{1}{rt - \alpha} - \frac{1}{t - \alpha} \right|$$

$$\leq \frac{1 - r}{|rt - \alpha||t - \alpha|} \leq \frac{1}{|t - \alpha|}.$$

Hence $|rt - \alpha|^{-1} \leq 2|t - \alpha|^{-1}$. Thus we can find a constant c_n such that

$$|f_\epsilon(t)| < \frac{c_n}{|\pi_n^*(t)|}.$$

Because $c_n / \pi_n^*(t) \in L_2(\mu)$, also $f_\epsilon \in L_2(\mu)$. Obviously, f_ϵ is analytic in $\overline{\mathbb{D}}$ so that $f_\epsilon \in \mathcal{P}$. Since there is also a constant d_n such that

$$\int |f(t)|^2 d\mu(t) \leq d_n \int \frac{d\mu(t)}{|\pi_n^*(t)|^2},$$

then $f \in L_2(\mu)$. It remains to show that in $L_2(\mu)$

$$\lim_{\epsilon \downarrow 0} \| f_\epsilon - f \| = 0.$$

Since by the Schwarz inequality

$$\| f_\epsilon - f \|^2 \leq \| f_\epsilon \|^2 + \| f \|^2 + 2 \| f_\epsilon \| \| f \|,$$

the right-hand side being uniformly bounded (not depending on ϵ), we can apply the dominated convergence theorem to find that

$$\lim_{\epsilon \downarrow 0} \| f_\epsilon - f \| = \| \lim_{\epsilon \downarrow 0} f_\epsilon - f \| = 0. \qquad \square$$

The analog for the real line is

Lemma 11.10.2. If $\mathring{\mu}$ is a finite measure on $\overline{\mathbb{R}}$, such that

$$\int |b_n(t)|^2 d\mathring{\mu}(t) < \infty, \qquad b_n(z) = \prod_{k=1}^{n} \frac{z}{1 - z/\alpha_k},$$

then $\mathcal{L}_n$, as a subspace of $L_2(\mathring{\mu})$, is in the closure $\mathcal{P}$ of

$$\text{span}\{1, t, t^2, \ldots\}.$$

Proof. Define

$$f(t) = \frac{p_n(t)}{\omega_n^*(t)}, \quad p_n \in \mathcal{P}_n, \ t \in \mathbb{R}, \qquad \omega_n(t) = \prod_{k=1}^{n} \left(1 - \frac{z}{\alpha_k}\right)$$

and

$$f_\epsilon(t) = \frac{p_n(t)}{\omega_n^*(t + \mathrm{i}\epsilon)}, \quad \text{with } 0 < \epsilon \leq 1.$$

For a fixed $\alpha \in \mathbb{R}$ and $t \in \overline{\mathbb{R}}$ arbitrary, we have

$$\frac{1}{\left|1 - \frac{t+\mathrm{i}\epsilon}{\alpha}\right|} - \frac{1}{|1 - t/\alpha|} \leq \left|\frac{1}{1 - \frac{t+\mathrm{i}\epsilon}{\alpha}} - \frac{1}{1 - 1/\alpha}\right| = \frac{|\alpha\epsilon|}{|\alpha - t - \mathrm{i}\epsilon||t - \alpha|}.$$

Since $|\alpha - t - \mathrm{i}\epsilon| \geq \epsilon$, the right-hand side is bounded by $1/|1 - t/\alpha|$ and thus is

$$\frac{1}{\left|1 - \frac{t+\mathrm{i}\epsilon}{\alpha}\right|} \leq \frac{2}{|1 - t/\alpha|}.$$

Thus we can find constant c_n and d_n such that

$$|f_\epsilon(t)| < c_n |f(t)| < d_n |b_n(t)|.$$

Because the right-hand side is in $L_2(\mathring{\mu})$, f and f_ϵ are also in $L_2(\mathring{\mu})$. Since $f_\epsilon \in H(\mathbb{U})$ we find that $f_\epsilon \in \mathcal{P}$. The rest of the proof is as in the circle case. $\square$

The previous arguments give, of course, that $\mathcal{L}_\infty$ is dense in the space $\mathcal{P}$ with respect to the $L_2(\mathring{\mu})$ metric both in the case of the circle and in the case of the line.

Note that the conditions

$$\int \frac{d\mu(t)}{|\pi_n^*(t)|^2} < \infty \quad \text{for } \mathbb{T} \quad \text{or} \quad \int |b_n(t)|^2 d\mathring{\mu}(t) < \infty \quad \text{for } \mathbb{R}$$

imply that the measure does not have a mass point at any of the $\alpha \in A_n = \{\alpha_1, \ldots, \alpha_n\}$. Consequently, defining as usual the Riesz–Herglotz–Nevanlinna transform

$$\Omega_\mu(z) = \int D(t, z) \, d\mathring{\mu}(t), \quad z \in \mathbb{O}$$

then the boundary function $\Omega_\mu(t)$, $t \in \overline{\partial\mathbb{O}}$, will not have a jump at α (see Ref. [62]). If $\alpha^\#$ is the number of times that α appears in A_n, then also the derivatives $\Omega_\mu^{(k)}(t)$ will not be discontinuous at α for $k = 0, 1, \ldots, \alpha^\# - 1$. In particular, for the case $\partial \mathbb{O} = \mathbb{R}$ and $\alpha = \infty$, we interpret continuity at ∞ of $f(t)$ to mean that $\lim_{t \to +\infty} f(t) = \lim_{t \to -\infty} f(t)$.

The kind of limits that were used in the last two lemmas will be needed frequently in this section. For the case of the circle, these limits will be radial limits; for the case of the real line, these are limits along vertical lines. We could call it orthogonal limits in general. We shall indicate this by a special notation defined as follows:

$$\lim_{z \mapsto \alpha} f(z) = \lim_{r \uparrow 1} f(r\alpha), \quad \alpha \in \mathbb{T}$$

and

$$\lim_{z \mapsto \alpha} f(z) = \lim_{\epsilon \downarrow 0} f(\alpha + i\epsilon), \quad \alpha \in \mathbb{R},$$

and

$$\lim_{z \mapsto \infty} f(z) = \lim_{z \mapsto 0} f(1/z).$$

If $\alpha \notin \overline{\partial \mathbb{O}}$, we can interpret the orthogonal limit as an ordinary limit.

We can now give an analog of Theorem 2.1.3, namely that the inner product in $\mathcal{L}_n$ is characterized by values of the class $\mathcal{C}$ function Ω_μ and possibly its derivatives in the points α_i. Consider the case of the circle $\mathbb{T}$ and let us use the basis $\{w_0, w_1, \ldots, w_n\}$ with

$$w_0 = 1 \quad \text{and} \quad w_k(z) = \frac{1}{\pi_k^*(z)}.$$

Recall that $\langle w_0, w_0 \rangle = \int d\mathring{\mu}(t) = 1 = \Omega_\mu(\beta)$. This is the left top element in the Gram matrix for these basis functions. The other entries in the Gram matrix involve integrals of the form $(k + l \geq 1)$

$$\langle w_k, w_l \rangle = \int \frac{\prod_{i=1}^{l}(-\alpha_i)t^l \, d\mu(t)}{\prod_{i=1}^{k}(t - \alpha_i) \prod_{i=1}^{l}(t - \alpha_i)}.$$

(A product $\prod_{i=1}^{j}$ with $j = 0$ is replaced by 1.) By partial fraction decomposition of the integrand, we see that this expression depends upon integrals of the form

$$\int \frac{t \, d\mu(t)}{(t - \alpha)^{k+1}}, \quad \alpha \in A_n = \{\alpha_1, \ldots, \alpha_n\}, \tag{11.46}$$

where $k = 0, 1, \ldots, 2\alpha^{\#} - 1$ with $\alpha^{\#}$ the number of times that α appears in A_n.

By Lemma 2.1.2, we know that for α replaced by $w \in \mathbb{D}$, such integrals are completely characterized by the values of $\Omega_\mu^{(k)}$, where $\Omega_\mu(w) = \int D(t, w) \, d\mu(t)$

and where the superscript means derivative. Using the estimates as in the proof of Lemma 11.10.1 for the integrand of the integral (11.46) above, it should be clear that we can apply the dominated convergence theorem so that

$$\lim_{w \mapsto \alpha} \Omega_\mu^{(k)}(w) = \lim_{w \mapsto \alpha} \int \frac{t \, d\mu(t)}{(t-w)^{k+1}} = \int \frac{t \, d\mu(t)}{(t-\alpha)^{k+1}}.$$

Thus we find that the integrals of the form (11.46), and hence also the inner product in $\mathcal{L}_n$, are completely characterized by the radial limit values

$$\lim_{w \mapsto \alpha} \Omega_\mu^{(k)}(w), \quad \alpha \in A_n, \; k = 0, 1, \ldots, 2\alpha^\# - 1$$

and by $\Omega_\mu(\beta)$, the latter assumed to be 1 by normalization.

For the real line, a similar argument can be used. We leave the technicalities to the reader. Thus we can formulate the following theorem.

Theorem 11.10.3. *Consider the inner product (11.45) and the Riesz–Herglotz–Nevanlinna transform*

$$\Omega_\mu(z) = \int D(t, z) \, d\mathring{\mu}(t).$$

Then in $\mathcal{L}_n$ this inner product is completely characterized by the values

$$\Omega_\mu(\beta) \quad and \quad \lim_{w \mapsto \alpha} \Omega_\mu^{(k)}(w), \quad \alpha \in A_n, \; k = 0, \ldots, 2\alpha^\# - 1,$$

where $\alpha^\#$ is the multiplicity of α in A_n.

This entails immediately the following result, which says as before that equality of the inner product in $\mathcal{L}_n$ corresponds to an interpolation result for the corresponding Riesz–Herglotz–Nevanlinna transforms. (Recall the definitions $A_n^\beta = \{\beta, \alpha_1, \ldots, \alpha_n\}$ and $\hat{A}_n^\beta = \{\hat{\beta}, \hat{\alpha}_1, \ldots, \hat{\alpha}_n\}$.)

Corollary 11.10.4. *Let μ and ν be two positive measures on $\overline{\partial \mathbb{O}}$ and define the sets (counting multiplicities)*

$$\tilde{A}_n^\beta = A_n^\beta \cup \hat{A}_n^\beta = \{\beta, \hat{\beta}, \alpha_1, \alpha_1, \alpha_2, \alpha_2, \ldots, \alpha_n, \alpha_n\}.$$

Then in $\mathcal{L}_n$ we have $\langle \cdot, \cdot \rangle_{\hat{\mu}} = \langle \cdot, \cdot \rangle_{\hat{\nu}}$ iff

$$\lim_{w \mapsto \alpha} \left[\Omega_\mu^{(k)}(w) - \Omega_\nu^{(k)}(w) \right] = 0, \quad \alpha \in \tilde{A}_n^\beta, \; k = 0, 1, \ldots, \alpha^\# - 1,$$

with $\alpha^\#$ the multiplicity of α in $\tilde{A}_n^\beta$.

Proof. In view of the foregoing, taking into account that the orthogonal limits are replaced by ordinary limits when $\alpha \notin \partial\mathbb{O}$, we need only to explain the $\hat{\beta}$. Therefore, we note that for any positive measure μ, we have, by definition, $[\Omega_\mu(z)]_* = -\Omega_\mu(z)$. Therefore, $\Omega_\mu(\hat{\beta}) = -\overline{\Omega_\mu(\beta)}$. Thus $\Omega_\mu(\beta) = \Omega_\nu(\beta)$ is equivalent with $\Omega_\mu(\hat{\beta}) = \Omega_\nu(\hat{\beta})$. $\square$

We first derive some results involving the quasi-orthogonal functions. Recall

$$Q_n(z, \tau) = \phi_n(z) + \tau \frac{Z_n(z)}{Z_{n-1}(z)} \phi_{n-1}(z), \quad \tau \in \overline{\mathbb{R}},$$

$$P_n(z, \tau) = \psi_n(z) + \tau \frac{Z_n(z)}{Z_{n-1}(z)} \psi_{n-1}(z), \quad \tau \in \overline{\mathbb{R}}.$$

It has also been remarked that, although we called $P_n(z, \tau)$ the functions of the second kind associated with $Q_n(z, \tau)$, it is _not_ true in general that

$$P_n(z, \tau) = \int D(t, z)[Q_n(t, \tau) - Q_n(z, \tau)] \, d\mathring{\mu}(t)$$

unless $\tau = 0$. However, we do have the following:

Lemma 11.10.5. Let $Q_n(z, \tau)$ be the quasi-orthogonal functions and $P_n(z, \tau)$ the associated functions of the second kind. Then for $n \geq 2$

$$\frac{Z_{n-1}(z)}{Z_n(z)} P_n(z, \tau) = \int D(t, z) \left[\frac{Z_{n-1}(t)}{Z_n(t)} Q_n(t, \tau) - \frac{Z_{n-1}(z)}{Z_n(z)} Q_n(z, \tau) \right] d\mathring{\mu}(t).$$

Proof. The right-hand side is

$$\int D(t, z) \left[\frac{Z_{n-1}(t)}{Z_n(t)} \phi_n(t) - \frac{Z_{n-1}(z)}{Z_n(z)} \phi_n(z) \right] d\mathring{\mu}(t)$$

$$+ \tau \int D(t, z) \left[\phi_{n-1}(t) - \phi_{n-1}(z) \right] d\mathring{\mu}(t).$$

The second of these integrals is ψ_{n-1} by definition. The first integral turns out to be

$$\frac{Z_{n-1}(z)}{Z_n(z)} \psi_n(z)$$

by Lemma 11.2.2. We need to check that, as a function of t, $D(t, z)[f(t) - f(z)] \in \mathcal{L}_{n-1}$, where

$$f(z) = \frac{Z_{n-1}(z)}{Z_n(z)} = \frac{(\alpha_\emptyset - \alpha_{n-1})(z - \alpha_n)}{(\alpha_\emptyset - \alpha_{n-1})(z - \alpha_{n-1})}.$$

This is indeed the case because

$$D(t, z)[f(t) - f(z)] = \frac{\alpha_\emptyset - \alpha_{n-1}}{\alpha_\emptyset - \alpha_n} D(t, z) \frac{(t - z)(\alpha_n - \alpha_{n-1})}{(t - \alpha_{n-1})(z - \alpha_{n-1})}.$$

This proves the lemma.   □

Now suppose that $Q_n(z, \tau)$ has precisely n simple zeros $\xi_{ni}(\tau) \in \overline{\partial\mathbb{O}} \setminus A_n$, $A_n = \{\alpha_1, \ldots, \alpha_n\}$, so that we can associate with this quasi-orthogonal function the quadrature formula as in Section 11.6

$$\int f(t)\, d\mathring{\mu}(t) \approx \sum_{j=1}^n f(\xi_{nj}(\tau))\lambda_{nj}(\tau) = \int f(t)\, d\mathring{\mu}_n(t, \tau)$$

with exact equality in $\mathcal{R}_{n-1} = \mathcal{L}_{n-1} \cdot \mathcal{L}_{n-1}$. If $\tau = 0$ is a regular value, then we can set $\tau = 0$ and the quadrature is even exact in $\mathcal{L}_{n-1} \cdot \mathcal{L}_n$ (see Theorem 11.6.2). Let us define the positive real function $\Omega_n(z, \tau)$ by the Riesz–Herglotz–Nevanlinna transform of $\mathring{\mu}_n$:

$$\Omega_n(z, \tau) = \int D(t, z)\, d\mathring{\mu}_n(t) = \sum_{j=1}^n \lambda_{nj}(\tau) D(\xi_{nj}(\tau), z).$$

We show the following:

Lemma 11.10.6. With $\Omega_n(z, \tau)$ as defined above we have

$$\Omega_n(z, \tau) = -\frac{P_n(z, \tau)}{Q_n(z, \tau)},$$

where $Q_n(z, \tau)$ is the quasi-orthogonal function and $P_n(z, \tau)$ is the associated function of the second kind.

Proof. Let us drop the argument τ for simplicity. By the previous Lemma 11.10.5, we have

$$\frac{Z_{n-1}(z)}{Z_n(z)} P_n(z) = \int D(t, z) \left[\frac{Z_{n-1}(t)}{Z_n(t)} Q_n(t) - \frac{Z_{n-1}(z)}{Z_n(z)} Q_n(z) \right] d\mathring{\mu}(t).$$

It is easily checked that this integrand is in $\mathcal{R}_{n-1}$ and therefore $d\mathring{\mu}$ can be replaced by $d\mathring{\mu}_n$. Recalling that $Q_n(\xi_i) = 0$, we have

$$\frac{Z_{n-1}(z)}{Z_n(z)} P_n(z) = - \left[\sum_{i=1}^n D(\xi_i, z)\lambda_i \right] \frac{Z_{n-1}(z)}{Z_n(z)} Q_n(z).$$

Since in the square brackets we recognize $\Omega_n(z)$, we have found the required equality.   □

Because the quadrature is exact in $\mathcal{R}_{n-1}$, which is equivalent to saying that the inner products with respect to $\mathring{\mu}$ and $\mathring{\mu}_n$ are the same for functions in $\mathcal{L}_{n-1}$, we have by Theorem 11.10.3 the following consequence.

Theorem 11.10.7. *Let $Q_n(z, \tau)$ be the quasi-orthogonal functions and $P_n(z, \tau)$ the associated functions of the second kind, and let $\Omega_\mu(z) = \int D(t, z) \, d\mathring{\mu}(t)$. Then, setting for a regular value τ*

$$\Omega_n(z, \tau) = -\frac{P_n(z, \tau)}{Q_n(z, \tau)},$$

we have for all $\alpha \in \tilde{A}^\beta_{n-1} = \{\beta, \hat{\beta}, \alpha_1, \alpha_1, \ldots, \alpha_{n-1}, \alpha_{n-1}\}$

$$\lim_{z \mapsto \alpha} \left[\Omega_\mu^{(k)}(z) - \Omega_n^{(k)}(z, \tau) \right] = 0, \quad k = 0, 1, \ldots, \alpha^\# - 1,$$

where $\alpha^\#$ is the multiplicity of α in $\tilde{A}^\beta_{n-1}$.
 If $\tau = 0$ is a regular value, then setting

$$B_n = \tilde{A}^\beta_{n-1} \cup \{\alpha_n\} = \{\beta, \hat{\beta}, \alpha_1, \alpha_1, \alpha_2, \alpha_2, \ldots, \alpha_{n-1}, \alpha_{n-1}, \alpha_n\},$$

we have (recall that $\Omega_n(z, 0) = -\psi_n(z)/\phi_n(z)$)

$$\lim_{z \mapsto \alpha} \left[\Omega_\mu^{(k)}(z) - \Omega_n^{(k)}(z, 0) \right] = 0, \quad k = 0, 1, \ldots, \alpha^\# - 1,$$

where $\alpha^\#$ is the multiplicity of α in B_n.

As a special interpretation of this theorem, we consider the example where $\partial \mathbb{O} = \mathbb{R}$ and where all $\alpha_k = \infty$ so that the orthogonal functions become the orthogonal polynomials on the real line. However, since we used the kernel $D(t, z)$ and not the simplified kernel $(t - z)^{-1}$ as Akhiezer does, we do not find the same result as given in Akhiezer [2, p. 22]. By the previous theorem we have that the first $2n - 2$ coefficients in the asymptotic expansion of $\Omega_\mu - \Omega_n$ at ∞ will vanish. Thus

$$\left\{ -\frac{P_n(z, \tau)}{Q_n(z, \tau)} - \mathbf{i} \left[s_0 + \frac{s_1}{z} + \frac{s_2}{z^2} + \cdots + \frac{s_{2n-3}}{z^{2n-3}} \right] \right\} z^{2n-3} \to 0, \quad z \mapsto \infty,$$

$$(11.47)$$

where

$$s_0 = \mu_{10}, \quad s_k = \mu_{k-1,0} + \mu_{k+1,0}, \quad k \geq 1, \quad \mu_{k0} = \int t^k d\mathring{\mu}(t), \quad k \geq 0.$$

Indeed,

$$D(t, z) = -i\frac{1 + tz}{t - z}$$

$$= i\left[t + \frac{1 + t^2}{z} + \frac{(1 + t^2)t}{z^2} + \cdots + \frac{(1 + t^2)t^{l-1}}{z^l} + \frac{(1 + t^2)t^l}{(z - t)z^l} \right],$$

so that

$$\int D(t, z)\, d\mathring{\mu}(t) = i\left[\int t\, d\mathring{\mu}(t) + \sum_{j=1}^{2n-3} \frac{1}{z^j} \int (1 + t^2)t^{j-1} d\mathring{\mu}(t) \right.$$

$$\left. + \int \frac{(1 + t^2)t^{2n-3}}{(z - t)z^{2n-3}}\, d\mathring{\mu}(t) \right].$$

Now

$$\int t^k d\mathring{\mu}_n(t) = \int t^k d\mathring{\mu}(t) = \mu_{k0}, \quad k = 0, 1, \ldots$$

and hence

$$\Omega_\mu(z) = \int D(t, z)\, d\mathring{\mu}(t)$$

$$= i\left[s_0 + \frac{s_1}{z} + \frac{s_2}{z^2} + \cdots + \frac{s_{2n-3}}{z^{2n-3}} + \int \frac{(1 + t^2)t^{2n-3}}{(z - t)z^{2n-3}}\, d\mathring{\mu}(t) \right].$$

Because

$$\lim_{z \mapsto \infty} \int \frac{(1 + t^2)t^{2n-3}}{z - t}\, d\mathring{\mu}(t) = 0,$$

(11.47) will follow. We also note that the result in Akhiezer is somewhat stronger since his results would translate as

$$-\frac{P_n(z, \tau)}{Q_n(z, \tau)} = i\left[s_0 + \frac{s_1}{z} + \frac{s_2}{z^2} + \cdots + \frac{s_{2n-3}}{z^{2n-3}} \right] + O\left(\frac{1}{z^{2n-2}}\right), \quad z \mapsto \infty,$$

that is,

$$\left| -\frac{P_n(z, \tau)}{Q_n(z, \tau)} - i\left[s_0 + \frac{s_1}{z} + \frac{s_2}{z^2} + \cdots + \frac{s_{2n-3}}{z^{2n-3}} \right] \right| \leq \frac{K}{z^{2n-2}},$$

as $z \mapsto \infty$, where K is some constant. However, imposing stronger conditions on the measure (or on the linear functional), it is possible to obtain such a result directly. This is done for example in Ref. [38]. We explain the tread that is followed there.

In the previous approach, we assumed that $\int f(t)\,d\mathring{\mu}(t)$ was finite for all $f \in \mathcal{L}_\infty \cdot \mathcal{L}_\infty$. Now we need these integrals to be defined for a somewhat larger space, namely the space $\mathcal{L}_\infty \cdot \mathcal{L}_\infty \cdot \mathcal{L}_\infty$. We note that then the integrals are automatically defined on the space $\tilde{\mathcal{R}}$, which consists of all the functions f having the form

$$f(t) = \frac{z^k t^l}{t-z} g(t), \quad g \in \mathcal{L}_\infty \cdot \mathcal{L}_\infty, \quad k,l = 0,1, \quad z \in (\overline{\mathbb{C}} \setminus \overline{\partial \mathbb{O}}) \cup A_\beta \cup \hat{A}_\beta,$$

with $A_\beta = \cup_n A_n^\beta$ and $\hat{A}_\beta = \cup_n \hat{A}_n^\beta$. Of course, this is needed for the general rational case. If, as in the previous example, we only consider the polynomial case on the real line, then obviously $\mathcal{L}_\infty = \mathcal{L}_\infty \cdot \mathcal{L}_\infty = \mathcal{L}_\infty \cdot \mathcal{L}_\infty \cdot \mathcal{L}_\infty$.

Recall the Newton expansion for $D(t,z)$ given in Lemma 6.1.5:

$$D(t,z) \sim 1 + 2\sum_{k=1}^\infty a_k(t)\varpi_\beta^*(z)\pi_{k-1}^*(z), \quad a_k(t) = \frac{\varpi_\beta(t)}{\varpi_\beta(\beta)\pi_k^*(t)}.$$

This is associated with successive interpolating polynomials (in z)

$$p_n(t,z) = 1 + 2\sum_{k=1}^n a_k(t)\varpi_\beta^*(z)\pi_{k-1}^*(z) \tag{11.48}$$

in the interpolation points $A_n^\beta = \{\beta, \alpha_1, \ldots, \alpha_n\}$.

The interpolation error for $p_n(t,z)$ is given by

$$e_n(t,z) = \frac{2\varpi_\beta(t)\varpi_\beta^*(z)\pi_n^*(z)}{\varpi_\beta(\beta)\pi_n^*(t)(t-z)} = 2\frac{a_n(t)}{t-z}(z-\beta)\pi_n^*(z). \tag{11.49}$$

Similarly, one may obtain successive interpolating "polynomials" in the points $\hat{A}_n^\beta = \{\hat{\beta}, \hat{\alpha}_1, \ldots, \hat{\alpha}_n\}$. For the unit circle, these are the points $\hat{\beta} = \infty$, and for $k \geq 1$, $1/\overline{\alpha}_k = \alpha_k$ since these are on $\mathbb{T}$. For the real line, these points correspond to $\hat{\beta} = -\mathbf{i}$ and for $k \geq 1$ again $\overline{\alpha}_k = \alpha_k$. The expansion is easily obtained by the relation $D(t,\hat{z})_* = -D(t,z)$, where the substar is with respect to t and recalling that $f_*(z) = \overline{f(\hat{z})}$. Thus we get

$$-D(t,z) \sim 1 + 2\sum_{k=1}^\infty [a_k(t)]_*[\varpi_\beta^*(z)\pi_{k-1}^*(z)]_*,$$

with the same notation as before. Note that for $t \in \overline{\partial \mathbb{O}}$, we have $[a_k(t)]_* = \overline{a_k(t)}$. For the other factor, we can write explicitly

$$[\varpi_\beta^*(z)\pi_{k-1}^*(z)]_* = \begin{cases} z^{-1}\left(z^{-1} - \alpha_1^{-1}\right)\cdots\left(z^{-1} - \alpha_{k-1}^{-1}\right) & \text{for } \mathbb{T}, \\ (z+\mathbf{i})(z-\alpha_1)\cdots(z-\alpha_{k-1}) & \text{for } \mathbb{R}. \end{cases}$$

Thus

$$\hat{p}_n(t, z) = 1 + 2 \sum_{k=1}^{n} [a_k(t)]_* [\varpi_\beta^*(z) \pi_{k-1}^*(z)]_* \qquad (11.50)$$

is a polynomial (in z for $\mathbb{R}$ and in z^{-1} for $\mathbb{T}$) that interpolates $-D(t, z)$ in the points $z \in \hat{A}_n^\beta$. The interpolation error is

$$\hat{e}_n(t, z) = 2 \left[\frac{a_n(t)}{t - \hat{z}} \right]_* [(z - \beta) \pi_n^*(z)]_* \qquad (11.51)$$

(the first substar w.r.t. t; the second one w.r.t. z).

The associated relations for Ω_μ are

$$\Omega_\mu(z) = P_n(z) + E_n(z)$$

$$\sim 1 + 2 \sum_{k=1}^{\infty} \mu_k (z - \beta) \pi_{k-1}^*(z)$$

and

$$-\Omega_\mu(z) = \hat{P}_n(z) + \hat{E}_n(z)$$

$$\sim 1 + 2 \sum_{k=1}^{\infty} \nu_k [\varpi_\beta^*(z) \pi_{k-1}^*(z)]_*,$$

where the generalized moments are

$$\mu_k = \int a_k(t) \, d\mathring{\mu}(t) \quad \text{and} \quad \nu_k = \int \overline{a_k(t)} \, d\mathring{\mu}(t) = \overline{\mu}_k.$$

The expressions

$$P_n(z) = \int p_n(t, z) \, d\mathring{\mu}(t) = 1 + 2 \sum_{k=1}^{n} \mu_k (z - \beta) \pi_{k-1}^*(z),$$

$$E_n(z) = \int e_n(t, z) \, d\mathring{\mu}(t) = (z - \beta) \pi_n^*(z) \int \frac{a_n(t)}{t - z} \, d\mathring{\mu}(t)$$

show that $P_n(z)$ interpolates $\Omega_\mu(z)$ in the point set A_n^β, whereas

$$\hat{P}_n(z) = \int \hat{p}_n(t, z) \, d\mathring{\mu}(t) = 1 + 2 \sum_{k=1}^{n} \nu_k [(z - \beta) \pi_{k-1}^*(z)]_*,$$

$$\hat{E}_n(z) = \int \hat{e}_n(t, z) \, d\mathring{\mu}(t) = [(z - \beta) \pi_n^*(z)]_* \int \left[\frac{a_n(t)}{t - \hat{z}} \right]_* d\mathring{\mu}(t)$$

show that $\hat{P}_n(z)$ interpolates $-\Omega_\mu(z)$ in the points of $\hat{A}_n^\beta$.

We want to show that ψ_n/ϕ_n interpolates Ω_μ linearly in Hermite sense in all the points of B_n as defined before and if $\tau = 0$ is a regular value then this holds in a nonlinear sense. We prove this by showing that this kind of interpolation holds for a truncated Newton expansion for Ω_μ as given above. We need a series of lemmas to come to this result.

We define

$$\Delta_{nl}(z) = \psi_n(z) + \phi_n(z)P_l(z),$$

$$\hat{\Delta}_{nl}(z) = \psi_n(z) - \phi_n(z)\hat{P}_l(z).$$

The polynomial numerators of ϕ_n and ψ_n are denoted by f_n and g_n respectively, that is, $\phi = f_n/\pi_n^*$ and $\psi_n = g_n/\pi_n^*$.

Lemma 11.10.8. Assume that $\int f \, d\mu$ exists for all $f \in \tilde{\mathcal{R}}_\infty$. Let $\phi_n = f_n/\pi_n^*$ and $\psi_n = g_n/\pi_n^*$ be the orthogonal functions and the functions of the second kind and let Δ_{nl} be as defined above. Furthermore, suppose $0 \leq m < n$ and let l be arbitrary. Then

$$\Delta_{nl}(z) = \frac{2\varpi_\beta^*(z)}{\varpi_\beta(\beta)\pi_n^*(z)} \int A_{n,m,l}(t, z) \, d\mathring{\mu}(t),$$

with

$$A_{n,m,l}(t, z) = \frac{\varpi_\beta(t)}{t - z} \cdot \frac{\pi_m^*(z)\pi_l^*(t)\pi_n^*(t)f_n(t) - \pi_m^*(t)\pi_l^*(z)\pi_n^*(t)f_n(z)}{\pi_m^*(t)\pi_l^*(t)\pi_n^*(t)}$$

for all $z \in (\overline{\mathbb{C}} \setminus \partial\mathbb{O}) \cup A_\beta \cup \hat{A}_\beta$.

Proof. Use Lemma 11.2.2 with $f = 1/\pi_m^*$ and the expression (11.49) for the error $e_n(t, z)$ to get

$$\Delta_{nl}(z) = \int \left\{ E(t, z)\frac{\pi_m^*(z)}{\pi_m^*(t)}\phi_n(t) - [D(t, z) - p_l(t, z)]\phi_n(z) \right\} d\mathring{\mu}(t)$$

$$= \int \left\{ E(t, z)\frac{\pi_m^*(z)}{\pi_m^*(t)}\phi_n(t) - \frac{2\varpi_\beta(t)\varpi_\beta^*(z)\pi_l^*(z)}{(t - z)\varpi_\beta(\beta)\pi_l^*(t)}\phi_n(z) \right\} d\mathring{\mu}(t)$$

$$= \int B_{n,m,l}(t, z) \, d\mathring{\mu}(t),$$

where $p_l(t, z)$ is as in (11.48). Furthermore,

$$E(t, z) = D(t, z) - 1 = 2\frac{\varpi_\beta^*(z)\varpi_\beta(t)}{\varpi_\beta(\beta)(t - z)}$$

so that

$$B_{n,m,l}(t,z) = \frac{2\varpi_\beta^*(z)\varpi_\beta(t)}{\varpi_\beta(\beta)(t-z)} \left[\frac{\pi_m^*(z)f_n(t)}{\pi_m^*(t)\pi_n^*(t)} - \frac{\pi_l^*(z)f_n(z)}{\pi_l^*(t)\pi_n^*(z)} \right].$$

Note that the two parts of $B_{n,m,l}(t,z)$ belong to $\tilde{\mathcal{R}}_\infty$ so that the integral is well defined for all $z \in (\overline{\mathbb{C}} \setminus \partial\mathbb{O}) \cup A_\beta \cup \hat{A}_\beta$. Hence the lemma follows. $\square$

Similarly we can obtain

Lemma 11.10.9. Assume that $\int f\,d\mathring{\mu}$ exists for all $f \in \tilde{R}$. Let $\phi_n = f_n/\pi_n^*$ and $\psi_n = g_n/\pi_n^*$ be the orthogonal functions and the functions of the second kind and let $\hat{\Delta}_{nl}$ be as above. Furthermore, suppose $0 \le m < n$ and let l be arbitrary. Then

$$\hat{\Delta}_{nl}(z) = -\frac{2[\varpi_\beta^*(z)]_*}{\varpi_\beta(\beta)\pi_n^*(z)} \int \hat{A}_{n,m,l}(t,z)\,d\mathring{\mu}(t),$$

with

$$\hat{A}_{n,m,l}(t,z) = \frac{\varpi_{\beta*}(t)}{(t-\hat{z})_*} \frac{[\pi_m^*(z)\pi_l^*(t)]_*\pi_n^*(t)f_n(t) - [\pi_m^*(t)\pi_l^*(z)]_*\pi_n^*(t)f_n(z)}{[\pi_m^*(t)\pi_l^*(t)]_*\pi_n^*(t)}$$

for all $z \in (\overline{\mathbb{C}} \setminus \partial\mathbb{O}) \cup A_\beta \cup \hat{A}_\beta$.

Proof. This proceeds along the same lines as the previous proof. Use Lemma 11.2.2 with $f = 1/[\pi_m^*]_*$ and the expression (11.51) for the error $\hat{e}_n(t,z)$ to get

$$\hat{\Delta}_{nl}(z) = \int \left\{ E(t,z)\frac{[\pi_m^*(z)]_*}{[\pi_m^*(t)]_*}\phi_n(t) - [D(t,z) + \hat{p}_l(t,z)]\phi_n(z) \right\} d\mathring{\mu}(t)$$

$$= \int \left\{ E(t,z)\frac{[\pi_m^*(z)]_*}{[\pi_m^*(t)]_*}\phi_n(t) + \frac{2\varpi_{\beta*}(t)[\varpi_\beta^*(z)\pi_l^*(z)]_*}{(t-\hat{z})_*\overline{\varpi_\beta(\beta)}[\pi_l^*(t)]_*}\phi_n(t) \right\} d\mathring{\mu}(t),$$

where $\hat{p}_l(t,z)$ is as in (11.50). In this expression, we can replace the kernel $E(t,z)$ by $E(t,z) - 2$ since the additional term is zero by orthogonality. Furthermore, using $D(t,\hat{z}) = -D(t,z)_*$ (substar with respect to t) we arrive at

$$E(t,z) - 2 = -E(t,\hat{z})_* = -2\frac{[\varpi_\beta^*(z)]_*\varpi_{\beta*}(t)}{\varpi_\beta(\beta)(t-\hat{z})_*},$$

so that $\hat{\Delta}_{nl}(z) = \int \hat{B}_{n,m,l}(t,z)\,d\mathring{\mu}(t)$ with

$$\hat{B}_{n,m,l}(t,z) = \frac{2[\varpi_\beta^*(z)]_*\varpi_{\beta*}(t)}{\varpi_\beta(\beta)(t-\hat{z})_*} \left[\frac{[\pi_m^*(z)]_*f_n(t)}{[\pi_m^*(t)]_*\pi_n^*(t)} - \frac{[\pi_l^*(z)]_*f_n(z)}{[\pi_l^*(t)]_*\pi_n^*(z)} \right].$$

Note that again the integral of $\hat{B}_{n,m,l}(t, z)$ will converge so that the lemma follows. $\qquad\square$

Note that

$$(t - \hat{z})_* = (t^{-1} - z^{-1}) \quad \text{for } \mathbb{T} \quad \text{and} \quad (t - \hat{z})_* = (t - z) \quad \text{for } \mathbb{R}$$

and

$$[\pi_k^*(z)]_* = \pi_k^*(z)/\prod_{i=1}^{k}(-z\alpha_i) \quad \text{for } \mathbb{T} \quad \text{while} \quad [\pi_k^*(z)]_* = \pi_k^*(z) \quad \text{for } \mathbb{R}.$$

Our next step is to construct Newton interpolants for Ω_μ. We define for each n two sets of interpolation points

$$C_{2n-1} = \{\gamma_0, \gamma_1, \ldots, \gamma_{2n-1}\} \quad \text{and} \quad \hat{C}_{2n-1} = \{\hat{\gamma}_0, \hat{\gamma}_1, \ldots, \hat{\gamma}_{2n-1}\},$$

where

$$\gamma_0 = \beta, \quad \gamma_k = \alpha_k \text{ for } 1 \le k \le n \quad \text{and} \quad \gamma_k = \alpha_{k-n} \text{ for } n < k < 2n.$$

Recall $\hat{\alpha}_k = \alpha_k$ and hence also $\hat{\gamma}_k = \gamma_k$ for all $k \ge 1$. Furthermore, we define the corresponding nodal polynomials

$$\tilde{\pi}_{0,n}^* = 1, \qquad \tilde{\pi}_{k,n}^*(z) = \prod_{i=1}^{k}(z - \gamma_i), \quad 1 \le k < 2n.$$

Note that

$$\tilde{\pi}_{k,n}^*(z) = \pi_n^*(z)\pi_{k-n}^*(z) \quad \text{for any } n < k < 2n.$$

The generalized moments are defined as follows:

$$\tilde{\mu}_{k,n} = \int \tilde{a}_k(t)\, d\mathring{\mu}(t), \qquad \tilde{a}_k(t) = \frac{\varpi_\beta(t)}{\varpi_\beta(\beta)\tilde{\pi}_{k,n}(t)}$$

while

$$\nu_{k,n} = \int \overline{\tilde{a}_k(t)}\, d\mathring{\mu}(t) = \overline{\tilde{\mu}}_{k,n}.$$

The interpolating "polynomials" for these nodes are

$$P_{2n-1}^{\sim}(z) = \tilde{\mu}_{0,k} + 2\sum_{k=1}^{2n-1} \tilde{\mu}_{k,n}(z - \beta)\tilde{\pi}_{k-1,n}^*(z)$$

and

$$\hat{P}^{\sim}_{2n-1}(z) = v_{0,n} + 2 \sum_{k=1}^{2n-1} v_{k,n}[(z-\beta)\tilde{\pi}^*_{k-1,n}(z)]_*.$$

We have the following theorem:

Theorem 11.10.10. *Let $\int f \, d\overset{\circ}{\mu}$ be defined for all $f \in \tilde{\mathcal{R}}_\infty$. Then the polynomial $P^{\sim}_{2n-1}$ as defined above interpolates Ω_μ in the following sense:*

$$\Omega_\mu(z) - P^{\sim}_{2n-1}(z) = (z-\beta)\pi^*_n(z)\pi^*_{n-1}(z)R_{2n-1}(z),$$

with R_{2n-1} given by

$$R_{2n-1}(z) = \frac{2}{\varpi_\beta(\beta)} \int \frac{\varpi_\beta(t) \, d\overset{\circ}{\mu}(t)}{(t-z)\pi^*_n(t)\pi^*_{n-1}(t)},$$

and this integral converges for $z \in A_n^\beta$.

A similar result holds for the other interpolation. The polynomials $\hat{P}^{\sim}_{2n-1}$ interpolate $-\Omega_\mu$ in the following sense:

$$-\Omega_\mu(z) - \hat{P}^{\sim}_{2n-1}(z) = [(z-\beta)\pi^*_n(z)\pi^*_{n-1}(z)]_* \hat{R}\hat{R}_{2n-1}(z),$$

with $\hat{R}_{2n-1}$ given by

$$\hat{R}_{2n-1}(z) = \frac{2}{\varpi_\beta(\beta)} \int \frac{\varpi_{\beta*}(t) \, d\overset{\circ}{\mu}(t)}{(t-\hat{z})_*[\pi^*_n(t)\pi^*_{n-1}(t)]_*},$$

and the integral converges for $z \in \hat{A}_n^\beta$.

Proof. This follows immediately from our general polynomial interpolation given above, and the integrals defining R_{2n-1} and $\hat{R}_{2n-1}$ converge because the integrands are in $\tilde{\mathcal{R}}_\infty$. □

We need another lemma

Lemma 11.10.11. Suppose the integrals $\int f \, d\overset{\circ}{\mu}$ converge for $f \in \tilde{\mathcal{R}}_\infty$. Let $\phi_n = f_n/\pi^*_n$ be the orthonormal functions and $\psi_n = g_n/\pi^*_n$ the functions of the second kind. The interpolating polynomials $P^{\sim}_{2n-1}$ and $\hat{P}^{\sim}_{2n-1}$ are as defined above. Then the following formulas hold:

$$g_n(z) - f_n(z)P^{\sim}_{2n-1}(z) = (z-\beta)\pi^*_n(z)\pi^*_{n-1}(z)r_{2n-1}(z)$$

and

$$-g_n(z) - f_n(z)\hat{P}^{\sim}_{2n-1}(z) = [(z - \beta)\pi_n^*(z)\pi_{n-1}^*(z)]_*\hat{r}_{2n-1}(z),$$

with $r_{2n-1}(z)$ bounded for $z \in A_n^\beta$ and $\hat{r}_{2n-1}(z)$ bounded for $z \in \hat{A}_n^\beta$.

Proof. We use Lemma 11.10.8 with $m = n - 1$ and $l = 2n - 1$ to get

$$\tilde{\Delta}_{n,2n-1}(z) = \frac{2\varpi_0(z)}{\varpi_0(\alpha_0)} \int \tilde{A}_{n,n-1,2n-1}(t, z)\, d\tilde{\mu}(t),$$

with (recall $\tilde{\pi}_{2n-1,n}^* = \pi_n^*\pi_{n-1}^*$)

$$\tilde{A}_{n,n-1,2n-1}(t, z) = \pi_{n-1}^*(z)\pi_n^*(z)B_n(t, z),$$

where

$$B_n(t, z) = \frac{f_n(t) - f_n(z)}{t - z}\frac{\varpi_0(t)}{\pi_{n-1}^*(t)\pi_n^*(t)}.$$

Hence the integrand will be in $\tilde{\mathcal{R}}_\infty$ and therefore the integral converges for $z \in A_n^\beta$.

A similar derivation can be given for the second part.   □

Now we have the final interpolation result.

Theorem 11.10.12. *Suppose $\int f\, d\tilde{\mu}$ converges for all $f \in \tilde{\mathcal{R}}_\infty$. Let Ω_μ be the Riesz–Herglotz–Nevanlinna transform of the measure μ. Let ϕ_n be the orthonormal functions and ψ_n the associated functions of the second kind. Then ψ_n/ϕ_n interpolates Ω_μ linearly in Hermite sense in the points*

$$B_n = \{\beta, \hat{\beta}, \alpha_1, \alpha_1, \ldots, \alpha_{n-1}, \alpha_{n-1}, \alpha_n\},$$

by which we mean the following: Let $\phi_n = f_n/\pi_n^*$ and $\psi_n = g_n/\pi_n^*$. Then

$$g_n(z) - f_n(z)\Omega_\mu(z) = (z - \beta)\pi_n^*(z)\pi_{n-1}^*(z)[(z - \beta)\pi_n^*(z)\pi_{n-1}^*(z)]_*R_n(z),$$

with $R_n(z)$ finite for $z \in A_n^\beta \cup \hat{A}_n^\beta$.

If $\tau = 0$ is a regular value, then we also have

$$\frac{\psi_n(z)}{\phi_n(z)} - \Omega_\mu(z) = (z - \beta)\pi_n^*(z)\pi_{n-1}^*(z)[(z - \beta)\pi_n^*(z)\pi_{n-1}^*(z)]_*r_n(z),$$

with $r_n(z)$ finite for $z \in A_n^\beta \cup \hat{A}_n^\beta$.

Proof. The linear interpolation follows immediately from a combination of Lemma 11.10.11 and Theorem 11.10.10.

If $\tau = 0$ is a regular value and $\phi_n = f_n / \pi_n^*$, then f_n has no zeros in $A_n^\beta \cup \hat{A}_n^\beta$, so that we can divide by f_n in the linear relations and obtain the result as stated. $\qquad\square$

We remark that for the case of the line and taking all $\alpha_i = \infty$, the previous theorem immediately yields the earlier mentioned result of Akhiezer [2, p. 22].

Our next move is to show interpolation properties involving reproducing kernels. These properties are obtained via our framework of orthogonality with respect to varying measures. We start, however, with a lemma not in terms of this framework. Set

$$k_n(z, w) = \sum_{k=0}^{n} \phi_k(z) \overline{\phi_k(w)},$$

the reproducing kernel for $\mathcal{L}_n$, and define as before the normalized kernels

$$K_n(z, w) = \frac{k_n(z, w)}{\sqrt{k_n(w, w)}}.$$

Then we have

Lemma 11.10.13. For the reproducing kernels $k_n(z, w)$ in $\mathcal{L}_n$, we have

$$\frac{k_n(z, w)}{\overline{b_n(w)}}\bigg|_{w=\alpha_n} = \kappa_n \phi_n(z) \quad \text{and} \quad K_n(z, \alpha_n) = \phi_n(z),$$

where $\phi_n(z) = \kappa_n b_n(z) + \kappa_n' n_{n-1}(z) + \cdots$ are the orthonormal functions with $\kappa_n > 0$.

Proof. The first part follows from

$$\frac{k_n(z, w)}{\overline{b_n(w)}} = \phi_0(z) \overline{\left(\frac{\phi_0(w)}{b_n(w)}\right)} + \phi_1(z) \overline{\left(\frac{\phi_1(w)}{b_n(w)}\right)} + \cdots + \phi_n(z) \overline{\left(\frac{\phi_n(w)}{b_n(w)}\right)}.$$

Because $\phi_k(w)/b_n(w)$ retains a factor $1/Z_n(w)$ for all $k < n$, and $1/Z_n(\alpha_n) = 0$, and since

$$\frac{\phi_n(w)}{b_n(w)}\bigg|_{w=\alpha_n} = \kappa_n,$$

it follows that the first equality holds.

For the second equality, we note that

$$\frac{k_n(z, w)}{b_n(z)\overline{b_n(w)}}\Bigg|_{z=w=\alpha_n} = \kappa_n^2,$$

and so

$$K_n(z, w) = \frac{k_n(z, w)/\overline{b_n(w)}}{\sqrt{k_n(w, w)/|b_n(w)|^2}}\eta(w), \quad \eta(w) = \frac{\overline{b_n(w)}}{|b_n(w)|} \in \mathbb{T},$$

with $\eta(w)$ real for $w \in \overline{\partial\mathbb{O}}$. This gives

$$K_n(z, \alpha_n) = \phi_n(z)\eta$$

with $\eta = \pm 1$. Because $[\phi_n(z)/b_n(z)]_{z=\alpha_n} = \kappa_n > 0$, it follows that $\eta = 1$. ☐

Let us now recall some basic facts concerning the varying measures considered in Section 9.5. Recall also that the α_0 from that section has to be replaced now by β. Define the spaces

$$\mathcal{L}_0^n = \mathrm{span}\{\zeta_\beta(z)^k : k = 0, 1, \dots, n\}$$

and the measure

$$d\mathring{\pi}_n(t) = |w_n(t)|^2 d\mathring{\mu}(t), \quad w_0 = 1, \quad w_n(z) = \prod_{k=1}^n \frac{\varpi_\beta(z)}{\varpi_k(z)}, \quad n \geq 1.$$

To avoid a lot of notational complication we shall assume that, in the case of the real line, a factor $\varpi_i(z) = z - \overline{\alpha}_i$ is replaced by 1 if $\alpha_i = \infty$.

Let ϕ_{nk}, $k = 0, 1, \dots, n$ be the orthonormal functions obtained by Gram–Schmidt orthogonalization of the basis $\{1, \zeta_\beta, \zeta_\beta^2, \dots, \zeta_\beta^n\}$ for $\mathcal{L}_0^n$. Setting $\mathsf{F}_{nk} = \phi_{nk}w_n \in \mathcal{L}_n$, we have that (see Section 9.5 for the nonboundary case, but the arguments are the same as in the boundary situation)

$$k_n(z, w) = \sum_{k=0}^n \mathsf{F}_{nk}(z)\overline{\mathsf{F}_{nk}(w)} = \sum_{k=0}^n \phi_k(z)\overline{\phi_k(w)}$$

is the reproducing kernel for $\mathcal{L}_n$. Next we show the following theorem.

Theorem 11.10.14. *Considered as a function of z, $k_n(z, w)$ is an outer function in H_2.*

Proof. We use the notation introduced above. Recall also that the functions in $\mathcal{L}_0^n$ have only poles in $\mathbb{O}^e$, so that we can apply the results of the previous chapters. Since

$$k_n(z, w) = k_\pi(z, w)w_n(z)\overline{w_n(w)}, \quad k_\pi(z, w) = \sum_{k=0}^n \phi_{nk}(z)\overline{\phi_{nk}(w)}$$

and because $k_\pi(z, w)$ is an outer function in H_2, it follows that $k_n(z, w)$ is a rational function without zeros in $\overline{\mathbb{O}}$ and without poles in $\mathbb{O}$. Hence, it is outer in H_2. □

By this, we can define the measure

$$d\mathring{\mu}_n(t) = \frac{d\mathring{\lambda}(t)}{|K_n(t, \beta)|^2},$$

where $K_n(z, w)$ is the normalized reproducing kernel for $\mathcal{L}_n$. It is a finite positive measure on $\overline{\partial\mathbb{O}}$. Now the proof of Theorem 6.3.2 can be repeated without changes and we immediately can formulate

Theorem 11.10.15. *Let $K_n(z, w)$ be the normalized reproducing kernel for $\mathcal{L}_n$ and define $d\mathring{\mu}_n(t) = d\mathring{\lambda}(t)/|K_n(t, \beta)|^2$. Then the inner product in $\mathcal{L}_n$ is the same for the measure μ and the measure μ_n, that is,*

$$\int f(t)\overline{g(t)}\,d\mathring{\mu}(t) = \int f(t)\overline{g(t)}\,d\mathring{\mu}_n(t), \quad \forall f, g \in \mathcal{L}_n.$$

As a consequence we have the following general interpolation result (see also Ref. [62]).

Corollary 11.10.16. *Let $\mathring{\mu}_n$ be defined as in the previous theorem. Then*

$$f(z)g_*(z)\int D(t, z)d[\mathring{\mu} - \mathring{\mu}_n](t) = \int D(t, z)f(t)\overline{g(t)}\,d[\mathring{\mu} - \mathring{\mu}_n](t).$$

Proof. Since

$$\int h(t)\,d[\mathring{\mu} - \mathring{\mu}_n](t) = 0, \quad \forall h \in \mathcal{R}_n,$$

we just have to check that, as a function of t,

$$D(t, z)[f(z)g_*(z) - f(t)g_*(t)] \in \mathcal{R}_n,$$

which is obviously true. □

When the recurrence relations of Section 11.1 are linked with the interpolation properties discussed in this section, one can set up a generalization of the theory of T-fractions and M-fractions. A (modified) T-fraction and a (modified) M-fraction arise as the even and odd contractions of a Perron–Carathéodory (PC-) fraction. These PC-fractions are related to two-point Padé approximants; that is, their convergents approximate two formal power series: one at 0 and one at ∞. The canonical denominators of T- and M-fractions are orthogonal Laurent polynomials obtained by orthogonalization of the basis $\{1, z^{-1}, z, z^{-2}, z^2, \ldots\}$ or $\{1, z, z^{-1}, z^2, z^{-2}, \ldots\}$ respectively. For this theory see Refs. [18], [123], [142], [118], [120], [121], [126], and [147].

The present situation is related to a multipoint generalization of this theory. The PC-fractions have to be replaced by Extended Multipoint Padé (EMP-) fractions. There, the two points 0 and ∞ are replaced by a sequence of points $\{\alpha_1, \alpha_2, \alpha_3, \ldots\}$ on the (extended) real line. Such an EMP-fraction defines two formal Newton series: one associated with the interpolation table $\{\infty, 0, \alpha_1, \alpha_1, \alpha_2, \alpha_2, \alpha_3, \alpha_3, \ldots\}$ and one with the interpolation table $\{\infty, 0, \alpha_2, \alpha_2, \alpha_1, \alpha_1, \alpha_3, \alpha_3, \ldots\}$. The even contractions of an EMP-fraction are called Multipoint Padé (MP-) fractions because their convergents are multipoint Padé approximants for the first Newton series. The canonical denominators of the convergents are orthogonal rational functions of the form studied in this chapter for the real line. An odd contraction of an EMP-fraction is not exactly an MP-fraction, but it is similar to an MP-fraction. This odd contraction is related to the second Newton series defined by the EMP-fraction just as the even contraction is related to the first one. The denominators of its convergents are now orthogonal rational functions related to a reordered set of basic points: $\{\alpha_2, \alpha_1, \alpha_4, \alpha_3, \ldots\}$. An extensive study of these properties and relations can be found in Ref. [41].

11.11. Convergence

We continue our treatment of convergence results in the same setting as in the previous section. In particular, it was already mentioned in Section 9.5 on these varying measures that the results for these functions were much more general than needed at that moment. There we applied these results for the case where all the α_k were in $\mathbb{O}$. However, they were formulated such that most of them also hold when the α_i are in $\overline{\mathbb{O}}$; thus, they can also be on the boundary (as they are here). The proofs of most of the results, as long as they do not need the divergence of the Blaschke products or the α_i to be compactly included in $\mathbb{O}$, can be applied without change in the present situation. We only draw attention to the following facts.

When $\alpha \in \mathbb{T}$, then

$$\zeta_i(z) = -\frac{\overline{\alpha}_i}{|\alpha_i|}\frac{z - \alpha_i}{1 - \overline{\alpha}_i z} = 1.$$

Similarly, we also get for the case of the real line that $\zeta_i(z) = 1$. Thus all these factors and also the Blaschke products B_n can be replaced by 1. We also have explained in Section 11.1 that $\phi_{n*} = \phi_n$; thus we get $\phi_n^* = \phi_n$. In the case of the line, and $\alpha_n = \infty$, then a factor $\varpi_n(z)$ has to be replaced by 1. Finally, the reader is advised to check the meaning of the condition $(A, \mu) \in \mathcal{AM}'$ given in Section 9.5. Taking these things into account, we can just reformulate the theorems given in Sections 9.6 and 9.8 in our new notation and refer for the proofs to the proofs given for the corresponding theorems in those sections. We give some illustrative examples.

Theorem 9.6.3 becomes

Theorem 11.11.1. *Let $(A, \mu) \in \mathcal{AM}'$ and $\sigma(z)$ the outer spectral factor for the measure, normalized with $\sigma(\beta) > 0$. Then*

$$\lim_{n \to \infty} \frac{\phi_n(z)\varpi_n(z)}{\phi_n(\beta)\varpi_n(\beta)} = \frac{\sigma(\beta)\varpi_\beta(z)}{\sigma(z)\varpi_\beta(\beta)},$$

locally uniformly in $\mathbb{O}$, and

$$\lim_{n \to \infty} \frac{\phi_n(z)\varpi_n^*(z)}{\phi_n(\beta)\varpi_n(\beta)} = \frac{\sigma(\beta)\varpi_\beta^*(z)}{\sigma_*(z)\overline{\varpi_\beta(\beta)}},$$

locally uniformly in $\mathbb{O}^e$.

Proof. The first part is by Theorem 9.6.3. The second part is by taking substar conjugates. $\square$

For Theorem 9.6.4 and its Corollary 9.6.5 we obtain

Theorem 11.11.2. *Suppose $(A, \mu) \in \mathcal{AM}'$ and let $k_n(z, w)$ be the reproducing kernel for $\mathcal{L}_n$ and $s_w(z)$ the Szegő kernel (9.3). Then we have*

$$\lim_{n \to \infty} k_n(z, w) = s_w(z), \quad (z, w) \in \mathbb{O} \times \mathbb{O},$$

while

$$\lim_{n \to \infty} \phi_n(z) = 0, \quad z \in \mathbb{C} \setminus \overline{\partial\mathbb{O}},$$

where convergence is locally uniform in the indicated regions.

Proof. The latter convergence is given in $\mathbb{O}$ by Corollary 9.6.5, but because $\phi_n(z) = \phi_{n*}(z) = \overline{\phi_n(\hat{z})}$ and $\hat{z} \in \mathbb{O}$ iff $z \in \mathbb{O}^e$, we get the convergence in the whole complex plane except on the boundary $\partial\mathbb{O}$. □

Note that for the first relation we can take substar conjugates with respect to z and with respect to w. The result for $k_n(z, w)$ is $\overline{k_n(z, w)}$. By a similar operation on $s_w(z)$, we obtain

$$\lim_{n\to\infty} k_n(z, w) = -\frac{\zeta_\beta(z)\overline{\zeta_\beta(w)}}{1 - \zeta_\beta(z)\overline{\zeta_\beta(w)}}\frac{1}{\sigma_*(z)\overline{\sigma_*(w)}} = -\frac{\varpi_\beta^*(z)\overline{\varpi_\beta^*(w)}}{\varpi_\beta(\beta)\varpi_w(z)}\frac{1}{\sigma_*(z)\overline{\sigma_*(w)}},$$

locally uniformly for $(z, w) \in \mathbb{O}^e \times \mathbb{O}^e$.

Corollary 11.11.3. *Under the same conditions as in the previous theorem, letting $K_n(z, w)$ be the normalized reproducing kernel,*

$$\lim_{n\to\infty} K_n(z, \beta) = \frac{1}{\sigma(z)}, \quad \text{locally uniformly in } \mathbb{O}.$$

Proof. Since by the previous theorem we have for $z \in \mathbb{O}$

$$\lim_{n\to\infty} k_n(z, \beta) = \frac{1}{\sigma(z)\sigma(\beta)}$$

(recall $\sigma(\beta) > 0$), thus also

$$\lim_{n\to\infty} k_n(\beta, \beta) = \frac{1}{\sigma(\beta)^2},$$

from which the first result follows by using the definition $K_n(z, \beta) = k_n(z, \beta)/\sqrt{k_n(\beta, \beta)}$. □

Concerning ratio asymptotics, we have from Lemma 9.8.1

Theorem 11.11.4. *Let $(A, \mu) \in \mathcal{AM}'$. Then*

$$\lim_{n\to\infty} \frac{\phi_{n+1}(z)}{\phi_n(z)}\frac{\varpi_{n+1}(z)\varpi_n(\beta)}{\varpi_n(z)\varpi_{n+1}(\beta)}\frac{\phi_n(\beta)}{\phi_{n+1}(\beta)} = 1, \quad z \in \mathbb{O}$$

and

$$\lim_{n\to\infty} \frac{\phi_{n+1}(z)}{\phi_n(z)}\frac{\varpi_{n+1}^*(z)\overline{\varpi_n(\beta)}}{\varpi_n^*(z)\overline{\varpi_{n+1}(\beta)}}\frac{\overline{\phi_n(\beta)}}{\overline{\phi_{n+1}(\beta)}} = 1, \quad z \in \mathbb{O}^e,$$

where convergence is locally uniform in the indicated regions.

For the quasi-orthogonal functions we get immediately from the proof of Theorem 11.8.2 that the following holds:

Theorem 11.11.5. *Let $Q_n(z, \tau_n)$ be the quasi-orthogonal functions and suppose that there is an infinite sequence of indices $n = n(h)$ for which ϕ_n is regular and τ_n is a regular value. We can then associate a quadrature formula with each of these indices characterized by the discrete measure $\mathring{\mu}_n(t, \tau_n)$. Define the class $\mathcal{C}$ functions*

$$\Omega_n(z) = \Omega_n(z, \tau_n) = -\frac{P_n(z, \tau_n)}{Q_n(z, \tau_n)} = \int D(t, z) \, d\mathring{\mu}_n(t, \tau_n).$$

Then there is a subsequence of $(n(h))$ such that $\mathring{\mu}_{n(h)}$ converges to a solution $\mathring{\mu}$ of the moment problem in $\mathcal{L}_\infty$ and we have

$$\lim_{s \to \infty} \Omega_{n(h(s))}(z) = \Omega_\mu(z) = \int D(t, z) \, d\mathring{\mu}(t),$$

locally uniformly in $\overline{\mathbb{C}} \setminus \overline{\partial \mathbb{O}}$.

As a special case we get convergence of ψ_n/ϕ_n, if we assume that there are infinitely many regular indices for which $\tau_n = 0$ is a regular value. Then we can select a subsequence of $\Omega_n(z, 0) = -\psi_n(z)/\phi_n(z)$ and this will also converge to $\Omega_\mu(z)$ locally uniformly in $\overline{\mathbb{C}} \setminus \overline{\partial \mathbb{O}}$.

12

Some applications

We give some applications of the previous theory. The theory of linear prediction of stationary stochastic processes is a classical application that dates back to the work of Kolmogorov and Wiener. By the work of Dewilde and Dym, this is known to be mathematically equivalent with Darlington synthesis and lossless inverse scattering. The basic connection is that the solution for these problems can be constructed recursively by the application of the Nevanlinna–Pick algorithm.

In the classical Nevanlinna–Pick problem one has to find a Schur function g that satisfies interpolation conditions $g(\alpha_i) = w_i$, $i = 1, \ldots, n$, where the $\alpha_i \in \mathbb{D}$. It is well known that this will give a solution if and only if the Pick matrix (we assume for simplicity that the α_i are all different from each other)

$$P_n = \left[\frac{1 - \overline{w}_i w_j}{1 - \overline{\alpha}_i \alpha_j} \right]_{i,j=1}^{n}$$

is positive definite. If one wants to find a Schur function that interpolates these data and satisfies $\|g\|_\infty \leq \gamma$, then this problem has a solution if and only if the matrix

$$P_n(\gamma) = \left[\frac{\gamma^2 - \overline{w}_i w_j}{1 - \overline{\alpha}_i \alpha_j} \right]_{i,j=1}^{n}$$

is positive definite. If this matrix has negative eigenvalues, then there can not be a solution. However, a solution with k poles in $\mathbb{D}$ can be found where k is the number of negative eigenvalues of $P_n(\gamma)$. This is the Nevanlinna–Pick–Takagi problem. The solution will therefore not be an H_∞ function anymore but we can still require that it is L_∞, that is, that $\|g\|_\infty \leq \gamma$.

Recalling our discussion of the Nevanlinna–Pick algorithm in Chapter 6, we know that if we start the recursion with a Schur function Γ, or equivalently from

342

data (α_i, w_i) drawn from a Schur function $[w_i = \Gamma(\alpha_i)$ with $\Gamma \in \mathcal{B}]$, then the Pick matrix will be positive definite and, by the Cayley transform of Γ, we can associate a Carathéodory function Ω with it and hence a positive measure on the unit circle. In that case we do have orthogonal rational functions and the whole machinery gets to work. If there are negative eigenvalues of the Pick matrix, then the Nevanlinna–Pick–Takagi algorithm can be used to solve a form of the Nehari problem. These kind of problems are related to several applications in H_∞ control and Hankel norm approximation problems in model reduction. We shall also outline the relation to the previous chapters.

12.1. Linear prediction

For the basic material of this section we used Refs. [202] and [203]. See also Refs. [174], [64], [60], and [62] for more details.

Let $\{x_t : t \in \mathbb{Z}\}$ be a complex-valued, zero-mean, finite variance, second-order stationary process. This means the following. Suppose E represents the expectation operator. Then

$$\mathrm{E}\{x_t\} = 0, \qquad \mathrm{E}\{|x_t|^2\} = \mu_0 < \infty.$$

The covariance is defined by

$$\mu_k = \mathrm{E}\{x_i \bar{x}_{i-k}\}, \quad k \in \mathbb{Z}.$$

Note that being *stationary* means that, in this definition, μ_k does not depend on i. Note also that $\mu_{-k} = \bar{\mu}_k$ and $|\mu_k| \le \mu_0$. The value $\mu_0 = \mathrm{E}\{|x_t|^2\}$ is positive. It is called the *energy* of the stochastic process. Since we assume that the energy is finite and nonzero, we can suppose without loss of generality that the process is normalized such that its energy is equal to 1.

With a stochastic process one can associate a (backward) *shift* or delay operator Z, which maps x_t to x_{t-1}. Since the index runs over all integers, the shift operator has an inverse Z^{-1}, which maps x_t to x_{t+1}. So we can write $x_t = Z^{-t}x_0, t \in \mathbb{Z}$. A stationary process is also said to be *shift invariant* because if we define $y_t = Z^i x_t$ for arbitrary integer i, then the process $\{x_t\}_{t=-\infty}^{\infty}$ and the process $\{y_t\}_{t=-\infty}^{\infty}$ have exactly the same first- and second-order statistics and are for our purposes indistinguishable.

We shall consider the pre-Hilbert space span$\{x_t : t \in \mathbb{Z}\}$ with inner product $\langle x, y \rangle_{\mathcal{X}} = \mathrm{E}\{x\bar{y}\}$. Completing this space with respect to the norm $\|x\|_{\mathcal{X}} = \langle x, x \rangle_{\mathcal{X}}^{1/2}$ turns it into a Hilbert space that we shall denote as

$$\mathcal{X} = \text{clos. span } \{x_t : t \in \mathbb{Z}\}.$$

Note that

$$\langle x_k, x_l \rangle_{\mathcal{X}} = \mathsf{E}\{x_k \bar{x}_l\} = \mu_{k-l}.$$

The space $\mathcal{X}$ is called the *time domain* for the process. The projection of $x \in \mathcal{X}$ onto a subspace $\mathcal{Y}$ is denoted by $(x|\mathcal{Y})$.

To describe the prediction problem in the time domain, we introduce the following subspaces:

$$\mathcal{X}_t^- = \text{clos. span } \{x_s : s = t, t-1, \ldots\} \quad \text{and}$$
$$\mathcal{X}_t^+ = \text{clos. span } \{x_s : s = t, t+1, \ldots\}.$$

If we consider x_t as being the *present* of the stochastic process then its *strict past* is defined as $\mathcal{X}_{t-1}^-$ and its *past* (in a weak sense since it includes the present) as $\mathcal{X}_t^-$. Similarly, we define the *strict future* and *future* of x_t as $\mathcal{X}_{t+1}^+$ and $\mathcal{X}_t^+$ respectively. The *remote past* is defined as

$$\mathcal{X}_{-\infty} = \bigcap_{t \in \mathbb{Z}} \mathcal{X}_t^-.$$

The stochastic process is called *regular* if $\mathcal{X}_{-\infty} = \{0\}$.

Let $\hat{x}_t = (x_t | \mathcal{X}_{t-1}^-)$ be the orthogonal projection of x_t onto its strict past. Then it is called the (one-step) *prediction* for x_t. The stochastic process $e_t = x_t - \hat{x}_t$ gives the error made by the prediction and it is called the forward *innovation process*.

Note that for a stationary process, the inner product is shift invariant and therefore

$$\left(x_t | \mathcal{X}_{t-1}^-\right) = \left(Z^{-t}x_0 | Z^{-t}\mathcal{X}_{-1}^-\right) = Z^{-t}\left(x_0 | \mathcal{X}_{-1}^-\right),$$
$$\hat{x}_t = Z^{-t}\hat{x}_0, \quad \text{and} \quad e_t = Z^{-t}e_0.$$

Thus the innovation process will be stationary too and the forward prediction error, which is defined as the energy of the forward innovation process, is $E = \mathsf{E}\{|e_t|^2\}$ and does not depend on t. Thus instead of finding the prediction $\hat{x}_t$ for each and every t, it is sufficient to find the prediction at time zero.

If the prediction error is nonzero, then it is impossible to predict x_0 (and for that matter also x_t) exactly from its strict past. The stochastic process is then called *unpredictable* or *nondeterministic*. If the prediction error is zero, then the process is called *predictable* or *deterministic*. If a stationary process is unpredictable, then its time domain $\mathcal{X}$ is infinite dimensional.

For a stationary deterministic process, x_0 belongs to $\mathcal{X}_{-1}^-$ since x_0 is perfectly predictable from its strict past. By stationarity, $x_{-1} \in \mathcal{X}_{-1}^-$ is again perfectly predictable from its strict past $\mathcal{X}_{-2}^-$, which implies that $x_0 \in \mathcal{X}_{-2}^-$. This argument can be repeated, which leads to the conclusion that, for a deterministic process, x_0 (and also any x_t) belongs to the remote past $\mathcal{X}_{-\infty}$ of the process.

Obviously, the prediction $\hat{x}_t$ will give the best possible estimate of x_t given its strict past. Thus it is a solution of the least-squares minimization problem

$$\inf \left\{ \|x_t - x\|_{\mathcal{X}}^2 : x \in \mathcal{X}_{t-1}^- \right\}.$$

The infimum is the prediction error E.

Note that if $\mathcal{E}_t^-$ represents the past for the innovation process e_t, then $\mathcal{X}_t^- = \mathcal{X}_{-\infty} + \mathcal{E}_t^-$ is an orthogonal decomposition. More generally, the Wold decomposition theorem (see Ref. [202, p. 137]) says

Theorem 12.1.1 (Wold decomposition). *Let* $\{x_t\}_{-\infty}^\infty$ *be a unpredictable stationary stochastic process with innovation process* $\{e_t\}_{-\infty}^\infty$. $\mathcal{X}_t^-$ *and* $\mathcal{E}_t^-$ *represent the past of* x_t *and* e_t *respectively. Then* $x_t = u_t + v_t$, *with* $u_t = (x_t|\mathcal{E}_t^-) \perp v_t = (x_t|\mathcal{X}_{-\infty})$. *The stochastic process* v_t *is predictable.*

The innovation process is orthogonal in the sense that $\langle e_s, e_t \rangle_\mathcal{X} = \delta_{st} E$. If the process is unpredictable, then $\{e_s : s = t, t-1, \ldots\}$ forms an orthogonal basis for $\mathcal{E}_t^-$ for any $t \in \mathbb{Z}$. If the process x_t is also regular then $\mathcal{X}_t^- = \mathcal{E}_t^-$ and then $\{e_s : s = t, t-1, \ldots\}$ is also an orthogonal basis for $\mathcal{X}_t^-$.

Note that, for a predictable process, $x_t \in \mathcal{X}_{-\infty}$ and thus in the Wold decomposition of x_t we get $v_t = (x_t|\mathcal{X}_{-\infty}) = x_t$ and $u_t = 0$.

We shall now consider the prediction problem for an unpredictable stationary process. This problem reduces to finding the projection $\hat{x}_0$ of x_0 onto its strict past $\mathcal{X}_{-1}^-$. Note that unpredictable means that $\mathcal{X}_{-1}^-$ is infinite dimensional. A possible way of solving this infinite-dimensional problem is by projecting x_0 onto finite-dimensional subspaces

$$\{0\} = \mathcal{X}_{0,0}^- \subset \cdots \subset \mathcal{X}_{0,n}^- \subset \mathcal{X}_{0,n+1}^- \subset \cdots \subset \mathcal{X}_{-1}^-$$

of its strict past and letting $\dim\mathcal{X}_{0,n}^- = n$ go to infinity, preferably such that $\cup_{n=0}^\infty \mathcal{X}_{0,n}^- = \mathcal{X}_{-1}^-$.

Denote the projection of x_0 onto $\mathcal{X}_{0,n}^-$ as $\hat{x}_{0,n} = (x_0|\mathcal{X}_{0,n}^-)$. If the process is regular, and if we arrange that $\cup_{n=0}^\infty \mathcal{X}_{0,n}^- = \mathcal{X}_{-1}^-$, then it is reasonable to expect that $\hat{x}_{0,n} \to \hat{x}_0$ for $n \to \infty$. If the process is not regular, then we might expect that at least $\hat{x}_{0,n} \to (x_0|\mathcal{E}_{-1}^-)$ as $n \to \infty$.

Let us now cast this problem into a function theoretic setting. This corresponds to the analysis of the process in the *frequency domain*. The reformulations of the previous concepts and prediction problem will then be very familiar in view of the analysis in the previous chapters.

With the correlation coefficients μ_k given above, we can define

$$\Omega(z) = \mu_0 + 2 \sum_{k=1}^{\infty} \mu_k z^k.$$

The series converges for $z \in \mathbb{D}$ to an analytic function and it has a positive real part. In other words, it is a Carathéodory function and by the Riesz–Herglotz representation theorem, there should be a positive measure on $\mathbb{T}$ (recall that we assumed $\mu_0 = 1$) such that

$$\Omega(z) = \Omega_\mu(z) = \int D(t, z) \, d\mu(t).$$

This measure μ is called the *spectral measure* of the stochastic process.

Suppose the Lebesgue decomposition of the spectral measure is $\mu = \mu_a + \mu_s$ with absolutely continuous part $\mu_a \ll \lambda$, that is, $d\mu_a = \mu' d\lambda$, where λ is the normalized Lebesgue measure, and singular part $\mu_s = \mu - \mu_a \perp \lambda$. The Radon–Nykodym derivative μ' is called the *spectral density* of the process. This decomposition is related to the Wold decomposition of the process. One has the following theorem (see Ref. [202]):

Theorem 12.1.2. *Let $\{x_t\}_{-\infty}^{\infty}$ be a stationary stochastic process with spectral measure μ. Then the following hold:*

1. *The process is unpredictable if and only if $\log \mu' \in L_1 = L_1(\mathbb{T}, \lambda)$, that is, μ satisfies the Szegő condition.*
2. *If the process is unpredictable and regular, then μ is absolutely continuous.*
3. *Assume the process is unpredictable and let $x_t = u_t + v_t$ be the Wold decomposition of the stationary stochastic process x_t. Let μ_u and μ_v be the spectral measures corresponding to the stochastic processes u_t and v_t respectively. Then $\mu = \mu_u + \mu_v$ corresponds to the Lebesgue decomposition of μ. μ_u is the absolutely continuous part and μ_v is the singular part w.r.t. the Lebesgue measure.*

The second conclusion is obvious since for a regular unpredictable process, the remote past is $\mathcal{X}_{-\infty} = \{0\}$, which means that the v_t component of the Wold decomposition is zero and thus also μ_v is zero, so that $\mu = \mu_u$ is absolutely continuous.

An intuitive explanation for the second statement can be given as follows. Via the relations

$$\langle x_k, x_l \rangle_\mathcal{X} = \mathsf{E}\{x_k \bar{x}_l\} = \mu_{k-l} = \int t^{l-k} \, d\mu(t) = \langle z^{-k}, z^{-l} \rangle,$$

where the last inner product is the inner product of $L_2(\mu) = L_2(\mathbb{T}, \mu)$, it can be shown that the mapping, $x_t \mapsto e^{-it\theta}$, called *Kolmogorov isomorphism*, will be an isomorphism between the Hilbert spaces $\mathcal{X}$ and $L_2(\mu)$, that is, between the time domain and the frequency domain. This explains why we have used the notation Z for the backward shift operator in $\mathcal{X}$ since a multiplication with z in $L_2(\mu)$ corresponds with applying the operator Z in $\mathcal{X}$.

By definition the stationary process x_t is unpredictable if $x_0 \notin \mathcal{X}_{-1}^-$, or equivalently, if $x_1 \notin \mathcal{X}_0^-$. It should be clear that the Kolmogorov isomorphism maps the past $\mathcal{X}_0^-$ onto the Hardy space $H_2(\mu)$. Thus for an unpredictable process, the time domain statement $x_1 \notin \mathcal{X}_0^-$ translates into the frequency domain the fact that z^{-1} is not in $H_2(\mu)$, which means that the polynomials are not dense in $L_2(\mu)$. By Theorem 7.2.1 we know that this is true iff Szegő's condition is satisfied, and this is an explanation for the second statement of Theorem 12.1.2 above.

It is now easy to formulate and, at least in principle, solve the prediction problem in the frequency domain. The projection of x_0 onto $\mathcal{X}_{-1}^-$ in the Hilbert space $\mathcal{X}$ corresponds to the projection of 1 onto the space $\{f \in H_2(\mu) : f(0) = 0\}$. The projecting vector $e_0 = x_0 - \hat{x}_0 \in \mathcal{X}$ corresponds to the projecting vector $f \in H_2(\mu)$, which is the function solving the optimization problem (compare with Theorem 1.4.2)

$$P^2(1, 0): \quad \inf\{\|f\|_\mu^2 : f \in H_2(\mu), f(0) = 1\}.$$

If the process is unpredictable, then $\log \mu' \in L_1$ and the Szegő kernel $s_w(z)$ will be defined in terms of the outer spectral factor σ as [see (1.43) and (2.18)]

$$s_w(z) = \frac{1}{\sigma(z)(1 - \overline{w}z)\overline{\sigma(w)}}, \qquad \sigma(z) = \exp\left\{\frac{1}{2}\int D(t, z) \log \mu'(t)\, d\lambda(t)\right\}.$$

From Lemma 2.3.6 we know that the Szegő kernel is reproducing for $H_2(\mu)$ and by Theorem 1.4.2 we know that the solution of problem $P^2(1, 0)$ will be given by $f(z) = s_0(z)/s_0(0) = \sigma(0)/\sigma(z)$ with minimum $s_0(0)^{-1} = |\sigma(0)|^2$. In other words, the optimal prediction is given by (Z is the backward shift and I is the identity)

$$\hat{x}_0 = x_0 - e_0 = x_0 - f(Z)x_0 = [I - f(Z)]x_0, \qquad f(z) = \sigma(0)/\sigma(z).$$

The operator $I - f(Z)$ is the orthogonal projection operator in $\mathcal{X}_0^-$ onto the subspace $\mathcal{X}_{-1}^-$. The prediction error is given by

$$E = |\sigma(0)|^2 = \exp\left\{\int \log \mu'(t)\, d\lambda(t)\right\}.$$

Thus the prediction problem (for an unpredictable process) is in fact equivalent with the Szegő problem (see Section 9.1), which is also a spectral factorization problem: Given a measure satisfying μ the Szegő condition, find its spectral factor σ.

The classical polynomial approach is to consider the problem $P^2(1, 0)$ in the finite-dimensional subspaces $\mathcal{P}_n \subset H_2(\mu)$ of all polynomials of degree at most n, hence $\dim \mathcal{P}_n = n + 1$. Denoting this problem as $P_n^2(1, 0)$ (see Theorem 2.3.2),

$$P_n^2(1, 0): \quad \inf\{\|f\|_\mu : f(0) = 1, f \in \mathcal{P}_n\},$$

we get the solution $f = \varphi_n^* = \kappa_n^{-1}\phi_n^*$, where ϕ_n is the nth orthonormal Szegő polynomial with leading coefficient $\kappa_n > 0$, and φ_n are the monic ones. The minimum is given by κ_n^{-2}.

Such finite-dimensional solutions approximate the prediction in the sense that one finds the best possible prediction of order n: $\hat{x}_{0,n} = (x_0|\mathcal{X}_{0,n}^-)$, where $\mathcal{X}_{0,n}^- = \text{span}\{x_{-1}, \ldots, x_{-n}\}$. Thus, setting

$$\hat{x}_{0,n} = -(a_1 x_{-1} + \cdots + a_n x_{-n}) = -(a_1 Z + \cdots + a_n Z^n)x_0,$$

we get for the nth-order forward innovation

$$e_{0,n} = x_0 - \hat{x}_{0,n} = (1 + a_1 Z + \cdots + a_n Z^n)x_0 = \varphi_n^*(Z)x_0.$$

Here φ_n^* is called the (forward) predictor (polynomial) of order n.

To give an interpretation for the Szegő polynomials φ_n and the Szegő–Levinson recurrence, we consider the nth-order *backward prediction* problem where it is required to "predict" the variable x_{-n} from its future values $x_{1-n}, \ldots, x_0$. Thus if $\mathcal{X}_{0,n}^+ = \text{span}\{x_{1-n}, \ldots, x_0\}$, we have to find the projection $\check{x}_{0,n} = (x_{-n}|\mathcal{X}_{0,n}^+)$. The projecting vector $f_{0,n} = x_{-n} - \check{x}_{-n}$ is called the nth-order *backward innovation*. In the frequency domain, this backward problem is translated into finding the solution of

$$\inf\{\|f\|_\mu^2 : f \in \mathcal{P}_n^M\},$$

where $\mathcal{P}_n^M$ is the set of monic polynomials of degree at most n. Because $\|\varphi_n\|_\mu = \|\varphi_n^*\|_\mu$, it is directly seen that the solution to this problem is the monic Szegő polynomial φ_n. We find $f_{0,n} = \varphi_n(Z)x_0$ and the error for this prediction problem is $\|f_{0,n}\|_{\mathcal{X}}^2 = \|\varphi_n\|_\mu^2 = \|\varphi_n^*\|_\mu^2 = \|e_{0,n}\|_{\mathcal{X}}^2 = \kappa_n^{-2} = E_n$. Thus it is the same as for the forward problem. Note that

$$E_n = \prod_{k=1}^n (1 - |\lambda_k|^2),$$

where the λ_k are the Szegő parameters or reflection coefficients that appear in the Szegő–Levinson recurrence [see (4.1)]

$$\begin{bmatrix} \varphi_n \\ \varphi_n^* \end{bmatrix} = \begin{bmatrix} 1 & \bar{\lambda}_n \\ \lambda_n & 1 \end{bmatrix} \begin{bmatrix} z & 0 \\ 0 & 1 \end{bmatrix} \begin{bmatrix} \varphi_{n-1} \\ \varphi_{n-1}^* \end{bmatrix}.$$

Replacing z by Z and applying this to x_t we find the time-domain relations

$$\begin{bmatrix} f_{t,n} \\ e_{t,n} \end{bmatrix} = \begin{bmatrix} 1 & \bar{\lambda}_n \\ \lambda_n & 1 \end{bmatrix} \begin{bmatrix} Z & 0 \\ 0 & 1 \end{bmatrix} \begin{bmatrix} f_{t,n-1} \\ e_{t,n-1} \end{bmatrix} = \begin{bmatrix} 1 & \bar{\lambda}_n \\ \lambda_n & 1 \end{bmatrix} \begin{bmatrix} f_{t-1,n-1} \\ e_{t,n-1} \end{bmatrix}.$$

The innovation prediction filter takes as input the stochastic process x_t and produces as output the nth-order innovations $e_{t,n}$ and $f_{t,n}$ as in Figure 12.1. By the Szegő–Levinson recursion, this can be realized in a cascade of elementary 2-ports as in Figure 12.2. Each 2-port is described by an elementary 2×2 matrix of the form

$$t_k = \begin{bmatrix} 1 & \bar{\lambda}_k \\ \lambda_k & 1 \end{bmatrix} \begin{bmatrix} Z & 0 \\ 0 & 1 \end{bmatrix},$$

which can be realized as a lattice in Figure 12.3. The Z-block represents a

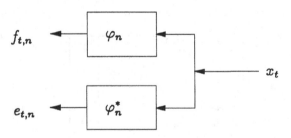

Figure 12.1. Innovation prediction filter.

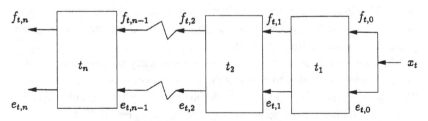

Figure 12.2. Cascade form of innovation prediction filter.

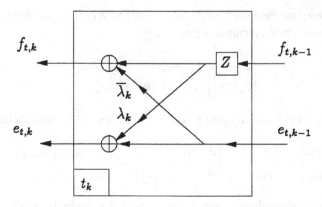

Figure 12.3. Lattice section of innovation prediction filter.

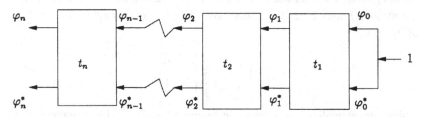

Figure 12.4. Frequency-domain formulation of innovation prediction filter.

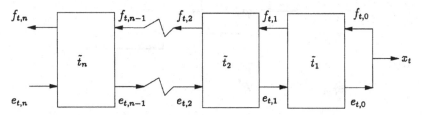

Figure 12.5. Modeling filter in the time domain.

delay. This innovation prediction filter is sometimes called a *whitening filter* because it can be shown that, under suitable conditions, the processes $e_{t,n}$ and $f_{t,n}$ become white noise as $n \to \infty$.

The same filter can be formulated in the frequency domain. In that case, the innovation prediction filter is just the Szegő–Levinson recurrence relation. The input is 1, the delay operator is replaced by a multiplication with z, and the output of the filter are the polynomials φ_n and φ_n^* as in Figure 12.4.

Inverting the filter gives a scheme like that shown in Figure 12.5. In the

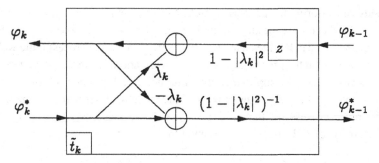

Figure 12.6. Modeling lattice section in the frequency domain.

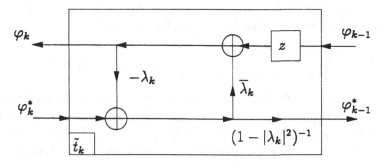

Figure 12.7. Modeling ladder section in the frequency domain.

frequency domain the effect can be described by the equations

$$\varphi_k(z) = (1 - |\lambda_k|^2)z\varphi_{k-1}(z) + \bar{\lambda}_k\varphi_k^*(z),$$
$$\varphi_{k-1}^*(z) = (1 - |\lambda_k|^2)^{-1}[\varphi_k^*(z) - \lambda_k\varphi_k(z)].$$

These are directly obtained from the Szegő–Levinson recursions. This elementary section is depicted in Figure 12.6. Such a filter section is equivalent to the ladder structure of Figure 12.7 since the previous relations can be rewritten as

$$\varphi_k(z) = z\varphi_{k-1}(z) + \bar{\lambda}_k\varphi_{k-1}^*(z),$$
$$(1 - |\lambda_k|^2)\varphi_{k-1}^*(z) = \varphi_k^*(z) - \lambda_k\varphi_k(z).$$

The filter of Figure 12.5 is called a *modeling filter* because, in the time domain, it gives the stochastic process as the output provided the innovation process $e_{t,n}$ is applied at the input. The backward innovation process $f_{t,n}$ comes as a bonus. Thus to generate the process x_t exactly from an nth-order modeling filter, we need to know the $e_{t,n}$, but storing this information is as difficult and expensive as storing the process x_t itself. However, in practical applications, it is assumed that during the analysis of the process, we obtained a whitening filter (predictor) that catches the characteristic information of the process quite well, which means that the process $e_{t,n}$ is almost informationless. In stochastic terms, this means that $e_{t,n}$ is approximately white noise and the only information it contains is its energy E_n. Thus, if we feed the modeling filter with a normalized white noise process w_t, that is, $E\{w_k \overline{w}_l\} = \delta_{kl}$ so that it has a perfectly flat spectrum $W(z) = 1$, then we can assume that

$$x_t \approx \tilde{x}_t = E_n^{1/2} \varphi_n^*(Z)^{-1} w_t = \phi_n^*(Z)^{-1} w_t.$$

Since the spectral measure of w_t is the normalized Lebesgue measure, we see that the autocorrelation coefficients of the process $\tilde{x}_t$ are given by (see Theorem 6.1.9 and recall that $E_n = \kappa_n^{-2}$ and $\phi_n = \kappa_n \varphi_n$)

$$\tilde{\mu}_{k-l} = E_n \int t^{l-k} \frac{d\lambda(t)}{|\varphi_n^*(t)|^2} = \int t^{l-k} \frac{d\lambda(t)}{|\phi_n^*(t)|^2} = \int t^{l-k} \, d\mu(t) = \mu_{k-l}$$

for all $0 \leq k, l \leq n$. Thus our model will match the autocorrelation coefficients μ_k of the given process for $k = 0, \pm 1, \ldots, \pm n$. The spectral density $\mu' = |\sigma|^2$ of the original process and the spectral density $\tilde{\mu}' = |\phi_n^*|^{-2}$ of the simulated process will interpolate in the sense that $\sigma(z) - 1/\phi_n^*(z)$ has a zero of order $n+1$ at the origin (see Theorem 6.3.1). This is equivalent with the Riesz–Herglotz transforms

$$\Omega(z) = \int D(t, z) \, d\mu(t) \quad \text{and} \quad \tilde{\Omega}(z) = \int D(t, z) \, d\tilde{\mu}(t) = \frac{\psi_n^*(z)}{\phi_n^*(z)}$$

interpolating each other such that $\Omega(z) - \tilde{\Omega}(z)$ has a zero of order $n + 1$ at the origin.

From Chapter 9 (in fact, as a special case, setting all $\alpha_k = 0$), we know how and when ϕ_n^* and κ_n (and also φ_n) converge as $n \to \infty$. Under suitable conditions we have convergence in some sense of $E_n = \kappa_n^{-2} \to |\sigma(0)|^2$ and $\varphi_n^*(z) \to \sigma(0)/\sigma(z)$. Thus we have convergence to the solution of the infinite-dimensional prediction problem. Note that if the process is unpredictable, then $\int \log \mu' d\lambda > -\infty$ and thus $|\sigma(0)|^2 > 0$. The prediction error E_n does *not* go to zero as $n \to \infty$. If it did, this would mean perfect prediction from the strict past, that is, a deterministic process.

The modeling filter $\phi_n^*(z)^{-1}$ has the advantage that it is stable. This means that all its poles (called transmission modes) are outside the closed unit disk. Also, its zeros (called transmission zeros) are outside the closed unit disk, since they are all at infinity. This is also a desirable property for implementation because it means that among all filters with the same amplitude characteristic, the range of the phase angle will be mimimized. Therefore, if the transmission zeros are outside the closed unit disk, the filter will be called *minimal phase*. A filter that has no finite transmission zeros is called an autoregressive filter. However, the fact that there is no freedom left in choosing the transmission zeros can be seen as a restriction. This is the point where the orthogonal rational functions have an advantage over the orthogonal polynomials. Let us motivate this by a simple example. If the spectral measure were

$$d\mu(t) = \left| \frac{t - \alpha}{t - \beta} \right|^2 d\lambda(t), \quad \alpha, \beta \in \mathbb{D},$$

then we are trying to approximate the predictor

$$\frac{\sigma(0)}{\sigma(z)} = \frac{1 - \overline{\beta}z}{1 - \overline{\alpha}z}$$

by a polynomial φ_n^*. Since

$$\frac{1 - \overline{\beta}z}{1 - \overline{\alpha}z} = 1 + (\overline{\alpha} - \overline{\beta})z[1 + \overline{\alpha}z + (\overline{\alpha}z)^2 + (\overline{\alpha}z)^3 + \cdots],$$

it is clear that if α is close to $\mathbb{T}$ and β significantly different from $\overline{\alpha}$, then the series between square brackets will converge very slowly. It forms a substantial part of the expansion since the factor $\overline{\alpha} - \beta$ in front will not be small if β is indeed significantly different from $\overline{\alpha}$. Thus we shall need a polynomial of a high degree to get a good approximation of the rational function on the left-hand side. If we had proposed an approximation of the form $p_1(z)/(1 - \overline{\alpha}z)$ with $p_1 \in \mathcal{P}_1$, then the optimal estimate would give the exact solution. Of course, this is only true if we had the correct value of the transmission zero $1/\overline{\alpha}$ of the modeling filter. Leaving the denominator free as well, thus considering p_1/q_1, where p_1 and q_1 are both in $\mathcal{P}_1$, would make the problem nonlinear, and if we do not want to give up the simplicity of computation, we should choose a compromise and assume that we have a method of estimating the value of α. If this estimate is α', then we may expect that approximating $(1 - \overline{\beta}z)/(1 - \overline{\alpha}z)$ by an approximant of the form $p_n(z)/(1 - \overline{\alpha}'z)$ with $p_n \in \mathcal{P}_n$ corresponds to an approximation of the rational function

$$(1 - \overline{\beta}z)\frac{1 - \overline{\alpha}'z}{1 - \overline{\alpha}z} = (1 - \overline{\beta}z)\{1 + (\overline{\alpha} - \overline{\alpha}')z[1 + \overline{\alpha}z + (\overline{\alpha}z)^2 + (\overline{\alpha}z)^3 + \cdots]\}$$

by a polynomial $p_n \in \mathcal{P}_n$. This will be much easier because the ratio $(1 - \overline{\alpha}'z)/(1 - \overline{\alpha}z)$ is almost 1, so that the rational function on the left-hand side is almost a polynomial of degree 1, which is reflected by the fact that on the right-hand side the part of degree 1 is dominant. Indeed, the slowly converging series, which also appeared in the previous expansion, is now multiplied by a small number $(\overline{\alpha} - \overline{\alpha}')$. Thus if this number is small enough, then an approximation with $n = 1$ will already give a fairly good fit.

This idea can of course be generalized, which leads to the setting of our orthogonal rational functions. Instead of working with the subspaces $\mathcal{P}_n$ of polynomials, we work with the subspaces $\mathcal{L}_n$, which are rational functions with given poles. If we want the modeling filter (which was $1/\phi_n^*$ in the polynomial case and is now of the form p_n/q_n with $p_n, q_n \in \mathcal{P}_n$) to be stable and minimum phase, then its poles and zeros (transmission modes and transmission zeros) should all be outside the closed unit disk. We fix the zeros by choosing them as estimates for the transmission zeros of the given process. Since we have only the spectral information of the given process, we can choose the transmission zeros to be either α or $1/\overline{\alpha}$ since indeed the spectral density is $|\sigma(t)|^2 = \sigma(t)\sigma_*(t)$. Thus if we want a minimal phase model, we should take care that they are chosen to be all outside the closed unit disk. Now it turns out that these zeros appear as the poles of the function spaces $\mathcal{L}_n$. So we are working in the spaces $\mathcal{L}_n$ of functions analytic in $\mathbb{D}$ as they were considered in the first chapters of this book. The stability of the optimal predictor will come as a bonus, just as in the polynomial case (see below).

One might expect that the optimal predictor is given by the rational function φ_n^*, but this is not true. Indeed, the finite-dimensional subspaces of the past on which we project x_0 have now a Kolmogorov image, which are the spaces $\mathcal{L}_n$. We know that the optimal nth-order predictor is found by solving the problem

$$\inf\{\|f\|_\mu : f(0) = 1, f \in \mathcal{L}_n\}.$$

The solution to this problem is given by $f(z) = k_n(z, 0)/k_n(0, 0)$, where $k_n(z, w)$ is the reproducing kernel for $\mathcal{L}_n$ (see Theorem 2.3.2). If $\kappa_n(z, w) = k_n(z, w)[k_n(w, w)]^{-1/2}$ is the normalized kernel, then, with $\kappa_n(z) = \kappa_n(z, 0)$, we can also describe the optimizing function as $f(z) = \kappa_n(z)/\kappa_n(0)$. The nth-order prediction error is given by $E_n = 1/k_n(0, 0) = \kappa_n(0)^{-2}$. See Theorem 1.4.2. By (3.14) we know that this error equals

$$E_n = \frac{1}{\kappa_n(0)^2} = \prod_{k=1}^{n} \frac{1 - |\rho_k(0)|^2}{1 - |\gamma_k(0)|^2} = \prod_{k=1}^{n} \frac{1 - |\rho_k(0)|^2}{1 - |\alpha_k \rho_k(0)|^2}, \qquad (12.1)$$

where ρ_k and γ_k are the coefficients that appear in the recurrence relation for the reproducing kernels. In analogy with the polynomial case, we find a modeling filter, which is given by

$$E_n^{1/2} \frac{K_n(0)}{K_n(z)} = \frac{1}{K_n(z)}.$$

We know that K_n is an outer function in H_2 [see Theorem 3.3.3(3)]. Thus its zeros, which are the transmission modes of the modeling filter, will all be outside the closed unit disk. Hence the modeling filter is stable. As in the polynomial case we have that the process

$$x_t \approx \tilde{x}_t = E_n^{1/2} K_n(0) K_n(Z)^{-1} w_t = K_n(Z)^{-1} w_t$$

has spectral measure $d\tilde{\mu} = d\lambda/|K_n|^2$ and autocorrelation coefficients

$$\tilde{\mu}_{k-l} = \int t^{l-k} \frac{d\lambda(t)}{|K_n(t)|^2}.$$

Since $\tilde{\mu}_0 = \mu_0$, the synthesized process has the same energy as the original one, but for the other autocorrelation coefficients we have in general $\tilde{\mu}_k \neq \mu_k, k > 1$. Instead, the approximation is such that the outer spectral factors $\tilde{\sigma} = 1/K(z)$ and σ interpolate each other in the points $0, \alpha_1, \ldots, \alpha_n$ (Theorem 6.3.1). Also, the Riesz–Herglotz transforms $\tilde{\Omega}(z) = L_n(z,0)/K_n(z,0)$ and Ω_μ interpolate in these points (Theorem 6.3.3). The inner products in $\mathcal{L}_n$ is the same with respect to $\tilde{\mu}$ and with respect to μ. This means that the spectral information of the original process x_t, which is contained in the "information space" $\mathcal{L}_n$, is matched completely by the simulated process $\tilde{x}_t$.

The realization of the filters (whitening and modeling) has the same cascade structure but the sections are somewhat more complicated because the recurrence relation for the kernels is more complicated. For example, section t_k, which is the frequency-domain formulation of a normalized innovation prediction filter, would be of the form given in Figure 12.8. The notation used

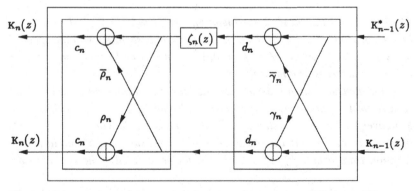

Figure 12.8. Section n of normalized innovation prediction filter in the frequency domain.

is the same as in Theorem 3.3.1. For practical applications, one can use the Nevanlinna–Pick algorithm as described in Section 6.4. The convergence of $K_n(z) = K_n(z, 0)$ to the inverse of the spectral factor was studied in Chapter 9.

12.2. Pisarenko modeling problem

This problem was originally considered in Ref. [179]. For other discussions of the problem see Refs. [53], [108], [11], and [55].

Let $\{x_t\}$ be a stationary process as before and let its $n + 1$ first autocorrelation coefficients

$$\mu_k^x = E\{x_t \overline{x}_{t-k}\}, \quad k = 0, \ldots, n$$

be given. Suppose we want to model this process as $x_t = y_t + w_t$, where y_t and w_t are uncorrelated processes and w_t is a white noise process with variance G. Then one has

$$\mu_k^x = \mu_k^y + \delta_{k0} G, \quad k = 0, 1, \ldots,$$

where $\mu_k^y = E\{y_t \overline{y}_{t-k}\}$ are the autocorrelation coefficients of the process $\{y_t\}$.
Define the covariance matrices

$$G_n^y = \left[\mu_{i-j}^y\right]_{i,j=0}^n \quad \text{and} \quad G_n^x = \left[\mu_{i-j}^x\right]_{i,j=0}^n.$$

Then $G_n^y = G_n^x - G I_n$ and since this is a covariance matrix it has to be positive semidefinite. Therefore, the maximal possible value of G is the smallest non-negative eigenvalue of G_n^x. For simplicity let us assume that this eigenvalue is simple. The Carathéodory–Toeplitz theorem says:

Theorem 12.2.1 (Carathéodory–Toeplitz). *Let $G_k = [\mu_{i-j}]_{i,j=0}^k, k = 0, 1, \ldots$ be the covariance matrices of size $k + 1$ that are associated with the spectral measure μ. Then $\det G_k \neq 0, k = 0, 1, \ldots, n - 1$ and $\det G_n = 0$ iff μ is positive measure with n mass points.*

If the measure has n mass points, then the nth orthogonal polynomial ϕ_n with respect to this measure is η-reciprocal, that is, $\phi_n^* = \eta \phi_n$ with $\eta \in \mathbb{T}$. Moreover, ϕ_n has n zeros that are all on $\mathbb{T}$ and coincide with the mass points of the measure. The outer spectral factor of the measure is $1/\phi_n^*$ and its Riesz–Herglotz transform is ψ_n^*/ϕ_n^*, where ψ_n is the polynomial of the second kind.

We apply this in the case of $G_n = G_n^y$ and $\mu = \mu^y$. Let the mass points be

$\xi_k = e^{-i\theta_k} \in \mathbb{T}$, and the corresponding masses $h_k > 0$. Then

$$\frac{\psi_n^*}{\phi_n^*} = \Omega_\mu(z) = \sum_{k=1}^{n} h_k D(\xi_k, z) = \sum_{k=1}^{n} h_k \frac{\xi_k + z}{\xi_k - z}, \quad h_k > 0. \quad (12.2)$$

Note that (see Chapter 5)

$$h_k = \frac{1}{k_{n-1}(\xi_k, \xi_k)} = \frac{1}{\sum_{k=0}^{n-1} |\phi_k(\xi_k)|^2}.$$

The above formula (12.2) for Ω_μ implies that

$$\Omega_\mu(z) = \mu_0 + 2 \sum_{l=1}^{\infty} \mu_l z^l \quad \text{with} \quad \mu_l = \sum_{k=1}^{n} h_k \xi_k^l.$$

Thus

$$y_j = \sum_{k=1}^{n} \sqrt{h_k} \exp\{i(j\theta_k + \gamma_k)\},$$

where the phase angles γ_k are uncorrelated zero-mean random variables. Such a process is generated by n sinusoidal wave generators. This is called the Pisarenko model [179] (see Figure 12.9). Thus the procedure to obtain this model goes as follows: First find the smallest eigenvalue G of G_n^x. Then define the moments $\mu_0 = \mu_0^x - G$ and $\mu_k = \mu_k^x$ for $k = 1, \ldots, n$. From these μ_k

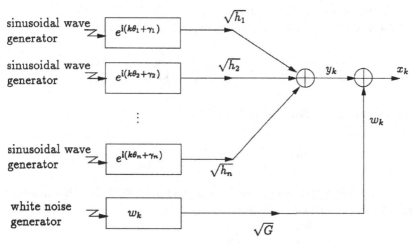

Figure 12.9. Pisarenko frequency model.

one can generate the Szegő polynomials $\phi_0, \ldots, \phi_n$. The zeros of ϕ_n are the $\xi_k = e^{-i\theta_k} \in \mathbb{T}$, $k = 1, \ldots, n$ and the weights are given by

$$h_k = \left[\sum_{j=0}^{n-1} |\phi_j(\xi_k)|^2 \right]^{-1}, \quad k = 1, \ldots, n.$$

In this classical model, one draws information about the spectrum of the process from the information space $\mathcal{P}_n$. This means that we use the matrix G_n^x, which is the Gram matrix of the basis $\{1, z, \ldots, z^n\}$ for $\mathcal{P}_n$ with respect to the measure μ.

Again, in this application, the use of orthogonal rational functions instead of orthogonal polynomials may have advantages. In this generalization to the rational case one draws information about the spectrum from the information space $\mathcal{L}_n$ instead of $\mathcal{P}_n$. Suppose we choose for $\mathcal{L}_n$ the Malmquist basis $\{b_k\}_{k=0}^n$ given by (2.13). Then if μ^x, μ^y, and μ^w are the spectral measures for the processes $\{x_k\}$, $\{y_k\}$, and $\{w_k\}$ respectively, then

$$\langle b_k, b_l \rangle_{\mu^x} = \langle b_k, b_l \rangle_{\mu^y} + \langle b_k, b_l \rangle_{\mu^w}.$$

Because μ^w is G times the normalized Lebesgue measure, and because the Malmquist basis is orthogonal for this measure, we find that the Gram matrices are given by

$$G_n^x = G_n^y + G I_n.$$

Because $G_n = G_n^y = G_n^x - G I_n$ has to be positive semidefinite, it follows that G is again the smallest eigenvalue of G_n^x.

For $\mu = \mu^y$, it follows from Theorem 2.2.4 and (3.14) that

$$\frac{1}{\kappa_n^2} = \frac{\det G_n}{\det G_{n-1}} = \frac{1}{k_n(\alpha_n, \alpha_n)} = \prod_{k=1}^{n} \frac{1 - |\rho_k(\alpha_n)|^2}{1 - |\gamma_k(\alpha_n)|^2}.$$

Thus if $\det G_n = 0$ then $\rho_n(\alpha_n) \in \mathbb{T}$. Thus

$$\rho_n(\alpha_n) = \frac{\overline{\phi_n(\alpha_n)}}{\phi_n^*(\alpha_n)} \in \mathbb{T} \quad \Rightarrow \quad \frac{\phi_n(\alpha_n)}{\phi_n^*(\alpha_n)} \in \mathbb{T}.$$

Because $\phi_n(z)/\phi_n^*(z)$ is an inner function, and $\alpha_n \in \mathbb{D}$, this implies that this function is a unimodular constant

$$\frac{\phi_n}{\phi_n^*} = \eta \in \mathbb{T}.$$

Thus $\phi_n(z) = \eta\phi_n^*(z)$ is η-reciprocal as in the polynomial case. The para-orthogonal function Q_n is thus given by

$$Q_n(z, \tau) = \phi_n(z) + \tau\phi_n^*(z) = (\eta + \tau)\phi_n^*(z).$$

Because Q_n has n simple zeros on $\mathbb{T}$, $\phi_n^*(z)$, and hence also $\phi_n(z)$, will have n simple zeros on $\mathbb{T}$.

Because the singularity of G_n also means that the Gram–Schmidt orthogonalization process breaks down in step $n + 1$ so that $H_2(\mu) = \mathcal{L}_n$ is $n + 1$ dimensional because μ is positive. This is only possible if μ is a discrete measure having n mass points with positive weights.

The rest of the solution goes as in the polynomial case. One computes the smallest eigenvalue G of G_n^x, sets $G_n = G_n^* - GI_n$, and from this Gram matrix generates the orthogonal rational functions $\phi_k(z)$. The zeros ξ_k of ϕ_n will all be on $\mathbb{T}$ and the weights h_k are given by $h_k = 1/k_{n-1}(\xi_k, \xi_k)$.

Again, as in the previous applications, when the information from the spaces $\mathcal{L}_n$ is available, or even approximately available via estimates of the α_k, then the rational Pisarenko model will give a better approximation of the signal than the polynomial one.

12.3. Lossless inverse scattering

The material of this section is mainly inspired by the work of Dewilde and Dym. For more detailed information refer to Refs. [64], [60], [58], [62], and [59].

Consider a dissipative scattering medium M as depicted in Figure 12.10. In inverse scattering one wants to find a model for the medium M, given an input signal $u(t)$ (incident wave) and an output signal $v(t)$ (reflected wave). For digital processing, the signals are sampled at discrete time intervals and we shall therefore consider discrete time signals w_k, $k \in \mathbb{Z}$. The energy of such a signal is defined as its ℓ_2-norm: $\sum_{k\in\mathbb{Z}} |w_k|^2$. We consider signals with finite energy, which are thus ℓ_2 sequences. The z-transform

$$W(z) = \sum_{k=-\infty}^{\infty} w_k z^k$$

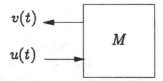

Figure 12.10. Scattering medium.

will converge to a function $W \in L_2(\mathbb{T})$. Obviously the energy is also given by

$$\|w\|_2^2 = \sum_{k=-\infty}^{\infty} |w_k|^2 = \|W\|_2^2 = \int |W(t)|^2 \, d\lambda(t),$$

where λ is the normalized Lebesgue measure on $\mathbb{T}$.

The medium is supposed to act as a linear system with transfer function S. This means that if we denote the z-transforms of the incident wave and reflected wave as $U(z)$ and $V(z)$ respectively, then $V(z) = S(z)U(z)$. The transfer function $S(z)$ is called the *scattering function* of the medium.

For physical reasons, it is plausible that the medium behaves as a causal system. This means that there can not be any output different from zero before there has been any input that was different from zero. Thus, when the medium is excited with a unit impulse at time zero, that is, with a signal $u_k = \delta_{k0}$ whose z-transform is $U(z) = 1$, then the output, with z-transform $V(z) = S(z)$ have Fourier coefficients $s_k = 0$ for $k < 0$. This means that S is analytic in $\mathbb{D}$. The system is supposed to be stable in the sense that bounded inputs are transformed in bounded outputs, and this means that $S \in H_\infty = H_\infty(\mathbb{T})$. Moreover, if the medium is *passive*, then it should add no energy to any signal while it is being scattered. Mathematically this forces $S(z)$ to be bounded by 1: $|S(z)| \le 1$ for all z in the closed unit disk. In other words, S is a Schur function: $S \in \mathcal{B}$. In what follows we use scattering function in the meaning of passive scattering function and this is a synonym for Schur function.

If the medium does not absorb energy, then the medium and its scattering function are called *lossless*, which means mathematically that $|S(t)| = 1$ a.e. for $t \in \mathbb{T}$. Thus a lossless scattering function is an inner function in H_∞.

We need a generalization of these concepts to square matrix valued functions. We say that an $n \times n$ matrix valued function $\Sigma(z)$ is a (passive) *scattering matrix* if it is analytic in $\mathbb{D}$ and if it is contractive in $\mathbb{D}$; thus $\Sigma(z)^H \Sigma(z) \le I_n$ for $z \in \mathbb{D}$. Here I_n stands for the $n \times n$ unit matrix and the inequality is understood in the sense of positive definiteness: $\Sigma^H \Sigma \le I_n$ iff $I_n - \Sigma^H \Sigma$ is positive semidefinite. Note that $\Sigma^H \Sigma \le I_n \Leftrightarrow \Sigma \Sigma^H \le I_n$. Thus Σ is a scattering matrix iff Σ^H is a scattering matrix.

A scattering matrix is called lossless if it is unitary a.e. on $\mathbb{T}$; thus $\Sigma(t)^H \Sigma(t) = I_n$ a.e., $t \in \mathbb{T}$.

For $n = 1$ we obtain the previous definition of a (lossless) scattering function. Thus a scattering function is a scattering matrix of size 1×1.

If we think of the scattering medium M as having a top surface at which the incident and reflected waves are observed and at the other end a bottom surface

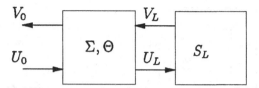

Figure 12.11. Scattering medium with load.

where another medium starts with some other scattering properties, then we can represent the whole system as in Figure 12.11. The scattering function of the underlying medium is S_L (the *load*). The scattering properties of the medium M are described by a scattering matrix Σ. We have

$$V_L(z) = S_L(z)U_L(z) \tag{12.3}$$

and

$$\begin{bmatrix} U_L(z) \\ V_0(z) \end{bmatrix} = \Sigma(z) \begin{bmatrix} U_0(z) \\ V_L(z) \end{bmatrix}, \tag{12.4}$$

where U_0 and V_0 now represent the incident wave and reflected wave at the top surface of the medium M, while U_L is the transmitted wave that emerges at the bottom surface of M and is incident for the underlying medium and V_L is the wave reflected by the underlying medium and is incident at the bottom surface of M.

Using circuit terminology, we say that the load is described by a 1-port (1 input, 1 output) and the medium itself is described by a 2-port (2 inputs, 2 outputs). The whole system is a cascade of the 2-port and the 1-port. We say that the 2-port is loaded by a passive load S_L.

The scattering matrix gives the relation between the incident waves (U_0 and V_L) and the output waves (V_0 and U_L). Although this is the most logical description, it has mathematically a number of disadvantages. For example, the overall scattering function for the combination of the medium M and its load is given by

$$S_0(z) = \frac{V_0(z)}{U_0(z)} = \Sigma_{21}(z) + \Sigma_{22}(z)S_L(z)[1 - \Sigma_{12}(z)S_L(z)]^{-1}\Sigma_{11}(z), \tag{12.5}$$

where $\Sigma = [\Sigma_{ij}]_{i,j=1,2}$. This is not a simple formula. Similarly, if we have two consecutive media, each having a scattering matrix, then the scattering matrix of the cascade of the two media is difficult to obtain. If we consider the cascade

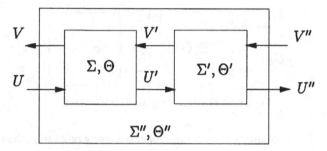

Figure 12.12. Cascade of two scattering media.

of Figure 12.12 then

$$\begin{bmatrix} U' \\ V \end{bmatrix} = \Sigma \begin{bmatrix} U \\ V' \end{bmatrix} \quad \text{and} \quad \begin{bmatrix} U'' \\ V' \end{bmatrix} = \Sigma' \begin{bmatrix} U' \\ V'' \end{bmatrix}.$$

For the cascade we have

$$\begin{bmatrix} U'' \\ V \end{bmatrix} = \Sigma'' \begin{bmatrix} U \\ V'' \end{bmatrix},$$

where Σ'' is given by the Redheffer product [183] of Σ and Σ':

$$\Sigma'' = \Sigma * \Sigma' = \begin{bmatrix} \Sigma'_{11} + \Sigma'_{12}\Sigma_{11}\Gamma\Sigma'_{12} & \Sigma'_{12}\Sigma_{11}\Gamma\Sigma'_{22}\Sigma_{12} + \Sigma'_{12}\Sigma_{12} \\ \Sigma_{21}\Gamma\Sigma'_{21} & \Sigma_{21}\Gamma\Sigma'_{22}\Sigma_{12} + \Sigma_{22} \end{bmatrix},$$

with $\Gamma = (1 - \Sigma'_{22}\Sigma_{11})^{-1}$. If $1 - \Sigma'_{22}\Sigma_{11}$ is not identically zero, this will exist for all values of z, except for at most a countable number of values in $\mathbb{D}$.

For our purposes, it is much easier to work with *chain scattering matrices*. Whereas the scattering matrix gives the relation between input and output waves for the 2-port, the chain scattering matrix gives the relation between the waves at the bottom and the waves at the top surface. Thus

$$\begin{bmatrix} U' \\ V \end{bmatrix} = \Sigma \begin{bmatrix} U \\ V' \end{bmatrix} \quad \Leftrightarrow \quad \begin{bmatrix} U' \\ V' \end{bmatrix} = \Theta \begin{bmatrix} U \\ V \end{bmatrix}. \tag{12.6}$$

The relation between Σ and Θ can be found as follows. Define the projection matrices

$$P = \begin{bmatrix} 1 & 0 \\ 0 & 0 \end{bmatrix} \quad \text{and} \quad P^{\perp} = \begin{bmatrix} 0 & 0 \\ 0 & 1 \end{bmatrix}.$$

Then from (12.6)

$$\begin{bmatrix} U' \\ V' \end{bmatrix} = (P\Sigma + P^\perp) \begin{bmatrix} U \\ V' \end{bmatrix} \quad \text{and} \quad \begin{bmatrix} U \\ V \end{bmatrix} = (P^\perp\Sigma + P) \begin{bmatrix} U \\ V' \end{bmatrix}.$$

Therefore,

$$\begin{bmatrix} U' \\ V' \end{bmatrix} = (P\Sigma + P^\perp)(P^\perp\Sigma + P)^{-1} \begin{bmatrix} U \\ V \end{bmatrix},$$

so that

$$\Theta = (P\Sigma + P^\perp)(P^\perp\Sigma + P)^{-1} = \begin{bmatrix} \Sigma_{11} - \Sigma_{12}\Sigma_{22}^{-1}\Sigma_{21} & \Sigma_{12}\Sigma_{22}^{-1} \\ -\Sigma_{22}^{-1}\Sigma_{21} & \Sigma_{22}^{-1} \end{bmatrix} \quad (12.7)$$

if Σ_{22} is not identically zero. Note that the inverse formulas, which express Σ in terms of Θ, are completely symmetric, that is,

$$\Sigma = (P\Theta + P^\perp)(P^\perp\Theta + P)^{-1} = \begin{bmatrix} \Theta_{11} - \Theta_{12}\Theta_{22}^{-1}\Theta_{21} & \Theta_{12}\Theta_{22}^{-1} \\ -\Theta_{22}^{-1}\Theta_{21} & \Theta_{22}^{-1} \end{bmatrix}.$$

These formulas connecting Σ and Θ are sometimes called the Mason rules [52, p. 100] or the Redheffer transformation rules [21, p. 58].

The chain scattering matrices have the enjoyable property that a cascade is described by an ordinary product, rather than a Redheffer product:

$$\Sigma'' = \Sigma * \Sigma' \quad \Leftrightarrow \quad \Theta'' = \Theta\Theta'.$$

Since Σ is passive (i.e., contractive in $\mathbb{D}$), it follows that Θ is also passive, which means that it is J-contractive in $\mathbb{D}$ because

$$\Theta(z)^H J\Theta(z) \leq J, \quad z \in \mathbb{D},$$

where

$$J = P - P^\perp = \begin{bmatrix} 1 & 0 \\ 0 & -1 \end{bmatrix}.$$

If Σ is lossless (i.e., unitary a.e. on $\mathbb{T}$), then Θ is J-lossless, which means J-unitary on $\mathbb{T}$ a.e.:

$$\Theta(t)^H J\Theta(t) = J, \quad \text{a.e. } t \in \mathbb{T}.$$

The chain scattering matrices are precisely the matrices discussed in Section 1.5. They showed up in the recurrence relations for the reproducing kernels and the orthogonal functions in Sections 3.3 and 4.1.

The following theorem is due to Dewilde and Dym [62, p. 647] (see also Theorem 1.5.1).

Theorem 12.3.1. *Every J-lossless chain scattering matrix is of the form*

$$\Theta(z) = \frac{1}{2} \begin{bmatrix} R_*^{-1}(1+\Omega_*) & R_*^{-1}(1-\Omega_*) \\ F^{-1}(1-\Omega) & F^{-1}(1+\Omega) \end{bmatrix},$$

with

$$F = (\Theta_{21} + \Theta_{22})^{-1} \in H_2,$$
$$R = (\Theta_{11*} + \Theta_{12*})^{-1} \in H_2,$$
$$\Omega = (\Theta_{22} + \Theta_{21})^{-1}(\Theta_{22} - \Theta_{21})$$
$$= (\Theta_{11*} - \Theta_{12*})(\Theta_{11*} - \Theta_{12*})^{-1} \in \mathcal{C},$$
$$\frac{1}{2}(\Omega + \Omega_*) = FF_* = R_*R$$

and where

$$T = (\Theta_{11} + \Theta_{12})^{-1}(\Theta_{21} + \Theta_{22}) = R_*F^{-1}$$

is inner.

The conclusion of the discussion is that a lossless inverse scattering problem can be formulated as follows. Given a lossless scattering function $S_0(z) = V_0(z)/U_0(z)$, find a J-lossless chain scattering matrix Θ and a passive load $S_L = V_L/U_L$ such that

$$\begin{bmatrix} 1 \\ S_L \end{bmatrix} U_L = \Theta \begin{bmatrix} 1 \\ S_0 \end{bmatrix},$$

where U_L is analytic in $\mathbb{D}$. Once this problem is solved, we have the relation

$$S_0 = -(\Theta_{22} - S_L\Theta_{12})^{-1}(\Theta_{21} - S_L\Theta_{11}). \tag{12.8}$$

If at the bottom surface of the medium we have perfect absorption, then $S_L = 0$ and in that case

$$S_0 = -\frac{\Theta_{21}}{\Theta_{22}} = \frac{\Omega - 1}{\Omega + 1}.$$

Since $V_0 = S_0 U_0$, we can rearrange this by an inverse Cayley transform as

$$\Omega = \frac{U_0 + V_0}{U_0 - V_0}.$$

If we consider U_0 and V_0 as incoming and outgoing waves for an electrical 1-port network of characteristic impedance 1, then $U_0 + V_0$ can be interpreted as a "voltage" and $U_0 - V_0$ as a "current" (see the next section or Ref. [23, pp. 160–161]). Therefore, Ω is sometimes called the *input impedance* with matched load (the latter referring to $S_L = 0$).

Once more, the orthogonal rational functions can be advantageous over the orthogonal polynomials. The original Schur algorithm is based on a recursive application of the Schur lemma (6.4.2) where each time α is taken to be 0. This algorithm checks whether a given function is a Schur function, but at the same time it constructs rational Schur functions that approximate (by repeated interpolation in the point $\alpha = 0$) the given Schur function. The Nevanlinna–Pick algorithm does basically the same thing but now it is allowed to choose for each of the successive applications of the Schur lemma arbitrary values for α, provided they are all in $\mathbb{D}$. Instead of orthogonal polynomials, one obtains orthogonal rational functions. To be more precise, the prominent approximating functions will be related to the reproducing kernels $k_n(z, 0)$ for the rational function spaces we have considered. Note that, in the polynomial case, these are just the reciprocal Szegő polynomials.

We recapitulate from Chapter 6 the following facts. Applying the Nevanlinna–Pick algorithm to the given Schur function S_0 computes the successive parameters ρ_k and $\gamma_k = \zeta_k \rho_k$ for a chosen sequence of points $\alpha_k \in \mathbb{D}$. As we know, this gives the elementary matrices θ_k of Theorem 3.3.1 of the recurrence for the normalized reproducing kernels. Let us assume without loss of generality that we apply the Nevanlinna–Pick algorithm with $w = 0$, that is, we assume that $S_0(0) = 0$. If this is not true, then a simple transformation

$$\frac{S_0(z) - S_0(0)}{1 - \overline{S_0(0)} S_0(z)}$$

will arrange for it. Then, setting

$$\Omega(z) = \frac{S_0(z) - 1}{S_0(z) + 1},$$

we find after application of the algorithm the relation (6.21), which can be rewritten as

$$\frac{1}{2} \begin{bmatrix} K_n^*(1 + \Omega_{n*}) & K_n^*(1 - \Omega_{n*}) \\ K_n(1 - \Omega_n) & K_n(1 + \Omega_n) \end{bmatrix} \begin{bmatrix} 1 \\ S_0 \end{bmatrix} = \frac{B_n}{\Omega + 1} \begin{bmatrix} \Delta_{n2} \\ -\Delta_{n1} \end{bmatrix}.$$

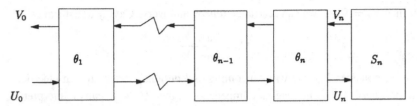

Figure 12.13. Layered medium.

Thus setting $U_n = B_n \Delta_{n2}/(\Omega + 1) \in H(\mathbb{D})$ and $S_n = \Delta_{n1}/\Delta_{n2} \in \mathcal{B}$, we see by comparing with Theorem 12.3.1 that we have solved an nth-order lossless inverse scattering problem with $S_L = -S_n$, $F(z) = 1/K_n(z,0)$, $\Omega = \Omega_n$, and $R_*(z) = 1/K_n^*(z,0)$. This solution corresponds to considering the medium as consisting of n layers. (See Figure 12.13.) Each layer is described by an elementary chain scattering matrix θ_k. The "unexplained" part of the medium is deferred to the load S_n. This corresponds to a parameterization of all Schur functions that match S_0 in all the interpolation points $\alpha_1, \dots, \alpha_n$. In our notation this is given by

$$-\frac{\Theta_{21}^{(n)} - \Theta_{11}^{(n)} S_n}{\Theta_{22}^{(n)} - \Theta_{12}^{(n)} S_n}, \quad S_n \in \mathcal{B}, \quad \Theta_n = \left[\Theta_{ij}^{(n)}\right]_{i,j=1,2},$$

where $\Theta_n = \theta_n \, \theta_{n-1} \, \cdots \, \theta_1$ is a factorization of Θ_n in elementary matrices.

The parallel with the prediction problem of the previous section should now also be obvious. If we neglect the unexplained part completely, that is, if we set $S_n = 0$, then the scattering function S_0 is approximated by

$$\Gamma_n(z) = -\frac{K_n(z,0) - L_n(z,0)}{K_n(z,0) + L_n(z,0)} \in \mathcal{B}.$$

Because we know that $L_n(z,0)/K_n(z,0)$ interpolates the input impedance Ω in the points $0, \alpha_1, \dots, \alpha_n$ (Theorem 6.3.3), it follows that Γ_n interpolates the scattering function S_0 at the same points. This solution (i.e., the one obtained by setting $S_n = 0$) can be interpreted as a maximal entropy approximant. Recall that the entropy integral for any positive $F \in L_1$ is given by

$$\int \log F(t) \, d\lambda(t).$$

Theorem 12.3.2. *Let* μ *be the Riesz–Herglotz measure for* $\Omega = (1 - S_0)/(1 + S_0)$ *and let* Θ_n *be the matrix that is constructed by n steps of the Nevanlinna–Pick algorithm applied with* $w = 0$ *(we assume $S_0(0) = 0$). Then*

$\Omega_n(z) = L_n(z, 0)/K_n(z, 0)$ *and the outer spectral factor of the Riesz–Herglotz measure for* Ω_n *is equal to* $\sigma_n(z) = 1/K_n(z, 0)$. *Moreover,*

$$\int \log \mu'(t)\, d\lambda(t) \le \int \log |\sigma_n(z)|^2 \, d\lambda(t)$$

with equality if $S_L = 0$, *that is,* $\mu' = |\sigma_n|^2$.

Proof. Since [116, p. 149]

$$\exp\left\{ \int \log \mu'(t)\, d\lambda(t) \right\} = \inf \left\{ \|f\|_\mu^2 : f \in H_2, f(0) = 1 \right\},$$

we shall not decrease the infimum if we replace H_2 by the subspace $\mathcal{L}_n$. Thus

$$\exp\left\{ \int \log \mu'(t)\, d\lambda(t) \right\} \le \inf \left\{ \|f\|_\mu^2 : f \in \mathcal{L}_n, f(0) = 1 \right\}$$

$$= \frac{1}{K_n(0, 0)} = \exp\left\{ \int \log |\sigma_n(t)|^2 \, d\lambda(t) \right\},$$

which proves the result. □

Thus setting $S_L = 0$ corresponds to picking from all solutions of the n-point Nevanlinna–Pick interpolation problem the one with maximal entropy.

Because of the factor ζ_k in θ_k, each section will add a transmission zero $1/\overline{\alpha}_k$ to the scattering medium. A transmission zero $e^{i\omega}$ would correspond to the fact that the frequency ω is completely absorbed. A transmission zero $re^{i\omega}$, with r close to 1, causes a significant reduction in the amplitudes that correspond to frequencies in the neighborhood of ω. Thus if we have good estimates of these transmission zeros, one might expect Γ_n to be a good approximation for S_0 for rather small values of n. In fact, if S_0 is rational of degree n and if the transmission zeros are estimated exactly, then we should have a perfect fit after n steps of the Nevanlinna–Pick algorithm.

Most inverse problems are notoriously ill conditioned. It is no different for the lossless inverse scattering problem. It is very difficult to recover the transmission zeros $1/\overline{\alpha}_k$ from S_0. The optimal choice for the α_k would correspond to letting the α_k vary freely, giving an approximation error that depends on these α_k. In the terminology of Section 12.1 we could use the prediction error E_n as a measure of how well the spectral measure is approximated, hence also as a measure of how well its Riesz–Herglotz transform and the associated scattering function are approximated. Therefore, we have to minimize this prediction error E_n with respect to the α_k if we want optimal locations of the transmission zeros. In

Ref. [28] some numerical examples are given that reveal that plots of E_n as a function of the α_k show a very flat behavior near the minimum. This means that this is an ill posed problem, which means that the α_k can not be pinned down with great accuracy when working in finite-precision arithmetic. However, by the same observation, this also means that the location of the points α_k (at least when moved in certain directions) does not influence the value of the approximation error E_n by large amounts. Thus if we are only interested in finding a good model giving a small approximation error, then it is of no crucial importance to find the location of these αs exactly. Rough estimates will do for approximating purposes.

So far, we have only considered transmission zeros that were chosen in $\mathbb{E}$, that is, α_k that were chosen inside $\mathbb{D}$. Each α_k gave rise to an elementary section in the cascade, which is described by a J-lossless chain scattering matrix θ_k. In Ref. [62], such an elementary section is called a Schur section in the cascade or, equivalently, θ_k is called a Schur factor of Θ. It is also possible to extract factors from Θ that have zeros on $\mathbb{T}$. Such sections are called Brune sections, referring to work of Brune related to network theory synthesis [25]. However, a Brune factor with transmission zero $\alpha \in \mathbb{T}$ is only possible if α is a *point of local losslessness* (PLL). Recalling the relation among the chain scattering matrix Θ, the scattering function S_0, the input impedance Ω, and its Riesz–Herglotz measure μ, one can define a PLL as a point where either $\mu(\{\alpha\}) > 0$ (there is a point mass in α) or $\mu(\{\alpha\}) = 0$, but then such that

$$\int \frac{d\mu(t)}{|t - \alpha|^2} < \infty.$$

It can be shown [62, Theorem 2.3] that $\alpha \in \mathbb{T}$ is a PLL iff $\lim_{r \uparrow 1} S_0(r\alpha) = c \in \mathbb{T}$ and

$$\lim_{r \uparrow 1} \frac{1 - \bar{c}S_0(r\alpha)}{1 - r} < \infty.$$

If one wants to extract several Brune sections at the same point $\alpha \in \mathbb{T}$, then one needs PLLs of higher order, which basically means the following. The point $\alpha \in \mathbb{T}$ will be a PLL of order k if either

$$\mu(\{\alpha\}) = 0 \quad \text{while} \quad \int \frac{d\mu(t)}{|t - \alpha|^{2k}} < \infty \quad \text{and} \quad \int \frac{d\mu(t)}{|t - \alpha|^{2k+2}} = \infty$$

or

$$\mu(\{\alpha\}) > 0 \quad \text{while} \quad \int \frac{d\mu^\alpha(t)}{|t - \alpha|^{2(k-1)}} < \infty \quad \text{and} \quad \int \frac{d\mu^\alpha(t)}{|t - \alpha|^{2k}} = \infty,$$

where μ^α is the measure μ with the mass point at α deleted (for more details see Ref. [62, pp. 649–650]).

Of course such conditions were also required for the study of the boundary situation in Chapter 11. However, our study was based on a three-term recurrence relation, whereas the form of the Brune factors θ_k in Ref. [62] is based on a coupled recurrence. Therefore the link is not immediate and we shall not elaborate on it here.

12.4. Network synthesis

The problem of Darlington synthesis of a passive lossless network is mathematically the same as the problem of lossless inverse scattering. In the previous section, we already used some terminology referring to that. We introduce this terminology here in a more systematic way. We essentially used the book by Belevitch [23].

An *electrical network* consists of a finite number of interconnected elements. Such elements involve, for example, resistances, capacitances, inductances, current or voltage generators, etc. Such a network can have several terminals to which other subnetworks can be connected. A network is called an n-port if it has $2n$ terminals that are paired in couples. Each couple is called a *port* and it is characterized by the port variables, which are, for example, a voltage and a current over that port. Figure 12.14 shows a 2-port. Note that the name variables is misleading since voltage and current are actually functions of time, which are related through a system of ordinary differential equations that describe the electrical properties of the composing elements. After taking the Laplace transform, the derivative with respect to time is transformed into a multiplication with the complex variable z. We shall only work in the Laplace transform domain, that is, the frequency domain. Thus voltages and currents will be represented by functions of a complex frequency variable z.

Suppose V_i and I_i are the voltage and current for port i. The complex *power* dissipated by the n-port is defined as $W = \sum_{i=1}^{n} \bar{I}_i V_i$, where the bar represents the complex conjugate. If we define the vectors $V = [V_1 \ V_2 \ \cdots \ V_n]^T$ and $I = [I_1 \ I_2 \ \cdots \ I_n]^T$, then the complex power is $W = I^H V$, where the superscript

Figure 12.14. A 2-port.

means complex conjugate transpose. If the n-port does not contain internal generators, then the relation between the vectors V and I can be described in homogeneous form by

$$V = RI \quad \text{or} \quad I = AV. \tag{12.9}$$

The $n \times n$ matrices R and A are complex rational matrix functions. R is called the *impedance matrix* of the network and A is called the *admittance matrix*.

An n-port is called *passive* if the active power (that is, the real part of W) is nonnegative in the right half plane: Re $W(z) \geq 0$ for all Re $z \geq 0$. If, moreover, it satisfies Re $W(z) = 0$ for Re $z = 0$ then it is said to be *lossless*. Consequently, if (12.9) holds, then Re $W = I^H(\text{Re } R)I = V^H(\text{Re } A)V$ so that the impedance and admittance matrices of a passive n-port have a real part that is nonnegative definite in the right half plane. A matrix satisfying Re $R(z) \geq 0$ in Re $z \geq 0$ is called a passive matrix. Similarly, for a lossless n-port, one has Re $R(z) = 0$ for Re $z = 0$ and such a matrix is called lossless.

If a 1-port contains internal (current or voltage) generators, then the relation between voltage V and current I is inhomogeneous: $V = RI + E$ or $I = AV + J$, where V and J are some combinations of internal variables. This means that it can be represented by a voltage generator in series with an impedance R (Thevenin's theorem) or as a current generator in parallel with an impedance A (Norton's theorem). We find $V = E$ when $I = 0$; thus E is the open-circuit voltage of the port. Similarly, J will be the short-circuit current of the port. Thus if all internal generators are put to zero (voltage generators replaced by short circuits and current generators by open circuits), then $V = RI$ and $I = AV$, so that we have the situation as above (12.9). The impedance $R = V/I$ is called the *internal impedance* of the 1-port.

A voltage generator can be seen as a 1-port that produces a voltage V and has some (internal) impedance Ω. If it is loaded with an impedance Ω_L, then there will be a current given by

$$I = \frac{V}{\Omega + \Omega_L}.$$

(See Figure 12.15.) If one makes the load Ω_L equal to $\overline{\Omega}$, then the total power dissipated in the load will be maximal [23, p. 159]. The current obtained for this matched load is

$$I_0 = \frac{V}{2 \operatorname{Re} \Omega},$$

and the relative difference for the currents with open circuit (load 0) and with

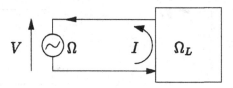

Figure 12.15. Generator with load.

load Ω_L,

$$s = \frac{I_0 - I}{I} = \frac{\Omega_L - \Omega}{\Omega_L + \Omega},$$

is called the *reflectance* of Ω_L relative to Ω. Note that if we match the load impedance Ω_L with the internal impedance Ω (i.e., $\Omega_L = \Omega$) then $s = 0$. Choosing the normalized variables $i = I\sqrt{\Omega}$, $v = V/\sqrt{\Omega}$, and $\omega = \Omega_L/\Omega$, this becomes

$$s = \frac{\omega - 1}{\omega + 1} \quad \text{or} \quad \omega = \frac{1 + s}{1 - s}.$$

Note $\omega = v/i$ because $\Omega = V/I$. These relations imply

$$\operatorname{Re}\omega = \frac{1 - |s|^2}{|1 - s|^2}.$$

This shows that Ω, and hence ω, being passive is equivalent with $|s(z)| \le 1$ for all $\operatorname{Re} z \ge 0$, and if ω is lossless, then $|s(z)| = 1$ for $\operatorname{Re} z = 0$. Replacing ω by v/i gives

$$s = \frac{v - i}{v + i}.$$

By another change of variables

$$x = \frac{v + i}{2} \quad \text{and} \quad y = \frac{v - i}{2},$$

one obtains the simple relation $y = sx$. The function x is called the *incoming wave* and y is the *outgoing wave*.

For an n-port, one can make the same change of variables for each port and thus obtain new wave variables (x_i, y_i) for port i. Defining the vectors $X = [x_1 \ x_2 \ \ldots \ x_n]^T = (V + I)/2$ and $Y = [y_1 \ y_2 \ \ldots \ y_n]^T = (V - I)/2$, one obtains a relation of the form $Y = SX$, which replaces (12.9). The matrix function S is called the *scattering matrix* of the n-port. It is related to the (internal) impedance matrix by

$$S = (R + I_n)^{-1}(R - I_n),$$

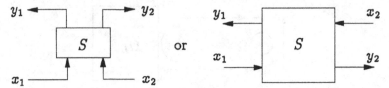

Figure 12.16. A 2-port with wave variables.

where I_n is the $n \times n$ unit matrix. For the 2-port of Figure 12.16, we have in terms of the wave variables

$$\begin{bmatrix} y_1 \\ y_2 \end{bmatrix} = S \begin{bmatrix} x_1 \\ x_2 \end{bmatrix}.$$

Note that for the dissipated power, we have

$$W = I^H V = i^H v = (x - y)^H (x + y) = x^H (I_n - S^H S)x.$$

Thus for a passive n-port $S^H S \leq I_n$ on Re $z \geq 0$, and for a lossless n-port, it is moreover true that $S^H S = 1$ on Re $z = 0$.

Now, modulo some minor adaptations and a change of notation, we are back in the situation of the previous section. There we had the special case of $n = 2$ and the scattering matrix was contractive in $\mathbb{D}$ instead of the right half plane, but the latter is easily dealt with by a bilinear transformation of the variable that maps the right half plane to the disk. Or one can rotate the right half plane to the upper half plane and then one has the theory of orthogonal rational functions on the real line. If we had sampled the electrical signals and used the z-transform instead of the Laplace transform, we would have obtained the disk situation directly.

By the same observations as made in the previous section, it is much more interesting for us to describe the cascade connection of two ports not by scattering matrices, but by chain scattering matrices.

Suppose we have a lossless 2-port where port 2 is replaced by an open circuit. Then as seen from port 1, we obtain a 1-port circuit with some impedance Ω. If this 2-port is loaded at port 2 by a passive 1-port with impedance Ω_L, then there will be a reflectance S_L (= scattering function) of Ω_L with respect to Ω: $S_L = (\Omega_L - \Omega)/(\Omega_L + \Omega)$. This S_L will be zero when we match the load impedance Ω_L with the internal impedance Ω of the 2-port (i.e., if we set $\Omega_L = \Omega$). This explains why the function Ω in the lossless inverse scattering framework is called the input impedance with matched load.

In the realization theory of these networks, the problem is to realize a certain complex network, which is considered as being a passive 1-port with possibly a high internal complexity. This means that the scattering matrix, or equivalently the chain scattering matrix Θ, is a rational function of a relatively high degree. The realization will be obtained as a cascade connection of simple sections represented by 2-ports, just like in the lossless inverse scattering framework. Mathematically this means that the rational matrix function Θ is factored as a product of elementary matrices. Elementary usually means of degree 1, but for practical reasons it is sometimes more interesting to put together two "complex conjugate" sections of degree one and combine them into a real section of degree two.

Since in practical situations the given scattering function is often rational, we know, at least in principle, the transmission zeros and we should be able to apply the machinery of the factorization and the orthogonal rational functions, which should give an exact realization after a finite number of steps. Thus certainly in such situations the rational, rather than the polynomial, approach is highly recommended.

12.5. H_∞ problems

12.5.1. The standard H_∞ control problem

Several problems considered in H_∞ control are tightly connected to inverse scattering. Again the machinery of J-unitary matrices and Nevanlinna–Pick interpolation can be set to work and that is the only point we want to make here. The reader who is interested in more details should consult the extensive literature on H_∞ control. A simple approach, which is close to ours, is given, for example, in Ref. [128]. See also Ref. [22, Part VI].

Suppose we consider a discrete time system; otherwise one should replace the unit circle by the real or the imaginary axis (depending on the formalism one wants to use). Let us redraw the picture given in Figure 12.11 as in Figure 12.17.

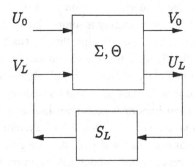

Figure 12.17. Plant with feedback.

This makes it easier to see this as a plant with input U_0 and output V_0 with some feedback loop, which contains a controller characterized by S_L. Suppose the plant is described by (12.3) and (12.4), or by the chain scattering equivalent as in (12.6). Then the closed loop transfer function of the plant is given by (12.5). Thus

$$S_0 = \frac{V_0}{U_0} = \Sigma_{21} + \Sigma_{22} S_L [1 - \Sigma_{12} S_L]^{-1} \Sigma_{11}$$

or

$$S_0 = -(\Theta_{22} - \Theta_{12} S_L)^{-1} (\Theta_{21} - \Theta_{11} S_L). \qquad (12.10)$$

The H_∞ control problem is to find a controller S_L such that the closed loop transfer function S_0 (from U_0 to V_0) satisfies

$$\|S_0\|_\infty \leq \gamma$$

for some given positive number $\gamma > 0$.

Of course, for physical reasons, one expects the system to be stable, which means in mathematical terms that S_0 is analytic in $\mathbb{D}$, and thus the norm $\|S_0\|_\infty$ that we used above is the norm in $H_\infty(\mathbb{D})$. This means that $\gamma^{-1} S_0$ should be a Schur function.

This problem can be given the following interpretation. We can consider V_0 as some error that can be observed. U_0 is some exogeneous input and U_L is an observed output of the system. The controller S_L uses these output observations to steer a control input V_L such that the effect of U_0 on the error signal V_0 is brought below a certain tolerance γ. Alternatively, one could consider U_0 as some noise that influences the output V_0 of the system, and a controller has to be designed to bring this influence below a certain level.

The problem of inverse scattering is indeed the inverse of this problem in the sense that, there, the scattering function S_0 is observed and one aims at constructing a model for the scattering medium, that is, one wants to compute Σ or Θ and S_L. Here, one knows in principle the plant as Σ or Θ and one wishes to construct a controller S_L such that the closed loop transfer function S_0 is a (scaled) Schur function. Let us assume for simplicity that the problem is scaled such that we can take $\gamma = 1$. Thus S_0 should be a genuine Schur function.

There is, however, an important constraint imposed by the physical realizability of the system. Namely, the closed loop system should be *internally stable*. This means that none of the signals that are generated somewhere internally in the system may become unbounded, whatever the input signals may be, provided of course that these input signals are bounded. This is physically obvious because otherwise the device could "explode" while operating with finite-energy input signals. It may well be that the system transfer function is stable, but that it is realized such that it is not internally stable. For example, if the transfer function is $1/(z - 5)$ but if this is implemented as a cascade of a transfer function $1/(1 - 2z)$ and a transfer function $(1 - 2z)/(z - 5)$, then it is not internally stable since the signal after the first transfer function is internally generated and it can become unbounded because the first transfer function is not stable (not in H_∞). Thus internal stability refers to a specific (state space) realization of the system, but we do not want to go into the details of state space realizations in this context.

There are several other problems in H_∞ control, including sensitivity minimization and robust stabilization, that can be reduced to this standard problem or variations thereof. See, for example, Refs. [19, Chap. 5], [84, Chap. 12], [128], [22, Part VI], [85], and [86] for much more on H_∞ control and interpolation.

Because the internal variable U_L of the closed loop system is

$$U_L = (1 - \Sigma_{12} S_L)^{-1} \Sigma_{11} U_0$$

and because of internal stability, it follows that $(1 - \Sigma_{12} S_L)^{-1} \Sigma_{11}$ should be stable. Furthermore, for internal stability reasons, the internal variable V_L should be bounded, and because $V_L = S_L U_L = S_L (1 - \Sigma_{12} S_L)^{-1} \Sigma_{11} U_0$, this means that S_L and $S_L (1 - \Sigma_{12} S_L)^{-1} \Sigma_{11}$ should be stable.

Now if α_i is an unstable zero of Σ_{22} (i.e., $\Sigma_{22}(\alpha_i) = 0$ with $\alpha_i \in \mathbb{D}$) then it can not be compensated by a pole of $S_L (1 - \Sigma_{12} S_L)^{-1} \Sigma_{11}$, since this is stable and thus has only stable poles. Thus, if we fill in α_i, we get

$$S_0(\alpha_i) = \Sigma_{21}(\alpha_i), \tag{12.11}$$

which should hold for all unstable zeros of Σ_{22}. Consequently, internal stability conditions impose Nevanlinna–Pick interpolation constraints. In fact, it will be shown below that, if a solution exists, then the solution for this Nevanlinna–Pick problem actually solves the control problem.

As in the inverse scattering problem, there is a considerable advantage in using the chain scattering matrix Θ instead of the scattering matrix Σ. These are related by (12.7), that is,

$$\Theta = (P\Sigma + P^\perp)(P^\perp \Sigma + P)^{-1}$$

if Σ_{22}^{-1} exists. The terminology "scattering" and "chain scattering" matrix is not exactly correct since there is no reason why the system should be described by a matrix Σ that has the properties of a scattering matrix. Moreover, in the multidimensional case, Σ is some $m \times n$ matrix that need not even be square. It is possible to generalize the ideas of J-contractive, J-unitary, J-lossless, etc. to nonsquare matrix functions, but since this was not discussed in the main part of this monograph, we shall not go into the details. The control problems where the signals are all scalar are of little practical importance. Usually, all the signals are vector valued and often of different dimensions, but since we only want to illustrate the simplest possible ideas, let us assume that Σ and Θ are 2×2 matrices, which means that all the inputs and outputs are scalar functions. Moreover, in all practical computations, they are assumed to be rational functions. Assuming that we use the "chain scattering matrix" Θ to describe the system, then

$$\begin{bmatrix} V_L \\ U_L \end{bmatrix} = \Theta \begin{bmatrix} V_0 \\ U_0 \end{bmatrix},$$

which means that (12.10) holds, or equivalently,

$$S_L = (\Theta_{21} + \Theta_{22} S_0)(\Theta_{11} + \Theta_{12} S_0)^{-1}. \tag{12.12}$$

If the controller is also described in the same way:

$$S' = (\Theta'_{21} + \Theta'_{22} S_L)(\Theta'_{11} + \Theta'_{12} S_L)^{-1},$$

that is,

$$\begin{bmatrix} V' \\ U' \end{bmatrix} = \Theta' \begin{bmatrix} V_L \\ U_L \end{bmatrix}, \qquad S' = \frac{V'}{U'},$$

then

$$S' = (\Theta''_{21} + \Theta''_{22} S_0)(\Theta''_{11} + \Theta''_{12} S_0)^{-1}, \quad \Theta'' = \Theta' \Theta.$$

Thus, if we have realized the system Θ as a cascade of elementary sections, then adding the controller just means that we have to extend the cascade with a few more sections. This illustrates once more the advantages of working with Θ instead of Σ.

Let us reformulate the control problem once again: Given some matrix Θ, we have to find a function S_L such that S_0 given by (12.10) is a Schur function and such that the system is internally stable.

It is obvious that if Θ is a J-lossless chain scattering matrix, then S_0 will be a Schur function for any choice of a Schur function S_L. This is of course a trivial system, which does not need a controller because setting $S_L = 0$ would solve

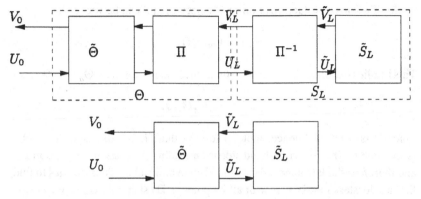

Figure 12.18. Lossless factorization.

the H_∞ control problem. By letting S_L range over all Schur functions, we get all the possible controllers.

However, for an arbitrary plant, there is no reason why Θ should be a J-lossless chain scattering matrix. In that case, one can hope to replace the system by another system as in Figure 12.18. There we replaced Θ by $\tilde\Theta \Pi$, where $\tilde\Theta$ is J-lossless and Π is outer. Recall that a matrix is J-lossless if it is (1) all-pass (i.e., J-unitary) on $\mathbb{T}$ and (2) passive (i.e., J-contractive) in $\mathbb{D}$. The matrix function Π is outer (in H_∞) when Π and its inverse Π^{-1} have entries that are analytic functions in $\mathbb{D}$ (it is a matrix version of an outer function). Such a factorization (if it exists) is called a J-lossless factorization of the matrix Θ. It can be seen from the figure that for the new system, where $\tilde\Theta$ is loaded (controlled) by $\tilde S_L$, we can choose as above $\tilde S_L$ to be any Schur function since $\tilde\Theta$ is J-lossless. Because (U_L, V_L) are the port values of a port described by Π^{-1} and terminated by $\tilde S_L$, it is clear that this corresponds to a controller S_L for the original system, which is given by

$$S_L = (\Pi_{21}\tilde S_L + \Pi_{22})^{-1}(\Pi_{11}\tilde S_L + \Pi_{12})$$

with $\tilde S_L$ an arbitrary Schur function. Equivalently, because of (12.10), the controller is given by

$$S_L = \frac{\Theta_{21} + \Theta_{22}S_0}{\Theta_{11} + \Theta_{12}S_0}.$$

Thus the problem of H_∞ control is reduced to a problem of J-lossless factorization.

Since $\tilde\Theta$ is J-lossless, it can not have poles on $\mathbb{T}$. Then it is always possible to factor it as $\tilde\Theta = \Theta_a\Theta_s$, with Θ_s J-lossless with all its poles in $\mathbb{E}$ (it is stable) and Θ_a J-lossless with all its poles in $\mathbb{D}$ (it is anti stable) [128, Lemma 4.9, p. 89].

So, if a J-lossless factorization of Θ exists, then it can be written as

$$\Theta = \Theta_a \Theta_s \Pi.$$

This implies (recall from Section 1.5 that, since Θ_a is J-lossless, $\Theta_a^{-1} = J\Theta_{a*}J$)

$$\Theta_* J \Theta_a = \Pi_* \Theta_{s*} J.$$

Since Π is outer, and hence stable, it follows that Π_* is anti stable. Also, Θ_{s*} is anti stable. Thus the right-hand side is anti stable (i.e., has all its poles in $\mathbb{D}$), and therefore the left-hand side should also be anti stable. Thus one has to find first a J-lossless matrix Θ_a (with all its poles in $\mathbb{D}$) such that $G_* = \Theta_* J \Theta_a$ is anti stable. If we can then find a J-lossless factorization for the stable matrix JG of the form $JG = \Theta_s \Pi$, then the problem has been solved because we then have $\Theta = \Theta_a \Theta_s \Pi$. Thus there remain two problems to be solved:

1. Given a rational matrix function H, find a J-lossless matrix Θ_a that makes $H\Theta_a$ anti stable.
2. Given a rational and stable matrix function G, find a J-lossless factorization $G = \Theta_s \Pi$.

The first problem is a problem of J-lossless anti stabilizing conjugation. One defines this problem as follows. Given a rational matrix H, then one says that Θ is a J-lossless stabilizing (anti stabilizing) conjugator for H if Θ is J-lossless and $H\Theta$ is stable (anti stable) and if the degree of Θ is equal to the number of anti stable (stable) poles of H. The last condition expresses some minimal degree condition for the conjugator. The Θ-matrix cancels the anti stable (stable) poles α_i and replaces them by their reflections $1/\overline{\alpha}_i$. The zeros and the stable (unstable) poles are left untouched. In the scalar case, this is a trivial matter. For example, the lossless stabilizer of

$$G(z) = \frac{z+3}{(1-2z)(z-5)}$$

is given by $\Theta(z) = (1 - 2z)/(z - 2)$ (a Blaschke factor, and thus lossless) because

$$G(z)\Theta(z) = \frac{z-3}{(z-2)(z-5)}$$

is stable.

We first note that problem 2 above can be solved by a J-lossless stabilizing conjugation. Indeed, suppose G is stable and let Θ be a J-lossless stabilizing conjugator for G^{-1}. Then $G^{-1}\Theta = G'$ with Θ J-lossless and with G' stable. Because

G is stable, G^{-1} has no anti stable zeros, and because J-lossless conjugation keeps all the zeros, G' can have no anti stable zeros. Thus G' is outer. Setting $\Pi^{-1} = G'$, we find that $G = \Theta \Pi$ and this is a J-lossless factorization of G.

Thus the only problem left to be solved is the problem of constructing a stabilizing (antistabilizing) J-lossless conjugator.

Such a construction is obtained step by step, where each elementary step eliminates an unstable pole by multiplying with an elementary J-lossless matrix θ_k as in the Nevanlinna–Pick algorithm. It is in fact equivalent with a Nevanlinna–Pick problem. In practical applications, where the signals are vector valued, the algorithm is often performed on a state space realization of the system. The details can, for example, be found in Ref. [128].

We prefer, however, to reformulate the problem as a Nehari problem for which we shall give a solution below. In the Nehari problem, one wants to do better than in the H_∞ control problem formulated so far. One wants to find a controller that is *optimal* in a certain sense. So, instead of just constructing a controller that arranges for $\|S_0\|_\infty \leq \gamma$, one wants to go further and find the best controller, that is, one that *minimizes* $\|S_0\|_\infty$. If such an optimal controller can be found for which $\|S_0\|_\infty \leq \gamma$, then the H_∞ problem as we defined it here has a solution; otherwise, there is no solution. This problem of optimal control is a so-called minimal norm problem. An alternative would be to solve the minimal degree problem, which will construct a (rational) controller that makes $\|S_0\|_\infty \leq \gamma$, but among all solutions finds the simplest possible one, that is, the rational function of lowest possible degree.

Nehari problems are usually formulated in terms of Hankel operators. Therefore, we shall give an introduction first.

12.5.2. Hankel operators

For an elementary introduction to Hankel operators see Ref. [175]. See also Ref. [182] for a more advanced text.

Let $\{x_k\}_{k=-\infty}^{\infty}$ be an input signal in ℓ_2 and suppose we apply this to a linear system with impulse response $\{h_k\}_{k=-\infty}^{\infty} \in \ell_\infty$ to give an output signal $\{y_k\}_{k=-\infty}^{\infty} \in \ell_2$. Then we can describe this as

$$
\begin{bmatrix} \vdots \\ y_{-1} \\ \boxed{y_0} \\ y_1 \\ \vdots \end{bmatrix} = \begin{bmatrix} \ddots & \ddots & & \ddots & \\ \cdots & h_0 & h_{-1} & h_{-2} & \cdots \\ \cdots & h_1 & \boxed{h_0} & h_{-1} & \cdots \\ \cdots & h_2 & h_1 & h_0 & \cdots \\ & & \ddots & \ddots & \ddots \end{bmatrix} \begin{bmatrix} \vdots \\ x_{-1} \\ \boxed{x_0} \\ x_1 \\ \vdots \end{bmatrix}.
$$

For causal systems, it will only be the past of the input that will define the future of the output. Thus the relevant part of the system will be the operator that maps the past of x_0 into the future of y_0. This operator can be described as

$$
\begin{bmatrix} y_0 \\ y_1 \\ y_2 \\ \vdots \end{bmatrix} = \begin{bmatrix} h_0 & h_1 & h_2 & \cdots \\ h_1 & h_2 & \cdots \\ h_2 & \cdots \\ \vdots \end{bmatrix} \begin{bmatrix} x_0 \\ x_{-1} \\ x_{-2} \\ \vdots \end{bmatrix}.
$$

To rewrite this in the frequency domain, we define some operators for the frequency domain. Let R be the reversion operator

$$
R : L_2 \to L_2 : f(z) \mapsto f(1/z).
$$

For given $h \in L_\infty$, let M_h be the (bounded) multiplication operator

$$
M_h : L_2 \to L_2 : f(z) \mapsto h(z)f(z).
$$

We note incidentally that

$$
\|M_h\|_2 = \|h\|_\infty.
$$

Furthermore, let Π_+ be the projection in L_2 onto H_2. Then a Hankel operator with *symbol* $h \in L_\infty$ is an operator on H_2 defined as $H_h = \Pi_+ M_h R|_{H_2}$, that is,

$$
H_h : H_2 \to H_2 : f(z) \mapsto \Pi_+ M_h R|_{H_2} f(z) = \Pi_+ h(z) f(1/z),
$$

where $M|_X$ means the restriction of operator M to the space X. Now we can transform our time domain input–output relation $(x_k)_0^{-\infty} \mapsto (y_k)_0^\infty$ to the frequency domain. Therefore we define the z-transforms $x(z) = \sum_{k=0}^\infty x_{-k} z^k \in H_2$ (note the minus sign in the index), $y(z) = \sum_{k=0}^\infty y_k z^k \in H_2$, and the transfer function $h(t) = \sum_{k=-\infty}^\infty h_k t^k \in L_\infty, t \in \mathbb{T}$. Thus

$$
x_k = \int t^k x(t) \, d\lambda(t), \qquad y_k = \int t^{-k} y(t) \, d\lambda(t), \quad k \in \mathbb{Z},
$$

where $x_k = y_{-k} = 0$, for $k = 1, 2, \ldots$, and

$$
h_k = \int t^{-k} h(t) \, d\lambda(t), \quad k \in \mathbb{Z}.
$$

Then the input–output relation simply reads $y(z) = H_h x(z)$. Obviously, with respect to the standard basis $\{1, z, z^2, \ldots\}$, the operator H_h is represented by the Hankel matrix $H_h = [h_{i+j}]_{i,j=0,1,2,\ldots}$. Note that we use the notation M_h for the operator in H_2 as well as for the corresponding matrix representation.

Given $h \in L_\infty$, this defines uniquely the Hankel operator H_h, but the converse is not true. If H is a Hankel operator and h is a symbol for it, that is, $H = H_h$, then any function $f = h + g$ with $g \in H_\infty^\perp$ arbitrary will also be a symbol for H. We used the notation $H_\infty^\perp$ for

$$H_\infty^\perp = \left\{ f \in L_\infty : \int t^k f(t)\, d\lambda(t) = 0, k = 0, 1, \ldots \right\}$$

$$= \left\{ f \in L_\infty : f(z) = \sum_{k=-1}^{-\infty} f_k z^k \right\}.$$

Obviously $H_f = H_h$ iff $f - h \in H_\infty^\perp$. The classical Nehari Theorem [151; 175, p. 31] says that it is always possible to find a symbol for a Hankel operator that is optimal in a certain sense.

Theorem 12.5.1 (Nehari). *Let H be a Hankel operator. Then there exists a symbol $h \in L_\infty$ such that $H = H_h$ and $\|H\|_2 = \|h\|_\infty$. This h is the solution of the optimization problem (the infimum is $\|H\|_2$)*

$$\inf\{\|h\|_\infty : h \in L_\infty, H = H_h\}.$$

Since we can write any symbol for $H = H_h$ as $f = h - h^\perp$ with $h^\perp \in H_\infty^\perp$, it follows that

$$\|H_h\|_2 = \inf\left\{ \|h - h^\perp\|_\infty : h^\perp \in H_\infty^\perp \right\}.$$

In other words, $\|H_h\|_2$ is equal to the L_∞-distance of $h \in L_\infty$ to $H_\infty^\perp$:

$$\|H_h\|_2 = \mathrm{dist}_\infty\left(h, H_\infty^\perp\right).$$

This optimization problem is also called a Nehari extension problem. That is defined as follows. Given a sequence $\{h_k : k = 0, 1, 2, \ldots\}$ with $\sum_{k=0}^\infty h_k z^k \in H_\infty$, one has to extend it with a sequence $\{h_k : k = -1, -2, \ldots\}$ such that $h(z) = \sum_{k=-\infty}^\infty h_k z^k \in L_\infty$. Then one has to find, among all the solutions of this extension problem, the one that has minimal norm. Sometimes another variant is formulated where the problem is to find among all the solutions of the extension problem some h that satisfies $\|h\|_\infty < 1$. Of course, if one can solve for the optimal h, then one can decide whether the second variant has a solution and if a solution exists, it is actually constructed.

Let us now see how the solution of this problem can help to solve the previous H_∞ control problem. First we recall that internal stability required that

$f = S_L(1 - \Sigma_{12}S_L)^{-1}\Sigma_{11}$ was stable, and hence in H_∞. Thus if we can find $f \in H_\infty$ that minimizes $\|\Sigma_{21} + \Sigma_{22}f\|_\infty$, then we have solved our problem, because we can easily compute S_L from f. Now let us assume for simplicity that our given system is stable and that Σ_{21} and Σ_{22} are in H_∞. Let $\Sigma_{22} = B_{22}O_{22}$ be an inner–outer factorization of Σ_{22}. Then because $OH_\infty = H_\infty$ for any outer function O, and because $\|Bf\|_\infty = \|f\|_\infty$ for any inner function B, it follows that our problem can be further reduced to finding the infimum of $\|h - f\|_\infty$, where f ranges over H_∞ and $h = -\Sigma_{21}/B_{22} \in L_\infty$ is given. Thus we have to find $\mathrm{dist}_\infty(h, H_\infty)$, and this can of course be formulated as a Nehari optimization problem of the previous type by a transformation $z \to 1/z$. Indeed

$$\inf_{f \in H_\infty} \|h - f\|_\infty = \inf_{\hat{f} \in H_\infty^\perp} \|\hat{h} - \hat{f}\|_\infty \quad \text{or} \quad \mathrm{dist}_\infty(h, H_\infty) = \mathrm{dist}_\infty\big(\hat{h}, H_\infty^\perp\big)$$

if

$$\hat{h}(z) = h(1/z)/z \quad \text{and} \quad \hat{f}(z) = f(1/z)/z.$$

For the Nehari optimization problem, Adamyan, Arov, and Kreĭn [6] gave a solution that is much more general, but for the solution of the Nehari problem, it simplifies to the following theorem. We first define an *all-pass* function $f \in L_\infty$ to be a function for which $|f(t)| = 1$ a.e. on $\mathbb{T}$. Note that an all-pass function is not necessarily inner because it need not be analytic in $\mathbb{D}$.

Theorem 12.5.2 (Adamyan–Arov–Kreĭn). *Let H be a Hankel operator with $\gamma = \|H\|_2$. Let $x \in H_2$ be nonzero such that $\|Hx\|_2 = \gamma\|x\|_2$. Set $y = \gamma^{-1}Hx$ and $\hat{x} = Rx$, and thus $\hat{x}(z) = x(1/z)$, and define $h = \gamma g = \gamma y/\hat{x}$. Then g is an all-pass function and $h \in L_\infty$ is the only function that solves the Nehari extension problem. Thus it is the only function for which $H = H_h$ and $\|H\|_2 = \|h\|_\infty$. Equivalently, if f is a symbol for H, then the function $h^\perp \in H_\infty^\perp$, that is closest to f in L_∞ is given by $h^\perp = f - h$, with h as constructed above and where the distance $\mathrm{dist}_\infty(f, H_\infty^\perp)$ is equal to $\|h\|_\infty$.*

Note that if the operator H is compact, then it has the singular value decomposition

$$Hf = \sum_{k=1}^\infty s_k \langle f, x_k \rangle y_k, \qquad f, x_k, y_k \in H_2, \qquad Hx_k = s_k y_k, \qquad (12.13)$$

where $\|H\|_2 = s_1 \geq s_2 \geq s_3 \geq \cdots$ are the singular values of H and (x_k, y_k) is a pair of singular vectors (Schmidt pairs) corresponding to s_k. The x and y

in the previous theorem are given by any Schmidt pair that corresponds to the maximal singular value γ.

If we return to our H_∞ control problem, then this theorem implies that there can only be a solution to the problem if the Hankel matrix H, defined by the symbol $\hat{h}$ with $\hat{h}(z) = h(1/z)/z$ and $h = \Sigma_{21}/B_{22}$, has a norm $\|H\|_2 \le \gamma$. If not, there is no solution.

If the given system Σ is rational, then it can be seen that the rank of this Hankel operator is finite and equal to the number of unstable zeros of Σ_{22} (multiplicities counted) that are the zeros of the Blaschke product B_{22}. This follows directly from the following classical theorem by Kronecker [132; 175, p. 37].

Theorem 12.5.3 (Kronecker). *Let* $H = H_h$ *be a Hankel matrix with symbol* $h \in H_\infty$. *Then the following statements are equivalent:*

1. *H has finite rank n.*
2. *$g(z) = z^{-1}h(z^{-1}) \in H_\infty^\perp$ is rational and has n poles (in $\mathbb{D}$).*
3. *$H = H_f$ with f of the form $f = B_n g$, where $g \in H_\infty^\perp$ and B_n is a Blaschke product with n zeros in $\mathbb{D}$.*

Now if the rank of the Hankel matrix is finite, then this may suggest that the problem is a finite-dimensional problem and that we do not have to compute the infinite-dimensional Schmidt pair as suggested in Theorem 12.5.2. This is indeed the case. It is in fact solved as a Nevanlinna–Pick interpolation problem. To see this, assume that H_f has finite rank. This means that the given symbol f has a finite number of poles $\{1/\bar{\alpha}_i\}_{i=1}^n$ that are in $\mathbb{E}$ (they are repeated according to their multiplicity). We can collect them in a Blaschke product B_n with zeros $\{\alpha_i\}_{i=1}^n$ so that f has the form $f = B_n \hat{f}$ with $\hat{f} \in H_\infty$. But because the approximant $h^\perp \in H_\infty^\perp$, it is obvious that the error $h = f - h^\perp$ will have the same poles as f in $\mathbb{E}$ and thus it will also be of the form $h = B_n \hat{h}$. Now the solution $h^\perp = f - h = B_n(\hat{f} - \hat{h})$ should be in $H_\infty^\perp$. By taking the substar, this is equivalent to saying that the solution $h^\perp$ should satisfy

$$\hat{f}_*(z) - \hat{h}_*(z) = B_n(z)h_*^\perp(z) \in z B_n(z) H_\infty.$$

Thus we have reduced the interpolation problem to the following one: Given $\Gamma = \hat{f}_* = B_n f_*$ and the numbers α_i (which are the unstable poles of f_*), find $\Gamma_n = \hat{h}_* = B_n h_*$ such that Γ_n interpolates Γ in the points $\{\alpha_i\}_{i=0}^n$ (where $\alpha_0 = 0$). Thus

$$\Gamma(z) - \Gamma_n(z) \in z B_n(z) H_\infty.$$

The solution of the original problem is then given by

$$h^{\perp} = B_n(\Gamma_* - \Gamma_{n*}) = f - \Gamma_n^*.$$

This interpolation problem can of course be solved by the Nevanlinna–Pick algorithm as described in Section 6.4 (where we take $w = 0$). The algorithm constructs J-unitary matrices with parameters ρ_k (and γ_k). These matrices are related to the recurrence of reproducing kernels as has been explained there. They can also be used to give all solutions to the interpolation problem. Indeed, this algorithm constructs transformations τ_k (see Theorem 6.4.1) and all solutions to the interpolation problem were given by $\Gamma_n = T_n(\Gamma_0)$ with Γ_0 arbitrary in $\mathcal{B}$ and where $T_n = (\tau_n \circ \tau_{n-1} \circ \cdots \circ \tau_1)^{-1}$ is defined by the associated Θ-matrix. If in the algorithm all the $\rho_k \in \mathbb{D}$, then Γ_n will be a Schur function for any choice of $\Gamma_0 \in \mathcal{B}$. A solution of minimal degree for Γ_n will be found if Γ_0 is a constant. Thus if we set $\Gamma_0 = \gamma$, then $\|\Gamma_n\|_{\infty} \leq \gamma$ and we have solved the minimal degree H_{∞} control problem.

However, the problem we had above is to find an all-pass function, which need not be a Schur function. Of course if $\Gamma \in \mathcal{B}$, then Γ_n will be in $\mathcal{B}$ too. However, if $\Gamma \notin \mathcal{B}$, then interpolation in the α_i may generate some Γ_n that is not in $\mathcal{B}$. This can of course be checked by the modulus of the numbers ρ_k. If some ρ_k falls in $\mathbb{E}$, then Γ_n will not be a Schur function. The H_{∞} control problem will not have a solution. However, it can well be that, if no degenerate situation occurs, the Nevanlinna–Pick algorithm can go on and construct some Γ_n. Of course the Pick matrix associated with this problem will not be positive definite anymore and there will not be a positive measure associated with the problem. This (rational) function Γ_n is, however, not completely useless. It is a solution of a Nevanlinna–Pick–Takagi problem. The function Γ_n will have unstable poles (poles in $\mathbb{D}$). The number of these unstable poles is equal to the number of negative eigenvalues of the (generalized) Pick matrix that can be associated with the interpolation problem. This number is equal to the number of sign changes in the sequence of determinants of the leading submatrices in this Pick matrix. By the determinant formulas for the kernels given in Theorem 2.2.2 and formula (12.1), this means that this is equal to the number of sign changes in the sequence of E_k, $k = 0, 1, \ldots, n$, where

$$E_k = \prod_{i=1}^{k} \frac{1 - |\rho_i|^2}{1 - |\alpha_i \rho_i|^2}$$

and the ρ_i are as produced in the Nevanlinna–Pick algorithm.

The Nevanlinna–Pick–Takagi problem has another important application in best rational approximation and in model reduction for linear systems. These problems are characterized by the criterion of Hankel norm approximation. Since almost all our results only involved positive measures, they do not apply to these kinds of applications. We therefore will only outline this problem very briefly in the next section.

12.5.3. Hankel norm approximation

Given a function $h \in H_\infty$, one can associate with it the Hankel operator H_h, and the norm $\|H_h\|_2$ is in fact a norm for h. It is known as the Hankel norm of h and we shall denote it as $\|h\|_H = \|H_h\|_2$. The Hankel norm lies somewhere between the L_2 and the L_∞ norm since one can show that [57, p. 184]

$$h(z) \in H_\infty \quad \Rightarrow \quad \|h\|_2 \leq \|h\|_H \leq \|h\|_\infty.$$

Note that for $h \in L_\infty$, this is not a norm, because the $H_\infty^\perp$-part of h does not contribute to the Hankel norm of h.

Suppose that $f \in H_\infty$ is the transfer function of a linear system. Then the corresponding input–output map is the Hankel operator $H = H_f$. The problem is to approximate the system by another system, where the approximating criterion is the Hankel norm, which corresponds to approximating the Hankel matrix $H = H_f$ by another Hankel matrix H'. Thus, if h' is the symbol of H' and if $g = f - h'$, then we consider

$$\|H - H'\|_2 = \|H_{f-h'}\|_2 = \|H_g\|_2 = \|g\|_H = \|f - h'\|_H$$

as the error of approximation. Either we want to bring this below a certain tolerance τ and if there is more than one solution to achieve this, we try to find the simplest possible one, that is, the one for which the rank of H' is as low as possible (this is the minimal degree problem), or we just want to find the optimal one of a certain degree, that is, we minimize the norm, given that the rank of H' is bounded by k (this is the minimal norm problem).

We first remark that approximating an operator by an operator of finite rank is a standard problem. If H is a Hankel operator with singular value decomposition (12.13), then

$$\inf \{\|H - K\|_2 : K \text{ has rank} \leq k\} = s_{k+1}.$$

An operator that is optimal is given by $K = \sum_{j=1}^{k} s_j \langle \cdot, x_j \rangle y_j$. However, this approximating operator is in general not Hankel.

The remarkable result of the Theorem 12.5.4 below, which generalizes Theorem 12.5.2, is that there is indeed an optimal approximating operator within the class of Hankel operators. Thus given a Hankel operator H, there exists a Hankel operator H' such that it solves

$$\inf\{\|H - H'\|_2 : H' \text{ has rank} \le k \text{ and } H' \text{ is Hankel}\}. \qquad (12.14)$$

The minimum is s_{k+1}.

We can reformulate this in terms of the symbols as follows. Let $H_\infty^{[k]}$ denote the class of functions that are symbols of Hankel operators of rank k. Thus $H_\infty^{[k]}$ represents functions that belong to classes of the form $B_k H_\infty^\perp$, where B_k is an arbitrary Blaschke product with k zeros in $\mathbb{D}$. Then the optimization problem (12.14) becomes: Given $f \in L_\infty$, find

$$\inf\left\{\|f - h'\|_H : h' \in H_\infty^{[k]}\right\}. \qquad (12.15)$$

Theorem 12.5.4 (Adamyan–Arov–Kreĭn). *Let $H = H_f$ be a compact Hankel operator with s_{k+1} the $(k+1)$st singular value and (x, y) an arbitrary Schmidt pair associated with s_{k+1}.*

A solution to the optimization problem (12.14) is obtained for K equal to a unique Hankel operator H'.

The Hankel operator $H - H'$ has a unique symbol h of minimal norm, that is, there is a function $h \in L_\infty$ such that $H - H' = H_h$ and

$$\|H_h\|_2 = s_{k+1} = \|h\|_\infty.$$

This symbol h can be constructed as follows: $h(z) = s_{k+1}x(z)/y(1/z)$. The function h/s_{k+1} is an all-pass function.

Equivalently, $h' = f - h$ is a solution to the Hankel norm problem (12.15).

Note that for $k = 0$ this theorem reduces to Theorem 12.5.2.

This solves the minimum norm problem. Given the Hankel operator H one finds the Hankel operator H' of rank k at most that minimizes $\|H - H'\|_2$. In the optimal degree problem, one is given the Hankel operator H and a tolerance τ. The problem is then to find the Hankel operator H' of minimal degree that satisfies $\|H - H'\|_2 \le \tau$. For the latter problem, it is seen from the previous theorem that if Hankel operators of rank k are allowed, then the minimal norm that can be obtained is s_{k+1}. This implies that the minimal rank required to satisfy $\|H - H'\|_2 \le \tau$ will be k if $s_k > \tau \ge s_{k+1}$.

The solution is constructed by the Nevanlinna–Pick algorithm as in the previous section. It is now allowed that there are unstable poles and the measure that is involved will not be positive definite anymore.

We note that this kind of problem really requires the Nevanlinna–Pick interpolation algorithm and thus in this case the problem can not be solved by a polynomial approach. When all $\alpha_i = 0$ in this kind of problem, then the corresponding Nehari problem is equivalent with a Carathéodory coefficient problem [52] and this of course can be solved by a polynomial approach.

In a recent paper [99], Gohberg and Landau discuss the linear prediction problem for two stationary stochastic processes. More precisely, one of these stochastic processes is predicted using the cross correlation between the past of the other process and the future of the predicted process. In their formulation of the problem, the authors obtain a unifying framework for prediction problems as we have studied them in Section 12.1 and the problems we have discussed in this section, but only for the polynomial case. Extensions to the rational case may be expected to be straightforward with the tools provided by this monograph.

Conclusion

We have given an introduction to the theory of orthogonal rational functions. Although it required a slightly more complicated notation, our treatment allowed us to discuss the case of the circle ($\partial \mathbb{O} = \mathbb{T}$) and the case of the real line ($\partial \mathbb{O} = \mathbb{R}$) simultaneously. The case where all the poles are in $\mathbb{O}^e$ and the so-called boundary case, where the poles of the rational functions are on the boundary of $\mathbb{O}^e$, were considered separately. If A represents the basic points $A = \{\alpha_1, \alpha_2, \ldots\}$, which fix the poles, then the first case (internal poles) corresponds to all the points A being interior to $\mathbb{O}$: $A \subset \mathbb{O}$ (see cases IT and IR in the table below) and the boundary case corresponds to $A \subset \partial \mathbb{O}$ (see cases BT and BR in the table below).

	$\mathbb{O} = \mathbb{D}$	$\mathbb{O} = \mathbb{U}$
$A \subset \mathbb{O}$	IT	IR
$A \subset \overline{\partial \mathbb{O}}$	BT	BR

The case $A \subset \mathbb{O}$ was extensively discussed in Chapters 2–10; the case $A \subset \overline{\partial \mathbb{O}}$ was discussed in Chapter 11.

The origin of these problems can be found in multipoint generalizations of classical moment problems and associated interpolation problems, which are usually related to one- or two-point rational approximants in a Padé-like sense.

The case IT from the previous scheme where $A \subset \mathbb{D}$ and orthogonality is considered for a measure on $\mathbb{T}$ is in a sense the most natural multipoint generalization of the theory of orthogonal polynomials on the unit circle and the associated trigonometric moment problem. The polynomial problem occurs as a special case by choosing all α_k equal to 0.

389

The case BR in that scheme corresponds to the multipoint generalization of the polynomials on the real line and the associated Hamburger moment problem. Here the polynomials appear as a special case when all α_k are chosen to be at ∞.

The cases IR and BT are obtained by conformal mapping of the cases IT and BR respectively. However, the polynomial situations for IT and BR are *not* mapped to polynomial situations in IR and BT. The polynomial situation in IT corresponds to a special rational situation in IR, in fact polynomials in $(z - \mathbf{i})/(z + \mathbf{i})$, and the polynomial case in BR is mapped to the case of special rationals functions, namely polynomials in $(1 - z)/(1 + z)$.

The band in between has been left open. There is much room left for considering the case where the points from A can be everywhere on the Riemann sphere $\overline{\mathbb{C}}$ or, maybe more realistically, when $A \subset \overline{\mathbb{O}} = \mathbb{O} \cup \partial\mathbb{O}$. Some remarks about $A \subset \mathbb{O} \cup \mathbb{O}^e$ were given in Section 4.5. The case $A \subset \overline{\mathbb{D}} = \mathbb{D} \cup \mathbb{T}$ was considered by Dewilde and Dym [62] in the context of lossless inverse scattering as was briefly mentioned in Section 12.3.

While the authors were lecturing about the topic of this monograph at several international conferences, they were often asked how this collaboration of people from four different countries came about. In fact, this was not just a coincidence. Coming from different directions – analytical, numerical, or applied – each of the authors was interested in the multipoint generalizations of the polynomial ideas as outlined above. Thus it was unavoidable that some day they should meet at a conference somewhere in the world. Once they had discovered their common interest, mutual visits and often participation in conferences, which sometimes took place in none of the countries where the authors reside, turned professional contacts into personal friendship. The collaboration has been quite intense during the past five years. For daily communication, the fax machine and the internet were indispensable tools. We finish with some historical notes about how the authors were brought together.

Olav Njåstad has been working for quite a while on moment problems, continued fractions, orthogonal polynomials, and related topics, both for the case of the real line and for the complex unit circle. Much of his work was in collaboration with W. B. Jones and W. Thron. For the trigonometric moment problem, the polynomials studied by Szegő are the natural building bricks to be used. As is well known, this theory naturally leads to rational approximants for positive real functions (Carathéodory functions), quadrature formulas, etc. For moment problems on the real line such as the Hamburger moment problem, the same kind of problems, tools, and solutions occur, but now using polynomials orthogonal on the real line. The discussion of strong Hamburger moment problem gave rise to a first generalization. The orthogonal polynomials have to be replaced by orthogonal Laurent polynomials and the rational approximants were replaced by approximants that did not approximate in one point, but in two points. So a step was made from one-point Padé-type approximation to two-point Padé approximation. The step from two to more than two is then only an unavoidable generalization. So the extended moment problems were born, which are related to multipoint Padé approximation. However, the idea of treating power series in more than one point somehow suggested to deal with only a finite number of points in which these power series were given. This explains why the earlier papers dealt with the so-called cyclic situation where the infinite sequence of points α_k was a cyclic repetition on only a finite number of different points.

Pablo González-Vera promoted in 1985 *Two-Point Padé Type Approximants*, which were closely related to orthogonal Laurent polynomials, quadrature, and strong moment problems.

It was in the fall of 1985 at a workshop organized by Claude Brezinski in Luminy in the South of France that Erik Hendriksen, Pablo González-Vera, and Olav Njåstad met. Erik visited Trondheim at the end of the year and Olav was invited to Amsterdam in 1986 and this started collaboration on the topic of orthogonal Laurent polynomials and multipoint generalizations, which give rise to the orthogonal rational functions. Erik's promotion of *Strong Moment Problems and Orthogonal Laurent Polynomials* was held in Trondheim in April 1989.

By their common interests in two-point Padé approximation, quadrature formulas, and more general multipoint Padé approximation, Pablo and Olav kept in touch. It was in May–June 1989 that Pablo spent some time in Trondheim. Olav and Pablo prepared some work to be presented at another Luminy workshop to be held in September of that year.

Adhemar Bultheel who earned his Ph.D. under the guidance of P. Dewilde had entered these topics motivated by the interests of his promoter in applied topics such as digital speech processing, lossless inverse scattering, etc. The algorithms of Schur and Nevanlinna–Pick were the tools par excellence to deal with these problems. The orthogonal rational functions are not of central interest though because the optimal predictors are given by reproducing kernels in the first place. His Ph.D. (1979) dealt with *Recursive Rational Approximation*, discussing both matrix versions of the Nevanlinna–Pick algorithm and Padé approximation in one and two points.

Olav and Adhemar met for the first time at the NATO Advanced Study Institute on *Orthogonal Polynomials and Their Applications* organized by Paul Nevai in Columbus, Ohio in May–June 1989.

It was only at the September meeting organized by Claude Brezinski on *Extrapolation and Rational Approximation* in Luminy, France, from September 24 to 30, 1989 that the present four authors met. Adhemar and Olav visited

La Laguna (Tenerife) in October–November of the same year and this really started a successful collaboration between the four authors in trying to extend the theory of rational functions, first in the unit circle and the α_k inside the disk, but later also for the boundary situation, that is, where the points are on the circle, which is the obvious choice when trying to get analogs for the extended moment problems on the line.

From December 1993 till December 1996, A. Bultheel, P. González-Vera, and O. Njåstad were collaborating in the framework of a European Human Capital and Mobility project ROLLS under contract number CHRX-CT93-0416. This project tried to unify topics in rational approximation, orthogonal functions, linear algebra, linear systems, and signal processing. The financial contribution from this project for the accomplishment of this monograph is greatly appreciated. The present book is one of the accomplishments of the project.

This historical note may explain how this introduction to the theory of orthogonal rational functions came about. The treatment is only introductory since we have not attempted to cover all possible generalizations of the corresponding Szegő theory. Many aspects of the theory were not discussed and there is much work to be done. There is the matrix case, which was, for example, discussed in Refs. [56], [26], [146], [3], and [165] and many other papers. For a matrix theory of orthogonal polynomials on the unit circle see Refs. [74] and [75] or even the case of operator-valued functions [186, 84, 19, 149]. This is related to many directional or tangential interpolation problems such as discussed in Refs. [80], [82], [81], [83], [164], [117], [100], and [10] or the theory of more general J-unitary matrices, a theory initiated by Potapov [181, 77, 9, 13, 14, 15, 16, 127, 89, 90, 88, 91]. More of the polynomial results in Ref. [19] could have been generalized. We could have stressed more the multipoint Padé aspect of this theory (see Refs. [101], [113], and [159]). Much more extensive results can be obtained for the asymptotics of the polynomials or the recursion coefficients and the convergence of Fourier series in these orthogonal functions. There is the theory of time-varying systems, which can be generalized. See Refs. [98], [197], [63], [20], and [198]. And there are of course the many beautiful applications, with their own terminology and their own problem settings. We have given but a brief introduction to some of these in the last chapter. One could consult various volumes [97, 84, 19, 52, 128]. And there is a very long bibliography that is related to all these topics. Citing them all would be a project on its own.

So we are well aware of the fact that the present discussion can only be an appetizing survey that may hopefully invoke some interest in the field, if that were necessary at all. We think it is a fascinating subject, even more fascinating than the theory of orthogonal polynomials, if that does not sound too much like a blasphemy.

Bibliography

[1] N. I. Achieser. *Theory of Approximation*. Frederick Ungar Publ. Co., New York, 1956.

[2] N. I. Akhiezer [Achieser]. *The Classical Moment Problem*. Oliver and Boyd, Edinburgh, 1969. Originally published Moscow, 1961.

[3] R. Ackner, H. Lev-Ari, and T. Kailath. The Schur algorithm for matrix-valued meromorphic functions. *SIAM J. Matrix Anal. Appl.*, 15:140–150, 1994.

[4] V. M. Adamjan, D. Z. Arov, and M. G. Kreĭn. Infinite Hankel matrices and generalized Carathéodory–Fejér and Riesz problems. *Functional Anal. Appl.*, 2:1–18, 1968.

[5] V. M. Adamjan, D. Z. Arov, and M. G. Kreĭn. Infinite Hankel matrices and generalized problems of Carathéodory–Fejér and I. Schur. *Functional Anal. Appl.*, 2:269–281, 1968.

[6] V. M. Adamjan, D. Z. Arov, and M. G. Kreĭn. Analytic properties of Schmidt pairs for a Hankel operator and the generalized Schur–Takagi problem. *Math. USSR-Sb.*, 15:31–73, 1971.

[7] V. M. Adamjan, D. Z. Arov, and M. G. Kreĭn. Infinite Hankel block matrices and related extension problems. *Izv. Akad. Nauk. Armjan SSR Ser. Mat.*, 6:87–112, 1971. See also *Am. Math. Soc. Transl.*, 111:133–156, 1978.

[8] A. C. Allison and N. J. Young. Numerical algorithms for the Nevanlinna–Pick problem. *Numer. Math.*, 42:125–145, 1983.

[9] D. Alpay, J. A. Ball, I. Gohberg, and L. Rodman. *j*-Unitary preserving automorphisms of rational matrix functions: State space theory, interpolation, and factorization. *Linear Algebra Appl.*, 197/198:531–566, 1994.

[10] D. Alpay and V. Bolotnikov. Two-sided interpolation for matrix functions with entries in the Hardy space. *Linear Algebra Appl.*, 223/224:31–56, 1995.

[11] G. S. Ammar and W. B. Gragg. Determination of Pisarenko frequency estimates as eigenvalues of an orthogonal matrix. In *Proc. SPIE, Int. Soc. for Optical Eng. Advanced Algorithms and Architectures for Signal Processing 2*, 826:143–145, 1987.

[12] N. Aronszajn. Theory of reproducing kernels. *Trans. Am. Math. Soc.*, 68:337–404, 1950.

[13] D. Z. Arov. γ-generating matrices, *J*-inner matrix-functions and related extrapolation problems. Part I. *Theory of Functions, Functional Analysis and Their Applications*, 51:61–67, 1989. (In Russian.)

393

[14] D. Z. Arov. γ-generating matrices, J-inner matrix-functions and related
      extrapolation problems. Part II. *Theory of Functions, Functional Analysis and
      Their Applications*, 52:103–109, 1989. (In Russian.)

[15] D. Z. Arov. γ-generating matrices, J-inner matrix-functions and related
      extrapolation problems. Part III. *Theory of Functions, Functional Analysis and
      Their Applications*, 53:57–64, 1990. (In Russian.)

[16] D. Z. Arov. Regular J-inner matrix-functions and related continuation
      problems. In G. Arsene et al., eds., *Linear Operators in Function Spaces*,
      vol. 43 of *Oper. Theory: Adv. Appl.*, pp. 63–87, 1990.

[17] G. A. Baker, Jr. and P. R. Graves-Morris. *Padé Approximants. Part II:
      Extensions and Applications*, vol. 14 of *Encyclopedia of Mathematics and Its
      Applications*. Addison-Wesley, Reading, MA, 1981.

[18] G. A. Baker, Jr. and P. R. Graves-Morris. *Padé Approximants*, vol. 59 of
      Encyclopedia of Mathematics and Its Applications. Cambridge University
      Press, Cambridge, 2nd ed., 1996.

[19] M. Bakonyi and T. Constantinescu. *Schur's Algorithm and Several
      Applications*, vol. 261 of *Pitman Research Notes in Mathematics*. Longman,
      Harlow, UK, 1992.

[20] J. A. Ball, I. Gohberg, and M. A. Kaashoek. Two-sided Nudelman interpolation
      for input–output operators of discrete time-varying systems. *Integral Equations
      Operator Theory*, 21:174–211, 1995.

[21] J. A. Ball, I. Gohberg, and L. Rodman. Realization and interpolation of rational
      matrix functions. In I. Gohberg, ed., *Topics in Interpolation Theory of Rational
      Matrix-Valued Functions*, vol. 33 of *Oper. Theory: Adv. Appl.*, pp. 1–72,
      Birkhäuser Verlag, Basel, 1988.

[22] J. A. Ball, I. Gohberg, and L. Rodman. *Interpolation of Rational Matrix
      Functions*, vol. 45 of *Oper. Theory: Adv. Appl.*, Birkhäuser Verlag, Basel, 1990.

[23] V. Belevitch. *Classical Network Theory*, pp. 93,136,141. Holden-Day,
      San Francisco, 1968.

[24] R. P. Brent and F. T. Luk. A systolic array for the solution of linear-time solution
      of Toeplitz systems of equations. *J. VLSI and Comp. Systems*, 1(1):1–23, 1983.

[25] O. Brune. Synthesis of finite two-terminal network whose driving point
      impedance is a prescribed function of frequency. *J. Math. Phys.*, 10:191–236,
      1931.

[26] A. Bultheel. Orthogonal matrix functions related to the multivariable
      Nevanlinna–Pick problem. *Bull. Soc. Math. Belg. Sér. B*, 32(2):149–170, 1980.

[27] A. Bultheel. On a special Laurent–Hermite interpolation problem. In L. Collatz,
      G. Meinardus, and H. Werner, eds., *Numerische Methoden der
      Approximationstheorie 6*, vol. 59 of *Int. Ser. of Numer. Math.*, pp. 63–79,
      Birkhäuser Verlag, 1981. Basel–New York–Berlin.

[28] A. Bultheel. On the ill-conditioning of locating the transmission zeros in least
      squares ARMA filtering. *J. Comput. Appl. Math.*, 11(1):103–118, 1984.

[29] A. Bultheel. *Laurent Series and Their Padé Approximations*, vol. OT-27 of
      Oper. Theory: Adv. Appl. Birkhäuser Verlag, Basel–Boston, 1987.

[30] A. Bultheel and P. Dewilde. Orthogonal functions related to the
      Nevanlinna–Pick problem. In P. Dewilde, ed., *Proc. 4th Int. Conf. on Math.
      Theory of Networks and Systems at Delft*, pp. 207–212, Western Periodicals,
      North Hollywood, CA, 1979.

[31] A. Bultheel, P. González-Vera, E. Hendriksen, and O. Njåstad. A Szegő theory
      for rational functions. Technical Report TW131, Department of Computer
      Science, K. U. Leuven, May 1990.

[32] A. Bultheel, P. González-Vera, E. Hendriksen, and O. Njåstad. Orthogonality and quadrature on the unit circle. In C. Brezinski, L. Gori, and A. Ronveaux, eds., *Orthogonal Polynomials and Their Applications*, vol. 9 of *IMACS Annals on Computing and Applied Mathematics*, pp. 205–210, J. C. Baltzer AG, Basel, 1991.

[33] A. Bultheel, P. González-Vera, E. Hendriksen, and O. Njåstad. The computation of orthogonal rational functions and their interpolating properties. *Numer. Algorithms*, 2(1):85–114, 1992.

[34] A. Bultheel, P. González-Vera, E. Hendriksen, and O. Njåstad. Orthogonal rational functions and quadrature on the unit circle. *Numer. Algorithms*, 3:105–116, 1992.

[35] A. Bultheel, P. González-Vera, E. Hendriksen, and O. Njåstad. Moment problems and orthogonal functions. *J. Comput. Appl. Math.*, 48:49–68, 1993.

[36] A. Bultheel, P. González-Vera, E. Hendriksen, and O. Njåstad. Asymptotics for orthogonal rational functions. *Trans. Am. Math. Soc.*, 346:331–340, 1994.

[37] A. Bultheel, P. González-Vera, E. Hendriksen, and O. Njåstad. Orthogonal rational functions with poles on the unit circle. *J. Math. Anal. Appl.*, 182:221–243, 1994.

[38] A. Bultheel, P. González-Vera, E. Hendriksen, and O. Njåstad. Orthogonality and boundary interpolation. In A. M. Cuyt, ed., *Nonlinear Numerical Methods and Rational Approximation II*, pp. 37–48. Kluwer, Dordrecht, 1994.

[39] A. Bultheel, P. González-Vera, E. Hendriksen, and O. Njåstad. Quadrature formulas on the unit circle based on rational functions. *J. Comput. Appl. Math.*, 50:159–170, 1994.

[40] A. Bultheel, P. González-Vera, E. Hendriksen, and O. Njåstad. On the convergence of multipoint Padé-type approximants and quadrature formulas associated with the unit circle. *Numer. Algorithms*, 13:321–344, 1996.

[41] A. Bultheel, P. González-Vera, E. Hendriksen, and O. Njåstad. Continued fractions and orthogonal rational functions. In W. B. Jones and A. S. Ranga, eds., *Orthogonal Functions, Moment Theory and Continued Fractions: Theory and Applications*, pp. 69–100, Marcel Dekker, New York, 1998.

[42] A. Bultheel, P. González-Vera, E. Hendriksen, and O. Njåstad. A Favard theorem for rational functions with poles on the unit circle. *East J. Approx.*, 3:21–37, 1997.

[43] A. Bultheel, P. González-Vera, E. Hendriksen, and O. Njåstad. Orthogonal rational functions and nested disks. *J. Approx. Theory*, 89:344–371, 1997.

[44] A. Bultheel, P. González-Vera, E. Hendriksen, and O. Njåstad. Rates of convergence of multipoint rational approximants and quadrature formulas on the unit circle. *J. Comput. Appl. Math.*, 77:77–102, 1997.

[45] A. Bultheel, P. González-Vera, E. Hendriksen, and O. Njåstad. A rational moment problem on the unit circle. *Methods Appl. Anal.*, 4(3):283–310, 1997.

[46] R. B. Burckel. *An Introduction to Classical Complex Analysis*. Birkhäuser Verlag, Basel, 1971.

[47] J. P. Burg. Maximum entropy spectral analysis. In D. G. Childers, ed., *Modern Spectral Analysis*, pp. 34–39, IEEE Press, New York, 1978. Originally presented at 37th Meet. Soc. Exploration Geophysicists, 1967.

[48] C. Carathéodory. Über den Variabilitätsbereich der Koeffizienten von Potenzreihen die gegebene Werte nicht annehmen. *Math. Ann.*, 64:95–115, 1907.

[49] C. Carathéodory. Über den Variabilitätsbereich der Fourier'schen Konstanten von positiven harmonischen Funktionen. *Rend. Circ. Mat. Palermo*, 32:193–217, 1911.

[50] C. Carathéodory and L. Fejér. Über den Zusammenhang der Extremen von harmonischen Funktionen mit ihren Koefficienten und über den Picard-Landauschen Satz. *Rend. Circ. Mat. Palermo*, 32:218–239, 1911.

[51] L. Cochran and S. C. Cooper. Orthogonal Laurent polynomials on the real line. In S. C. Cooper and W. J. Thron, eds., *Continued Fractions and Orthogonal Functions*, pp. 47–100, Marcel Dekker, New York, 1994.

[52] T. Constantinescu. *Schur Analysis, Factorization and Dilation Problems*, vol. 82 of *Oper. Theory: Adv. Appl.* Birkhäuser Verlag, Basel, 1996.

[53] G. Cybenko. Computing Pisarenko frequency estimates. In *Proc. 1984 Conf. Inform. Syst. Sci.*, pp. 587–591. Princeton Univ., 1984.

[54] P. J. Davis and P. Rabinowitz. *Methods of Numerical Integration*. Academic Press, 2nd ed., 1984.

[55] Ph. Delsarte and Y. Genin. A survey of the split approach based techniques in digital signal processing applications. *Phillips J. Res.*, 43:346–374, 1988.

[56] Ph. Delsarte, Y. Genin, and Y. Kamp. The Nevanlinna–Pick problem for matrix-valued functions. *SIAM J. Appl. Math.*, 36:47–61, 1979.

[57] Ph. Delsarte, Y. Genin, and Y. Kamp. On the role of the Nevanlinna–Pick problem in circuit and system theory. *Int. J. Circuit Th. Appl.*, 9:177–187, 1981.

[58] P. Dewilde. Stochastic modeling with orthogonal filters. In *Outils et Modèles Mathématiques pour l'Automatique, l'Analyse de Systèmes et le Traitement du Signal, Vol. 2*, pp. 331–398, Editions du CNRS, Paris, 1982.

[59] P. Dewilde. The lossless inverse scattering problem in the network-theory context. In H. Dym and I. Gohberg, eds., *Topics in Operator Theory, Systems and Networks*, vol. 12 of *Oper. Theory: Adv. Appl.*, pp. 109–128, Birkhäuser Verlag, Basel, 1984.

[60] P. Dewilde and H. Dym. Schur recursions, error formulas, and convergence of rational estimators for stationary stochastic sequences. *IEEE Trans. Inf. Th.*, IT-27:446–461, 1981.

[61] P. Dewilde and H. Dym. Lossless inverse scattering with rational networks: theory and applications. Technical Report 83-14, Delft University of Technology, Dept. of Elec. Eng. Network Theory Section, December 1982.

[62] P. Dewilde and H. Dym. Lossless inverse scattering, digital filters, and estimation theory. *IEEE Trans. Inf. Th.*, IT-30:644–662, 1984.

[63] P. Dewilde, M. A. Kaashoek, and M. Verhaegen, eds. *Challenges of a Generalized System Theory*. Essays of the Dutch Academy of Arts and Sciences. Dutch Acad. Arts Sci., Amsterdam, 1993.

[64] P. Dewilde, A. Viera, and T. Kailath. On a generalized Szegő–Levinson realization algorithm for optimal linear predictors based on a network synthesis approach. *IEEE Trans. Circuits and Systems*, CAS-25:663–675, 1978.

[65] M. M. Djrbashian. Expansions in systems of rational functions on a circle with a given set of poles. *Doklady Akademii Nauk SSSR*, 143:17–20, 1962. (In Russian. Translation in *Soviet Mathematics Doklady*, 3:315–319, 1962.)

[66] M. M. Djrbashian. Orthogonal systems of rational functions on the unit circle with given set of poles. *Doklady Akademii Nauk SSSR*, 147:1278–1281, 1962. (In Russian. Translation in *Soviet Mathematics Doklady*, 3:1794–1798, 1962.)

[67] M. M. Djrbashian. Orthogonal systems of rational functions on the circle. *Izv. Akad. Nauk Armyan. SSR*, 1:3–24, 1966. (In Russian.)

[68] M. M. Djrbashian. Orthogonal systems of rational functions on the unit circle. *Izv. Akad. Nauk Armyan. SSR*, 1:106–125, 1966. (In Russian.)

[69] M. M. Djrbashian. Expansions by systems of rational functions with fixed poles. *Izv. Akad. Nauk Armyan. SSR*, 2:3–51, 1967. (In Russian.)

[70] M. M. Djrbashian. A survey on the theory of orthogonal systems and some open problems. In P. Nevai, ed., *Orthogonal Polynomials: Theory and Practice*, vol. 294 of *Series C: Mathematical and Physical Sciences* pp. 135–146, NATO-ASI, Kluwer Academic Publishers, Boston, 1990.

[71] W. F. Donoghue Jr. *Monotone Matrix Functions and Analytic Continuation*. Springer-Verlag, Berlin, 1974.

[72] R. G. Douglas. *Banach Algebra Techniques in Operator Theory*. Academic Press, New York, 1972.

[73] R. G. Douglas, H. S. Shapiro, and A. L. Shields. Cyclic vectors and invariant subspaces for the backward shift operator. *Ann. Inst. Fourier*, 20:37–76, 1970.

[74] V. K. Dubovoj, B. Fritzsche, and B. Kirstein. On a class of matrix completion problems. *Math. Nachr.*, 143:211–226, 1989.

[75] V. K. Dubovoj, B. Fritzsche, and B. Kirstein. *Matricial Version of the Classical Schur Problem*, vol. 129 of *Teubner-Texte zur Mathematik*. Teubner Verlagsgesellschaft, Stuttgart, Leipzig, 1992.

[76] P. L. Duren. *The Theory of H^p Spaces*, vol. 38 of *Pure and Applied Mathematics*. Academic Press, New York, 1970.

[77] H. Dym. *J-Contractive Matrix Functions, Reproducing Kernel Hilbert Spaces and Interpolation*, vol. 71 of *CBMS Regional Conf. Ser. in Math.* Am. Math. Soc., Providence, RI, 1989.

[78] T. Erdélyi, P. Nevai, J. Zhang, and J. S. Geronimo. A simple proof of "Favard's theorem" on the unit circle. *Atti. Sem. Mat. Fis. Univ. Modena*, 29:551–556, 1991. Proceedings of the Meeting "Trends in Functional Analysis and Approximation Theory," 1989, Italy.

[79] J. Favard. Sur les polynomes de Tchebicheff. *C. R. Acad. Sci. Paris*, 200:2052–2053, 1935.

[80] I. P. Fedčina. A criterion for the solvability of the Nevanlinna–Pick tangent problem. *Mat. Issled. (Kishinev)*, 7:213–227, 1972. (In Russian.)

[81] I. P. Fedčina. A description of the solutions of the Nevanlinna–Pick tangent problem. *Dokl. Akad. Nauk Armjan SSR, Ser. Mat.*, 60:37–42, 1975. (In Russian.)

[82] I. P. Fedčina. The tangential Nevanlinna–Pick problems with multiple points. *Dokl. Akad. Nauk Armjan SSR, Ser. Mat.*, 61:214–218, 1975.

[83] I. P. Fedčina. The Schur problem for vector valued functions. *Ukrain Mat. Z.*, 30(6):797–805, 861, 1978.

[84] C. Foiaş and A. Frazho. *The Commutant Lifting Approach to Interpolation Problems*, vol. 44 of *Oper. Theory: Adv. Appl.* Birkhäuser Verlag, Basel, 1990.

[85] C. Foiaş, J. W. Helton, H. Kwakernaak, and J. B. Pearson. H_∞-*Control Theory*. Number 1496 in Lecture Notes in Math. Springer-Verlag, Berlin, 1991.

[86] B. A. Francis. *A Course in H^∞ Control Theory*. Springer-Verlag, Berlin, Heidelberg, 1987.

[87] G. Freud. *Orthogonal Polynomials*. Pergamon Press, Oxford, 1971.

[88] B. Fritzsche, B. Fuchs, and B. Kirstein. Schur sequence parametrizations of Potapov-normalized full rank j_{pq}-elementary factors. *Linear Algebra Appl.*, 191:107–150, 1994.

[89] B. Fritzsche and B. Kirstein. Darlington synthesis with Arov-singular j_{pq}-inner functions. *Analysis*, 13:215–228, 1993.

[90] B. Fritzsche and B. Kirstein. On the Weyl balls associated with nondegenerate matrix-valued Carathéodory functions. *Z. Anal. Anw.*, 12:239–261, 1993.

[91] B. Fritzsche and B. Kirstein. Carathéodory sequence parametrizations of Potapov-normalized full rank j_q-elementary factors. *Linear Algebra Appl.*, 214:145–186, 1995.

[92] J. B. Garnett. *Bounded Analytic Functions*. Academic Press, New York, 1981.

[93] Ya. Geronimus. *Polynomials Orthogonal on a Circle and Their Applications*, vol. 3 of *Transl. Math. Monographs*, pp. 1–78. Am. Math. Soc., 1954.

[94] Ya. Geronimus. *Polynomials Orthogonal on a Circle and Interval*. International Series of Monographs in Pure and Applied Mathematics. Pergamon Press, Oxford, 1960.

[95] Ya. Geronimus. *Orthogonal Polynomials*. Consultants Bureau, New York, 1961.

[96] L. Gillman and M. Jerison. *Rings of Continuous Functions*, vol. 43 of *Graduate Texts in Mathematics*. Van Nostrand, Princeton, NJ, 1976.

[97] I. Gohberg, ed. *I. Schur Methods in Operator Theory and Signal Processing*, vol. 18 of *Oper. Theory: Adv. Appl.* Birkhäuser Verlag, Basel, 1986.

[98] I. Gohberg, ed. *Time-Variant Systems and Interpolation*, vol. 56 of *Oper. Theory: Adv. Appl.* Birkhäuser Verlag, Basel, 1992.

[99] I. Gohberg and H. J. Landau. Prediction of two processes and the Nehari problem. *J. Fourier Anal. Appl.*, 3:43–62, 1997.

[100] I. Gohberg and L. A. Sakhnovich, eds. *Matrix and Operator Valued Functions*, vol. 72 of *Oper. Theory: Adv. Appl.* Birkhäuser Verlag, Basel, 1994.

[101] P. González-Vera and O. Njåstad. Szegő functions and multipoint Padé approximation. *J. Comput. Appl. Math.*, 32:107–116, 1990.

[102] U. Grenander and G. Szegő. *Toeplitz Forms and Their Applications*. University of California Press, Berkeley, 1958.

[103] T. H. Gronwall. On the maximum modulus of an analytic function. *Ann. of Math.*, 16(2):77–81, 1914–15.

[104] H. Hamburger. Ueber eine Erweiterung des Stieltjesschen Moment Problems I. *Math. Ann.*, 81:235–319, 1920.

[105] H. Hamburger. Ueber eine Erweiterung des Stieltjesschen Moment Problems II. *Math. Ann.*, 82:120–164, 1921.

[106] H. Hamburger. Ueber eine Erweiterung des Stieltjesschen Moment Problems III. *Math. Ann.*, 82:168–187, 1921.

[107] G. Hamel. Eine Charakteristische Eigenschaft beschränkter analytischer Funktionen. *Math. Ann.*, 78:257–269, 1918.

[108] M. H. Hayes and M. A. Clements. An efficient algorithm for computing Pisarenko's harmonic decomposition using Levinson's algorithm. *IEEE Trans. Acoust. Speech Signal Process.*, ASSP-34:485–491, 1986.

[109] G. Heinig and K. Rost. *Algebraic Methods for Toeplitz-Like Matrices and Operators*. Akademie Verlag, Berlin, 1984. Also Birkhäuser Verlag, Basel.

[110] H. Helson. *Lectures on Invariant Subspaces*. Academic Press, New York, 1964.

[111] J. W. Helton. Orbit Structure of the Möbius Transformation Semi-Group Acting on H^∞ (Broadband Matching), vol. 3 of *Adv. in Math. Suppl. Stud.*, pp. 129–197. Academic Press, New York, 1978.

[112] E. Hendriksen and O. Njåstad. A Favard theorem for rational functions. *J. Math. Anal. Appl.*, 142(2):508–520, 1989.

[113] E. Hendriksen and O. Njåstad. Positive multipoint Padé continued fractions. *Proc. Edinburgh Math. Soc.*, 32:261–269, 1989.

[114] E. Hendriksen and H. van Rossum. Orthogonal Laurent polynomials. *Proc. of the Kon. Nederl. Akad. Wetensch, Proceedings A*, 89(1):17–36, 1986.

[115] P. Henrici. *Applied and Computational Complex Analysis. Volume 2: Special Functions, Integral Transforms, Asymptotics, Continued Fractions*, vol. II of *Pure and Applied Mathematics, a Wiley-Interscience Series of Texts, Monographs and Tracts*. John Wiley & Sons, New York, 1977.

[116] K. Hoffman. *Banach Spaces of Analytic Functions.* Prentice-Hall, Englewood Cliffs, 1962.
[117] T. S. Ivanchenko and L. A. Sakhnovich. An operator approach to the Potapov scheme for the solution of interpolation problems. In I. Gohberg and L. A. Sakhnovich, eds., *Matrix and Operator Valued Functions,* vol. 72 of *Oper. Theory: Adv. Appl.,* pp. 48–86. Birkhäuser Verlag, Basel, 1994.
[118] W. B. Jones, O. Njåstad, and W. J. Thron. Two-point Padé expansions for a family of analytic functions. *J. Comput. Appl. Math.,* 9:105–124, 1983.
[119] W. B. Jones, O. Njåstad, and W. J. Thron. Orthogonal Laurent polynomials and the strong Hamburger moment problem. *J. Math. Anal. Appl.,* 98:528–554, 1984.
[120] W. B. Jones, O. Njåstad, and W. J. Thron. Continued fractions associated with the trigonometric moment problem and other strong moment problems. *Constr. Approx.,* 2:197–211, 1986.
[121] W. B. Jones, O. Njåstad, and W. J. Thron. Perron–Carathéodory continued fractions. In J. Gilewicz, M. Pindor, and W. Siemaszko, eds., *Rational Approximation and Its Applications in Mathematics and Physics,* vol. 1237 of *Lecture Notes in Math.,* pp. 188–206, Springer-Verlag, Berlin, 1987.
[122] W. B. Jones, O. Njåstad, and W. J. Thron. Moment theory, orthogonal polynomials, quadrature and continued fractions associated with the unit circle. *Bull. London Math. Soc.,* 21:113–152, 1989.
[123] W. B. Jones and W. J. Thron. *Continued Fractions. Analytic Theory and Applications.* Addison-Wesley, Reading, MA, 1980.
[124] T. Kailath. A view of three decades of linear filtering theory. *IEEE Trans. Inf. Th.,* IT-20:146–181, 1974. Reprinted in [125], 10–45.
[125] T. Kailath et al., ed. *Linear Least-Squares Estimation,* vol. 17 of *Benchmark Papers in Electrical Engineering and Computer Science.* Dowden, Hutchinson and Ross, Stroudsburg, PA, 1977.
[126] J. Karlsson. Rational interpolation and best rational approximation. *J. Math. Anal. Appl.,* 52:38–52, 1976.
[127] V. E. Katsnelson. Left and right Blaschke–Potapov products and Arov-singular matrix-valued functions. *Integral Equations Operator Theory,* 13:836–848, 1990.
[128] M. Kimura. *Chain Scattering Approach to H-Infinity-Control.* Birkhäuser Verlag, Basel, 1997.
[129] A. N. Kolmogorov. Stationary sequences in Hilbert's space. *Bull. Moscow State Univ.,* 2(6):1–40, 1940. Reprinted in [125], 66–89.
[130] P. Koosis. *Introduction to H^p Spaces,* vol. 40 of *London Mathematical Society Lecture Notes.* Cambridge University Press, Cambridge, 1980.
[131] M. G. Kreĭn and A. A. Nudel'man. *The Markov Moment Problem and Extremal Problems,* vol. 50 of *Transl. Math. Monographs.* Am. Math. Soc., Providence, RI, 1977.
[132] L. Kronecker. Algebraische Reduction der Schaaren bilinearer Formen. *S.-B. Akad. Berlin,* pp. 763–776, 1890.
[133] H. J. Landau. Maximum entropy and the moment problem. *Bull. Am. Math. Soc. (N.S.),* 16(1):47–77, 1987.
[134] N. Levinson. The Wiener rms (root mean square) error criterion in filter design and prediction. *J. Math. Phys.,* 25:261–278, 1947.
[135] X. Li and K. Pan. Strong and weak convergence of rational functions orthogonal on the unit circle. *J. London Math. Soc.,* 53:289–301, 1996.

[136] X. Li and E. B. Saff. On Nevai's characterization of measures with almost everywhere positive derivative. *J. Approx. Theory*, 63:191–197, 1990.

[137] G. L. Lopes [López-Lagomasino]. Conditions for convergence of multipoint Padé approximants for functions of Stieltjes type. *Math. USSR-Sb.*, 35:363–376, 1979.

[138] G. L. Lopes [López-Lagomasino]. On the asymptotics of the ratio of orthogonal polynomials and convergence of multipoint Padé approximants. *Math. USSR-Sb.*, 56:207–219, 1985.

[139] G. L. López [López-Lagomasino]. Szegő's theorem for orthogonal polynomials with respect to varying measures. In M. Alfaro et al., eds., *Orthogonal Polynomials and Their Applications*, vol. 1329 of *Lecture Notes in Math.*, pp. 255–260. Springer-Verlag, Berlin, 1988.

[140] G. L. López [López-Lagomasino]. Asymptotics of polynomials orthogonal with respect to varying measures. *Constr. Approx.*, 5:199–219, 1989.

[141] G. L. López [López-Lagomasino]. Convergence of Padé approximants of Stieltjes type meromorphic functions and comparative asymptotice for orthogonal polynomials. *Math. USSR-Sb.*, 64:207–227, 1989.

[142] L. Lorentzen and H. Waadeland. *Continued Fractions with Applications*, vol. 3 of *Studies in Computational Mathematics*. North-Holland, Dordrecht, 1992.

[143] J. Makhoul. Linear prediction: a tutorial review. *Proc. IEEE*, 63:561–580, 1975.

[144] J. D. Markel and A. H. Gray Jr. *Linear Prediction of Speech*. Springer-Verlag, New York, 1976.

[145] A. Máté, P. Nevai, and V. Totik. Strong and weak convergence of orthogonal polynomials. *Am. J. Math.*, 109:239–281, 1987.

[146] R. Mathias. Matrices with positive Hermitian part: inequalities and linear systems. *SIAM J. Matrix Anal. Appl.*, 13(2):640–654, 1992.

[147] J. H. McCabe and J. A. Murphy. Continued fractions which correspond to power series expansions at two points. *J. Inst. Math. Appl.*, 17:233–247, 1976.

[148] H. Meschkowski. *Hilbertsche Räume mit Kernfunktion*. Springer-Verlag, Berlin, 1962.

[149] K. Müller. *Arov–Dewilde–Dym–Parametrization of j_{qq}-inner functions*. PhD thesis, Univ. Leipzig, 1995.

[150] K. Müller and A. Bultheel. Translation of the Russian paper "Orthogonal systems of rational functions on the unit circle" by M. M. Džrbašian. Technical Report TW253, Department of Computer Science, K. U. Leuven, February 1997.

[151] Z. Nehari. On bounded bilinear forms. *Ann. of Math.*, 65:153–162, 1957.

[152] P. Nevai. Géza Freud, orthogonal polynomials and Christoffel functions. A case study. *J. Approx. Theory*, 48:3–167, 1986.

[153] R. Nevanlinna. Über beschränkte Funktionen die in gegebenen Punkten vorgeschriebene Werte annehmen. *Ann. Acad. Sci. Fenn. Ser. A.*, 13(1):71, 1919.

[154] R. Nevanlinna. Asymptotische Entwickelungen beschränkter Funktionen und das Stieltjessche Momentenproblem. *Ann. Acad. Sci. Fenn. Ser. A.*, 18(5):53, 1922.

[155] R. Nevanlinna. Kriterien für die Randwerte beschränkter Funktionen. *Math. Z.*, 13:1–9, 1922.

[156] R. Nevanlinna. Über beschränkte analytische Funktionen. *Ann. Acad. Sci. Fenn. Ser. A.*, 32(7):75, 1929.

[157] O. Njåstad. An extended Hamburger moment problem. *Proc. Edinburgh Math. Soc.*, 28:167–183, 1985.

[158] O. Njåstad. Unique solvability of an extended Hamburger moment problem. *J. Math. Anal. Appl.*, 124:502–519, 1987.

[159] O. Njåstad. Multipoint Padé approximation and orthogonal rational functions. In A. Cuyt, ed., *Nonlinear Numerical Methods and Rational Approximation*, pp. 258–270, D. Reidel, Dordrecht, 1988.

[160] O. Njåstad. A modified Schur algorithm and an extended Hamburger moment problem. *Trans. Am. Math. Soc.*, 327(1):283–311, 1991.

[161] O. Njåstad. Classical and strong moment problems. *Comm. Anal. Th. Continued Fractions*, 4:4–38, 1995.

[162] O. Njåstad and W. J. Thron. The theory of sequences of orthogonal L-polynomials. In H. Waadeland and H. Wallin, eds., *Padé Approximants and Continued Fractions*, Det Kongelige Norske Videnskabers Selskab Skrifter (No. 1), pp. 54–91, 1983.

[163] O. Njåstad and W. J. Thron. Unique solvability of the strong Hamburger moment problem. *J. Austral. Math. Soc. (Series A)*, 40:5–19, 1986.

[164] A. A. Nudelman. Matrix versions of interpolation problems of Nevanlinna–Pick and Loewner type. In U. Helmke, R. Mennicken, and J. Saurer, eds., *Systems and Networks: Mathematical Theory and Applications, Vol. I (Regensburg, 1993)*, number 77 in Mathematical Research, pp. 291–309, Akademie Verlag, Berlin, 1994.

[165] A. A. Nudelmann. Multipoint matrix moment problem. *Dokl. Acad. Nauk.*, 298:812–815, 1988.

[166] K. Pan. On characterization theorems for measures associated with orthogonal systems of rational functions on the unit circle. *J. Approx. Theory*, 70:265–272, 1992.

[167] K. Pan. On orthogonal systems of rational functions on the unit circle and polynomials orthogonal with respect to varying measures. *J. Comput. Appl. Math.*, 47(3):313–322, 1993.

[168] K. Pan. Strong and weak convergence of orthogonal systems of rational functions on the unit circle. *J. Comput. Appl. Math.*, 46:427–436, 1993.

[169] K. Pan. On orthogonal polynomials with respect to varying measures on the unit circle. *Trans. Am. Math. Soc.*, 346:331–340, 1994.

[170] K. Pan. On the convergence of rational interpolation approximant of Carathéodory functions. *J. Comput. Appl. Math.*, 54:371–376, 1994.

[171] K. Pan. Extensions of Szegő's theory of rational functions orthogonal on the unit circle. *J. Comput. Appl. Math.*, 62:321–331, 1995.

[172] K. Pan. On the orthogonal rational functions with arbitrary poles and interpolation properties. *J. Comput. Appl. Math.*, 60:347–355, 1995.

[173] K. Pan. On the convergence of rational functions orthogonal on the unit circle. *J. Comput. Appl. Math.*, 76:315–324, 1996.

[174] A. Papoulis. Levinson's algorithm, Wold decomposition, and spectral estimation. *SIAM Rev.*, 27(3):405–441, 1985.

[175] J. R. Partington. *An Introduction to Hankel Operators*, vol. 13 of *London Mathematical Society Student Texts*. Cambridge University Press, Cambridge, 1988.

[176] G. Pick. Über die Beschränkungen analytischen Funktionen welche durch vorgegebene Funktionswerte bewirkt werden. *Math. Ann.*, 77:7–23, 1916.

[177] G. Pick. Über die Beschränkungen analytischen Funktionen durch vorgegebene Funktionswerte. *Math. Ann.*, 78:270–275, 1918.

[178] G. Pick. Über beschränkte Funktionen mit vorgeschriebenen Wertzuordnungen. *Ann. Acad. Sci. Fenn. Ser. A*, 15(3):17, 1920.

[179] V. P. Pisarenko. The retrieval of harmonics from a covariance function. *Geophys. J. R. Astron. Soc.*, 33:347–366, 1973.

[180] V. P. Potapov. *The Multiplicative Structure of J-Contractive Matrix Functions*, vol. 15 of *Am. Math. Soc. Transl. Ser. 2*, pp. 131–243. Am. Math. Soc., Providence, RI, 1960.

[181] V. P. Potapov. *Linear Fractional Transformations of Matrices*, vol. 138 of *Am. Math. Soc. Transl. Ser. 2*, pp. 21–35. Am. Math. Soc., Providence, RI, 1988.

[182] S. C. Power. *Hankel Operators on Hilbert Space*. Pitman Advanced Public Program, Boston, 1982.

[183] R. Redheffer. On the relation of transmission-line theory to scattering and transfer. *J. Math. Phys.*, 41:1–41, 1962.

[184] F. Riesz. Über ein Problem des Herrn Carathéodory. *J. Reine Angew. Math.*, 146:83–87, 1916.

[185] F. Riesz. Über Potenzreihen mit vorgeschriebenen Anfangsgliedern. *Acta Math.*, 42:145–171, 1918.

[186] M. Rosenblum and J. Rovnyak. *Hardy Classes and Operator Theory*. Oxford University Press, New York, 1985.

[187] W. Rudin. *Real and Complex Analysis*. McGraw-Hill, New York, 2nd ed., 1974.

[188] D. Sarason. Generalized interpolation in H^∞. *Trans. Am. Math. Soc.*, 127:179–203, 1967.

[189] I. Schur. Über ein Satz von C. Carathéodory. *S.-B. Preuss. Akad. Wiss. (Berlin)*, pp. 4–15, 1912.

[190] I. Schur. Über Potenzreihen die im Innern des Einheitskreises Beschränkt sind I. *J. Reine Angew. Math.*, 147:205–232, 1917. See also [97, pp. 31–59].

[191] I. Schur. Über Potenzreihen die im Innern des Einheitskreises Beschränkt sind II. *J. Reine Angew. Math.*, 148:122–145, 1918. See also [97, pp. 36–88].

[192] J. A. Shohat and J. D. Tamarkin. *The Problem of Moments*, volume 1 of *Math. Surveys*. Am. Math. Soc., Providence, RI, 1943.

[193] H. Stahl and V. Totik. *General Orthogonal Polynomials*. Encyclopedia of Mathematics and Its Applications. Cambridge University Press, Cambridge, 1992.

[194] T. J. Stieltjes. Recherches sur les fractions continues. *Ann. Fac. Sci. Toulouse*, 8:J.1–122, 1894, 9:A.1–47, 1895. English transl.: Oeuvres Complètes, Collected Papers, Vol. 2, 609–745, Springer-Verlag, Berlin, 1993.

[195] M. H. Stone. *Linear Transformations in Hilbert Space and Their Applications to Analysis*, vol. 15 of *Am. Math. Soc. Colloq. Publ.* Am. Math. Soc., New York, 1932.

[196] G. Szegő. *Orthogonal Polynomials*, vol. 33 of *Am. Math. Soc. Colloq. Publ.* Am. Math. Soc., Providence, RI, 3rd ed., 1967. First edition 1939.

[197] A.-J. van der Veen. *Time-variant system theory and computational modeling. Realization, approximation and factorization*. PhD thesis, Technical University Delft, The Netherlands, June 1993.

[198] A.-J. van der Veen and P. Dewilde. Embedding time-varying contractive systems in lossless realizations. *Math. Control Signals Systems*, 7:306–330, 1995.

[199] J. L. Walsh. Interpolation and functions analytic interior to the unit circle. *Trans. Am. Math. Soc.*, 34:523–556, 1932.

[200] J. L. Walsh. *Interpolation and Approximation*, vol. 20 of *Am. Math. Soc. Colloq. Publ.* Am. Math. Soc., Providence, RI, 3rd ed., 1960. First edition 1935.

[201] N. Wiener. *Extrapolation, Interpolation and Smoothing of Time Series*. Wiley, New York, 1949.

[202] N. Wiener and P. Masani. The prediction theory of multivariate stochastic processes, I. The regularity condition. *Acta Math.*, 98:111–150, 1957.
[203] N. Wiener and P. Masani. The prediction theory of multivariate stochastic processes, II. The linear predictor. *Acta Math.*, 99:93–139, 1958.
[204] D. C. Youla and M. Saito. Interpolation with positive-real functions. *J. Franklin Inst.*, 284(2):77–108, 1967.

Index

admissible, 142
admittance matrix, 370
all-pass function, 377
autoregressive filter, 353

backward shift, 343
Banach space, 20, 36
Bessel inequality, 160
Beurling theorem, 45
Beurling–Lax theorem, 34
Blaschke factor, 42, 65
Blaschke product, 31, 43, 53, 84, 91, 110, 122,
    135, 142, 149, 151, 153, 157, 164, 176,
    179, 181, 184, 241, 244, 257
Blaschke–Potapov factor, 38
boundary situation, 101
Brune section, 368

Carathéodory class, 11, 12, 15, 23
Carathéodory coefficient problem, 104, 387
Carathéodory–Toeplitz theorem, 356
Carleman condition, 194, 195
Cauchy integral, 22, 46, 62, 138, 153
Cauchy kernel, 23
Cauchy–Stieltjes integral, 22, 122
causal system, 360
Cayley transform, 15, 16, 25, 111, 144, 182
chain scattering matrix, 40, 362
Christoffel function, 118, 175
Christoffel–Darboux relation, 12, 64, 67, 78, 93,
    101, 137, 185, 191, 192, 200, 204, 243,
    245, 246, 272
compactification, 253, 302
continued fraction, 96, 103, 105
    approximants, 95
    convergents, 95
control problem, 373
convergence factor, 32
covariance, 343

Darlington synthesis, 369
density, 34, 149, 155, 174, 254
determinant expression, 55, 57
determinant formula, 89, 90, 92, 118, 243, 246
dissipated power, 369

EMP-fraction, 338
energy of stochastic process, 343
Erdős–Turán condition, 173, 194, 209
expectation operator, 343
extended multipoint Padé fraction, 338
extended recurrence relation, 309
extremal problem, 35, 36, 42, 58, 174

Favard theorem, 13, 121, 161, 307
Fourier transform, 21
Fourier–Stieltjes transform, 21
frequency domain, 345, 369
functions of second kind, 92, 100, 104, 105,
    111, 114, 117, 121, 123, 145, 181, 241,
    242, 267, 269, 331

Gram matrix, 46, 48, 49, 53, 56, 57
Gram–Schmidt orthogonalization, 56
Green's formula, 93, 277

Hamburger representation, 29
Hankel norm approximation, 385
Hardy class, 15, 17
harmonic function, 186
harmonic majorant, 17, 40
Helly's theorems, 252
Hurwitz theorem, 185, 204

impedance matrix, 370
incident wave, 359
incoming wave, 371
inner function, 31, 34, 156
inner-outer factorization, 31, 190, 382

405